Lecture Notes in Mathematics

Editors:
A. Dold, Heidelberg
B. Eckmann, Zürich
F. Takens, Groningen

H. Komatsu (Ed.)

Functional Analysis and Related Topics, 1991

Proceedings of the International Conference in
Memory of Professor Kôsaku Yosida held at RIMS,
Kyoto University, Japan, July 29-Aug. 2, 1991

Springer-Verlag
Berlin Heidelberg New York
London Paris Tokyo
Hong Kong Barcelona
Budapest

Editor

Hikosaburo Komatsu
Department of Mathematical Sciences
University of Tokyo
Hongo, Tokyo, 113 Japan

Mathematics Subject Classification (1991): 46-06, 46N20, 47D06, 47H20, 47A40, 46L37

ISBN 3-540-56471-3 Springer-Verlag Berlin Heidelberg New York
ISBN 0-387-56471-3 Springer-Verlag New York Berlin Heidelberg

Typesetting: Camera-ready by author/editor
46/3140-543210 - Printed on acid-free paper

Preface

These notes are a collection of papers presented at the International Conference on Functional Analysis and Related Topics, 1991, held at the Research Institute for Mathematical Sciences, Kyoto University, Japan, on July 29 - August 2, 1991. Approximately 180 mathematicians from 7 countries attended the conference.

The conference was organized by the Research Institute for Mathematical Sciences as a special conference of the institute, and supported by the ICM-90 Commemorative Meeting Fund of the Mathematical Society of Japan, and by the Inoue Foundation. It was held in memory of Professor Kôsaku Yosida who passed away a year before, on June 20, 1990 after a short illness. The Organizing Committee, consisting of M. Sato, H. Fujita, Y. Kōmura and H. Komatsu, invited 27 speakers who carry on Professor Yosida's research tradition.

As the attached list of publications shows, Professor Yosida had a very wide interest in analysis, and guided many students, not only through personal contact but also through his numerous books, including his famous textbook "Functional Analysis" of which six distinct editions appeared. His collected works in English will soon be published by Springer Verlag.

In 1969 an international conference of the same title was held in Tokyo on the occasion of his 60th anniversary. It covered Partial Differential Equations, Differential Equations on Manifolds, Hyperfunctions, Markov Processes and Potentials, and Ergodic Theory. This time the range of topics is more restricted, but we hope that some of these topics have deepened considerably since then. Professor Yosida was essentially a theorist but he always had applications in mind. Whenever he created a beautiful abstract theory, he was also the first to apply it to more concrete problems. The organizers of the conference will be delighted if the reader would recognize this flavor in the following pages.

Hikosaburo Komatsu

Contents

Program

July 29 (Monday)

11:00 - 12:00   Jacques-Louis Lions (Collège de France)*
Distributed systems with incomplete data and uniqueness theorems
13:30 - 14:25   Kiyosi Itô (Kyoto Univ.)
Semigroups in probability theory
14:30 - 15:25   Daisuke Fujiwara (Tokyo Inst. of Tech.)
Some Feynman path integrals as oscillatory integrals over a Sobolev space
15:30 - 16:25   Hikosaburo Komatsu (Univ. of Tokyo)
Operational calculus and semigroups of operators
16:30 - 16:45
Donation of the Yosida Library
17:30 -
Reception at Kyodai Kaikan

July 30 (Tuesday)

9:50 - 10:50   Haïm Brezis (Univ. of Paris VI)
Mathematical problems of liquid crystals
11:00 - 12:00   Yukio Kōmura - Kiyoko Furuya (Ochanomizu Univ.)
Wave equations in non-reflexive spaces
13:30 - 14:25   Ken-iti Sato (Nagoya Univ.)
Stochastic processes of Ornstein-Uhlenbeck type on Euclidean spaces
14:30 - 15:25   Takashi Suzuki (Tokyo Metropolitan Univ.)
Symmetry breaking: a variational approach
15:30 - 16:25   Hisashi Okamoto (RIMS, Kyoto Univ.)
Computer-assisted analysis of 2D Navier-Stokes equations
16:30 - 17:00   Yasuyuki Kawahigashi (Univ. of Tokyo)
Solvable orbifold models and operator algebras

July 31 (Wednesday)

9:50 - 10:50   Tosio Kato (Univ. of California, Berkeley)
Abstract evolution equations, linear and quasilinear, revisited
11:00 - 12:00   Alberto Venni (Univ. of Bologna)
Complex powers of operators and related problems of operator theory
13:30 - 14:25   Yoshikazu Giga (Hokkaido Univ.)
L^p estimates for the Navier-Stokes system
14:30 - 15:25   Hitoshi Kitada (Univ. of Tokyo)
Completeness of N-body wave operators — long-range quantum systems
15:30 - 16:25   Minoru Murata (Kumamoto Univ.)
Nonnegative solutions of linear parabolic equations

16:30 - 17:00   Teruo Ushijima - Mihoko Matsuki (Univ. of Electro-Communications)
Fully discrete approximation of a second order linear evolution equation related to the water wave problem

August 1 (Thursday)
9:50 - 10:50   P. P. Narayanaswami (Memorial Univ. of Newfoundland)
The separable quotient problem for Banach and (LF)-spaces
11:00 - 12:00   D. C. Struppa (Univ. of Calabria) - T. Kawai (RIMS, Kyoto Univ.)
Interpolating varieties and the Fabry-Ehrenpreis-Kawai gap theorem
13:30 - 14:25   Shinnosuke Oharu (Hiroshima Univ.)
Characterization of nonlinearly perturbed semigroups
14:30 - 15:25   Yoshikazu Kobayashi (Niigata Univ.)
Semigroups of locally Lipschitzian operators and applications
15:30 - 16:25   Mitsuharu Otani (Waseda Univ.)
A priori estimates for some nonlinear parabolic equations via Lyapunov functions
16:30 - 17:00   Hiroko Morimoto (Meiji Univ.)
Asymptotic behavior of solutions of the convection equation

August 2 (Friday)
9:50 - 10:50   Philippe Clément (Delft Univ. of Technology)
Maximal regularity L^p - L^q for a class of integro-differential equations
11:00 - 12:00   Giovanni Dore (Univ. of Bologna)
L^p-regularity for abstract differential equations
13:30 - 14:25   Atsushi Yoshikawa (Kyushu Univ.)
Quasilinear oscillations and geometric optics
14:30 - 15:25   Atsushi Yagi (Himeji Inst. of Tech.)
Global solution to some quasilinear parabolic system in mathematical biology
15:30 - 16:25   Shigetake Matsuura (RIMS, Kyoto Univ.)
On non-convex curves of constant angle

*) Professor Lions was unable to attend the conference because of a minor accident.

List of Publications of Kôsaku Yosida

Books

A. 連続群論 (= *Theory of Continuous Groups*), 岩波数学講座, Iwanami (岩波書店), Tokyo, 1934, iii+180 pp.

B. りい環論 (= *Theory of Lie Rings*), 大阪帝国大学数学講演集 IV, Iwanami, Tokyo, 1939, 3+48 pp.

C. 線型作用素 (= *Linear Operators*), 現代数学叢書, Iwanami, Tokyo, 1943, 2+118 pp.

D. スペクトル解析 (= *Spectral Analysis*), 近代数学全書, Kyoritsu (共立出版), Tokyo, 1947, 1+2+125 pp.

E. エルゴード諸定理 (= *Ergodic Theorems*), 確率統計叢書, Chubunkan (中文館書店), Tokyo, 1948, iii+82 pp.

F. 物理数学概論 (= *Topics in Mathematical Physics*), Nihon Hyoron (日本評論社), Tokyo, 1949, 6+202+6 pp.

G. 積分方程式論 (= *Theory of Integral Equations*), 岩波全書 117, Iwanami, Tokyo, 1950, 8+234 pp.

H. 位相解析 I (= *Topological Analysis I*), 現代数学 8, Iwanami, Tokyo, 1951, 2+2+339 pp.

I. ヒルベルト空間論 (= *Theory of Hilbert Spaces*), 共立全書 49, Kyoritsu, Tokyo, 1953, 2+4+215 pp.

J. 微分方程式の解法 (= *Methods of Differential Equations*), 岩波全書 189, Iwanami, Tokyo, 1954, 9+263 pp.

K. (With A. Amemiya, K. Itô, T. Kato, Y. Matsushima et al.) 応用数学便覧 (=*Handbook of Applied Mathematics*), Maruzen (丸善), Tokyo, 1954, 2+8+518 pp.

L. 近代解析 (= *Modern Analysis*), 基礎数学講座 20, Kyoritsu, Tokyo, 1956, 1+3+121 pp.

M. 超函数論 (= *Theory of Distributions*), 現代数学講座 13, Kyoritsu, Tokyo, 1956, v+169 pp.

N. *Lectures on Semi-group Theory and its Application to Cauchy's Problem in Partial Differential Equations*, Lectures on Mathematics and Physics 8, Tata Inst. Fund. Research, Bombay, 1957, iv+127+iv pp.

O. 位相解析 (= *Topological Analysis*), 岩波講座現代応用数学 A4, Iwanami, Tokyo, 1957, iv+234 pp.

L$_{II}$. 近代解析 第2版 (= *Modern Analysis, 2nd ed.*), 基礎数学講座 20, Kyoritsu, Tokyo, 1958, 2+4+219+4 pp.

P. (With K. Kunugui, S. Nakanishi and S. Itô) 積分論・位相解析 (= *Theory of Integrals, Topological Analysis*), 数学演習講座 15, Kyoritsu, Tokyo, 1958, Part 2, 2+105 pp.

G$_E$. *Lectures on Differential and Integral Equations*, Pure and Applied Mathematics Vol. X, Interscience, New York-London, 1960, ix+220 pp.

O$_{\text{II}}$. (With Y. Kawada and T. Iwamura) 位相解析の基礎 (= *Fundamental of Topological Analysis*), Iwanami, Tokyo, 1960, pp. 95–330.

Q. (With T. Kato) 大学演習応用数学 I (= *Exercises in Applied Mathematics I*), 大学演習新書, Shokabo (裳華房), Tokyo, 1961, vii+347 pp.

R. *Functional Analysis*, Grundlehren Math. Wiss. Bd. 123, Springer, Berlin-Göttingen-Heidelberg, 1965, XI+458 pp.

R$_{\text{R}}$. *Funkčional'nyi Analiz*, Izdat. MIR, Moscow, 1967, 624 pp.

K$_{\text{II}}$. (With A. Amemiya, K. Itô, T Kato, Y. Matsushima, S. Furuya et al.) 応用数学便覧 新版 (= *Handbook of Applied Mathematics, New ed.*), Maruzen, Tokyo, 1967, 2+9+603 pp.

R$_{\text{II}}$. *Functional Analysis, 2nd ed.*, Grundlehren Math. Wiss. Bd. 123, Springer, Berlin-Heidelberg-New York, 1968, XI+465 pp.

R$_{\text{III}}$. *Functional Analysis, 3rd ed.*, Grundlehren Math. Wiss. Bd. 123, Springer, Berlin-Heidelberg-New York, 1971, XI+475 pp.

G$_{\text{F}}$. *Equations Différentielles et Intégrals*, Dunod, Paris, 1971, xv+230 pp.

F$_{\text{II}}$. 物理数学概論 (= *Topics in Mathematical Physics*), 数理解析とその周辺 5, Sangyo Tosho (産業図書), 1974, 6+198 pp.

R$_{\text{IV}}$. *Functional Analysis, 4th ed.*, Grundlehren Math. Wiss. Bd. 123, Springer, Berlin-Heidelberg-New York, 1974, XI+496 pp.

S. 測度と積分 (= *Measures and Integrals*), 岩波講座基礎数学, 解析学 (I) iii, Iwanami, Tokyo, 1976, vii+172 pp.

T. (With S. Itô, A. Orihara and T. Muramatsu) 函数解析と微分方程式 (= *Functional Analysis and Differential Equations*), 現代数学演習叢書 4, Iwanami, Tokyo, 1976, xi+474 pp.

R$_{\text{V}}$. *Functional Analysis, 5th ed.*, Grundlehren Math. Wiss. Bd. 123, Springer, Berlin-Heidelberg-New York, 1978, XII+501 pp.

G$_{\text{II}}$. 積分方程式論 第2版 (= *Theory of Integral Equations, 2nd ed.*), 岩波全書 117, Iwanami, Tokyo, 1978, vi+292 pp.

J$_{\text{II}}$. 微分方程式の解法 第2版 (= *Methods of Differential Equations, 2nd ed.*), 岩波全書 189, Iwanami, Tokyo, 1978, xiv+307 pp.

R$_{\text{VI}}$. *Functional Analysis, 6th ed.*, Grundlehren Math. Wiss. Bd. 123, Springer, Berlin-Heidelberg-New York, 1980, XII+501 pp.

U. 私の微分積分法 — 解析入門 (= *My Calculus — An Introduction to Analysis*), Kodansha (講談社), Tokyo, 1981, 250 pp.

V. 演算子法 一つの超函数論 (= *Operational Calculus A Theory of Hyperfunctions*), UP 応用数学選書 5, Univ. of Tokyo Press (東京大学出版会), 1982, viii+171 pp.

V$_{\text{E}}$. *Operational Calculus A Theory of Hyperfunctions*, Applied Mathematical Sciences Vol. 55, Springer, New York-Berlin-Heidelberg-Tokyo, 1984, x+170 pp.

W. 19世紀の数学 解析学 I (= *Analysis I Mathematics in the 19th Century*), 数学の歴史 9, Kyoritsu, Tokyo, 1986, xii+256 pp.

S$_{\text{II}}$. (With H. Fujita) 現代解析入門 (= *Introduction to Modern Analysis*), 岩波基礎数学選書, Iwanami, Tokyo, 1991, pp. 243–456.

Papers

1. *On the asymptotic property of the differential equation $y'' + H(x)y = f(x, y, y')$,* Japan. J. Math. **9** (1932), 145–152.
2. *On the asymptotic property of the differential equation $y'' + H(x)y = f(x, y, y')$, II,* Japan. J. Math. **9** (1932), 227–230.
3. *A remark to a theorem due to Halphen,* Japan. J. Math. **9** (1932), 231–232.
4. *A generalisation of a Malmquist's theorem,* Japan. J. Math. **9** (1932), 253–256.
5. *Some remarks on the theory of Fredholm's integral equations,* Proc. Phys.-Math. Soc. Japan **14** (1932), 381–384.
6. *On the distribution of α-points of solutions for linear differential equation of the second order,* Proc. Imp. Acad. Tokyo **8** (1932), 335–336.
7. *A note on Riccati's equation,* Proc. Phys.-Math. Soc. Japan **15** (1933), 227–232.
8. *On the characteristic function of a transcendental meromorphic solution of an algebraic differential equation of the first order and of the first degree,* Proc. Phys.-Math. Soc. Japan **15** (1933), 337–338.
9. *On algebroid-solutions of ordinary differential equations,* Japan. J. Math. **10** (1933), 199–208.
10. *On a class of meromorphic functions,* Proc. Phys.-Math. Soc. Japan **16** (1934), 227–235.
11. *Wronskian ニ就テ (= On Wronskians),* Zenkoku Sizyo Sugaku Danwakai (全国紙上数学談話会) **3** (1934), 3–5.
12. *Algebroid function ニ就テ (= On algebroid functions),* Zenkoku Sizyo Sugaku Danwakai **6** (1934), 2–6 and **10** (1934), d-e.
13. *Picard ノ定理ニツイテ (= On Picard's theorem),* Zenkoku Sizyo Sugaku Danwakai **18** (1934), 11–12.
14. *有理型函数ノ derivative ニ就テ (= On derivatives of meromorphic functions),* Zenkoku Sizyo Sugaku Danwakai **21** (1934), 7–10.
15. (With T. Shimizu and S. Kakutani) *On meromorphic functions. I,* Proc. Phys.-Math. Soc. Japan **17** (1935), 1–10.
16. *Beurling ノ定理ノ應用例 (= An application of Beurling's theorem),* Zenkoku Sizyo Sugaku Danwakai **24** (1934), 28–30.
17. *Cartan-角谷-Selberg ノ定理ニ就テ (= On the Cartan-Kakutani-Selberg theorem),* Zenkoku Sizyo Sugaku Danwakai **25** (1935), 11–14 and **30** (1935), 9–11.
18. (With T. Shimizu and S. Kakutani) *Function group ニ就テ (=On Function groups),* Zenkoku Sizyo Sugaku Danwakai **28** (1935), 2–8.
19. *Stone ノ定理ニ就テ (= On Stone's theorem),* Zenkoku Sizyo Sugaku Danwakai **35** (1935), 9–12.
20. *A theorem concerning the derivatives of meromorphic functions,* Proc. Phys.-Math. Soc. Japan **17** (1935), 170–173.
21. (With S. Kakutani) *in kleinen affine 寫像ニ就テ (= On locally affine (quasi-conformal) mappings),* Zenkoku Sizyo Sugaku Danwakai **40**(1935), 5–8 and **41** (1935), 10–18.
22. *Fatou ノ定理ニ對スル小サナ注意 (= A short remark on Fatou's theorem),* Zenkoku Sizyo Sugaku Danwakai **45** (1935), 6–8.

23. 等角寫像ニ於ケル *metrical* ナ一定理 (= *A metrical theorem on conformal mappings*), Zenkoku Sizyo Sugaku Danwakai **47** (1935), 5–8.

24. *Picard-Vessiot ノ 理論ニ就テ* (= *On the theory of Picard-Vessiot*), Zenkoku Sizyo Sugaku Danwakai **51** (1935), 8–14, **52** (1935), 1–3, **53** (1935), 9–10, **63** (1935), 28–33 and **67** (1935), 19–22.

25. *Closure ニ関スル一問題への Stone ノ 定理ノ 應用* (= *An application of Stone's theorem to a problem of closure*), Zenkoku Sizyo Sugaku Danwakai **60** (1935), 15–17.

26. *On the groups of rationality for linear differential equations*, Proc. Phys.-Math. Soc. Japan **17** (1935), 498–510.

27. 距離付ケラレタ環ニ於テ閉ヂタ連續群 (= *Closed continuous groups in metrical rings*), Zenkoku Sizyo Sugaku Danwakai **68** (1935), 1–9, **69** (1935), 15–18, **70** (1935), 5–7 and **77** (1935), 4–9.

28. "det $A \neq 0$ ナル *Matrix* ハ $A = \exp B$" ノ 正田教授ニヨル証明其ノ他 (= *A matrix* $A = \exp B$ *if* $\det A \neq 0$; *Prof. Shoda's proof and other topics*), Zenkoku Sizyo Sugaku Danwakai **72** (1935), 1–6.

29. *On the group embedded in the metrical complete ring*, Japan J. Math. **13** (1936), 7–26.

30. *Locally compact ナ topological group ノ 連続表現* (= *Continuous representations of locally compact topological groups*), Zenkoku Sizyo Sugaku Danwakai **87** (1936), 1–8, **88** (1936), 6–8, **99** (1936), 1–3 and **101** (1936), 8–9.

31. *On the group embedded in the metrical complete ring, II*, Japan J. Math. **13** (1936), 459–472.

32. 距離付ケラレタ環ニ付イテ (= *On metrical rings*), Zenkoku Sizyo Sugaku Danwakai **101** (1936), 6–8.

33. *Topological group ノ 連續表現* (= *Continuous representations of topological groups*), Zenkoku Sizyo Sugaku Danwakai **107** (1936), 5–7.

34. *H. Auerbach ノ 定理ニツイテ* (= *On a theorem of H. Auerbach*), Zenkoku Sizyo Sugaku Danwakai **110** (1936), 5–6.

35. *Homomorphie ニヨル次元ノ関係* (= *Dimension relations under homomorphisms*), Zenkoku Sizyo Sugaku Danwakai **111** (1936), 16–18.

36. *A note on the continuous representation of topological groups*, Proc. Imp. Acad. Tokyo **12** (1936), 329–331.

37. *Riemann 空間ノ 等長変換ノ 解析性* (= *Analyticity of isometric transformations of Riemannian spaces*), Zenkoku Sizyo Sugaku Danwakai **120** (1937), 33–35.

38. 完閉群ニ於ケル線状移動可能微分演算子 (= *Linear translatable differential operators in compact groups*), Zenkoku Sizyo Sugaku Danwakai **123** (1937), 87–91 and **124** (1937), 118–119.

39. *A remark on a theorem of B. L. van der Waerden*, Tôhoku Math. J. **43** (1937), 411-413.

40. 単純群ノ一ツノ *class* ニ就テ (= *On a class of simple groups*), Zenkoku Sizyo Sugaku Danwakai **126** (1937), 143–146.

41. *Lie ノ 第二基本定理ニ関聯シタ一ツノ 問題* (= *A problem concerning the second fundamental theorem of Lie*), Zenkoku Sizyo Sugaku Danwakai **128** (1937), 179–185.

42. *A problem concerning the second fundamental theorem of Lie*, Proc. Imp. Acad. Tokyo **13** (1937), 152–155.

43. 準單純りい群ニ関スル一定理 (= *A theorem on semi-simple Lie groups*), Zenkoku Sizyo Sugaku Danwakai **133** (1937), 267–272.

44. *Locally bicompact* ナ *topological group* ノ 連續表現 (= *Continuous representations of locally bicompact topological groups*), Zenkoku Sizyo Sugaku Danwakai **135** (1937), 37–43.

45. *Lie* ノ第二基本定理ニ就テ (= *On the second fundamental theorem of Lie*), Zenkoku Sizyo Sugaku Danwakai **137** (1937), 75–78.

46. *A theorem concerning the semi-simple Lie groups*, Tôhoku Math. J. **44** (1938), 81–84.

47. 豊田浩七氏ノ論文ヲ読ミテ (= *On a paper of Toyoda*), Zenkoku Sizyo Sugaku Danwakai **139** (1937), 138–140.

48. *Topological group* ノ 微分可能性ニ就テ (= *On the differentiability of topological groups*), Zenkoku Sizyo Sugaku Danwakai **141** (1937), 185–189.

49. *On the exponential-formula in the metrical complete ring*, Proc. Imp. Acad. Tokyo **13** (1937), 301–304.

50. *A note on the differentiability of the topological group*, Proc. Phys.-Math. Soc. Japan **20** (1938), 6–10.

51. *A characterisation of the adjoint representations of the semi-simple Lie-rings*, Japan J. Math. **14** (1938), 169–173.

52. 單純且準單純ナ環ノ表現ノ *reduction* ニ就テ (= *On the reduction of representations of simple and semi-simple Lie rings*), Zenkoku Sizyo Sugaku Danwakai **153** (1937), 56–60.

53. *Lie* 環ノ *derivation* (= *Derivations of Lie rings*), Zenkoku Sizyo Sugaku Danwakai **156** (1938), 125–129 and **157** (1938), 167–170.

54. *On the fundamental theorem of the tensor calculus*, Proc. Imp. Acad. Tokyo **14** (1938), 211–213.

55. 確率論ヘノ積分方程式ノ應用 (= *Applications of integral equations to probability theory*), Zenkoku Sizyo Sugaku Danwakai **160** (1938), 245–254, **161** (1938), 282–295, **162** (1938), 296–306, **163** (1938), 358–364, **164** (1938), 394–397 and **165** (1938), 429–441.

56. *Birkhoff-Khintchine* ノ *ergodic theorem* (= *The ergodic theorem of Birkhoff-Khintchine*), Zenkoku Sizyo Sugaku Danwakai **166** (1938), 476–485.

57. *Abstract integral equations and the homogeneous stochastic process*, Proc. Imp. Acad. Tokyo **14** (1938), 286–291.

58. *Mean ergodic theorem in Banach spaces*, Proc. Imp. Acad. Tokyo **14** (1938), 292–294.

59. *Mean ergodic theorem* ノ 應用 (= *Applications of mean ergodic theorems*), Zenkoku Sizyo Sugaku Danwakai **167** (1938), 543–549.

60. (With Y. Mimura and S. Kakutani) 有界且ツ可測ナ核ニヨル積分 *operator* ニ就イテ (= *On integral operators with bounded measurable kernels*), Zenkoku Sizyo Sugaku Danwakai **168** (1938), 631–637 and **169** (1938), 681–684.

61. (With S. Kakutani) *Application of mean ergodic theorem to the problems of Markoff's process*, Proc. Imp. Acad. Tokyo **14** (1938), 333–339.

62. *Doeblin* ノ結果ノ積分方程式的取扱 (= *A treatment of Doeblin's results by integral equations*), Zenkoku Sizyo Sugaku Danwakai **169** (1938), 656–666.

63. (With Y. Mimura and S. Kakutani) *Integral operator with bounded kernel*, Proc. Imp. Acad. Tokyo **14** (1938), 359–362.

64. *Operator-theoretical treatment of the Markoff's process*, Proc. Imp. Acad. Tokyo **14** (1938), 363–367.

65. 完全連續ナ對稱作用素ノ固有値存在ノ証明 (= *A proof of the existence of eigenvalues for completely continuous symmetric operators*), Zenkoku Sizyo Sugaku Danwakai **171** (1938), 756–758.

66. *Quasi-completely-continuous linear functional operations*, Japan. J. Math. **15** (1939), 297–301.

67. 可附番無限個ノ可能ナ状態ニ関スル *Markoff* 過程 (= *Markoff processes with countably infinite possible states*), Zenkoku Sizyo Sugaku Danwakai **172** (1939), 11–19, **173** (1939), 31–37 and **176** (1939), 170–173.

68. *F. Riesz* ノ *mean ergodic theorem* (= *F. Riesz's mean ergodic theorem*), Zenkoku Sizyo Sugaku Danwakai **176** (1939), 166–170.

69. (With S. Kakutani) *Markoff process with an enumerable infinite number of possible states*, Japan. J. Math. **16** (1940), 47–55.

70. 準完全連續線型作用素に就いて (= *On quasi-completely continuous linear operators*), Isô Sûgaku (位相数学) **1** No. 2 (1939), 32–34.

71. *Markoff* 過程ニ於ケルーツノ固有値問題 (= *An eigenvalue problem in Markoff processes*), Zenkoku Sizyo Sugaku Danwakai **177** (1939), 187–193.

72. *Operator-theoretical treatment of Markoff's process, II*, Proc. Imp. Acad. Tokyo **15** (1939), 127–130.

73. (With S. Kakutani) *Birkhoff ergodic theorem* ト *maximal ergodic theorem* (= *Birkhoff's ergodic theorem and the maximal ergodic theorem*), Zenkoku Sizyo Sugaku Danwakai **179** (1939), 216–221 and **181** (1939), 267–291.

74. (With S. Kakutani) *Birkhoff's ergodic theorem and the maximal ergodic theorem*, Proc. Imp. Acad. Tokyo **15** (1939), 165–168.

75. 漸近的概週期性ト *ergodic theorems* (= *Asymptotic almost periodicities and ergodic theorems*), Zenkoku Sizyo Sugaku Danwakai **185** (1939), 407–414 and **186** (1939), 450–461.

76. *Asymptotic almost periodicities and ergodic theorems*, Proc. Imp. Acad. Tokyo **15** (1939), 255–259.

77. *Markoff chain* ノーツノ抽象化 (= *An abstraction of Markoff chains*), Zenkoku Sizyo Sugaku Danwakai **187** (1939), 481–490.

78. *P. Lévy* ノ定理ノ直接ナ証明 (*by G. Ottaviani*) (= *The direct proof of P. Lévy's theorem by G. Ottaviani*), Zenkoku Sizyo Sugaku Danwakai **188** (1939), 524–527.

79. (With S. Kakutani) *Operator-theoretical treatment of Markoff's Process and mean ergodic theorem*, Ann. Math. (2) **42** (1941), 188–228.

80. *Markoff process with stationary uniform distribution*, Zenkoku Sizyo Sugaku Danwakai **189** (1939), 534–541 and **191** (1939), 640–647.

81. *Markoff chain* と *H*-定理 (= *Markoff chains and the H-theorem*), Isô Sûgaku **2** No. 1 (1939), 41–42.

82. *An ergodic theorem of Birkhoff-Khintchine type*, Zenkoku Sizyo Sugaku Danwakai **193** (1940), 64–66 and **195** (1940), 100–107.

83. (With S. Kakutani) *Compact* ナ空間ニ於ケル *transition process* (= *Transition processes in compact spaces*), Zenkoku Sizyo Sugaku Danwakai **196** (1940), 139–147.

84. *Ergodic theorems of Birkhoff-Khintchine's type*, Japan. J. Math. **17** (1940), 31–36.

85. *The Markoff process with a stable distribution*, Proc. Imp. Acad. Tokyo **16** (1940), 43–48.

86. *Individual ergodic theorem* ニ就テ (= *On the individual ergodic theorem*), Zenkoku Sizyo Sugaku Danwakai **199** (1940), 263–272.

87. *An abstract treatment of the individual ergodic theorem*, Proc. Imp. Acad. Tokyo **16** (1940), 280–284.

88. *Pythagorian ring* ニ就テ (= *On Pythagorian rings*), Zenkoku Sizyo Sugaku Danwakai **200** (1940), 293–299, **201** (1940), 306–314 and **203** (1940), 368–374.

89. *On the theory of spectra*, Proc. Imp. Acad. Tokyo **16** (1940), 378–383.

90. *Hilbert* 空間の *Hermite* 作用素の作る "環" に就て (*Pythagorian ring*) (= *On "rings" of Hermitian operators in Hilbert spaces (Pythagorian rings)*), Isô Sûgaku **3** No.1 (1940), 64–66.

91. *Regularly convex set* ニ就テ (= *On regularly convex sets*), Zenkoku Sizyo Sugaku Danwakai **207** (1940), 473–478.

92. (With M. Fukamiya) *On regularly convex sets*, Proc. Imp. Acad. Tokyo **17** (1941), 49–52.

93. *On vector lattice with a unit*, Proc. Imp. Acad. Tokyo **17** (1941), 121–124.

94. 単位ヲ有スル *vector-lattice* ニ就テ (= *On vector lattices with unit*), Zenkoku Sizyo Sugaku Danwakai **211** (1941), 94–98 and **212** (1941), 119.

95. *Spectral theorem* に就いて (= *On the spectral theorem*), Isô Sûgaku **3** No. 2 (1941), 47–49 and **4** No. 1 (1942), 62.

96. *Radon-Nikodym* ノ定理ニ就テ (= *On the Radon-Nikodyn theorem*), Zenkoku Sizyo Sugaku Danwakai **217** (1941), 274–282 and **218** (1941), 328–330.

97. *Vector lattices and additive set functions*, Proc. Imp. Acad. Tokyo **17** (1941), 228–232.

98. アルキメデス的 *vector lattice* ノ表現 (= *A representation of Archimedean vector lattices*), Zenkoku Sizyo Sugaku Danwakai **225** (1941), 499–502.

99. (With M. Fukamiya) 単位ヲ有スル *vector* 束ニ就テ, III (= *On vector lattices with unit*), Zenkoku Sizyo Sugaku Danwakai **226** (1941), 574–578 and **227** (1941), 643–644.

100. 平均エルゴード定理及び個別エルゴード定理 (= *The mean and the individual ergodic theorems*), Tokyo Buturi Gakko Zassi (東京物理學校雜誌) **50** (1941), 463–465.

101. (With M. Fukamiya) *On vector lattice with a unit, II*, Proc. Imp. Acad. Tokyo **17** (1941), 479–482.

102. *Vector* 束の表現に就て (= *On representations of vector lattices*), Isô Sûgaku **4** No. 1 (1942), 1–7 and **4** No. 2 (1942), 54–55.

103. *On the representation of the vector lattice*, Proc. Imp. Acad. Tokyo **18** (1942), 339–342.

104. *Spectral theorem* ノ証明ニ就テ (= *On a proof of the spectral theorem*), Zenkoku Sizyo Sugaku Danwakai **240** (1942), 1232–1237.

105. (With T. Nakayama) 準順序環及ビソノ應用ニツイテ (= *On semi-ordered rings and their application*), Zenkoku Sizyo Sugaku Danwakai **242** (1942), 1309–1320 and **250** (1943), 158–163.

106. (With T. Nakayama) *On the semi-ordered ring and its application to the spectral theorem*, Proc. Imp. Acad. Tokyo **18** (1942), 555–560.

107. 淡中氏相對定理ノ証明ニ就テ (= *On a proof of Tannaka's duality theorem*), Zenkoku Sizyo Sugaku Danwakai **246** (1942), 1591–1595.

108. (With T. Nakayama) *On the semi-ordered ring and its application to the spectral theorem. II*, Proc. Imp. Acad. Tokyo **19** (1943), 144–147.

109. *On the duality theorem of non-commutative compact groups*, Proc. Imp. Acad. Tokyo **19** (1943), 181–183.

110. *L. Pontrjagin* ノ双對定理ノ証明ニツイテ (= *On a proof of L. Pontrjagin's duality theorem*), Zenkoku Sizyo Sugaku Danwakai **254** (1943), 318–321.

111. のるむ環トスペクトル定理 (= *Normed rings and spectral theorems*), Zenkoku Sizyo Sugaku Danwakai **255** (1943), 362–364.

112. *Normed rings and spectral theorems*, Proc. Imp. Acad. Tokyo **19** (1943), 356–359.

113. 線型函数方程式の可解性 (= *Solvability of linear functional equations*), Isô Sûgaku **5** No. 1 (1943), 23.

114. *Normed rings and spectral theorems, II*, Proc. Imp. Acad. Tokyo **19** (1943), 466–470.

115. *Normed rings and spectral theorems, III*, Proc. Imp. Acad. Tokyo **20** (1944), 71–73.

116. *Normed rings and spectral theorems, IV*, Proc. Imp. Acad. Tokyo **20** (1944), 183–185.

117. *Normed rings and spectral theorems, V*, Proc. Imp. Acad. Tokyo **20** (1944), 269–273.

118. (With T. Iwamura) *Equivalence of two topologies of Abelian groups*, Proc. Imp. Acad. Tokyo **20** (1944), 451–453.

119. *Normed rings and spectral theorems, VI*, Proc. Imp. Acad. Tokyo **20** (1944), 580–583.

120. *On the representation of functions by Fourier integrals*, Proc. Imp. Acad. Tokyo **20** (1944), 655–660.

121. *On the unitary equivalence in general Euclid space*, Proc. Japan Acad. **22** (1946), 242–245.

122. *Unitary equivalence* について (= *On the unitary equivalence*), Sûgaku (数学) **1** (1948), 88–89.

123. *Simple Markoff process with a locally compact phase space*, Math. Japon. **1** (1948), 99–103.

124. *On the differentiability and the representation of one-parameter semi-group of linear operators*, J. Math. Soc. Japan **1** No. 1 (1948), 15–21.

125. 線型作用素の作る *1-parameter* 準群 (= *One parameter semi-groups of linear operators*), Sûgaku **1** (1948), 201–203.

126. *An operator-theoretical treatment of temporally homogeneous Markoff process*, J. Math. Soc. Japan **1** No. 3 (1949), 244–253.

127. 三次元球面上の *Brown* 運動 (= *Brownian motion on the sphere in the 3-space*), Sûgaku **1** (1949), 327–329.

128. *Brownian motion on the surface of the 3-sphere*, Ann. Math. Statist. **20** (1949), 292–296.

129. *Compact Riemann* 空間の上での *Fokker-Planck* 偏微分方程式の積分 (= *Integration of the Fokker-Planck equation on compact Riemannian spaces*), Sûgaku **2** (1949), 166–168.

130. *Integration of Fokker-Planck's equation in a compact Riemannian space*, Ark. Mat. **1** No. 9 (1949), 71–75.

131. *An extension of Fokker-Planck's equation*, Proc. Japan Acad. **25** No. 9 (1949), 1–3.

132. *On Titchmarsh-Kodaira's formula concerning Weyl-Stone's eigenfunction expansion*, Nagoya Math. J. **1** (1950), 49–58, Correction. ibid. **6** (1953), 187–188.

133. *Stochastic processes built from flows*, Proc. Japan Acad. **26** No. 8 (1950), 1–3.

134. *Integration of Fokker-Planck's equation with a boundary condition*, J. Math. Soc. Japan **3** (1951), 69–73.

135. *Integrability of the backward diffusion equation in a compact Riemannian space*, Nagoya Math. J. **3** (1951), 1–4.

136. (With E. Hewitt) *Finitely additive measures*, Trans. Amer. Math. Soc. **72** (1952), 46–66.

137. *Fokker-Planck 方程式およびその積分について* (= *On the Fokker-Planck equation and its integral*), Sûgaku **3** (1951), 129–136.

138. *A theorem of Liouville's type for meson equation*, Proc. Japan Acad. **27** (1951), 214–215.

139. *On Brownian motion in a homogeneous Riemannian space*, Pacific J. Math. **2** (1952), 263–270.

140. *On the existence of the resolvent kernel for elliptic differential operator in a compact Riemann space*, Nagoya Math. J. **4** (1952), 63–72.

141. *An ergodic theorem associated with harmonic integrals*, Proc. Japan Acad. **27** (1951), 540–543.

142. *Homogeneous space の上の Brown 運動の定義* (= *A definition of Brownian motions on homogeneous spaces*), Sûgaku **4** (1952), 32–34.

143. *On the integration of diffusion equations in Riemannian spaces*, Proc. Amer. Math. Soc. **3** (1952), 864–873.

144. *Fokker-Planck 方程式およびその積分について, II*, Sûgaku **4** (1952), 145–150.

145. *On Cauchy's problem in the large for wave equations*, Proc. Japan Acad. **28** (1952), 396–403.

146. *On the fundamental solution of the parabolic equation in a Riemannian space*, Osaka Math. J. **5** (1953), 65–74.

147. *Titchmarsh-Kodaira の固有函数による展開定理の証明について* (= *On a proof of the eigenfunction expansion theorem of Titchmarsh-Kodaira*), Sûgaku **5** (1952), 228.

148. *On the integration of the temporally inhomogeneous diffusion equation in a Riemannian space*, Proc. Japan Acad. **30** (1954), 19–23.

149. *On the integration of the temporally inhomogeneous diffusion equation in a Riemannian space, II*, Proc. Japan Acad. **30** (1954), 273–275.

150. *Semi-group theory and the integration problem of diffusion equations*, Proc. of Internat. Congress of Mathematicians, Amsterdam, 1954, Vol. **1**, pp. 405–420.

151. *On the generating parametrix of the stochastic processes*, Proc. Nat. Acad. Sci. U.S.A. **41** (1955), 240–244.

152. *A characterization of the second order elliptic differential operators*, Proc. Japan Acad. **31** (1955), 406–409.

153. *An operator-theoretical integration of the wave equation*, J. Math. Soc. Japan **8** (1956), 79–92.

154. *Semi-group* の理論による波動方程式の積分 (= *Integration of the wave equation by the theory of semi-groups*), Sûgaku **8** (1956), 65–71.

155. *An operator-theoretical integration of the temporally inhomogeneous wave equation*, J. Fac. Sci. Univ. Tokyo Sect. I, **7** (1957), 463–466.

156. *On the reflexivity of the space of distribution*, Sci. Papers Coll. Gen. Ed. Univ. Tokyo **7** (1957), 151–155.

157. *On the differentiability of semi-groups of linear operators*, Proc. Japan Acad. **34** (1958), 337–340.

158. 発展方程式に関連して (= *Regarding the evolution equation*), Sûgaku **10** (1959), 205–211.

159. *An abstract analyticity in time for solutions of a diffusion equation*, Proc. Japan Acad. **35** (1959), 109–113.

160. *Fractional powers of infinitesimal generators and the analyticity of the semi-groups generated by them*, Proc. Japan Acad. **36** (1960), 86–89.

161. *On a class of infinitesimal generators and the integration problem of evolution equations*, Proc. Fourth Berkeley Sympos. on Math. Stat. and Prob., 1961, Vol. **II**, pp. 623–633.

162. *Ergodic theorems for pseudo-resolvents*, Proc. Japan Acad. **37** (1961), 422–425.

163. *Abelian ergodic theorems in locally convex spaces*, Ergodic Theory, Proc. Internat. Sympos. Tulane Univ. 1961, pp. 293–299.

164. *On the integration of the equation of evolution*, J. Fac. Sci. Univ. Tokyo Sect. I, **9** (1963), 397–402.

165. *Holomorphic semi-groups in a locally convex linear topological space*, Osaka Math. J. **15** (1963), 51–57.

166. *Holomorphic semi-groups*, Séminaire sur les Équations aux Dérivées Partielles (1963–1964), II, 68–76.

167. *Positive pseudo-resolvents and potentials*, Proc. Japan Acad. **41** (1965), 1–5.

168. *Time dependent evolution equations in a locally convex space*, Math. Ann. **162** (1965), 83–86.

169. *A perturbation theorem for semi-groups of linear operators*, Proc. Japan Acad. **41** (1965), 645–647.

170. *On holomorphic Markov processes*, Proc. Japan Acad. **42** (1966), 313–317.

171. *Positive resolvents and potentials (An operator-theoretical treatment of Hunt's theory of potentials)*, Z. Wahrscheinlichkeitstheorie und Verw. Gebiete **8** (1967), 210–218.

172. (With T. Watanabe and H. Tanaka) *On the pre-closedness of the potential operator*, J. Math. Soc. Japan **20** (1968), 419–421.

173. *The existence of the potential operator associated with an equicontinuous semigroup of class* (C_0), Studia Math. **31** (1968), 531–533.

174. *On the potential operators associated with Brownian motions*, J. Analyse Math. **23** (1970), 461–465.

175. *On the pre-closedness of Hunt's potential operators and its applications*, Proc. Intern. Conf. on Functional Analysis and Related Topics, Tokyo, 1969, Univ. of Tokyo Press, 1970, pp. 324-331..

176. *Abel* 型エルゴード定理と *Hunt* のポテンシャル論 (= *Abelian ergodic theorem and Hunt's theory of potentials*), Sûgaku **22** (1970), 81–91.

177. *On the existence and a characterization of abstract potential operators*, Proc. Troisième Colloq. sur l'Analyse Fonctionnelle, Liège, 1970, pp. 129–136.

178. *Abstract potential operators on Hilbert space*, Publ. Res. Inst. Math. Sci. **8** (1972), 201–205.

179. ブラウン運動と等速度運動 拡散方程式の一面 (= *Brownian motions and uniform motions   An aspect of diffusion equations*), Sûrikagaku (数理科学) **146** (1975), 9–14.

180. *A note on Malmquist's theorem on first order algebraic differential equations*, Proc. Japan Acad. **53** (1977), 120–123.

181. (With S. Okamoto) *A note on Mikusiński's operational calculus*, Proc. Japan Acad. Ser. A Math. **56** (1980), 1–3.

182. *A brief biography on Takakazu Seki (1642?–1708)*, Math. Intelligencer **3** (1980/81), no.3, 121–122.

183. *A note on the fundamental theorem of calculus*, Proc. Japan Acad. Ser. A Math. **57** (1981), 241.

184. *Some aspects of E. Hille's contribution to semi-group theory*, Integral Equations Operator Theory **4** (1981), 311–329.

185. 函数解析５０年 (= *50 years of functional analysis*), Sûgaku **34** (1982), 354–364.

186. *A simple complement to Mikusiński's operational calculus*, Studia Math. **77** (1983), 95–98.

187. 角谷静夫氏の学士院賞・恩賜賞の受賞に際して (= *Award of Academy-Imperial Prize to Shizuo Kakutani*), Sûgaku **34** (1982), 351–353.

188. (With S. Matsuura) *A note on Mikusiński's proof of the Titchmarsh convolution theorem*, Conference in Modern Analysis and Probability, New Haven, Conn., 1982, Contemp. Math., **26** (1984), 423–425.

189. *The algebraic derivative and Laplace's differential equation*, Proc. Japan Acad. Ser. A Math. **59** (1983), 1–4.

190. *Chaos からの距離* (= *A metric from the chaos*), Sûrikagaku Suppl. *Entropy and Chaos* (1984), 111–114.

191. *Sato, a perfectionist*, Algebraic Analysis Vol. **1**, Academic Press, Boston, 1988, pp. 17–18.

Prepared by Hikosaburo Komatsu with the cooperation of Akira Kaneko and Nobuhisa Iwasaki.

Interpolation Theorems in Several Complex Variables and Applications

C.A. Berenstein [*]   T. Kawai [†]   D.C. Struppa [‡]

Dedicated to the memory of the late Professor Kôsaku Yosida

1   Introduction.

In this report we present a new approach to the treatment of classical gap theorems, which relies on a rather deep theory of interpolation varieties in spaces of analytic functions satisfying growth conditions at infinity.

Our first source of inspiration has been a classical result known as the Fabry gap theorem [17], for which we have already provided several different proofs, interpretations and extensions [2],[11],[12]. The major interest of our approach lies in an intermediate result (Theorem 4.1) which shows that, for a large class of holomorphic functions, every translation operator can be represented by a constant coefficients linear differential operator of infinite order, which acts upon the sheaf of holomorphic functions as a sheaf homomorphism. Otherwise stated, the germ of such a function at a point far away from, say, the origin, can be found by the local study of the function at the origin. This result, which applies, in particular, to the Riemann theta-zerovalue function $\theta(t)$, has a rather long, and interesting history, which may be worthwhile relating. In [20] Sato, Kashiwara and one of us (T.K.) proved the $\mathbf{R}$-constructibility of the solution complex of an $\mathbf{R}$-holonomic complex, a simple example of which is given by the equations that $\theta(t)$ solves. An explicit cohomology calculation [13] together with the above mentioned result, guarantees that, in the one-dimensional case, for any real number a, there exists a linear differential operator G_a such that $\theta(t + a) = G_a\theta(t)$. In the present work, a simple and explicit construction for G_a is given, which applies to a rather wide class of functions, including the higher-dimensional Riemann theta-zerovalue. An explicit construction of the operator G_a in the one-dimensional case was also done in [9], which is based on a prototype of the interpolation theorem for holomorphic functions of one complex variable (e.g. Boas [5]).In view of the highly technical nature of some of the arguments concerning interpolating varieties, we have given, in this paper, the main philosophy and ideas of the work, and we refer the reader to the forthcoming [1] for fuller details.

---

[*]Department of Mathematics, University of Maryland, College Park, MD 20742, USA

[†]RIMS, Kyoto University, Kyoto 606-01, Japan

[‡]Department of Mathematics, University of Calabria,87036 Rende (CS), Italy and Department of Mathematics, George Mason University, Fairfax, Virginia 22030 USA

The plan of the paper is as follows: section 2 is devoted to the discussion of the classical Fabry gap theorem, [11] [12], which will help the reader's understanding of this article, in terms of hyperfunctions and infinite order differential operators. Section 3 contains the relevant material on interpolating varieties (and [1] is the key reference). Finally section 4 contains our main results and applications.

Acknowledgments. The authors are grateful to Japan Association for Mathematical Sciences for supporting the preparation of this report on several occasions.

2 The Fabry Gap Theorem.

Let us begin by giving one of the versions under which the original Fabry gap theorem can be stated. We fix the following notations:

$$\Pi_+ = \{z \in \mathbf{C} : \operatorname{Im} z > 0\},$$

$$\Delta_\delta = \{z \in \mathbf{C} : |z| < \delta\},$$

and for Ω a subset of $\mathbf{C}$ (or of $\mathbf{C}^n$ later in this paper), $H(\Omega)$ will denote the ring of holomorphic functions on Ω. We can then state the Fabry gap theorem as follows:

Theorem 2.1 *Suppose that a given series $\sum_{j=1}^{+\infty} c_j \exp(ia_j z)$, $a_j \in \mathbf{R}, c_j \in \mathbf{C}$, converges uniformly, on compact subsets of Π_+, to a function $f \in H(\Pi_+)$. Suppose, moreover, that f extends holomorphically to $\Pi_+ \cup \Delta_\delta$ and that the sequence $\{a_j\}$ is lacunary, in the sense that the number of elements of the set*

$$\{j : c_j \neq 0, |a_j| \leq n\}$$

is $o(n)$, when n goes to infinity .

Under these hypotheses, the series $\sum_{j=1}^{+\infty} c_j \exp(ia_j z)$ converges locally uniformly on $\Pi_+ - i\delta = \{z \in \mathbf{C} : \operatorname{Im} z > -\delta\}$, to a holomorphic continuation of f.

We note, incidentally, that the hypothesis requires that $|a_n|/n \to +\infty$ as n diverges to $+\infty$; as a consequence if σ_c and σ_a denote, respectively, the abscissa of convergence and the abscissa of absolute convergence of the series, one has that $\sigma_c = \sigma_a$. We also note that for Dirichlet series it is well known that uniform convergence on compact sets is equivalent to uniform convergence on half planes, but our method also applies to the case in which the coefficients c_j could be polynomials and the frequencies a_j may not be confined to lie on a single direction.

Remark 2.1 We wish to point out two key elements in the statement of Theorem 2.1 which are somehow essential for the understanding of our ideas. The first such element is the fact that the exponential series representation for $f(z)$ entails that f is the "solution" of some kind of equation

$$T(f) = 0,$$

where the a_j are the elements of the "spectrum" of T. The second key element is the lacunarity (i.e. the sparsity) of the sequence $\{a_j\}$, which implies that T must satisfy some "special" conditions that guarantee the locality of T.

Remark 2.2 A few words should be said on the history of the Fabry Gap Theorem, after its original formulation and proof by Fabry (who gave a theorem of overconvergence of power series, which only after an exponential change of variables becomes Theorem 2.1). The first fresh look at Fabry's theorem was given by Ehrenpreis, in his fundamental work [6], which has been the source of inspiration for all of our works in this direction. As it is well known, in [6] a substantially new approach to linear constant coefficients partial differential operators is introduced. As a consequence, Ehrenpreis proposed to replace the series of exponentials of Theorem 2.1 with a series $\sum_{j=0}^{+\infty} f_j$ of solutions to suitable systems of partial differential equations, $\vec{D}_j(f_j) = 0$; in the particular case of Fabry's theorems, one has $D_j = \frac{d}{dz} - ia_j$. This approach, while quite interesting, leads to many difficulties, and Ehrenpreis himself suggested that one should try to look at convolution equations instead. Later on, one of us (T.K.) exploited this idea in [11], [12], and provided an approach based on a special class of convolution operators, namely infinite order differential operators. Soon after the publication of these works, two of us (C.B. and D.S.) extended these results to the case of more general convolution operators, and introduced the notion of interpolating variety as a tool for this kind of problems, [2], [3]. Finally, in [1], whose results are for the first time presented here, a more subtle interpolation theory is developed, and deeper results can finally be obtained, even when returning to the case of infinite order differential operators.

Before we get to explain these ideas, we wish to set the stage by providing some basic notations on convolution operators and infinite order differential operators, and by giving the general idea in [11],[12], emphasizing its relation to this article.

To begin with, let us recall that if f is an entire function, $f \in H(\mathbf{C}^n)$, and μ is an analytic functional, i.e. $\mu \in H'(\mathbf{C}^n)$, the strong dual of $H(\mathbf{C}^n)$, then their convolution can be defined by:

$$\mu * f(z) := < \mu, \zeta \to f(z + \zeta) >,$$

where $<,>$ denotes the duality bracket beetwen H and H'. It is easily seen that, with this definition, $\mu * f$ is an entire function. A crucial tool for the study of convolution operators is the so-called Fourier-Borel transformation which is defined on the elements of H' as follows:

$$\mathcal{F}\mu(z) = \hat{\mu}(z) := < \mu, \zeta \to \exp(iz\zeta) > .$$

For any $\mu \in H'(\mathbf{C}^n)$, $\hat{\mu}$ is an entire function, and if μ is carried by a compact set K (see, e.g.,[7] for more details on these matters), then one can prove that for any $\varepsilon > 0$ there exists $A_e > 0$ such that:

$$\mid \mu(z) \mid \le A_\varepsilon \exp(H_K(z) + \varepsilon \mid z \mid),$$

where $H_K(z)$ is the supporting function of the convex hull of K. A very special (but very important) case occurs when the carrier K is the origin,i.e. $K = \{0\}$. In this case we say that $\mu*$ is an infinite order differential operator and one can show that, via Fourier-Borel transformation, we have a (topological) isomorphism between the ring of infinite order differential operators and the ring $Exp_0(\mathbf{C}^n)$ of functions of infraexponential type, defined as

$$Exp_0(\mathbf{C}^n) = \{f \in H(\mathbf{C}^n) :\mid f(z) \mid \le a_\varepsilon \exp(\varepsilon \mid z \mid)\}$$

Let us note that the reason for the name that these special convolution operators have, is due to the fact that if μ is carried by the origin and if

$$\hat{\mu}(z) = \sum a_{i_1\ldots i_n} z_1^{i_n} \ldots z_n^{i_n},$$

then for any entire function f,

$$\mu * f(z) = \sum a_{i_1\ldots i_n} \frac{\partial^{i_1+\ldots+i_n} f}{\partial z_1^{i_1} \ldots \partial z_n^{i_n}}(z),$$

The traditional example for an infraexponential type function in one variable is

$$\cosh(\sqrt{z}) := \sum_{n=0}^{+\infty} \frac{z^n}{(2n)!},$$

and if $\hat{\mu}(z) = \cosh(\sqrt{z})$ one has

$$\mu * f(z) = \sum_{n=0}^{+\infty} \frac{1}{(2n)!} \frac{d^n f}{dz^n}(z)$$

Let us conclude this section by providing the interpretation given in [11] of the Fabry gap theorem as a propagation of analiticity theorem for hyperfunctions. Its relation with our approach will become most manifest in Corollary 4.1. For simplicity we discuss the one-dimensional case. Let

$$f_+(z) = \sum_{a_n \geq 0} c_n \exp(ia_n z),$$

and

$$f_-(z) = \sum_{a_n < 0} c_n \exp(ia_n z),$$

define two holomorphic functions, respectively $f_+ \in H(\Pi_+)$ and $f_- \in H(\Pi_-)$. Assume, moreover, that the sequence $\{a_n\}$ is lacunary in the sense of Theorem 2.1. Then if we denote by f the hyperfunction defined on $\mathbf{R}$ by the pair (f_+, f_-) of holomorphic functions, we can restate the Fabry gap theorem as follows: *if f is real analytic in a neighborhood of the origin then f is real analytic everywhere.*

To prove so, just note that if f is real analytic in a neighborhood of the origin, then both f_+ and f_- must extend holomorphically to a neighborhood of the origin. At this point, by the Fabry gap theorem, both f_+ and f_- must extend holomorphically to a full neighborhood of the real line, and therefore f is real analytic everywhere: even more, f is holomorphic in the open set $\{z \in \mathbf{C} :| \operatorname{Im} z |< \lambda\}$. A generalization of this discussion to the higher-dimensional case can be found in [11]. The results given there are, however, not very suited for practical applications, mainly because subsidiary finite order equations, which are important in analyzing concrete objects such as the Riemann theta-zerovalue (cf.[10],[13],[19],[20]) are not taken into account. Our approach in this article, as will be seen in Corollary 4.1, naturally resolves this trouble.

3 Interpolating Varieties.

Because of the highly technical content of this section, we will try to have to convey the main ideas, while referring the reader to [1], [4] for the precise definitions, statements and proofs. Let us begin by fixing some notations: Let $p(z)$ be a non-negative plurisubharmonic function on $\mathbf{C}^n$ such that:

(i) $\log(1+\mid z \mid) = O(p(z))$,

(ii) There exists $D_1, D_2 > 0$ such that $\mid z_1 - z_2 \mid \leq 1$ implies $p(z_1) \leq D_1 p(z_2) + D_2$.

We then define two important algebras of entire functions: $A_p = A_p(\mathbf{C}^n) = \{f \in H(\mathbf{C}^n) :$ there exist $A, B > 0$ such that $\mid f(z) \mid \leq A \exp(Bp(z))\}$, $A_{p,0} = A_{p,0}(\mathbf{C}^n) = \{f \in H(\mathbf{C}^n) :$ for any $\varepsilon > 0$ there exists A_ε such that $\mid f(z) \mid \leq A_\varepsilon \exp(\varepsilon p(z))\}$.

The main example we are concerned with is $p(z) = \mid z \mid$, in which case $A_p = Exp(\mathbf{C}^n)$ is the space of functions of exponential type, and $A_{p,0} = Exp_0(\mathbf{C}^n)$ is the space of functions of infraexponential type. Other interesting and important examples arise for

$$p(z) = \mid z \mid^\rho, \rho > 1,$$

see [3], or for

$$p(z) = \mid \operatorname{Im} z \mid + \log(1+\mid z \mid),$$

in which case A_p is the space of Fourier transforms of compactly supported distributions.

Let us give a very rough description of what the interpolation problem in these spaces might be (the reader is referred to [4], where these thechniques and ideas were first introduced).

Let $f_1, ..., f_m \in A_p(\mathbf{C}^n)$, and suppose that the variety

$$V = \{z \in \mathbf{C}^n : f_1(z) = ... = f_m(z) = 0\}$$

is discrete, say

$$V = \{z_k\}_{k=1}^{+\infty}.$$

The following natural question can be asked:

(a) Given a sequence $\{a_k\}_{k=1}^{+\infty}$ of complex numbers such that

$$\mid a_k \mid \leq A \exp(Bp(z_k)),$$

does there exist $f \in A_p(\mathbf{C}^n)$ such that

$$f(z_k) = a_k ?$$

(b) Define

$$H(V) := \frac{H(\mathbf{C}^n)}{(f_1, ..., f_m)},$$

with $(f_1, ..., f_m)$ the closed ideal generated by $f_1, ..., f_m$ in $H(\mathbf{C}^n)$, and define $A_p(V)$ in some "natural" way (see e.g., [4]). Consider the canonical restriction

$$\rho = \rho_V : H(\mathbf{C}^n) \to H(V) :$$

how can one describe $\rho_V(A_p(\mathbf{C}^n))$?

(c) When is

$$\rho_V(A_p(\mathbf{C}^n)) = A_p(V)?$$

When this happens, we will say that the variety V is **interpolating**.

(d) Ask the same questions for $A_{p,0}$ and its related spaces.

The key notion, when dealing with such interpolation problems, is that of **slowly decreasing ideal**. This notion was originally developed,for the case of principal ideals, by Ehrenpreis in the late fifties (and Sato independently), but, in the form we need, it is essentially due to Berenstein and Taylor [4]. We provide here a simplified formulation of what is defined and used in our forthcoming [1].

Let $f_1, ..., f_n \in A_p(\mathbf{C}^n)$, and suppose $V = V(\vec{f})$ is discrete; given two positive constants ε, C, we define the set

$$S(\vec{f}; \varepsilon, C) := \{z \in \mathbf{C}^n : \sum_{j=1}^{n} \mid f_j(z) \mid \leq \varepsilon \exp(-Cp(z))\}.$$

Definition 3.1 *We say that* $\vec{f} = (f_1, ..., f_m)$ *is* **slowly decreasing** *if there are positive constants* ε, C, C_1, C_2 *such that the set* $S(\vec{f}; \varepsilon, C)$ *has bounded connected components, which are such that if* z, ζ *belong to the same component, then*

$$p(z) \leq C_1 p(\zeta) + C_2.$$

The main result in our forthcoming [1] is the following:

Theorem 3.1 *If a variety* V *is defined by a slowly decreasing n-tuple* $\vec{f}$, *then* $\rho_V(A_p(\mathbf{C}^n))$ *can be explicitly described. Moreover if, for a suitable choice of* $\varepsilon, C > 0$, *every component of* $S(\vec{f}; \varepsilon, C)$ *contains at most one point of* V *(counting multiplicities), then* $\rho_V(A_p(\mathbf{C}^n)) = A_p(V)$.

Proof The proof of this result, which can also be stated for non discrete varieties, is fairly complicated, and comprises a large part of [1]. We content ourselves by saying that it is based on the so-called Jacobi interpolation formula (first used in this connection by Delsarte), and on Hörmander's L^2 techniques.

Remark 3.1 A completely parallel result can be stated and proved for the space $A_{p,0}$ and, in particular, for the space Exp_0 of infraexponential type functions. To do so, two of us (T.K. and D.S.) employed in [15], the natural projective limit structure of Exp_0 together with some results due to Kaneko [8] which permit the application of Hörmander's thechniques to this situation. In the forthcoming [1], on the other hand, we manage to express Exp_0 as a sort of inductive limit of A_p spaces, and we get, in this way, a more natural approach to the problem.

Because the definition of slow decrease would seem rather misterious, let us provide here a typical example of a slowly decreasing n-tuple, at least for the case $p(z) = \mid z \mid$. Fix $r \leq n$ in such a way that $f_1(z_1), f_2(z_2), ..., f_r(z_r)$ define a discrete variety in the space $\mathbf{C}^r$ whose variables are $z_1, ..., z_r$. Consider now, for $j > r$, functions such as

$$f_j(z) = z_j^{d_j} - P_j(z_1, ..., z_r),$$

with d_j being positive integers and P_j being polynomials. One can show that the family $(f_1, ..., f_n)$ is slowly decreasing according to Definition 3.1. More general classes of examples will be given in [1].

4 Applications.

We now conclude our report by showing how Theorem 3.1 can play a role in the study of the questions addressed in the introduction.

Theorem 4.1 *Let a_l, $l = 1, ..., r$ be r sequences of positive real numbers satisfying the following conditions:*

$$\lim_{n \to +\infty} \frac{n}{a_l(n)} = 0 \quad (l = 1, ..., r);$$

there exists a positive C such that

$$| a_l(n) - a_l(n') | \geq C | n - n' | \quad (l = 1, ..., r; n, n' \in \mathbf{N});$$

and let $p_{r+1}, ..., p_n$ be polynomials such that the variety

$$V = \{z \in \mathbf{C}^n : p_{r+1}(z) = ... = p_n(z) = 0\} \cap$$

$$\{z \in \mathbf{C}^n : \text{ for any } l \text{ there exists } n(l) \in \mathbf{N} : z_l^2 = -a_l(n(l))\}$$

is discrete. Let $c(\nu)$, for $\nu \in V$, be complex numbers such that, for any $\varepsilon > 0$, there exists $C_\varepsilon > 0$ such that

$$| c(\nu) | \leq C_\varepsilon \exp(\varepsilon | \nu |), \nu \in V \cap i\mathbf{R}^n,$$

and set

$$f(x) = \sum_{\nu \in V \cap i\mathbf{R}^n} c(\nu) \exp(< \nu, x >).$$

Then $f(x)$ defines a hyperfunction on $\mathbf{R}^n$ and, for any real vector $a \in \mathbf{R}^n$, there exists a linear differential operator with constant coefficients

$$G(a; \frac{\partial}{\partial x_1}, ..., \frac{\partial}{\partial x_n})$$

such that

$$f(x + a) = G(a; \frac{\partial}{\partial x_1}, ..., \frac{\partial}{\partial x_n}) f(x).$$

<u>Proof</u> The fact that $f(x)$ defines a hyperfunction on $\mathbf{R}^n$ is well known. Define now the following sequence, indexed on the variety V:

$$\tilde{c}(\nu; a) = \exp(< \nu, a >), \text{ if } \nu \in V \cap i\mathbf{R}^n$$

and

$$\tilde{c}(\nu; a) = 0, \text{ if } \nu \in V \cap (\mathbf{C}^n \setminus i\mathbf{R}^n)$$

Note that, trivially,

$$| \tilde{c}(\nu; a) | \leq C_\varepsilon \exp(\varepsilon | \nu |),$$

hence, by Theorem 3.1 and Remark 3.1 there exists $G \in Exp_0$ such that

$$\rho_V(G) = \tilde{c}.$$

If we now consider the infinite order differential operator whose symbol is the function G, we have:

$$G(f) = \sum c(\nu)G(\exp(<\nu, x>)) =$$

$$\sum c(\nu)G(\nu; a)\exp(<\nu, x>) = \sum c(\nu)\exp(<\nu, a + x>) = f(a + x).\square$$

An immediate consequence is the following:

Corollary 4.1 *Let f be the hyperfunction as in Theorem 4.1. Then if f is real analytic in a neighborhood of the origin, it is actually real analytic everywhere.*

<u>Proof</u> It is sufficient to notice that the local nature of G preserves and propagates analyticity.$\square$

As we have anticipated, Theorem 4.1 applies immediately to the Riemann theta-zerovalue $\theta(t)$,or, to be more precise, to its boundary value taken from the domain $\text{Im } t >> 0$, for t an $m \times m$ symmetric matrix. Indeed, if we identify the totality of $m \times m$ complex symmetric matrices $t = (t_{jk})_{1 \leq j,k \leq m}$ with $\mathbf{C}^{m(m+1)/2}$, we see (cf.[10],[20]) that the variety

$$V = \cap_{l=1}^m \{\zeta \in \mathbf{C}^{m(m+1)/2} : \zeta_{ll} = in_l^2 \text{ for some } n_l \in \mathbf{N}\}$$

$$\cap(\cap_{1 \leq j,k \leq m} \{\zeta \in \mathbf{C}^{m(m+1)/2} : 4\zeta_{jj}\zeta_{kk} = \zeta_{jk}^2\})$$

satisfy the conditions for being interpolating; therefore if $s = (s_{jk})_{1 \leq j,k \leq m}$ is a real symmetric matrix and $\mu \in \mathbf{Z}^m$, then, for any bounded sequence $c(\mu)$ the series

$$f(s) = \sum_{\mu \in \mathbf{Z}^m} c(\mu)\exp(\pi i < s\mu, \mu >)$$

satisfies the hypothesis of Theorem 4.1. In the particular case in which, $c(\mu) \equiv 1$, we have the m-dimensional Riemann theta-zerovalue function

$$\theta(t) = \sum_{\mu \in \mathbf{Z}^m} \exp(\pi i < t\mu, \mu >)$$

to which, therefore, Theorem 4.1 applies.

REFERENCES

[1]C.A. Berenstein, T. Kawai and D.C. Struppa, Interpolating varieties and the Fabry-Ehrenpreis-Kawai gap theorem, forthcoming.

[2]C.A. Berenstein and D.C. Struppa, On the Ehrenpreis-Kawai gap Theorem, Publ R.I.M.S. Kyoto University 23 (1987),565-574.

[3]C.A. Berenstein and D.C. Struppa, Convolution equations and Dirichlet series, Publ. R.I.M.S. Kyoto University 24 (1988), 783-810

[4]C.A. Berenstein and B.A. Taylor, Interpolations problems in $\mathbf{C}^n$ with applications to harmonic analysis, J. Analyse Math. 38 (1980),188-254.

[5]R.P. Boas, Entire Functions, Academic Press, 1959.

[6]L. Ehrenpreis, Fourier Analysis in Several Complex Variables, Wiley-Interscience, New York, 1970.

[7]L. Hörmander, An Introduction to Complex Analysis in Several Variables, Van Nostrand, Princeton, 1966.

[8]A. Kaneko, On continuation of regular solutions of partial differential equations to compact convex sets II, J. Fac. Sci. Univ. Tokyo 18 (1971),415-433.

[9]M. Kashiwara and T. Kawai, A differential relation between $\theta(t+a)$ and $\theta(t)$, RIMS Technical Report No 485, RIMS, Kyoto University. (1984)

[10]M. Kashiwara, T. Kawai and Y. Takei, The structure of cohomology groups associated with the theta-zerovalues, in Proc. of "Algebraical and Geometrical Aspects in Several Complex Variables", Cetraro 1989, C.A. Berenstein and D.C. Struppa eds.

[11]T. Kawai, The Fabry-Ehrenpreis gap Theorem for hyperfunctions, Proc Japan Acad. 60 A (1984), 276-278.

[12]T. Kawai, The Fabry-Ehrenpreis gap theorem and linear differential equations of infinite order, Amer. J. Math. 109 (1987), 57-64.

[13]T. Kawai, An example of a complex of linear differential operators of infinite order, Proc. Japan Acad. 59A(1983),113-115.

[14]T. Kawai, Some remarks on microlocal analysis of θ-functions, Suken-kokyuroku, no. 410, RIMS, Kyoto University, 76-87,(1980), (in Japanese).

[15]T.Kawai and D.C. Struppa, On the existence of holomorphic solutions of systems of linear differential equations of infinite order and with constant coefficients, Int. J. Math. 1 (1990), 83-82.

[16]T. Kawai and Y. Takei, "Fundamental Principle" and θ-zerovalue, Sukenkokyuroku, no. 675, RIMS, Kyoto University, 79-86 (1988), (in Japanese).

[17]N. Levinson, Gap and Density Theorems, A.M.S.,1940

[18]M. Sato, Pseudo-differential equations and theta function, Astérisque 2-3 (1973), 286-291.

[19]M. Sato, M. Kashiwara and T. Kawai, Linear differential equations of infinite order and theta functions, Advances in Math. 47 (1983),300-325.

[20]M. Sato, M. Kashiwara and T. Kawai, Microlocal analysis of theta functions, Adv. Studies in Pure Math. 4 (1984) 267-289.

[21]M. Sato, T. Kawai and M. Kashiwara, Microfunctions and pseudo-differential equations, Lecture Notes in Math. 286, Springer, Berlin-Heidelberg- New York, 1973, 265-529.

New energies for harmonic maps and liquid crystals

HAÏM BREZIS

Université P. et M. Curie
and
Rutgers University

Dedicated to the memory of Professor K. Yosida

Let Ω be a smooth bounded domain in $\mathbf{R}^3$. Consider the vector-valued Sobolev space

$$H^1(\Omega; \mathbf{R}^3) = \{u \in L^2(\Omega; \mathbf{R}^3); \frac{\partial u}{\partial x_i} \in L^2(\Omega; \mathbf{R}^3), \ i = 1, 2, 3\}$$

and the subset

$$H^1(\Omega; S^2) = \{u \in H^1(\Omega; \mathbf{R}^3); \ |u(x)| = 1 \ \text{a.e.}\}.$$

We shall often fix a boundary condition $g : \partial\Omega \to S^2$ and set

$$H_g^1 = H_g^1(\Omega; S^2) = \{u \in H^1(\Omega; S^2); \ u = g \ \text{on} \ \partial\Omega\}.$$

Note that for *any* (smooth) g the set H_g^1 is not empty. For example, if Ω is the unit ball then

$$u(x) = g(x/|x|) \in H_g^1$$

since $\int_\Omega \left| \nabla \left[\frac{x}{|x|} \right] \right|^2 < \infty.$

Similarly, we may consider

$$C_g^1 = C_g^1(\bar\Omega; S^2) = \{u \in C^1(\bar\Omega; \mathbf{R}^3); \ |u(x)| = 1 \ \forall x \ \text{and} \ u = g \ \text{on} \ \partial\Omega\}.$$

This set is nonempty if and only if $\deg(g, \partial\Omega) = 0$. Here $\deg(g, \partial\Omega)$ denotes the Brouwer degree (or winding number) of g.

A natural quantity to study is

(1) $$\min_{u \in H_g^1} \int_\Omega |\nabla u|^2$$

where $|\nabla u|^2 = \sum_{i,j=1}^3 \left[\frac{\partial u_i}{\partial x_j} \right]^2$. With the help of standard functional analysis methods (see e.g. Yosida [25]) it is easy to see that the minimum in (1) is achieved (it suffices to consider a minimizing sequence and pass to the limit using the lower-semicontinuity of the integral under weak convergence).

Every solution of (1) satisfies the corresponding Euler equation

$$
(2) \quad
\begin{cases}
-\Delta u = u|\nabla u|^2 & \text{in} \quad \Omega \\
|u(x)| = 1 & \text{in} \quad \Omega \\
u = g & \text{on} \quad \partial\Omega
\end{cases}
$$

which is a nonlinear system of 3 equations coupled through the Lagrange multiplier $|\nabla u|^2$.

Equation (2) is easily derived by writing that

$$
\int_\Omega |\nabla u|^2 \leq \int_\Omega \left| \nabla \left[\frac{u + \varepsilon v}{|u + \varepsilon v|} \right] \right|^2
$$

with $v \in C_0^\infty(\Omega; \mathbf{R}^3)$ and then taking $\varepsilon \to 0$. Weak H^1 solutions of (2) are called *weakly harmonic maps*.

The important question of regularity is quite delicate and has two different aspects:

Question 1: How regular are the minimizers of (1)?

Question 2: How regular are weak H^1 solutions of the system (2)?

The answer to Question 1 is now rather complete. A result of Schoen-Uhlenbeck [23] asserts that if u is a minimizer of (1), then u is smooth except on a finite set of points. The behavior of u near the singularities is fully understood. In a joint work with Coron and Lieb [9] we proved that if u is a minimizer and x_0 is a singularity then there is a rotation R of S^2 such that

$$
u(x) \simeq \pm R \frac{x - x_0}{|x - x_0|} \quad \text{as} \quad x \to x_0.
$$

Of course R depends on x_0 and in general we have different rotations for the various singularities.

By contrast, the answer to Question 2 is far from being clear. T. Rivière [21], [22] has constructed remarkable examples of weak H^1 solutions of (2). More precisely let ω be the largest open set in Ω on which u (or a representative in the same equivalence a.e. class) is continuous. It is well-known that a weak solution of (2) which is continuous must be C^∞ and thus u is C^∞ on ω. The complement of ω is the singular set S. One example of Rivière [21] shows that S could be a line. Another striking example of Rivière [22] shows that one may even have weak H^1 solutions of (2) such that $S = \Omega$. The construction of Rivière relies on the use of a variant of the relaxed energy described below.

Question 3: It would be very interesting to decide whether any *arbitrary* closed set in Ω can be the singular set of some weakly harmonic map.

Let us mention that there is a subset of the weakly harmonic maps called stationary harmonic maps: they are also stationary with respect to variations of the domain, i.e.

$$\frac{d}{dt}\int_\Omega |\nabla(u \circ \varphi_t)|^2_{|t=0} = 0$$

where φ_t is the solution of

$$\frac{d\varphi}{dt} = X(\varphi), \quad \varphi(0) = \text{id}$$

and X is an arbitrary smooth vector-field from Ω into $\mathbf{R}^3$ such that $X = 0$ on $\partial\Omega$. An interesting result of C. Evans [14] asserts that if u is a stationary harmonic map then $\mathcal{H}^1(S) = 0$ where $\mathcal{H}^1$ denotes the one dimensional Hausdorff measure.

Question 4: It is not known whether this result is optimal or whether stationary harmonic maps have only a finite number of point singularities.

Let us now return to the minimizers of (1), also called minimizing harmonic maps. One may wonder whether they really do have singularities. For example, if g is a constant then the minimizer is a constant and it has no singularity. When g is "close" to a constant it is reasonable to believe that the corresponding minimizers have no singularities:

Question 5: It would be interesting to prove that there is an explicit universal constant ε_0 (for example 8π) such that

$$\int_{\partial\Omega} |\nabla_T g|^2 < \varepsilon_0 \Rightarrow u \text{ has no singularity}$$

where ∇_T denotes the tangential gradient.

For a general (smooth) boundary condition g the corresponding minimizers *do have singularities*:

1) If $\deg(g, \partial\Omega) \neq 0$ then singularities are forced to exist from *topological considerations*. Indeed any element in H_g^1 must have at least one point singularity; otherwise we could homotopy g continuously to a constant and then $\deg(g, \partial\Omega)$ would vanish.
2) If $\deg(g, \partial\Omega) = 0$, singularities are not forced by the topology. Minimizers may still have singularities for *energy reasons*. This is a consequence of the following:

Theorem 1. *There exist smooth boundary conditions* $g : \partial\Omega \to S^2$ *such that* $\deg(g, \partial\Omega) = 0$ *and*

$$(3) \qquad \min_{H_g^1}\int_\Omega |\nabla u|^2 < \inf_{C_g^1}\int_\Omega |\nabla u|^2.$$

The conclusion of Theorem 1 is called a *gap phenomenon* and was originally discovered by Hardt-Lin [18] for special (dumbbell-shaped) domains. However this phenomenon is quite general and does not depend on the special shape of Ω. In [7] I gave a construction of such g when Ω is a ball (and the same holds in a general domain). In fact, given any $\varepsilon > 0$ one can construct a g such that $\deg(g, \partial\Omega) = 0$,

$$\min_{H_g^1} \int_\Omega |\nabla u|^2 < \varepsilon \quad \text{and} \quad \inf_{C_g^1} \int |\nabla u|^2 > \frac{1}{\varepsilon}.$$

The gap phenomenon is extremely interesting; it has many consequences and raises some challenging problems:

1) Suppose we take a g given by Therem 1. Then the minimizers in (1) must have singularities; otherwise we would have a contradiction with (3). This means that the system creates singularities in order to *reduce its energy* — not because of a topological constraint. This fact is somewhat surprizing because one usually thinks of singularities as carrying much energy. Note, however, that in our situation a singularity like $x/|x|$ has limited energy since

$$\int_{B_R} \left| \nabla \left[\frac{x}{|x|} \right] \right|^2 = 8\pi R.$$

Of course, the system also decides where it is most "advantageous" to place the singularities and also how many singularities. It is very difficult to estimate the number of singularities (in terms of g and Ω). A result of Almgren-Lieb [1] asserts that

$$\#(\text{singularities}) \leq C \int_{\partial\Omega} |\nabla_T g|^2,$$

where C depends on Ω. However their argument is quite involved and indirect and does not yield an explicit bound for C.

2) Another consequence of (3) is that C_g^1 is not dense in H_g^1. Here in fact the boundary condition is irrelevant; $C^1(\bar{\Omega}; S^2)$ is not dense in $H^1(\Omega; S^2)$ as was first proved in [23]. For example, if $0 \in \Omega$ the map $u(x) = x/|x|$ belongs to $H^1(\Omega; S^2)$ but there is no sequence (u_n) in $C^1(\bar{\Omega}; S^2)$ such that $u_n \to u$ in H^1. This follows from a simple degree argument. Note that the usual technique of regularizing by convolution can not be applied since $\rho_n * u$ does not take its values in S^2.

One may formulate the same question in a more general framework following Eells and Lemaire [12]. Let M and N be two compact manifolds with $\partial N = \phi$ (we may, for example, assume that M is a domain in $\mathbf{R}^N$). Consider the Sobolev class $W^{1,p}(M, N)$ which consists of all maps u from M into N such that $\nabla u \in L^p$ with $1 \leq p < \infty$.

Question 6: Is $C^1(M, N)$ dense in $W^{1,p}(M, N)$?

a) If $p > \dim M$ the answer is yes for every target N. Indeed, if $u \in W^{1,p}(M, N)$ then $u \in C(M, N)$ by the Sobolev imbedding theorem. Thus $\rho_n * u \to u$ uniformly. Thus $\rho_n * u$ takes its values in some ε-neighbourhood N_ε of N and then we may take $u_n = P(\rho_n * u)$ where P is the projection from N_ε onto N.

b) If $p = \dim M$ the above technique still applies but it is a little more delicate to check that $\rho_n * u$ takes its values in N_ε (see [23]).

c) If $p < \dim M$ the answer is provided by a deep result of Bethuel [3] (see also an earlier work of Bethuel-Zheng [6]):

Theorem 2. *Assume $p < \dim M$. Then the answer to Question 6 is positive if and only if $\pi_k(N) = 0$ where $k = [p]$ is the integer part of p.*

In particular for $p = 2$ we have:
— if $\dim M = 1$ or 2 then $C^1(M, N)$ is always dense in $H^1(M, N)$;
— if $\dim M \geq 3$ then $C^1(M, N)$ is dense in $H^1(M, N)$ if and only if $\pi_2(N) = 0$.

In the cases where the answer to Question 6 is negative there are still many other questions of interest:

Question 7: Can one approximate any map u in $W^{1,p}(M, N)$ by a sequence of maps (u_n) which are smooth except on small sets, for example, points or lines or smooth manifolds of lower dimension?

Question 8: Given u in $W^{1,p}(M, N)$, is there a sequence (u_n) in $C^1(M, N)$ such that $u_n \rightharpoonup u$ weakly in $W^{1,p}$?

Question 9: What is the closure of $C^1(M, N)$ in $W^{1,p}(M, N)$ for the strong $W^{1,p}$ topology? In other words can one find simple criteria to decide whether a given map u in $W^{1,p}$ is approximable by smooth maps?

Some partial answers to Questions 7, 8 and 9 are given in [2], [3] and [17] (see also Lemmas 1 and 3 below) but many problems are still open.

3) A very natural question related to (3) is:

Question 10: Given a smooth $g : \partial\Omega \to S^2$ with $\deg(g, \partial\Omega) = 0$, is

$$\inf_{C^1_g} \int_\Omega |\nabla u|^2 \quad \text{achieved?}$$

This seems to be a difficult problem but the answer would be quite interesting, both for mathematical and physical reasons:

A) **Mathematical motivation:** If the answer to Question 10 is positive then we would have a smooth harmonic map (i.e., a smooth solution of (2)) with a prescribed boundary condition g. This is already an open problem:

Question 11: Given a smooth $g : \partial\Omega \to S^2$ with $\deg(g, \partial\Omega) = 0$, is there a C^1 solution of (2)?

In principle, it could happen that the answer to Question 10 is negative but that the answer to Question 11 is positive. This would be a rather intriguing situation.

B) **Physical motivation**: This comes from the theory of liquid crystals (see e.g. [10], [11], [13] and [20]). A liquid crystal consists of long molecules with some orientational ordering. At every point $x \in \Omega$ (Ω = the container) the optical axis of the molecule has an orientation which we denote $u(x)$ — called the director — so that $|u(x)| = 1$. Physicists observe that u depends smoothly on x, except at the defects of the liquid crystals which consists of points (and sometimes lines). Every configuration u has a distortion energy — the simplest model being

$$E(u) = \int_\Omega |\nabla u|^2$$

(there are more complicated models, but they all correspond to *integrals over* Ω of expressions which are quadratic in ∇u; see e.g. [11] and [13]).

Principles of physics assert that stable configurations correspond to minimizers of the energy, but the class of testing maps is not clearly specified. Should one choose

$$\inf_{H^1_g} \int_\Omega |\nabla u|^2$$

or rather

$$\inf_{C^1_g} \int_\Omega |\nabla u|^2 \ ?$$

At first it would seem that

$$\inf_{H^1_g} \int_\Omega |\nabla u|^2$$

is the right choice because it yields the lowest energy among all finite energy configurations (possibly with singularities).

However in many experiments singularities move towards each other and eventually they coalesce — sometimes after 2 or 3 days (see e.g. [10]). This phenomenon of "collapse of singularities" is not well understood. It suggests that configurations with singularities are not stable and then it seems more reasonable to search for stable configuration by minimizing

$$\inf_{C^1_g} \int_\Omega |\nabla u|^2.$$

But this yields to a paradoxical situation in view of the gap phenomenon. Why would the system choose to end up in a configuration without singularities at a high energy level when it could reduce its energy by introducing singularities.

The explanation of this paradox that we propose is that the usual energy

$$E(u) = \int_\Omega |\nabla u|^2$$

does not describe accurately the energy of a configuration with singularities. The new energy that we propose was first introduced by the author in 1988 (see [8]) and its properties are investigated in a joint work with Bethuel and Coron [5]. It comes from a standard mathematical procedure called *relaxation* which has been considered by many author (one of the earliest systematic treatments may be found in Serrin [24]).

Let us first illustrate this procedure on a very elementary example. Consider the expression

$$E(u) = \int_0^1 [1 + (u')^2]^{1/2}$$

which makes sense for functions $u \in C^1([0,1]; \mathbf{R})$. The relaxed energy of a general function $u : [0,1] \to \mathbf{R}$ (which need not be C^1) is defined by

$$E_{\text{rel}}(u) = \inf\{\liminf E(v_n)\}$$

where the infimum is taken over all sequences (v_n) in $C^1([0,1]; \mathbf{R})$ such that $v_n \to u$ a.e. For example if u is a piecewise smooth function then it is easy to prove that

$$E_{\text{rel}}(u) = \int_0^1 [1 + (u')^2]^{1/2} + |\text{jumps}|.$$

We shall follow a similar approach in our setting. We first fix some boundary condition $g : \partial\Omega \to S^2$ such that $\deg(g, \partial\Omega) = 0$. Then we define

(4)
$$E_{\text{rel}}(u) = \inf\{\liminf \int_\Omega |\nabla v_n|^2\}$$

where the infimum is taken over all sequences (v_n) in $C_g^1(\bar{\Omega}, S^2)$ such that $v_n \to u$ a.e.

It is a remarkable fact that $E_{\text{rel}}(u)$ has a rather *explicit* form, at least in the case where u has just a finite number of singularities. In order to describe E_{rel} we have to introduce some notations and preliminary facts.
Set
$$R_g = \{u \in H_g^1; u \text{ is smooth except at a finite number of points}\}$$

(of course these points are not prescribed).

Lemma 1. (Bethuel-Zheng [6]). *R_g is dense in H_g^1 (for the strong H^1 topology).*

This result is related to Question 7 and is extremely useful — especially since C_g^1 is not dense in H_g^1. A number of results are first proved for maps in R_g and then extended to all maps in H_g^1 by density.

For a given $u \in R_g$ we now define $L(u)$, the *length of a minimal connection connecting the singularities of u* (this notion was first introduced in [9]). Let x_0 be a singularity of u. Let $\deg(u, x_0)$ be a degree of u restricted to any small sphere centered at x_0 (clearly it is independent of its radius r, provided r is sufficiently small). We disregard the singularities of degree zero. We assume first, for simplicity, that all remaining singularities have degree ± 1. We denote by $p_1, p_2, \cdots, p_k$ the singularities of degree $+1$ and $n_1, n_2, \cdots, n_k$ the singularities of degree -1. Note that the number of positive singularities is the same as the number of negative singularities since the total degree is $\deg(g, \partial\Omega) = 0$. Set

$$L(u) = \min_\sigma \sum_{i=1}^{k} d(p_i, n_{\sigma(i)})$$

where $d(a, b)$ denotes the geodesic distance of a to b in Ω and the minimum is taken over all permutations σ of the integers $\{1, 2, \cdots, k\}$. In the case of degrees d_i which are not ± 1 we repeat the corresponding point $|d_i|$ times in the list $(p_i), (n_i)$.

A very useful representation formula for $L(u)$, due to Coron, Lieb and myself [9] is given in the following

Lemma 2. *Assume $u \in R_g$ then*
(5)

$$L(u) = \sup\left\{ \frac{1}{4\pi} \int_\Omega D(u) \cdot \nabla\zeta - \frac{1}{4\pi} \int_{\partial\Omega} (\text{Jac } g)\zeta;\ \zeta \in C^1(\Omega; \mathbf{R}) \text{ and } \|\nabla\zeta\|_{L^\infty} \le 1 \right\}$$

where the D-field is defined by

$$D(u) = (u \cdot u_y \times u_z,\ u \cdot u_z \times u_x,\ u \cdot u_x \times u_y).$$

The proof of Lemma 2 is based on the fact that, in the sense of distributions,

(6)
$$\operatorname{div} D(u) = 4\pi \sum_i \deg(u, a_i)\delta_{a_i}$$

where the summation is taken over all singularities of u and δ_a denotes the Dirac mass at a.

The proof of Lemma 2 also relies on the following general lemma.

Lemma 3. *Assume M is a metric space with $2k$ points denoted $p_1, p_2, \cdots, p_k$ and $n_1, n_2, \cdots, n_k$. Then*

$$L = \min_\sigma \sum_{i=1}^{n} d(p_i, n_{\sigma(i)}) = \sup\left\{ \sum_{i=1}^{k} \zeta(p_i) - \zeta(n_i);\ \zeta : M \to \mathbf{R} \text{ with } \|\zeta\|_{\text{Lip}} \le 1 \right\}.$$

For an elementary proof of Lemma 3 we refer to [7].

The conclusion of Lemma 2 suggests that $L(u)$ makes sense even for maps in H_g^1 (not just R_g) since $D(u) \in L^1(\Omega; \mathbf{R}^3)$. This is indeed the case. More precisely we have proved in [5]:

Lemma 4: *There is a universal constant C such that*

(7)
$$|L(u) - L(v)| \le C\|u - v\|_{H^1}(\|u\|_{H^1} + \|v\|_{H^1}) \quad \forall u, v \in R_g.$$

Using Lemma 1 and Lemma 4 we may now extend the function L by continuity and density to all of H_g^1. The extended L is still given by formula (5) for every $u \in H_g^1$. Note that $L(u) = 0$ if $u \in C_g^1$.

We now return to the relaxed energy E_{rel} defined in (4) and we may now give an exlicit formula for E_{rel} due to Bethuel, Coron and myself [5]:

Theorem 3. *We have*

$$E_{\text{rel}}(u) = \int_\Omega |\nabla u|^2 + 8\pi\, L(u) \quad \forall u \in H_g^1.$$

In other words, the relaxed energy corresponds to the usual energy plus a *nonlocal* term involving the *interaction of singularities*. The additional term is a price that the system has to pay for creating a $(+1)$ and a (-1) singularities far apart. Let us mention that Giaquinta, Modica and Soucek, using their theory of Cartesian currents have also reached the conclusion that it is quite natural to add the term $8\pi\, L(u)$ to the usual energy (see [15] and [16]).

The proof of Theorem 3 is given in [5] and relies on several steps:

Step 1: The function

$$u \longmapsto \int_\Omega |\nabla u|^2 + 8\pi\, L(u)$$

is lower semicontinuous for the weak H^1 topology.

This follows easily from the representation formula (5) and the fact that for each *fixed* ζ the function

$$u \longmapsto \int_\Omega |\nabla u|^2 + 2\int_\Omega D(u) \cdot \nabla\zeta$$

is lower semicontinuous for the weak H^1 topology. Note that the functions $u \longmapsto \int_\Omega |\nabla u|^2$ is lower semicontinuous for the weak H^1 topology but the funtion $u \longmapsto L(u)$ is *not* weakly lower semicontinuous. Any combination $\int_\Omega |\nabla u|^2 + 8\pi\lambda\, L(u)$ is weakly lower semicontinuous for $0 \le \lambda \le 1$ but it is not weakly lower continuous for $\lambda > 1$.

Step 2: Given any $u \in R_g$ there is a sequence (v_n) in C_g^1 such that $v_n \to u$ a.e. and $v_n \rightharpoonup u$ weakly in H_g^1 and such that

$$\limsup \int_\Omega |\nabla v_n|^2 \le \int_\Omega |\nabla u|^2 + 8\pi\, L(u).$$

The proof of Step 2 is a beautiful construction of Bethuel [2]. A pair of singularities (p, n) is eliminated by modifying u only in some ε-tubular neighbourhood of the geodesic line joining p to n. The resulting map u_ε has now a higher energy but the increase may be controlled; more precisely

$$\int_\Omega |\nabla u_\varepsilon|^2 \leq \int_\Omega |\nabla u|^2 + 8\pi \, d(p, n) + \varepsilon.$$

This construction is an elaborate version of the "dipole construction" of Coron, Lieb and myself (see [9] and also [7]).

Step 3: Proof of Theorem 3 concluded.

From Step 2 we deduce that

$$E_{rel}(u) \leq \int_\Omega |\nabla u|^2 + 8\pi \, L(u) \quad \forall u \in R_g.$$

Using Lemma 1 and the fact that L is continuous from the strong H^1 topology we see that

$$(8) \qquad E_{rel}(u) \leq \int_\Omega |\nabla u|^2 + 8\pi \, L(u) \quad \forall u \in H^1_g.$$

On the other hand if u is any map from Ω into S^2 and (v_n) is any sequence in C^1_g such that $v_n \to u$ a.e. and

$$\int_\Omega |\nabla v_n| \leq C$$

we deduce that (for a subsequence) $v_n \rightharpoonup u$ weakly in H^1. Hence by Step 1

$$\int_\Omega |\nabla u|^2 + 8\pi \, L(u) \leq \liminf \left\{ \int_\Omega |\nabla v_n|^2 + 8\pi \, L(v_n) \right\} = \liminf \int_\Omega |\nabla v_n|^2.$$

Hence, by definition of E_{rel},

$$(9) \qquad \int_\Omega |\nabla u|^2 + 8\pi \, L(u) \leq E_{rel}(u).$$

The conclusion of Theorem 3 follows from (8) and (9).

We may now return to the quantity we started with, namely

$$\inf_{C^1_g} \int_\Omega |\nabla u|^2$$

and relate it to the relaxed energy in the following:

Theorem 4: *We have*

$$\min_{H^1_g} E_{rel}(u) \ \text{ is achieved}$$

and any minimizer of E_{rel} satisfies (2).

In addition,

(10) $$\inf_{C_g^1} \int_\Omega |\nabla u|^2 = \min_{H_g^1} E_{\text{rel}}(u).$$

Moreover **if**

$$\inf_{C_g^1} \int_\Omega |\nabla u|^2 \quad \text{is achieved by some } u_0 \in C_g^1$$

then u_0 is a minimizer for E_{rel} on H_g^1.

Conversely if $u_1 \in H_g^1$ is a minimizer for

$$\min_{H_g^1} E_{\text{rel}}(u)$$

and **if** $u_1 \in C_g^1$ then u_1 is a minimizer for

$$\inf_{C_g^1} \int_\Omega |\nabla u|^2.$$

Proof. The fact that $\min_{H_g^1} E_{\text{rel}}$ is achieved follows from Step 1 above and standard functional analysis (see e.g. Yosida [25]). If $u \in C_g^1$ then

$$\int_\Omega |\nabla u|^2 = E_{\text{rel}}(u)$$

and thus

$$\inf_{C_g^1} \int_\Omega |\nabla u|^2 = \inf_{C_g^1} E_{\text{rel}}(u) \geq \min_{H_g^1} E_{\text{rel}}(u).$$

Conversely, it follows from the definition of the relaxed energy that

$$\inf_{C_g^1} \int |\nabla u|^2 \leq \min_{H_g^1} E_{\text{rel}}(u)$$

and thus equality in (10) holds.

The fact that any minimizer of E_{rel} satisfies the Euler equation is a direct consequence of the following property of L:

(11) $$L\left[\frac{u + \varepsilon v}{|u + \varepsilon v|}\right] = L(u) \quad \forall u \in H_g^1, \quad \forall v \in C_0^\infty(\Omega; \mathbf{R}^3), \quad \forall \varepsilon \quad \text{small.}$$

Note that (11) is obvious when $u \in R_g$ since $\frac{u + \varepsilon v}{|u + \varepsilon v|}$ and u have the same singularities. For a general $u \in H_g^1$ it suffices to argue by density with the help of Lemma 1 and Lemma 4.

Suppose $u_0 \in C_g^1$ satisfies

$$\int_\Omega |\nabla u_0|^2 = \inf_{C_g^1} \int |\nabla u|^2.$$

Then clearly

$$E_{\text{rel}}(u_0) = \min_{H_g^1} E_{\text{rel}}(u).$$

Suppose $u_1 \in C_g^1$ satisfies

$$E_{\text{rel}}(u_1) = \min_{H_g^1} E_{\text{rel}}(u).$$

Then

$$\int_\Omega |\nabla u_1|^2 = \min_{H_g^1} E_{\text{rel}}(u) = \inf_{C_g^1} \int |\nabla u|^2.$$

Concluding remarks:

1) Let us return to Question 10. We still do not know whether

$$\inf_{C_g^1} \int_\Omega |\nabla u|^2 \quad \text{is achieved.}$$

But at least we have replaced this problem by a weaker formulation *which does have a solution*. This is a standard procedure in the calculus of variations. We may consider the minimizers of E_{rel} as "generalized solutions" of Question 10. In order to solve Question 10 we need a regularity result (which is still missing) for the minimizers of E_{rel}. One tentative approach is to introduce a family E_λ of energies connecting $E(u) = \int_\Omega |\nabla u|^2$ to $E_{\text{rel}}(u)$:

$$E_\lambda(u) = \int_\Omega |\nabla u|^2 + 8\pi\lambda\, L(u), \quad u \in H_g^1, \ \lambda \in [0, 1].$$

One can prove (see [4]) that for each $\lambda \in [0, 1)$ the minimizers of E_λ on H_g^1 exist and they are smooth except at a finite number of singularities. As λ increases from 0 to 1 we suspect that these singularities move towards each other and coalesce at $\lambda = 1$. Indeed, as λ increases the "penalty" term $L(u)$ becomes more "visible" and in order to decrease E_λ the system has an "advantage" to reduce the distance between singularities. What happens as $\lambda = 1$ is unclear. Residual singularities of degree zero could appear (a similar phenomenon occurs in [19]).

The fact that minimizers of the relaxed energy also yields weakly harmonic maps is a very powerful tool for constructing new harmonic maps (see e.g. [5], [21] and [22]).

2) Let us return to the physics. It would be very interesting to devise experiments in order to decide which energy is "prefered" by liquid crystals, E or E_{rel}? In general they lead to different minimizers. For example, if we are in the situation of a boundary condition g with a gap, such as in Theorem 1, let u be a minimizer for E on H_g^1. Then u has a finite number of singularities. We prove (see [5]) that u is *not* a minimizer for E_{rel}: one can lower a little E_{rel} by moving singularities closer to each other. If indeed E_{rel} is a prefered energy then the observed collapse of singularities would become a natural consequence.

3) Here is some further evidence in favor of E_{rel}:

23

a) E_{rel} has better "stability" properties with respect to smoothing. For example given $u \in H_g^1$ there is always a sequence (v_n) in C_g^1 such that $v_n \to u$ a.e. and

$$E_{\text{rel}}(v_n) \to E_{\text{rel}}(u).$$

This is not true for E.

b) If we consider higher order terms in the energy, for example

$$E_\varepsilon(u) = \int_\Omega |\nabla u|^2 + \varepsilon \int_\Omega |\nabla u|^p \quad \text{with} \quad p \geq 3$$

then one can prove that, for every g with $\deg(g, \partial\Omega) = 0$,

$$\min_{W_g^{1,p}} E_\varepsilon(u) \to \min_{H_g^1} E_{\text{rel}} \quad \text{as} \quad \varepsilon \to 0.$$

References

[1] F. Almgren–E. Lieb, Singularities of energy minimizing maps from the ball to the sphere: examples, counterexamples and bounds, Annals of Math. **129** (1988), 483–530.
[2] F. Bethuel, A characterization of maps in $H^1(B^3, S^2)$ which can be approximated by smooth maps, Ann. IHP Analyse Nonlinéaire **7** (1990), 269–286.
[3] F. Bethuel, The approximation problem for Sobolev maps between manifolds, Acta Math. **167** (1991), 153–206.
[4] F. Bethuel–H. Brezis, Regularity of minimizers of relaxed problems for harmonic maps, J. Funct. Anal. **101** (1991), 145–161.
[5] F. Bethuel–H. Brezis–J.M. Coron, Relaxed energies for harmonic maps, in *Variational Problems* (H. Berestycki, J.M. Coron and I. Ekeland ed.), Birkhauser (1990).
[6] F. Bethuel–X. Zheng, Density of smooth functions between two manifolds in Sobolev spaces, J. Funct. Anal. **80** (1988), 60–75.
[7] H. Brezis, Liquid crystals and energy estimates for S^2-valued maps, in *Theory and Applications of Liquid Crystals* (J. Ericksen and D. Kinderlehrer ed.), Springer (1987).
[8] H. Brezis, S^k-valued maps with singularities, in *Topics in the Calculus of Variations* (M. Giaquinta ed.), Lecture Notes in Math. vol. 1365, Springer (1989), 1–30.
[9] H. Brezis–J.M. Coron–E. Lieb, Harmonic maps with defects, Comm. Math. Phys. **107** (1986), 649–705.
[10] W. Brinkman–P. Cladis, Defects in liquid crystals, Physics Today, May 1982, 48–54.
[11] P. DeGennes, *The Physics of Liquid Crystals*, Clarendon Press, Oxford (1974).
[12] J. Eells–L. Lemaire, A report on harmonic maps, Bull. London Math. Soc. **10** (1978), 1–68.
[13] J. Ericksen–D. Kinderlehrer ed., *Theory and Applications of Liquid Crystals*, IMA Series vol. 5, Springer (1987).

[14] C. Evans, Partial regularity for stationary harmonic maps into the sphere, Archive Rat. Mech. Anal. **116** (1991), 101–113.

[15] M. Giaquinta–G. Modica–J. Soucek, Cartesian currents and variational problems for mappings into spheres, Ann. Sc. Norm. Sup. Pisa **16** (1989), 393–485.

[16] M. Giaquinta–G. Modica–J. Soucek, The Dirichlet energy of mappings with values into the sphere, Manuscripta Math. **65** (1989), 489–507.

[17] P. Hajlasz, Approximation of Sobolev mapping, Diff. and Int. Eq. (to appear).

[18] R. Hardt–F.H. Lin, A remark on H^1 mappings, Manuscripta Math. **56** (1986), 1–10.

[19] R. Hardt–F.H. Lin–C.C. Poon, Axially symmetric harmonic maps minimizing a relaxed energy (to appear).

[20] D. Kinderlehrer, Recent developments in liquid crystal theory, Proc. Conf. in honor of J.L. Lions (R. Dautray ed.), North Holland (1991).

[21] T. Rivière, Applications harmoniques de B^3 dans S^2 ayant une ligne de singularités, CRAS **313** (1991), 583–587.

[22] T. Rivière, Applications harmoniques de B^3 dans S^2 partout discontinues, CRAS **314** (1992), 719–723 and detailed paper to appear.

[23] R. Schoen–K. Uhlenbeck, A regularity theory for harmonic maps, J. Diff. Geom. **17** (1982), 307–335 and Boundary regularity and the Dirichlet problem for harmonic maps, J. Diff. Geom. **18** (1983), 253–268.

[24] J. Serrin, On the definition and properties of certain variational integrals, Trans. Amer. Math. Soc. **101** (1961), 139–167.

[25] K. Yosida, *Functional Analysis*, Springer (1965).

L^p Regularity for
Abstract Differential Equations

Giovanni Dore

Dipartimento di Matematica, Università di Bologna,
Piazza di Porta S.Donato 5, 40127 Bologna, Italy

Dedicated to the memory
of the late professor Kôsaku Yosida

1 Introduction

This paper is a survey of results about L^p regularity for a Cauchy problem for a differential equation in a Banach space.

We restrict ourself to autonomous equations. Results about non-autonomous equations can be found in [3], [8], [10] th. 7.22, [11], [14], [15], [33], [34] and [35].

From now on we denote with X a complex Banach space, with A a linear closed densely defined operator in X and with p a number in $]1, \infty[$.

We say that there is L^p regularity on the interval I (with $I = [0, T]$ or $I = [0, +\infty[$) for the Cauchy problem

(CP)
$$\begin{cases} u'(t) = Au(t) + f(t) & t \in I \\ u(0) = 0 \end{cases}$$

if for every $f \in L^p(I, X)$ there exists one and only one $u \in W^{1,p}(I, X) \cap L^p(I, \mathcal{D}(A))$ satisfying the Cauchy problem (we consider $\mathcal{D}(A)$ as a Banach space endowed with the graph norm).

From the closed graph theorem it follows easily that if there is L^p regularity then there exists $C \in \mathbf{R}^+$ such that

$$\|u\|_{L^p} + \|u'\|_{L^p} + \|Au\|_{L^p} \le C\|f\|_{L^p}.$$

We denote with $\mathcal{M}$ the operator in $L^p(I, X)$ such that $\mathcal{M}f = u$.

Every solution in the L^p sense is a weak solution in the sense of Ball (see [2]) and therefore if A generates a strongly continuous semigroup then this solution is given by the variation of constants formula

(VC)
$$u(t) = \int_0^t e^{(t-s)A} f(s)\, ds \ .$$

If A is bounded, then from (VC) it follows easily that there is L^p regularity for every $p \in]1, +\infty[$ on every bounded interval. Moreover if the semigroup e^{tA} has negative exponential type (i.e. there exist $M, m \in \mathbf{R}^+$ such that $\|e^{tA}\| \leq Me^{-mt}$ for $t \in [0, +\infty[)$ then there is also regularity on $[0, +\infty[$. Indeed to get regularity on $[0, +\infty[$ the summability of $\|e^{tA}\|$ would be sufficient, but this condition implies that the exponential type of the semigroup is negative ([22] ch. 4 th. 4.1).

If A generates a strongly continuous semigroup, but is unbounded, then (VC) gives a "mild" solution of (CP) whenever $f \in L^1_{\text{loc}}$. In this case if $f \in L^p$ then $u \in L^p$, but in general it is not true that $u(t) \in \mathcal{D}(A)$ for a.e. $t \in I$.

If the semigroup e^{tA} is analytic then $e^{tA}x \in \mathcal{D}(A)$ for every $t \in \mathbf{R}^+$ and $x \in X$, so that Au would be the convolution of $t \mapsto Ae^{tA}$ with f. However if A is unbounded then $\|e^{tA}\|$ behaves exactly as t^{-1} as $t \to 0^+$ (see [13] th. 4.1.8), so that it is not obvious that the operator defined through that convolution is bounded in L^p. Hence the problem of L^p regularity is a problem of singular integrals.

The theory of strongly continuous semigroups could suggest that it is more natural to study the continuous regularity for the Cauchy problem (CP), i.e. the existence and uniqueness of a solution $u \in C^1([0, T], X) \cap C([0, T], \mathcal{D}(A))$ for any continuous f. But Baillon ([1] th. 1) proved that if there is continuous regularity for an unbounded operator A that generates a strongly continuous semigroup, then the space X must contain a subspace isomorphic to c_0 (the space of sequences converging to 0). This fact implies that X cannot be reflexive. On the other hand there are good results of L^p regularity in some reflexive spaces.

2 Necessary conditions

As a first consequence of L^p regularity one can obtain estimates of the resolvent operator $(\lambda - A)^{-1}$.

Theorem 2.1 *Let X be a complex Banach space and A a linear closed densely defined operator in X such that there is L^p regularity on the interval $[0, +\infty[$ for (CP).*
Then

$$\{\lambda \in \mathbf{C} : Re\,\lambda \geq 0\} \subset \rho(A)$$

and

$$\exists C \in \mathbf{R}^+ \quad Re\,\lambda \geq 0 \implies \|(\lambda - A)^{-1}\| \leq \frac{C}{1 + |\lambda|} \,.$$

Let $\lambda \in \mathbf{C}$ such that $Re\,\lambda > 0$. Let $f_\lambda \in L^p([0, +\infty[, \mathbf{C})$ such that

$$f_\lambda(t) = \begin{cases} e^{\lambda t} & \text{for } 0 \leq t \leq \frac{1}{Re\,\lambda} \\ 0 & \text{for } t > \frac{1}{Re\,\lambda} \end{cases}$$

and, for $x \in X$, put

$$R_\lambda x = Re\,\lambda \int_0^\infty e^{-\lambda t} \mathcal{M}(f_\lambda x)(t)\, dt = \frac{Re\,\lambda}{\lambda} \int_0^\infty e^{-\lambda t} \frac{d\mathcal{M}(f_\lambda x)}{dt}(t)\, dt \,.$$

Since $\mathcal{M}$ is bounded from $L^p([0, +\infty[, X)$ to $W^{1,p}([0, +\infty[, X) \cap L^p([0, +\infty[, \mathcal{D}(A))$ it is easily seen that R_λ is a bounded linear operator in $L^p([0, +\infty[, X)$ whose norm is bounded by $C(1+|\lambda|)^{-1}$, with C that doesn't depend on λ; moreover a simple computation shows that R_λ is a right inverse of $\lambda - A$.

The fact that R_λ is also a left inverse requires only some technicalities (i.e. the fact that $\mathcal{M}(f_\lambda Ax) = A\mathcal{M}(f_\lambda x)$).

In this way the theorem is proved for $Re\ \lambda > 0$, but the estimate of $\|(\lambda - A)^{-1}\|$ doesn't blow up as $Re\ \lambda$ approaches to 0 so that it holds also on the imaginary axis.

This estimate shows that A generates a strongly continuous analytic semigroup e^{tA} of negative exponential type (see [36], [32] th. 3.3.1 and rem. 3.3.2).

If we have L^p regularity on a bounded interval, then there is no information on the behaviour of the semigroup for large values of t, so that the previous theorem become false. However we have:

Theorem 2.2 *Let X be a complex Banach space and A a linear closed densely defined operator in X such that there is L^p regularity on the interval $[0, T]$ for (CP).*

Then $\exists\ \delta \in \mathbf{R}^+$ such that

$$\{\lambda \in \mathbf{C} : Re\ \lambda \geq \delta\} \subset \rho(A)$$

and

$$\exists C \in \mathbf{R}^+ \quad Re\ \lambda \geq \delta \implies \|(\lambda - A)^{-1}\| \leq \frac{C}{1 + |\lambda|}\ .$$

In particular, A generates an analytic semigroup.

In this case we choose $\lambda \in \mathbf{C}$ such that $Re\ \lambda > \frac{1}{T}$ and we denote with f_λ the function in $L^p([0, T], \mathbf{C})$ such that

$$f_\lambda(t) = \begin{cases} e^{\lambda t} & \text{for } 0 \leq t \leq \frac{1}{Re\ \lambda} \\ 0 & \text{for } \frac{1}{Re\ \lambda} < t \leq T \end{cases}$$

and, for $x \in X$, we put

$$R_\lambda x = Re\ \lambda \int_0^T e^{-\lambda t} \mathcal{M}(f_\lambda x)(t)\ dt\ .$$

In this case R_λ is no more the inverse of $\lambda - A$, but the difference between $R_\lambda(\lambda - A)$ and the identity operator is small if $Re\ \lambda$ is large and so we obtain the theorem.

L^p regularity doesn't depend on the interval on which we consider (CP).
First of all we have:

Theorem 2.3 *Let X be a complex Banach space and A a linear closed densely defined operator in X.*

If there is L^p regularity on the interval $[0, +\infty[$ for (CP) then $\forall T \in \mathbf{R}^+$ there is L^p regularity on $[0, T]$.

If there is L^p regularity on $[0, +\infty[$ then A generates a strongly continuous analytic semigroup, so that there is uniqueness of the solution on $[0, T]$.

To get existence, if $f \in L^p([0, T], X)$, then let $\tilde{f}$ be the extension of f with 0 outside $[0, T]$; the restriction to $[0, T]$ of the solution of (CP) on $[0, +\infty[$ with non-homogeneous term $\tilde{f}$ is the required solution.

On the other hand we have the following theorem whose proof is due to T. Kato (personal communication):

Theorem 2.4 *Let X be a complex Banach space and A a linear closed densely defined operator in X.*

If there is L^p regularity on the interval $[0, T]$ for (CP) and the semigroup generated by A has negative exponential type, then there is L^p regularity on $[0, +\infty[$.

A generates an analytic semigroup (by theorem 2.2) so that, as we have already observed, if $f \in L^p([0, +\infty[, X)$ and u is an L^p solution of (CP), then it is

$$u(t) = \int_0^t e^{(t-s)A} f(s) \, ds \, .$$

This ensures the uniqueness of the solution.

Moreover from L^p regularity on $[0, T]$ and the fact that e^{tA} is analytic and has negative exponential type it follows easily that $u \in L^p([0, +\infty[, X)$ and $u(t) \in \mathcal{D}(A)$ for almost every $t \in \mathbf{R}^+$. Since the distributional derivative of u equals $Au + f$, it remains only to prove that Au belongs to $L^p([0, +\infty[, X)$.

The function $t \mapsto \|Ae^{tA}\|$ is summable on $[T, +\infty[$, so that $t \mapsto \int_0^{t-T} Ae^{(t-s)A} f(s) \, ds$ is the convolution of a function in L^1 with one in L^p hence it belongs to L^p.

It remains to estimate:

$$\int_0^T \left\| A \int_0^t e^{(t-s)A} f(s) \, ds \right\|^p dt + \int_T^{+\infty} \left\| A \int_{t-T}^t e^{(t-s)A} f(s) \, ds \right\|^p dt \, .$$

To this end for $k \in \mathbf{N} \cup \{0\}$ put $f_k = \chi_{[kT,(k+1)T[} f$ (we denote with χ the characteristic function of a set).

The above-mentioned quantity is equal to

$$\int_0^T \left\| A \int_0^t e^{(t-s)A} f_0(s) \, ds \right\|^p dt \, +$$

$$\sum_{j=1}^{\infty} \int_{jT}^{(j+1)T} \left\| A \int_{t-T}^{jT} e^{(t-s)A} f_{j-1}(s) \, ds + A \int_{jT}^t e^{(t-s)A} f_j(s) \, ds \right\|^p dt \le$$

$$2^{p-1} \left(\sum_{j=0}^{\infty} \int_{jT}^{(j+1)T} \left\| A \int_{jT}^t e^{(t-s)A} f_j(s) \, ds \right\|^p dt + \sum_{j=1}^{\infty} \int_{jT}^{(j+1)T} \left\| A \int_{t-T}^{jT} e^{(t-s)A} f_{j-1}(s) \, ds \right\|^p dt \right) =$$

$$2^{p-1} \left(\sum_{j=0}^{\infty} \int_0^T \left\| A \int_0^t e^{(t-s)A} f_j(s+jT) \, ds \right\|^p dt + \sum_{j=1}^{\infty} \int_0^T \left\| A \int_0^{T-t} e^{(t+s)A} f_{j-1}(jT-s) \, ds \right\|^p dt \right) .$$

From the regularity on $[0, T]$ it follows that the first sum is bounded by

$$C_1 \sum_{j=0}^{\infty} \|f_j\|_{L^p}^p$$

while the second is bounded by

$$\sum_{j=1}^{\infty} \int_0^T \left(\int_0^T \left\| A e^{(t+s)A} f_{j-1}(jT-s) \right\| ds \right)^p dt \ .$$

Now the semigroup e^{tA} is analytic, therefore there exists $C_2 \in \mathbf{R}^+$ such that for $t, s \in [0, T]$ $\|A e^{(t+s)A}\| \le C_2(t+s)^{-1}$ (see [13] th. 4.1.5), but the kernel $(t+s)^{-1}$ defines a bounded operator on $L^p([0,T], \mathbf{R})$, so that the last term is dominated by

$$C_3 \sum_{j=1}^{\infty} \|f_{j-1}\|_{L^p}^p \ .$$

Therefore we have

$$\int_0^T \left\| \int_0^t A e^{(t-s)A} f(s)\, ds \right\|^p dt + \int_T^{+\infty} \left\| \int_{t-T}^t A e^{(t-s)A} f(s)\, ds \right\|^p dt \le$$

$$C \sum_{j=0}^{\infty} \|f_j\|_{L^p}^p = C \|f\|_{L^p}^p$$

since the functions f_j have disjoint supports. This proves the theorem.

We proved that for the generator of a semigroup of negative exponential type L^p regularity is independent on the interval I on which we consider (CP). However if A generates a strongly continuous semigroup then, for a suitable $\lambda \in \mathbf{C}$, the operator $A + \lambda$ generates a semigroup of negative exponential type and L^p regularity (on a bounded interval) for the operator A is equivalent to L^p regularity for $A + \lambda$. This observation allows to get a result of L^p regularity on $[0, +\infty[$ from every result of L^p regularity on $[0, T]$.

As a consequence of this remark (and of theorems 2.3 and 2.4) we have:

Theorem 2.5 *Let X be a complex Banach space and A a linear closed densely defined operator in X.*

If there exists $T_0 \in \mathbf{R}^+$ such that there is L^p regularity on the interval $[0, T_0]$ for (CP) then $\forall T \in \mathbf{R}^+$ there is L^p regularity on $[0, T]$.

3 An example

One can ask if for every generator of an analytic semigroup there is L^p regularity. The answer is negative as the following example shows.

Let P be the Poisson semigroup in $L^2(\mathbf{R}, \mathbf{C})$, that is for $\phi \in L^2(\mathbf{R}, \mathbf{C})$, $t \in \mathbf{R}^+$ put

$$(P(t)\phi)(x) = \int_{\mathbf{R}} \frac{t}{\pi(y^2 + t^2)} \phi(x-y)\, dy \ .$$

P is an analytic semigroup whose generator is essentially the Hilbert transform of the derivative.

Next choose a Banach space E and consider the semigroup $\tilde{P}$ in $X = L^2(\mathbf{R}, E)$ defined by $\tilde{P}(t) = P(t) \otimes I_E$, i.e.

$$(\tilde{P}(t)f)(x) = \int_{\mathbf{R}} \frac{t}{\pi(y^2 + t^2)} f(x - y) \, dy .$$

Coulhon and Lamberton (see [9] § 5) proved that there is L^p regularity for the generator $\tilde{A}$ of the semigroup $\tilde{P}$ in X on $[0, T]$ if and only if E is a UMD space.

A Banach space X has the UMD property if martingale differences are unconditional in $L^p(\mathbf{R}, X)$ for $p \in \,]1, +\infty[$.

A space X is UMD if and only if it is ζ-convex, namely if it exists a symmetric function $\zeta : X \times X \to \mathbf{R}$, convex in each variable and such that $\zeta(0, 0) > 0$ and $\zeta(x, y) \leq \|x + y\|$ whenever $\|x\| \leq 1 \leq \|y\|$.

The interest of ζ-convex spaces in this context is due to the fact that a Banach space X is ζ-convex if and only if the Hilbert transform behaves well in X; i.e. if and only if the Hilbert transform is a bounded operator in $L^p(\mathbf{R}, X)$ for every $p \in \,]1, +\infty[$ or, equivalently, for some $p \in \,]1, +\infty[$.

If X is ζ-convex then it is super-reflexive and hence reflexive.

The class of of ζ-convex spaces is rather wide: every Hilbert space is ζ-convex, if X is ζ-convex and $p \in \,]1, +\infty[$ then every L^p space of X-valued functions is ζ-convex; in particular every L^p space of scalar-valued functions, with $1 < p < \infty$, is ζ-convex.

ζ-convexity is preserved by Banach space isomorphisms; closed subspaces and cartesian products of ζ-convex spaces are ζ-convex. If (X, Y) is an interpolation couple of ζ-convex spaces, then the complex interpolation spaces $[X, Y]_\theta$ and the real interpolation spaces $(X, Y)_{\theta, p}$, with $1 < p < \infty$, are ζ-convex spaces.

For more informations about ζ-convex spaces see Rubio de Francia's survey article [25] or Burkholder's one [6].

4   Sufficient conditions

As we observed in section 2, there is no difference in studying L^p regularity in bounded or unbounded intervals. In this section the results are quoted in the one or in the other case according to the original papers.

The first result on sufficient conditions for L^p regularity concerns the Hilbert space case (de Simon [29]).

Theorem 4.1 *Let H be a Hilbert space and A a closed linear operator in H that generates a strongly continuous analytic semigroup with negative exponential type.*

Then there is L^p regularity for (CP) on $[0, +\infty[$.

This theorem can be proved in three steps.

I. By means of the Fourier transform one obtains the regularity in L^2.

II. A multiplier theorem of Schwartz ([27] th. 2) ensures the boundedness of the operator $f \mapsto Au$ from L^1 to L^1_w, so that Marcinkiewicz interpolation theorem guarantees that such operator is bounded in L^p for $1 < p < 2$, therefore there is L^p regularity for $1 < p < 2$.

III. An argument of duality shows that there is regularity also for $2 < p < +\infty$.

This theorem completely solves the problem in the Hilbert space case.

The proof of this theorem doesn't work for operators in Banach spaces, since step I is based on the fact that the Fourier transform is an isomorphism in $L^2(\mathbf{R}, X)$ and this is true only if X is a Hilbert space.

Moreover step III requires the reflexivity of X. But this step can be modified so that it works with no reflexivity assumption; in fact it can be proved that the operator $f \mapsto Au$ is bounded from L^∞ to BMO and so by Stampacchia interpolation theorem (see [31] and [7]) it is bounded in L^p for $p > 2$.

Therefore in the case that X is a Banach space, one must have some substitute for step I, and so what can be proved is the following (Cannarsa-Vespri [8] th. 1.7, Vespri [33] th. 1.1, Coulhon-Lamberton [9] § 4):

Theorem 4.2 *Let X be a Banach space and A a closed linear operator in X that generates a strongly continuous analytic semigroup.*

If there exists $\tilde{p} \in]1, +\infty[$ such that there is $L^{\tilde{p}}$ regularity on $[0, T]$ for (CP) then $\forall p \in]1, +\infty[$ there is L^p regularity.

This theorem was already stated by Sobolevskiĭ (see [30] th. 3) who suggested that the proof would be based on the multiplier theorem of Benedek, Calderón and Panzone (see [4]).

Here is the proof as reconstructed by T. Kato (personal communication). For another example of application of the idea of Sobolevskiĭ see [34].

In view of the results of section 2 it is sufficient to consider the case of regularity on $[0, +\infty[$ (for the generator of a semigroup with negative exponential type).

We have to prove that the operator K defined by:

$$f \mapsto Au = A \int_0^t e^{(t-s)A} f(s)\, ds$$

is bounded in L^p. This operator is defined through the convolution with the kernel k such that:

$$k(t) = \begin{cases} Ae^{tA} & \text{for } t > 0 \\ 0 & \text{for } t \le 0 \end{cases}$$

From the $L^{\tilde{p}}$ regularity it follows that K is bounded in $L^{\tilde{p}}([0, +\infty[, X)$ so that, by translation, it is bounded in $L^{\tilde{p}}([a, +\infty[, X)$, for any $a \in \mathbf{R}$ with the same bound. But $\cup_{a \in \mathbf{R}} L^{\tilde{p}}([a, +\infty[, X)$ is dense in $L^{\tilde{p}}(\mathbf{R}, X)$ so that K is bounded in $L^{\tilde{p}}(\mathbf{R}, X)$.

Moreover $k(t)$ is locally summable away from 0.

Finally, taking into account the inequality $\|A^2 e^{tA}\| \le Ct^{-2}$ ([36], [32] th. 3.3.1), we have:

$$\int_{|t| > 2|s|} \|k(t-s) - k(t)\|\, dt =$$

$$\int_{t>2|s|} \|Ae^{(t-s)A} - Ae^{tA}\| \, dt = \int_{t>2|s|} \left\| \int_t^{t-s} A^2 e^{\tau A} \, d\tau \right\| dt \le$$

$$C \int_{t>2|s|} \left| \int_t^{t-s} \frac{1}{\tau^2} \, d\tau \right| dt = C \int_{t>2|s|} \left| \frac{1}{t} - \frac{1}{t-s} \right| dt$$

and this is bounded by a constant independent on s, so that k satisfies a Hörmander type condition.

Therefore K fulfills the assumptions of [26] part II th. 1.3 (i.e. the multiplier theorem of Benedek, Calderón and Panzone with no assumption about the summability of the kernel near 0) and it is bounded in $L^p(\mathbf{R}, X)$ for every $p \in \,]1, +\infty[$. This implies the boundedness of K in $L^p([0, +\infty[, X)$ and thus the L^p regularity.

In real interpolation spaces there is L^p regularity (Grisvard [19] th. 6.1, Da Prato-Grisvard [10] th. 4.7).

Theorem 4.3 *Let Y be a Banach space and A a closed linear operator in Y that generates a strongly continuous analytic semigroup with negative exponential type. Choose $\theta \in \,]0, 1[$ and $p \in \,]1, +\infty[$ and put $X = (Y, \mathcal{D}(A))_{\theta, p}$.*
Then there is L^p regularity on $[0, +\infty[$ for (CP) in the space X.

This theorem is a consequence of a general theorem on the sum of commuting operators.

In the space $L^p([0, +\infty[, Y)$ one considers the operators

$$\mathcal{A} : L^p([0, +\infty[, \mathcal{D}(A)) \to L^p([0, +\infty[, Y)$$

such that

$$(\mathcal{A}u)(t) = -A(u(t))$$

and

$$\mathcal{B} : \left\{ u \in W^{1,p} : u(0) = 0 \right\} \to L^p([0, +\infty[, Y)$$

such that

$$\mathcal{B}u = \frac{du}{dt} .$$

In this way the Cauchy problem is equivalent to the equation $(\mathcal{A} + \mathcal{B})u = f$, in the sense that there is L^p regularity for the Cauchy problem if and only if $\forall f \in L^p([0, +\infty[, Y)$ there is one and only one $u \in \mathcal{D}(\mathcal{A}) \cap \mathcal{D}(\mathcal{B})$ such that $(\mathcal{A} + \mathcal{B})u = f$. So that one has to prove that $\mathcal{A} + \mathcal{B}$ is invertible.

To this end one writes the operator

$$S = \frac{1}{2\pi i} \int_\gamma (\lambda - \mathcal{A})^{-1} (\lambda + \mathcal{B})^{-1} \, d\lambda$$

with γ a suitable path, separating the spectra of $\mathcal{A}$ and of $-\mathcal{B}$.

It can be easily proved that under suitable hypotheses on $\mathcal{A}$ and $\mathcal{B}$ (satisfied in our case) S is a left inverse of $\mathcal{A} + \mathcal{B}$; in general it is not a right inverse, since its range is not contained in $\mathcal{D}(\mathcal{A}) \cap \mathcal{D}(\mathcal{B})$. However the restriction of S to a real interpolation space between the space in which the operators $\mathcal{A}$ and $\mathcal{B}$ act and $\mathcal{D}(\mathcal{A})$ or $\mathcal{D}(\mathcal{B})$ is an inverse of the restriction of $\mathcal{A} + \mathcal{B}$.

The fact that

$$(L^p([0,+\infty[\,,Y),L^p([0,+\infty[\,,\mathcal{D}(\mathcal{A})))_{\theta,p} = L^p([0,+\infty[\,,(Y,\mathcal{D}(\mathcal{A}))_{\theta,p})$$

completes the proof.

We now consider the case in which X is neither a Hilbert space nor an interpolation one. One first result is (Lamberton [20] th. 1):

Theorem 4.4 *Let Ω be a measured space and A the generator of a strongly continuous semigroup e^{tA} in $L^2(\Omega)$ such that:*

a) *e^{tA} is analytic and bounded in $L^2(\Omega)$*

b) *$\forall p \in [1,+\infty]\ \forall f \in L^p(\Omega) \cap L^2(\Omega)\ \forall t \in \mathbf{R}^+\ \|e^{tA}f\|_{L^p} \le \|f\|_{L^p}$.*

Then $\forall p \in\,]1,+\infty[$ there is L^p regularity for (CP) in the space $X = L^p(\Omega)$ on $[0,T]$.

The proof of this theorem is carried out as follows.

Let $\psi \in\,]\frac{\pi}{2},\pi[$ such that the sector $\{z \in \mathbf{C} : |\arg z| < \psi\}$ is contained in the resolvent set of A and $\|\lambda(\lambda - A)^{-1}\|$ is bounded on it.

Choose $\psi_0 \in\,]0,\frac{\pi}{2}[$ and $\psi_1 \in\,]\frac{\pi}{2},\psi[$. For $z \in [0,1] + i\mathbf{R}$ and $\xi \in \mathbf{R}$ put

$$\rho(z,\xi) = |\xi|\exp\left(i(z\psi_1 + (1-z)\psi_0)\mathrm{sgn}(\xi)\right).$$

Now for $z \in [0,1] + i\mathbf{R}$ define a linear bounded operator U_z in $L^2(\mathbf{R} \times \Omega) = L^2(\mathbf{R}, L^2(\Omega))$ by

$$\widehat{U_z f}(\xi) = \rho(z,\xi)\,(\rho(z,\xi) - A)^{-1}\,\hat{f}(\xi).$$

Then $z \mapsto U_z$ is continuous and bounded from $[0,1] + i\mathbf{R}$ to $\mathcal{L}(L^2)$, holomorphic on $]0,1[+ i\mathbf{R}$. Moreover if θ is such that $\theta\psi_1 + (1-\theta)\psi_0 = \frac{\pi}{2}$ then $U_\theta f$ is the derivative of the solution of the Cauchy problem (CP) with datum f.

Next by the transference method one shows that if $q \in\,]1,+\infty[$ then $\exists C_q \in \mathbf{R}^+$ such that $\forall t \in \mathbf{R}\ \forall f \in L^q \cap L^2\ \|U_{it}f\|_{L^q} \le C_q\|f\|_{L^q}$. From this fact and Stein interpolation theorem it follows that $\|U_\theta f\|_{L^p} \le D_p\|f\|_{L^p}$ and this implies L^p regularity in the space $L^p(\Omega)$.

The assumptions of this theorem allow the application to parabolic problems only in case of second order operators, since in general elliptic operators (and systems) don't generate contraction semigroups on L^p.

A more widely applicable theorem is the following one (Dore-Venni [12] th. 3.2, see also [23] and [18]):

Theorem 4.5 *Let X be a ζ-convex Banach space and A a closed linear densely defined operator in X such that:*

a) *$-A$ is positive, i.e. $]-\infty,0] \subset \rho(-A)$ and $\|(\lambda + A)^{-1}\| \le C(1 + |\lambda|)^{-1}$ for every $\lambda \in\,]-\infty,0]$*

b) *The operator $(-A)^{it}$ is bounded and $\exists \theta \in \left[0,\frac{\pi}{2}\right[\ \exists C \in \mathbf{R}^+\ \forall t \in \mathbf{R}\ \|(-A)^{it}\| \le Ce^{\theta|t|}$.*

Then for $p \in \,]1, +\infty[$ there is L^p regularity on $[0, T]$ for (CP).

This theorem, like theorem 4.3, is obtained as a consequence of a theorem on the sum of commuting operators ([12] th. 2.1).

Theorem 4.6 *Suppose that $\mathcal{A}$ and $\mathcal{B}$ are linear closed densely defined operators in the ζ-convex Banach space X and that:*

(H1) $\mathbf{R}^- \cup \{0\} \subseteq \rho(\mathcal{A}) \cap \rho(\mathcal{B})$ *and there exists $M \in \mathbf{R}^+$ such that $\forall t \in \mathbf{R}^+ \cup \{0\}$*
$$\|(\mathcal{A} + t)^{-1}\| \le M(1 + t)^{-1}, \quad \|(\mathcal{B} + t)^{-1}\| \le M(1 + t)^{-1}$$

(H2) *if $\lambda \in \rho(\mathcal{A})$, $\mu \in \rho(\mathcal{B})$, then $(\lambda - \mathcal{A})^{-1}(\mu - \mathcal{B})^{-1} = (\mu - \mathcal{B})^{-1}(\lambda - \mathcal{A})^{-1}$*

(H3) *the operators $\mathcal{A}^{is}$ and $\mathcal{B}^{is}$ are bounded for every $s \in \mathbf{R}$ and there exist $K, \theta_\mathcal{A}, \theta_\mathcal{B} \in \mathbf{R}^+$, with $\theta_\mathcal{A} + \theta_\mathcal{B} < \pi$, such that for $s \in \mathbf{R}$*
$$\|\mathcal{A}^{is}\| \le K e^{\theta_\mathcal{A}|s|}, \quad \|\mathcal{B}^{is}\| \le K e^{\theta_\mathcal{B}|s|}.$$

Then $\mathcal{A} + \mathcal{B}$ is closed, invertible, with bounded inverse.

For $c \in \,]0, 1[$ put
$$S = \frac{1}{2i} \int_{c-i\infty}^{c+i\infty} \frac{\mathcal{A}^{-z}\mathcal{B}^{z-1}}{\sin(\pi z)} \, dz \,.$$

From (H3) it follows that the operator S is bounded in X and doesn't depend on c, moreover it is easy to show that it is a left inverse of $\mathcal{A} + \mathcal{B}$. We want to prove that it is also a right inverse. For this it is sufficient to prove that the range of S is contained in $\mathcal{D}(\mathcal{A}) \cap \mathcal{D}(\mathcal{B})$.

Now we have
$$Sx = \frac{1}{2i} \int_{\Gamma_\varepsilon} \frac{\mathcal{A}^{-z}\mathcal{B}^{z-1}x}{\sin(\pi z)} \, dz$$

where Γ_ε is the curve, oriented with increasing imaginary part, composed by the half-lines $\{is : |s| \ge \varepsilon\}$ and by the half-circle $\{z \in \mathbf{C} : |z| = \varepsilon, Re\,z \ge 0\}$.

Thus
$$Sx = \frac{1}{2} \int_{|s| \ge \varepsilon} \frac{\mathcal{A}^{-is}\mathcal{B}^{-1+is}x}{\sin(i\pi s)} \, ds + \frac{1}{2} \int_{-\pi/2}^{\pi/2} \frac{\varepsilon e^{i\theta}}{\sin(\pi \varepsilon e^{i\theta})} \mathcal{A}^{-\varepsilon e^{i\theta}} \mathcal{B}^{\varepsilon e^{i\theta}-1}x \, d\theta$$

and we denote with $I_{1,\varepsilon}$ and $I_{2,\varepsilon}$ the first and second summands respectively.

As $\varepsilon \to 0^+$, $I_{2,\varepsilon}$ converges to $\frac{1}{2}\mathcal{B}^{-1}x$. Moreover for every $\varepsilon \in \,]0, 1[$ $I_{1,\varepsilon}$ belongs to $\mathcal{D}(\mathcal{B})$, so that if we prove that $\mathcal{B}I_{1,\varepsilon}$ converges as $\varepsilon \to 0^+$ then we have
$$Sx - \frac{1}{2}\mathcal{B}^{-1}x = \lim_{\varepsilon \to 0^+} I_{1,\varepsilon}x \in \mathcal{D}(\mathcal{B})$$

and hence $Sx \in \mathcal{D}(\mathcal{B})$.

Now
$$\mathcal{B}I_{1,\varepsilon} = \frac{1}{2} \int_{|s| \ge \varepsilon} \frac{\mathcal{A}^{-is}\mathcal{B}^{is}x}{\sin(i\pi s)} \, ds =$$

$$\frac{1}{2}\int_{|s|\geq 1}\frac{\mathcal{A}^{-is}\mathcal{B}^{is}x}{\sin(i\pi s)}\,ds + \frac{1}{2}\int_{\varepsilon\leq|s|\leq 1}\frac{\mathcal{A}^{-is}\mathcal{B}^{is}x}{i\pi s}\,ds + \frac{1}{2}\int_{\varepsilon\leq|s|\leq 1}\mathcal{A}^{-is}\mathcal{B}^{is}x\left(\frac{1}{\sin(i\pi s)}-\frac{1}{i\pi s}\right)ds\,.$$

The first summand does not depend on ε and the third one is convergent as $\varepsilon\to 0^+$ since $\frac{1}{\sin(i\pi s)}-\frac{1}{i\pi s}$ is bounded on $[-1,1]$. The second summand equals $(2i)^{-1}(\mathcal{H}_\varepsilon F)(0)$ where $\mathcal{H}_\varepsilon$ is the truncated Hilbert transform, i.e.

$$(\mathcal{H}_\varepsilon f)(t) = \frac{1}{\pi}\int_{|s|\geq\varepsilon}\frac{f(t-s)}{s}\,ds\,,$$

and $F(s) = \chi_{]-1,1[}(s)\,\mathcal{A}^{is}\mathcal{B}^{-is}x$.

F belongs to $L^p(\mathbf{R};X)$, $\forall p\in\,]1,\infty[$. Since X is ζ-convex, the truncated Hilbert transform of F converges in the norm of X for almost every t (see [5] th. 2). Therefore we fix such a t (with $|t| < 1$) and we get

$$\int_{\varepsilon\leq|s|\leq 1}\frac{\mathcal{A}^{-is}\mathcal{B}^{is}x}{i\pi s}\,ds =$$

$$\frac{1}{\pi i}\mathcal{A}^{-it}\mathcal{B}^{it}\left(\int_{|s|\geq\varepsilon}\frac{F(t-s)}{s}\,ds - \int_1^{t+1}\frac{\mathcal{A}^{i(t-s)}\mathcal{B}^{i(s-t)}x}{s}\,ds + \int_{-1}^{t-1}\frac{\mathcal{A}^{i(t-s)}\mathcal{B}^{i(s-t)}x}{s}\,ds\right)$$

which converges as $\varepsilon\to 0^+$ since $\mathcal{A}^{-it}\mathcal{B}^{it}\in\mathcal{L}(X)$.

Analogously one obtains that $\mathcal{S}x\in\mathcal{D}(\mathcal{A})$ and the proof is complete.

Theorem 4.5 is a consequence of theorem 4.6 and of the following one concerning the imaginary powers of the operator of derivation in $L^p([0,T],X)$ ([12] th. 3.1).

Theorem 4.7 *Let X be a ζ-convex Banach space, $\mathcal{B}$ the operator in $L^p([0,T],X)$ such that*

$$\mathcal{D}(\mathcal{B}) = \left\{u\in W^{1,p}([0,T],X):u(0)=0\right\}$$

and

$$\mathcal{B}u = \frac{du}{dt}\,.$$

Then $\forall s\in\mathbf{R}$ $\mathcal{B}^{is}$ is bounded and there exist K such that for $s\in\mathbf{R}$

$$\|\mathcal{B}^{is}\|\leq K(1+s^2)e^{\frac{\pi}{2}|s|}\,.$$

The proof of this theorem is based on an extension of the Mihlin multiplier theorem to the case of vector valued functions that holds in ζ-convex spaces (see [21] th. 1.1 or [37] prop. 3).

The assumptions on the imaginary powers of the operator A required to apply theorem 4.5, or some generalization of it, are satisfied by several operators. We mention:

- The realization in L^p of an elliptic (in the sense of Agmon, Douglis and Nirenberg) system with C^∞ coefficients under general boundary conditions (see [28])

- The realization in L^p of a second order elliptic operator with Hölder continuous coefficients under Dirichlet boundary conditions (see [24])

- The Stokes operator in $(L^p(\Omega))^n$ under Dirichlet boundary conditions when Ω is a bounded domain (see [16]) or an exterior domain (see [17]).

5 Open problems

We conclude with some open problems concerning L^p regularity suggested by the results exposed in section 4.

Problem 1 (see [9] § 1) *As we have seen (th. 4.1) if X is a Hilbert space, then for every operator A in X that generates a strongly continuous analytic semigroup there is L^p regularity for (CP).*
Is the same true when X is a ζ-convex Banach space?

Problem 2 *Let X be a Banach space, in which there are strongly continuous analytic semigroups with unbounded generator. Suppose that for every generator of a strongly continuous analytic semigroup there is L^p regularity for (CP).*
Is X necessarily ζ-convex? Or (in case the answer to Problem 1 is no) is X necessarily a Hilbert space?

One can also ask if in this case is true a theorem similar to the already quoted one that holds if there is continuous regularity (see [1]).

Problem 3 *Let X be a Banach space such that there exists an unbounded operator A in X for which there is L^p regularity for (CP).*
Is it always possible to find an infinite-dimensional ζ-convex subspace of X?

We cannot expect that X itself is ζ-convex, as the following example shows.
Let X_1 be a Hilbert space and X_2 a Banach space that is not ζ-convex. Let A_1 be a closed unbounded operator in X_1 that generates a strongly continuous analytic semigroup and A_2 a bounded operator in X_2.
Put $X = X_1 \times X_2$ and $A : \mathcal{D}(A_1) \times X_2 \to X$ such that $A(x_1, x_2) = (A_1 x_1, A_2 x_2)$. Then A is an unbounded operator in X for which there is L^p regularity, but X is not ζ-convex.

References

[1] J. B. Baillon: Caractère borné de certains générateurs de semigroupes linéaires dans les espaces de Banach; C. R. Acad. Sci. Paris **290** (1980) 757–760.

[2] J. M. Ball: Strongly continuous semigroups, weak solutions, and the variation of constants formula; Proc. Amer. Math. Soc. **63** (1977) 370–373.

[3] C. Bardos: A regularity theorem for parabolic equations; J. Funct. Anal. **7** (1971) 311–322.

[4] A. Benedek, A. P. Calderón and R. Panzone: Convolution operators on Banach space valued functions; Proc. Nat. Acad. Sci. U.S.A. **48** (1962) 356–365.

[5] D. L. Burkholder: A geometrical condition that implies the existence of certain singular integrals of Banach-space-valued functions; in: W. Beckman, A. P. Calderón, R. Fefferman and P. W. Jones (Eds.) *Conference on harmonic analysis in honor of Antoni Zygmund* (Proceedings, Chicago 1981); Wadsworth, Belmont, 1983, 270–286.

[6] D. L. Burkholder: Martingales and Fourier analysis in Banach spaces; in: G. Letta and M. Pratelli (Eds.) *Probability and analysis* (Proceedings, Varenna 1985); Lect. Notes Math. **1206**, Springer, Berlin Heidelberg New York, 1986, 61–108.

[7] S. Campanato: Su un teorema di interpolazione di G. Stampacchia; Ann. Scuola Norm. Sup. Pisa **20** (1966) 649–652.

[8] P. Cannarsa and V. Vespri: On maximal L^p regularity for the abstract Cauchy problem; Boll. Un. Mat. It. B (6) **5** (1986) 165–175.

[9] T. Coulhon and D. Lamberton: Régularité L^p pour les équations d'évolution; in: B. Beauzaumy, B. Maurey and G. Pisier (Eds.) *Séminaire d'Analyse Fonctionelle 1984/1985*; Publ. Math. Univ. Paris VII **26**, Université Paris VII, 1986, 155–165.

[10] G. Da Prato and P. Grisvard: Sommes d'opérateurs linéaires et équations différentielles opérationnelles; J. Math. Pures Appl. (9) **54** (1975) 305–387.

[11] G. Di Blasio: L^p regularity for solutions of non autonomous parabolic equations in Hilbert spaces; Boll. Un. Mat. Ital. C (6) **1** (1982) 395–407.

[12] G. Dore and A. Venni: On the closedness of the sum of two closed operators; Math. Z. **196** (1987) 189–201.

[13] H. O. Fattorini: *The Cauchy problem*; Addison-Wesley, Reading, 1983.

[14] M. Fuhrman: Sums of linear operators of parabolic type in a Hilbert space: strict solutions and maximal regularity; preprint.

[15] M. Giga, Y. Giga and H. Sohr: L^p estimate for abstract linear parabolic equations; Proc. Japan Acad. Ser. A Math. Sci. **67** (1991) 197–202.

[16] Y. Giga: Domains of fractional powers of the Stokes operator in L_r Spaces; Arch. Rational Mech. Anal. **89** (1985) 251–265.

[17] Y. Giga and H. Sohr: On the Stokes operator in exterior domains; J. Fac. Sci. Univ. Tokyo Sect. IA Math. **36** (1989) 103–130.

[18] Y. Giga and H. Sohr: Abstract L^p estimates for the Cauchy problem with applications to the Navier-Stokes equations in exterior domains; J. Funct. Anal. **102** (1991) 72–94.

[19] P. Grisvard: Equations différentielles abstraites; Ann. Sci. École Norm. Sup. (4) **2** (1969) 311–395.

[20] D. Lamberton: Equations d'évolutions linéaires associées á des semi-groupes de contractions dans les espaces L^p; J. Funct. Anal. **72** (1987) 252–262.

[21] T. R. McConnell: On Fourier multiplier transformations of Banach-valued functions; Trans. Amer. Math. Soc. **285** (1984) 739–757.

[22] A. Pazy: *Semigroups of linear operators and applications to partial differential equations*; Springer, New York Berlin Heidelberg Tokyo, 1983.

[23] J. Prüss and H. Sohr: On operators with bounded imaginary powers in Banach spaces; Math. Z. **203** (1990) 429–452.

[24] J. Prüss and H. Sohr: Imaginary powers of elliptic second order differential operators in L^p spaces; preprint.

[25] J. L. Rubio de Francia: Martingale and integral transform of Banach space valued functions; in: J. Bastero and M. San Miguel (Eds.) *Probability and Banach spaces* (Proceedings, Zaragoza 1985); Lect. Notes Math. **1221**, Springer, Berlin Heidelberg New York, 1986, 195–222.

[26] J. L. Rubio de Francia, F. J. Ruiz and J. L. Torrea: Calderón-Zygmund theory for operator-valued kernels; Adv. in Math. **62** (1986) 7–48.

[27] J. T. Schwartz: A remark on inequalities of Calderon-Zygmund type for vector valued functions; Comm. Pure Appl. Math. **14** (1961) 785–799.

[28] R. Seeley: Norms and domains of the complex powers A_B^z; Amer. J. Math. **93** (1971) 299–309.

[29] L. de Simon: Un'applicazione della teoria degli integrali singolari allo studio delle equazioni differenziali lineari astratte del primo ordine; Rend. Sem. Mat. Univ. Padova **34** (1964) 547–558.

[30] P. E. Sobolevskiĭ: Coerciveness inequalities for abstract parabolic equations (in russian); Dokl. Acad. Nauk SSSR **157** (1964) 52–55; translated in: Soviet Math. Dokl. **5** (1964) 894–897.

[31] G. Stampacchia: The spaces $\mathcal{L}^{(p,\lambda)}$, $\mathcal{N}^{(p,\lambda)}$ and interpolation; Ann. Scuola Norm. Sup. Pisa **19** (1965) 443–462.

[32] H. Tanabe: *Equations of evolution*; Pitman, London San Francisco Melbourne, 1979.

[33] V. Vespri: Regolarità massimale in L^p per il problema di Cauchy astratto e regolarità $L^p(L^q)$ per operatori parabolici; in: L. Modica (Ed.) *Atti del convegno su equazioni differenziali e calcolo delle variazioni*; Pisa, 1985, 205–213.

[34] W. von Wahl: The equation $u' + A(t)u = f$ in a Hilbert space and L^p-estimates for parabolic equations; J. London Math. Soc. (2) **25** (1982) 483–497.

[35] Y. Yamamoto: On L^p-solutions for nonautonomous parabolic equations in Hilbert spaces; preprint.

[36] K. Yosida: On the differentiability of semi-groups of linear operators; Proc. Japan Acad. **34** (1958) 337–340.

[37] F. Zimmermann: On vector-valued Fourier multiplier theorems; Studia Math. **93** (1989) 201–222.

Some Feynman Path Integrals As Oscillatory Integrals Over A Sobolev Manifold

BY

DAISUKE FUJIWARA

Department of Mathematics
Tokyo Institute of Technology
2-12-1 Ohokayama, Meguro-ku
Tokyo 152, Japan

§1. Introduction

Let

$$L(\dot{x}, x) = \frac{1}{2}|\dot{x}|^2 - V(x), \quad x \in \mathbf{R},$$

be a Lagrangian with real potential $V(x)$. For any integer $k \geq 2$, let $[V]_k = \max_{2 \leq j \leq k} \sup_{x \in \mathbf{R}} |V^{(j)}(x)|$. We assume that

(ASS-I) $[V]_k < \infty$ for any $k \geq 2$.

Let $\gamma : [0, T] \to \mathbf{R}$ be a path satisfying $\frac{d}{dt}\gamma \in L^2(0, T)$, $\gamma(0) = y$, and $\gamma(T) = x$. The action $S(\gamma)$ of γ is the functional

$$(1) \qquad S(\gamma) = \int_0^T L\left(\frac{d}{dt}\gamma, \gamma\right) dt.$$

The Feynman path integral is a heuristic integral

$$(2) \qquad \int_\Omega e^{i\hbar^{-1}S(\gamma)} \mathcal{D}[\gamma]$$

over a formal path space Ω, where $\hbar$ is a small positive constant.

Many people discussed mathematically rigorous meaning of the Feynman path integral. Feynman [6] (cf. also [7]) introduced the path integral by using finite dimensional approximations of path space, which we now call the time slicing approximation. At the same time he suggested the possibility that his path integral may be considered as integration by a complex-valued measure over a path space. Unfortunately, after some effort, it was proved impossible. Cf. [5], [13]. Most successful approach to give meaning to Feynman path integral is the analytic continuation method. Cf. Nelson [15]. On the other hand, some people tried to define the path integral of Feynman as an improper integral over a Hilbert space of paths. Among them Itô's work [14] is the most beautiful. Later Albeverio and Høegh-Krohn [1], [2] applied his idea and discussed many problems. Cf. also [3].

In [8], [9] and [10] the present author proved that the time slicing approximation of Feynman path integral actually converges in a very strong topology to the fundamental solution of Schrödinger equation if the potential satisfies our assumption (ASS-I) and succeeded in discussing classical limit of Feynman path integrals.

The present report is concerned with Itô's definition of Feynman path integrals. Cf. [11]. By the way, the author gives a new simple proof of convergence of the time slicing approximation, which is based on the stationary phase method on a space of large dimension [12].

Remark. 1)   For the sake of simplicity of notation we assumed that x-space is of dimension one. The same results hold in the case of higher dimension.

2)   Pauli [16] discussed Feynman path integrals. His assumption about potential there was close to ours.

§2   Time slicing approximation.

2.1.   Let T_0 be so small that

$$(1) \qquad\qquad T_0^2[V]_2 < 2^{-4} \quad \text{and} \quad T_0^3[V]_2^2 < 2^{-2}.$$

We always assume in the following that $|T| \leq T_0$. Then under assumption (ASS-I) the classical path starting from y at time 0 and reaching x at time T is unique. We denote it by γ^{cl} and the action along γ^{cl} by $S(T, x, y)$.

Let $\Delta : 0 = \tau_0 < \tau_1 < \cdots < \tau_{L-1} < \tau_L = T$ be an arbitrary division of the interval $[0, T]$ into subintervals. Let $\Delta \tau_j$ denote $\tau_j - \tau_{j-1}$ and $|\Delta|$ denote the size of division $\max_{0 \leq j \leq L-1} \Delta \tau_j$. Let $x_1, \cdots, x_{L-1}$ be arbitrary points of the configuration space $\mathbb{R}$, let $x_0 = y$ and $x_L = x$. Then we denote by $\gamma_\Delta^{\text{cl}}$ the piecewise classical orbit joining (τ_j, x_j), $j = 0, \cdots, L$, i.e., $\gamma_\Delta^{\text{cl}}(t)$ satisfies Euler's equation

$$(2) \qquad\qquad \frac{d^2}{dt^2} \gamma_\Delta^{\text{cl}}(t) - V(\gamma_\Delta^{\text{cl}}(t)) = 0 \qquad \text{for } \tau_{j-1} < t < \tau_j$$

and at $t = \tau_j$ it satisfies $\gamma_\Delta^{\text{cl}}(\tau_j) = x_j$, $j = 0, 1, \cdots, L$. If $|\Delta| \leq T_0$, then such orbit $\gamma_\Delta^{\text{cl}}$ is unique. When we wish to emphasize the dependency of $\gamma_\Delta^{\text{cl}}$ on $X = (x_1, \cdots, x_{L-1})$, we denote it by $\gamma_\Delta^{\text{cl}}(X)$. The action along $\gamma_\Delta^{\text{cl}}$ can be written

$$(3) \qquad\qquad S(\gamma_\Delta^{\text{cl}}) = S(\Delta \tau_L, x_L, x_{L-1}) + \cdots + S(\Delta \tau_1, x_1, x_0).$$

The time slicing approximation of Feynman path integral is

$$(4) \qquad \begin{aligned} &K(\Delta; \hbar, T, x, y) \\ &= \prod_{j=1}^{L} \left(\frac{1}{2\pi i \hbar \Delta \tau_j} \right)^{\frac{1}{2}} \int_{\mathbb{R}^{L-1}} e^{i\hbar^{-1} \sum_{j=1}^{L} S(\Delta \tau_j, x_j, x_{j-1})} \prod_{j=1}^{L-1} dx_j. \end{aligned}$$

In [10] we have proved the existence of $\lim_{|\Delta| \to 0} K(\Delta; \hbar, T, x, y)$. We give a sketch of a simpler new proof based on results in [12].

2.2. Proposition. If $|T| \leq T_0$, we can write

$$(1) \qquad S(T, x, y) = \frac{|x - y|^2}{2T} + T\phi(T, x, y).$$

For any integer $m \geq 2$ there exists a positive constant κ_m such that if $2 \leq \alpha + \beta \leq m$, we have

$$(2) \qquad |\partial_x^\alpha \partial_y^\beta \phi(T, x, y)| \leq \kappa_m,$$

We can choose $\kappa_2 = [V]_2$.

Proof is ommitted.

2.3. Let Δ be an arbitrary division of $[0, T]$ as above. Let $K(\Delta; \hbar, T, x, y)$ be the corresponding time-slicing approximation of Feynman path integral. In order to prove that $\lim_{|\Delta| \to 0} K(\Delta; \hbar, T, x, y)$ exists we require some more preparations.

Let $H(\Delta)$ be the following $(L-1) \times (L-1)$ matrix

$$(1) \qquad H(\Delta) = \begin{pmatrix} \frac{1}{\Delta\tau_1} + \frac{1}{\Delta\tau_2}, & -\frac{1}{\Delta\tau_2}, & 0, & 0, \\ -\frac{1}{\Delta\tau_2}, & \frac{1}{\Delta\tau_2} + \frac{1}{\Delta\tau_3}, & -\frac{1}{\Delta\tau_3}, & 0, \\ 0, & -\frac{1}{\Delta\tau_3}, & \frac{1}{\Delta\tau_3} + \frac{1}{\Delta\tau_4}, & -\frac{1}{\Delta\tau_4}, & 0, \\ 0, & 0, & \cdots\cdots\cdots\cdots\cdots\cdots \end{pmatrix}.$$

Let $G(\Delta)$ be the inverse of $H(\Delta)$. Then its (i, j) entry is

$$(2) \qquad G(\Delta)_{ij} = \begin{cases} \frac{\tau_i(T-\tau_j)}{T}, & \text{if } \tau_i \leq \tau_j, \\ \frac{\tau_j(T-\tau_i)}{T}, & \text{if } \tau_j < \tau_i. \end{cases}$$

Denoting $\phi(\Delta\tau_j, x_j, x_{j-1})$ by ϕ_j and ∂_{x_j} by ∂_j, $W_{jk}(\Delta; T, x, y)$ is the i, j entry of the following matrix

$$W(\Delta; T, x, y)$$
$$= \begin{pmatrix} \Delta\tau_1\partial_1^2\phi_1 + \Delta\tau_2\partial_1^2\phi_2, & \Delta\tau_2\partial_1\partial_2\phi_2, & 0, & 0, \\ \Delta\tau_2\partial_2\partial_1\phi_2, & \Delta\tau_2\partial_2^2\phi_2 + \Delta\tau_3\partial_2^2\phi_3, & \Delta\tau_3\partial_2\partial_3\phi_3, & 0, \\ 0, & \Delta\tau_3\partial_3\partial_2\phi_3, & \Delta\tau_3\partial_3^2\phi_3 + \Delta\tau_4\partial_4^2\phi, & 0, \\ 0, & 0, & \cdots\cdots\cdots\cdots \end{pmatrix}.$$

2.4. Proposition. Let $|T| \leq T_0$. Then the critical point X^* of $S(\gamma_\Delta^{\mathrm{cl}}(X))$ exists uniquely. The orbit $\gamma_\Delta^{\mathrm{cl}}(X^*)$ corresponding to the critical point coincides with the classical orbit γ^{cl}.

2.5. Theorem. There exists a positive constant T_1 not greater than T_0 such that if $|T| \leq T_1$ then $K(\Delta; \hbar, T, x, y)$ is of the form

$$(1) \qquad K(\Delta; \hbar, T, x, y) = \left(\frac{1}{2\pi i\hbar T}\right)^{\frac{1}{2}} e^{i\hbar^{-1}S(T,x,y)} D(\Delta; x, y)(1 + r(\Delta; \hbar, x, y)),$$

where

(2)
$$D(\Delta; x, y) = (\det [I + G(\Delta)W^*(\Delta; T, x, y)])^{-\frac{1}{2}}$$

with $W^*(\Delta; T, x, y) = W(\Delta; T, x, y)|_{X=X^*}$. $D(\Delta; x, y)$ is of the form

(3)
$$D(\Delta; x, y) = 1 + T^2 d(\Delta; x, y).$$

For any α, β there exists a constant $C_{\alpha\beta}$ such that

(4)
$$|\partial_x^\alpha \partial_y^\beta d(\Delta; x, y)| \leq C_{\alpha\beta},$$

and

(5)
$$|\partial_x^\alpha \partial_y^\beta r(\Delta; \hbar, x, y)| \leq C_{\alpha\beta}\hbar T^3.$$

Proof. We apply the stationary phase method Theorem 2 of [12] to the right hand side of (2.1.4). Then Proposition follows.

2.6. In order to prove the existence of $\lim_{|\Delta|\to 0} K(\Delta; \hbar, T, x, y)$ we have to prove the existence of two limits $\lim_{|\Delta|\to 0} D(\Delta; x, y)$ and $\lim_{|\Delta|\to 0} r(\Delta; \hbar, T, x, y)$.

First we discuss $\lim_{|\Delta|\to 0} D(\Delta; x, y)$. The first variation of the functional $\gamma \to S(\gamma_0 + \gamma)$ vanishes at the classical path γ^{cl}. Its second variation at γ^{cl} is

$$\mathcal{H} \ni u \longrightarrow \int_0^T \left|\frac{d}{dt}u(t)\right|^2 dt - \int_0^T V''(\gamma_0(t) + \gamma^{\text{cl}}(t))u(t)^2 dt.$$

The associated differential operator (Jacobi's differential operator) is

(1)
$$J = -\frac{d^2}{dt^2} - V''(\gamma_0(t) + \gamma^{\text{cl}}(t)).$$

Let $g(t, s)$ be the Green function of Dirichlet problem in $[0, T]$, i.e., $g(t, s) = t(T-s)$ for $0 \leq t \leq s \leq T$ and $s(T-t)$ for $0 \leq s \leq t \leq T$. Let G be the Green operator i.e., for any $f \in L^2(0, T)$

$$Gf(t) = \int_0^T g(t, s)f(s)ds.$$

The operator G is of trace class.

2.7. Theorem. Let G be the Green operator. Then we have

(1)
$$\lim_{|\Delta|\to 0} D(\Delta; x, y) = \det [I - GV''(\gamma_0 + \gamma^{\text{cl}})]^{-\frac{1}{2}} = [\det(GJ)]^{-\frac{1}{2}}.$$

Here det means Fredholm's determinant.

Proof. It is clear that $H(\Delta) + W(\Delta; T, x, y)$ is a finite element approximation of Jacobi's operator J. Hence a standard argument of finite element method gives a proof to Theorem.

2.8. The next theorem shows that $\{r(\Delta; \hbar, x, y)\}_\Delta$ forms a Cauchy net.

Theorem. Let $|T| \leq T_1$ and let Δ' be any refinement of Δ. Then
(i) We introduce $d(\Delta', \Delta; x, y)$ by equality

$$(1) \qquad \frac{D(\Delta'; x, y)}{D(\Delta; x, y)} = 1 + |\Delta| T d(\Delta', \Delta; x, y).$$

Then for any α and β there exists a constant $C_{\alpha\beta} > 0$ such that

$$(2) \qquad \left| \partial_x^\alpha \partial_y^\beta d(\Delta', \Delta; x, y) \right| \leqq C_{\alpha\beta}.$$

(ii) For any α and β there exists a constant $C_{\alpha\beta} > 0$ such that

$$(3) \qquad \left| \partial_x^\alpha \partial_y^\beta (r(\Delta; \hbar, x, y) - r(\Delta'; \hbar, x, y)) \right| \leqq C_{\alpha\beta} T^2 |\Delta|.$$

2.9. Proof. Let j be any integer $1 \leqq j \leqq L$. Since Δ' is a refinement of Δ, Δ' divides the interval $[\tau_{j-1}, \tau_j]$. We denote this division of $[\tau_{j-1}, \tau_j]$ by Δ'_j, i.e.,

$$(1) \qquad \Delta'_j : \quad \tau_{j-1} = \tau_{j,0} < \tau_{j,1} < \cdots < \tau_{j,k_j} = \tau_j, \quad (k_j > 0).$$

Corresponding to this subdivision we set $x_{j,l} = \gamma_{\Delta'}^{\mathrm{cl}}(\tau_{j,l})$, $l = 1, \cdots, k_j$. And $\Delta\tau_{j,l} = \tau_{j,l} - \tau_{j,l-1}$.

Now we can prove (i) of Theorem 2.8. Proposition 2.6 of [12] yields that

$$(2) \qquad D(\Delta'; x, y) = D(\Delta; x, y) \prod_{j=1}^{L} D(\Delta'_j; x_j^{\#}, x_{j-1}^{\#}),$$

where $x_j^{\#} = \gamma^{\mathrm{cl}}(\tau_j)$. This together with (2.5.3) proves (2.8.1) and (2.8.2).

Next we prove (2.8.3). We compare $K(\Delta; \hbar, T, x, y)$ with

$$(3) \qquad K(\Delta'; \hbar, T, x, y) = \int_{\mathbb{R}^{L-1}} \prod_{j=1}^{L} J(\Delta'_j; \hbar, x_j, x_{j-1}) \prod_{j=1}^{L-1} dx_j,$$

where

$$(4) \qquad \begin{aligned} J(\Delta'_j; \hbar, x_j, x_{j-1}) &= \prod_{m=1}^{k_j} \left(\frac{1}{2\pi i \hbar \Delta\tau_{j,m}} \right)^{\frac{1}{2}} \\ &\times \int_{\mathbb{R}^{k_j-1}} e^{i\hbar^{-1} \sum_{m=1}^{k_j} S(\Delta\tau_{j,m}, x_{j,m}, x_{j,m-1})} \prod_{m=1}^{k_j-1} dx_{j,m}. \end{aligned}$$

We apply the stationary phase method, Theorem 2 of [12] to $J(\Delta'_j; \hbar, x_j, x_{j-1})$. Then we have

$$
\begin{aligned}
(5) \qquad & J(\Delta'_j; \hbar, x_j, x_{j-1}) \\
&= \left(\frac{1}{2\pi i \hbar \Delta\tau_j} \right)^{\frac{1}{2}} e^{i\hbar^{-1} S(\Delta\tau_j, x_j, x_{j-1})} (1 + \Delta\tau_j^2 p_j(\Delta'_j; \hbar, x_j, x_{j-1})),
\end{aligned}
$$

where we have set

$$
(6) \qquad 1 + \Delta\tau_j^2 p_j(\Delta'_j; \hbar, x_j, x_{j-1}) = D(\Delta'_j; x_j, x_{j-1})(1 + r(\Delta'_j; \hbar, x_j, x_{j-1})).
$$

For any α and β there exists a constant $C_{\alpha\beta} > 0$ such that we have

$$
(7) \qquad \left| \partial_{x_j}^\alpha \partial_{x_{j-1}}^\beta p_j(\Delta'_j; \hbar, x_j, x_{j-1}) \right| \leq C_{\alpha\beta}.
$$

This gives that

$$
\begin{aligned}
K(\Delta'; \hbar, x, y) &= \prod_{j=1}^{L} \left(\frac{1}{2\pi i \hbar \Delta\tau_j} \right)^{\frac{1}{2}} \\
&\times \int_{\mathbf{R}^{L-1}} e^{i\hbar^{-1} \sum_{j=1}^{L} S(\Delta\tau_j, x_j, x_{j-1})} \prod_{j=1}^{L} (1 + \Delta\tau_j^2 p_j(\Delta'_j; \hbar, x_j, x_{j-1})) \prod_{j=1}^{L-1} dx_j.
\end{aligned}
$$

Therefore,

$$
\begin{aligned}
(8) \qquad & K(\Delta'; \hbar, T, x, y) - K(\Delta; \hbar, T, x, y) \\
&= \prod_{j=1}^{L} \left(\frac{1}{2\pi i \hbar \Delta\tau_j} \right)^{\frac{1}{2}} \\
&\times \int_{\mathbf{R}^{L-1}} e^{i\hbar^{-1} \sum_{j=1}^{L} S(\Delta\tau_j, x_j, x_{j-1})} b(\Delta', \Delta; \hbar, x_L, \cdots, x_0) \prod_{j=1}^{L-1} dx_j.
\end{aligned}
$$

Here

$$
(9) \qquad b(\Delta', \Delta; \hbar, x_L, x_{L-1}, \cdots, x_0) = \prod_{j=1}^{L} (1 + \Delta\tau_j^2 p_j(\Delta'_j; \hbar, x_j, x_{j-1})) - 1.
$$

We can prove that $b(\Delta', \Delta; \hbar, x_L, \cdots, x_0)$ fulfills assumptions, with $A_K = C_K |\Delta| T$ and $X_K = 1$, of Theorem 1 in [12]. It follows from this that

$$
\begin{aligned}
(10) \qquad & K(\Delta'; \hbar, T, x, y) - K(\Delta; \hbar, T, x, y) \\
&= \left(\frac{1}{2\pi i \hbar T} \right)^{\frac{1}{2}} e^{i\hbar^{-1} S(T, x, y)} D(\Delta; x, y) q(\Delta', \Delta; \hbar, x, y)
\end{aligned}
$$

and that for any α and β there exist positive constants $C_{\alpha\beta}$ and $C'_{\alpha\beta}$ such that

$$(11) \qquad |\partial_x^\alpha \partial_y^\beta q(\Delta', \Delta; \hbar, T, x, y)| \leq \prod_{j=1}^{L}(1 + C_{\alpha\beta}\Delta\tau_j^2) - 1 \leq C'_{\alpha\beta}|\Delta||T|.$$

On the other hand, by (2.5.1) we obtain

$$(12) \qquad q(\Delta', \Delta; \hbar, T, x, y) = \frac{D(\Delta'; x, y)}{D(\Delta; x, y)}(1 + r(\Delta'; \hbar, x, y)) - (1 + r(\Delta; \hbar, x, y)).$$

(2.8.3) follows from (2.8.1), (10), (11) and (12).

2.10. As a consequence of Theorems 2.5, 2.7 and 2.8, we have

Theorem. If $|T| \leq T_1$, then

$$K(\hbar, T, x, y) = \lim_{|\Delta| \to 0} K(\Delta; \hbar, T, x, y)$$

exists. $K(\hbar, T, x, y)$ is of the form

$$(1) \qquad K(\hbar, T, x, y) = \left(\frac{1}{2\pi i\hbar T}\right)^{\frac{1}{2}} e^{i\hbar^{-1}S(T,x,y)} (\det(GJ))^{-\frac{1}{2}} (1 + r(\hbar, T, x, y)).$$

For any α and $\beta \geq 0$ there exists a constant $C_{\alpha\beta}$ such that

$$(2) \qquad |\partial_x^\alpha \partial_y^\beta r(\hbar, T, x, y)| \leq C_{\alpha\beta}\hbar T^3.$$

Moreover $K(\hbar; T, x, y)$ is the fundamental solution of the Schrödinger equation.

2.11. **Proof.** We have only to prove that $K(\hbar, T, x, y)$ is the fundamental solution of the Schrödinger equation. Let us introduce the operator $K(t)$, $|t| \leq T_1$, by

$$(1) \qquad K(t)f(x) = \int_{\mathbf{R}} K(\hbar; t, x, y)f(y)dy.$$

We can apply the result of [4] and prove that $\{K(t)\}_{T_1 \geq t > 0}$ forms a family of uniformly bounded operators on $L^2(\mathbf{R})$. cf. [9]. It is clear that $K(t)$ has the semi-group property. So we can extend the definition of $K(t)$ to an arbitrary $t > 0$ by the semi-group property. Moreover the stationary phase method proves that $s\text{-}\lim_{t \to 0} K(t) = I$ in $L^2(\mathbf{R})$. Therefore $\{K(t)\}_{t > 0}$ forms a Hille-Yosida semi-group in $L^2(\mathbf{R})$. Since $S(t, x, y)$ satisfies the Hamilton-Jacobi equation

$$(2) \qquad \frac{\partial}{\partial t}S + \frac{1}{2}\left(\frac{\partial s}{\partial x}\right)^2 + V(x) = 0,$$

a direct calculation shows that

(3)
$$\left(\frac{\hbar\partial}{i\partial t} + \frac{1}{2}\left(\frac{\hbar\partial}{i\partial x}\right)^2 + V(x)\right)K(\hbar, t, x, y)$$
$$= \left(\frac{1}{2\pi i t\hbar}\right)^{\frac{1}{2}} e^{i\hbar^{-1}S(t,x,y)} te(\hbar, t, x, y).$$

The function $e(\hbar, t, x, y)$ satisfies the estimate

(4)
$$|\partial_x^\alpha \partial_y^\beta e(\hbar; t, x, y)| \leqq C_{\alpha\beta}.$$

Theorem 2.1 of [4] shows that the operator $E(t)$ defined below has a norm bounded by some constant $N(E) > 0$:

(5)
$$E(t) = \left(\frac{1}{2\pi i t\hbar}\right)^{\frac{1}{2}} \int_{\mathbf{R}} e^{i\hbar^{-1}S(t,x,y)} e(\hbar, t, x, y)f(y)dy.$$

Therefore, the L^2 norm of $\left(\frac{\hbar\partial}{i\partial s} + \frac{1}{2}\left(\frac{\hbar\partial}{i\partial x}\right)^2 + V(x)\right)K(s)f$ is bounded by $N(E)s\|f\|_{L^2}$. This and the semi-group property show that

$$\left(\frac{\hbar\partial}{i\partial t} + \frac{1}{2}\left(\frac{\hbar\partial}{i\partial x}\right)^2 + V(x)\right)K(t)f$$
$$= s\text{-}\lim_{s\to 0} \left(\frac{\hbar\partial}{i\partial s} + \frac{1}{2}\left(\frac{\hbar\partial}{i\partial x}\right)^2 + V(x)\right)K(s+t)f$$
$$= s\text{-}\lim_{s\to 0} sE(s)K(t)f = 0.$$

Furthermore, if $f \in C_0^\infty(\mathbf{R})$, then $\left(\frac{1}{2}\left(\frac{\hbar\partial}{i\partial x}\right)^2 + V(x)\right)K(s)f$ belongs to $L^2(\mathbf{R})$. Therefore Theorem has been shown.

§3. Oscillatory integral over a Sobolev manifold.

3.1. We regard the Feynman path integral as an oscillatory integral over a Hilbert space. We follow Itô's formulation [14]. First we give meaning to the path space Ω of introduction as a Sobolev manifold. Let $\gamma_0(t)$ be the straight line segment starting from y at time 0 and reaching x at time T, i.e.,

(1)
$$\gamma_0(t) = \frac{t}{T}x + \frac{T-t}{T}y, \quad 0 \leqq t \leqq T.$$

Let $\mathcal{H}$ be the Sobolev space $H_0^1(0, T) = \left\{\gamma \,|\, \gamma \in L^2(0, T), \dfrac{d}{dt}\gamma \in L^2(0, T) \text{ and } \gamma(0) = \gamma(T) = 0\right\}$. The space $\mathcal{H}$ is a Hilbert space equipped with the inner product

$$(2) \qquad (\gamma_1, \gamma_2)_{\mathcal{H}} = \int_0^T \frac{d}{ds}\gamma_1(s)\frac{d}{ds}\gamma_2(s)ds.$$

We set $\Omega = \gamma_0 + \mathcal{H} = \{\gamma_0 + \gamma : \gamma \in \mathcal{H}\}$. Ω is a Sobolev (affine) manifold.

3.2. Next we explain how we consider the Feynman path integral as an oscillatory integral over Ω. For that purpose we introduce a quadratic form Q on $\mathcal{H}$, using a special complete ortho-normal system (c.o.n.s. in short) of $\mathcal{H}$ consisting of indefinite integrals of Haars' functions. For each $n = 1, 2, 3, \cdots$, and $k = 1, 2, \cdots, 2^{n-1}$, we define a finite binary fraction

$$(1) \qquad q = q(n, k) = 2^{-n}(2k - 1).$$

Since n and k are uniquely determined by q, we denote $n = n(q)$ and $k = k(q)$. We define

$$(2) \qquad \delta_q = 2^{-n(q)}T$$

and

$$(3) \qquad m(q) = 2^{n(q)-1} + k(q) - 1.$$

For each q, we define the function $e_q(s)$ by

$$(4) \qquad e_q(s) = \begin{cases} 0 & \text{for} \quad |s - qT| \geq \delta_q, \\ (2\delta_q)^{-1/2}(s - qT + \delta_q) & \text{for} \quad qT - \delta_q \leqq s \leqq qT, \\ (2\delta_q)^{-1/2}(qT + \delta_q - s) & \text{for} \quad qT \leqq s \leqq qT + \delta_q. \end{cases}$$

The system $\left\{\dfrac{d}{ds}e_q(s)\right\}_q$ and $\{T^{-1/2}\}$ form Haars' c.o.n.s. in $L^2(0, T)$. So $\{e_q(s)\}$ forms a c.o.n.s. of $\mathcal{H}$. For any $\gamma \in \mathcal{H}$ we denote its e_q component by $y_q = (\gamma, e_q)_{\mathcal{H}}$.

Let $\lambda > 1$ be fixed. Then we define a quadratic form Q on $\mathcal{H}$:

$$(5) \qquad Q(\gamma) = \sum_q \lambda^{-m(q)}|y_q|^2.$$

We denote $Q^{-1}(\gamma) = \sum_q \lambda^{m(q)}|y_q|^2$, if the right hand side is finite. Let $\mathcal{D}(Q^{-1}) = \{\gamma \in \mathcal{H} : Q^{-1}(\gamma) < \infty\}$. Then $\mathcal{D}(Q^{-1})$ is a Hilbert space with norm $[Q^{-1}(\gamma)]^{1/2}$. The natural injection $\mathcal{D}(Q^{-1}) \subset \mathcal{H}$ is of trace class. We denote by $N(d\gamma, b, nQ)$ the Gaussian measure over $\mathcal{H}$ with the mean vector $b \in \mathcal{H}$ and the variance nQ for $n = 1, 2, \cdots$.

3.3. Definition. We can now define, following Itô [14],

$$(1) \qquad \int_\Omega e^{i\hbar^{-1} S(\gamma)} \mathcal{D}[\gamma] = \lim_{n \to \infty} I_n(\hbar; T, x, y),$$

if the right hand side exists. Here

$$(2) \qquad I_n(\hbar, T, x, y) = (2\pi i\hbar)^{-\frac{1}{2}} \prod_q \left[1 + \frac{n\lambda^{-m(q)}}{i\hbar} \right]^{\frac{1}{2}} \int_{\mathcal{H}} e^{i\hbar^{-1} S(\gamma_0 + \gamma)} N(d\gamma, b, nQ).$$

We wish to show $\lim_{n \to \infty} I_n(\hbar, T, x, y)$ exists and equals $K(\hbar, T, x, y)$.

It should be remarked that Itô's original formulation has more symmetry, because he uses the full net of all positive quadratic forms Q of trace class.

3.4. We discuss Itô's integral $I_n(\hbar; T, x, y)$ over the Hilbert space $\mathcal{H}$. If $n\lambda^{-m(q)}$ is small the measure over y_q-space is concentrated on a small neighbourhood of $y_q = b_q$. On the other hand, if $n\lambda^{-m(q)}$ is large, then the variance of the Gaussian measure is large and the oscillatory character of $e^{i\hbar^{-1} S(\gamma_0 + \gamma)}$ is dominant in the y_q-space. So we are forced to distinguish various parts of space $\mathcal{H}$.

3.5. We classify binary fractions $\{q\}$ according to magnitude of $n\lambda^{-m(q)}$. We introduce natural numbers $m_0(n) < m_1(n) < m_2(n)$ in the following way: We fix a constant ρ satisfying $2 > \rho > 1$. Let N_0 be an integer so large that for any $N \geq N_0$

$$(1) \qquad 2^{-N} < \hbar^{2\rho}, \qquad 2^N \lambda^{-2^{N-12}} < 1.$$

Choose any integer n so large that

$$(2) \qquad n^{-1} \lambda^{2^{N_0}} < \hbar^{-1}.$$

Then we define the positive integer $m_2(n)$ by

$$(3) \qquad n^{-1} \lambda^{m_2(n)} \leqq \hbar^{-1} < n^{-1} \lambda^{m_2(n)+1}.$$

Let N_n be the largest integer not exceeding $\log_2 m_2(n)$. Then we define $m_1(n)$ and $m_0(n)$ by inequalities

$$(4) \qquad n^{-1} \lambda^{m_1(n)} \leqq 2^{-\frac{1}{4\rho} N_n} < n^{-1} \lambda^{m_1(n)+1},$$

$$(5) \qquad n^{-1} \lambda^{m_0(n)} \leqq 2^{-\rho N_n} < n^{-1} \lambda^{m_0(n)+1}.$$

We shall often denote $m_i(n)$ by $m_i, i = 0, 1, 2$, for the sake of brevity. We define $Q\{1\} = \{q : m(q) < m_0(n)\}$, $Q\{2\} = \{q : m_0(n) \leqq m(q) < m_1(n)\}$, $Q\{3\} = \{q : m_1(n) \leqq m(q) < m_2(n)\}$ and $Q\{4\} = \{q : m_2(n) \leqq m(q)\}$.

3.6. We decompose $\mathcal{H}$ into four subspaces: Let $\mathcal{H}_j = $ span of $\{e_q | q \in Q\{j\}\}$, $j = 1, 2, 3, 4$. Then we have an orthogonal decomposition: $\mathcal{H} = \mathcal{H}_1 + \mathcal{H}_2 + \mathcal{H}_3 + \mathcal{H}_4$. According to this decomposition any element γ can be decomposed into four parts $\gamma = \gamma_1 + \gamma_2 + \gamma_3 + \gamma_4$. This decomposition reduces the quadratic form $Q(\gamma)$, i.e., $Q(\gamma) = Q_1(\gamma_1) + Q_2(\gamma_2) + Q_3(\gamma_3) + Q_4(\gamma_4)$. Therefore, the Gaussian measure $N(d\gamma, b, nQ)$ is represented as the product measure

$$N(d\gamma, b, nQ)$$
$$= N(d\gamma_1, b_1, nQ_1) \times N(d\gamma_2, b_2, nQ_2) \times N(d\gamma_3, b_3, nQ_3) \times N(d\gamma_4, b_4, nQ_4),$$

where $N(d\gamma_i, b_i, nQ_i)$, $i = 1, 2, 3, 4$, is the Gaussian measure on $\mathcal{H}_i$ with mean b_i, the $\mathcal{H}_i$ component of b, and variance nQ_i. In calculating the right hand side of (3.3.2), we can apply Fubini's theorem and we can integrate consecutively over $\mathcal{H}_j$, $j = 1, 2, 3, 4$.

3.7. Oscillatory character of the integral becomes manifest in the integration over the space $\mathcal{H}_1$, because the variance nQ_1 of the Gaussian measure $N(d\gamma_1, b_1, nQ_1)$ is large. So we wish to apply the stationary phase method, which was studied in [12]. For that purpose the coordinate system $\{y_q\}$ is not convenient. We will make a change of variables in $\mathcal{H}_1$ and use a more convenient coordinate system. $\mathcal{H}_1$ is the space of all piecewise linear continuous functions with vertices corresponding to the division

(1) $$\Delta_n : 0 = \tau_0 < \tau_1 < \tau_2 < \cdots < \tau_{m_0 - 1} < \tau_{m_0} = T,$$

of the interval $[0, T]$, where $\{\tau_j\}_{1 \le j \le m_0 - 1}$ is the rearrangement of $\{qT\}_{1 \le m(q) \le m_0 - 1}$ in the order of magnitude. We introduce new basis (not orthogonal) $w_j(s)$, $1 \le j \le m_0 - 1$, of $\mathcal{H}_1$: For any $j = 1, 2, \ldots, m_0 - 1$, let $w_j(t)$ be the continuous piecewise linear function with support in the interval $[\tau_{j-1}, \tau_{j+1}]$, with one peak of hight 1 at τ_j, i.e.,

$$w_j(t) = \begin{cases} 0 & \text{outside } [\tau_{j-1}, \tau_{j+1}], \\ \Delta\tau_j^{-1}(t - \tau_{j-1}) & \text{for } t \in [\tau_{j-1}, \tau_j] \\ \Delta\tau_{j+1}^{-1}(\tau_{j+1} - t) & \text{for } t \in [\tau_j, \tau_{j+1}]. \end{cases}$$

By the way we introduce

$$w_0(t) = \begin{cases} \Delta\tau_1^{-1}(\tau_1 - t) & \text{for } t \in [0, \tau_1], \\ 0 & \text{for } t \text{ outside } [0, \tau_1], \end{cases}$$

$$w_L(t) = \begin{cases} \Delta\tau_L^{-1}(t - \tau_{L-1}) & \text{for } t \in [\tau_{L-1}, t], \\ 0 & \text{outside } [\tau_{L-1}, t]. \end{cases}$$

For any $\gamma_1 \in \mathcal{H}_1$ let $z_j = \gamma_1(\tau_j)$, $j = 1, 2, \ldots, m_0 - 1$. Then $\{z_j\}$ is the coordinates of γ_1 with respect to the basis $\{w_j\}_{j=1}^{m_0 - 1}$. i.e., $\gamma_1(t) = \sum_{j=1}^{m_0 - 1} z_j w_j(t)$. We can use functions $\{w_j(t)\}_{j=1}^{m_0}$ to express $\gamma_0 + \gamma_1$. Let $x_j = \gamma_0(\tau_j) + \gamma_1(\tau_j)$. Then

$$\gamma_0(t) + \gamma_1(t) = \sum_{j=0}^{L} x_j w_j(t).$$

We use the coordinates $\{x_j\}_{j=1}^{m_0-1}$ of $\gamma_1 \in \mathcal{H}_1$.

3.8. It is easy to show that

(1)
$$\prod_{m(q)=1}^{m_0-1} dy_q = \prod_{j=1}^{m_0} \left[\frac{1}{\Delta \tau_j}\right]^{\frac{1}{2}} \prod_{j=1}^{m_0-1} dx_j.$$

Therefore,

(2)
$$\left[\frac{1}{2\pi i \hbar}\right]^{\frac{1}{2}} \prod_{m(q)<m_0} \left[1 + \frac{n\lambda^{-m(q)}}{i\hbar}\right]^{\frac{1}{2}} N(d\gamma_1, b_1, nQ_1)$$
$$= \prod_{j=1}^{m_0} \left[\frac{1}{2\pi i \hbar \Delta \tau_j}\right]^{\frac{1}{2}} \prod_{j=1}^{m_0-1} \left[1 + \frac{i\hbar \lambda^j}{n}\right]^{\frac{1}{2}}$$
$$\times e^{-\frac{1}{2n} \sum_{m(q)=1}^{m_0-1} \lambda^{m(q)}(y_q-b_q)^2} \prod_{j=1}^{m_0-1} dx_j.$$

Consequently, Fubini's theorem gives that

(3)
$$I_n(\hbar, T, x, y) = \prod_{j=1}^{m_0-1} \left[1 + \frac{i\hbar \lambda^j}{n}\right]^{\frac{1}{2}} \prod_{j=1}^{m_0} \left[\frac{1}{2\pi i \hbar \Delta \tau_j}\right]^{\frac{1}{2}}$$
$$\times \int_{\mathbf{R}^{m_0-1}} e^{i\hbar^{-1}S(\gamma_0+\gamma_1+\gamma_2^{\#}+b_3+b_4)} p(\gamma_1) \prod_{j=1}^{m_0-1} dx_j,$$

where

(4)
$$p(\gamma_1) = e^{-\frac{1}{2n} \sum_{m(q)<m_0} \lambda^{m(q)}(y_q-b_q)^2} \prod_{m(q)\geqq m_0}^{\infty} \left[1 + \frac{n\lambda^{-m(q)}}{i\hbar}\right]^{\frac{1}{2}}$$
$$\times \int_{\mathcal{H}_2+\mathcal{H}_3+\mathcal{H}_4} e^{i\hbar^{-1}(S(\gamma_0+\gamma)-S(\gamma_0+\gamma_1+\gamma_2^{\#}+b_3+b_4))} \prod_{j=2}^{4} N(d\gamma_j, b_j, nQ_j),$$

and $\gamma_2^{\#}$ is the critical point of the map $\gamma_2 \to S(\gamma_0 + \gamma_1 + \gamma_2 + b_3 + b_4)$, which is unique under assumption (2.1.1).

We regard (3.8.3) as an oscillatory integral over $\mathbf{R}^{m_0-1}$ with the phase function $S(\gamma_0 + \gamma_1 + \gamma_2^{\#} + b_3 + b_4)$ and the amplitude function $p(\gamma_1)$.

3.9. It is clear that for each e_q there exists only one j, which is denoted by j_q, such that $(\tau_{j-1}, \tau_j) \cap \operatorname{spt} e_q \neq 0$, where $\operatorname{spt} e_q$ stands for the support of function e_q. For $q \in Q\{2\}$ we define

(1)
$$\delta_q \omega_q(y_q; x_{j_q}, x_{j_q-1}) = \int_{\tau_{j_q-1}}^{\tau_{j_q}} V((\gamma_0 + \gamma_1 + y_q e_q + b_3 + b_4)(t)) \, dt.$$

We assumed (2.1.1). Let $\gamma_2^{\#}$ be the stationary point of the function

$$\mathcal{H}_2 \ni \gamma_2 \longrightarrow S(\gamma_0 + \gamma_1 + \gamma_2 + b_3 + b_4).$$

Clearly $\gamma_2^{\#}$ is a function of γ_1. Let γ_1^* be the critical point of the function

$$\mathcal{H}_1 \ni \gamma_1 \longrightarrow S(\gamma_0 + \gamma_1 + \gamma_2^{\#} + b_3 + b_4).$$

We denote $\gamma_2^* = \gamma_2^{\#}(\gamma_1^*)$. Then (γ_1^*, γ_2^*) is the critical point of

$$\mathcal{H}_1 + \mathcal{H}_2 \ni \gamma_1 + \gamma_2 \longrightarrow S(\gamma_0 + \gamma_1 + \gamma_2 + b_3 + b_4).$$

We use the notation $y_q^* = (\gamma_1^* + \gamma_2^*, e_q)_{\mathcal{H}}$ and $x_j^* = \gamma_1^*(\tau_j)$.
   Let $\delta_n = 2^{-N_n T}$, $\Delta\tau_j = \tau_j - \tau_{j-1}$ and $\nu = \hbar^{-1}$. We have $\Delta\tau_j < 2\delta_n$.
   Let

$$(2) \qquad \xi_q^* = \int_0^T V'\left((\gamma_0 + \gamma_1^* + \gamma_2^* + b_3 + b_4)(t)\right) e_q(t)dt.$$

3.10. Theorem. Under hypothesis (ASS-I) there exists a positive constant T_1 such that if $0 < T \leq T_1$ then we have

$$(1) \quad I_n(\hbar, T, x, y) = I(\Delta_n, b|\hbar, T, x, y)$$

$$\times \prod_{m(q)=1}^{m_1-1} \left[1 + \frac{i\hbar\lambda^{m(q)}}{n}\right]^{\frac{1}{2}} \prod_{q \in Q\{2\}} \left[1 + \delta_q \partial_q^2 \omega_q(y_q^*, x_{j_q}^*, x_{j_q-1}^*)\right]^{\frac{1}{2}}$$

$$\times \exp -\frac{1}{2}\left[\sum_{m(q)<m_1} \frac{\lambda^{m(q)}}{n}(y_q^* - b_q)^2 + \sum_{m(q)\geq m_1} \frac{n\lambda^{-m(q)}}{1 - in\nu\lambda^{-m(q)}}(\xi_q^* - b_q)^2\right]$$

$$+ \left[\frac{1}{2\pi i\hbar T}\right]^{\frac{1}{2}} e^{i\nu S(\gamma_0 + \gamma_1^* + \gamma_2^* + b_3 + b_4)} r_n(\hbar; T, x, y),$$

where

$$(2) \qquad I(\Delta_n, b|\hbar, T, x, y)$$

$$= \prod_{j=1}^{m_0-1} \left[\frac{1}{2\pi i\hbar\Delta\tau_j}\right]^{\frac{1}{2}} \int_{\mathbf{R}^{m_0-1}} e^{i\nu S(\gamma_0 + \gamma_1 + \gamma_2^{\#} + b_3 + b_4)} \prod_{j=1}^{m_0-1} dx_j,$$

and amplitude $r_n(\hbar, T, x, y)$ of the remainder term satisfies the following estimate: There exists a positive constant ε such that

$$(3) \qquad |\partial_x^\alpha \partial_y^\beta r_n(\hbar; T, x, y)| \leq C_{\alpha\beta} \hbar \delta_n^\varepsilon,$$

where $C_{\alpha\beta}$ is a positive constant independent of n, x, y and $\hbar$.

3.11. When n goes to ∞, δ_n tends to 0. So $r_n(\hbar, T, x, y)$ together with its derivatives goes to 0. The exponential factors of the main term goes to 1 uniformly with respect to $\hbar$. In fact we can show

Theorem. There exist positive constants ε and C such that

$$
(1) \qquad \left| \sum_{m(q) \leqq m_1} \frac{n\lambda^{-m(q)}}{1 - i\nu n\lambda^{-m(q)}} (\xi_q^* - b_q)^2 \right|
$$
$$
\leqq C\hbar^\rho \delta_n \left(1 + |x|^2 + |y|^2 + \|b_3 + b_4\|_{\mathcal{H}}^2 \right).
$$

Furthermore

$$
(2) \qquad \sum_{m(q) < m_1} \frac{\lambda^{m(q)}}{n} (y_q^* - b_q)^2 \leqq C\hbar^\rho \delta_n^\varepsilon \left[1 + |x|^2 + |y|^2 + \|b_3 + b_4\|_{\mathcal{H}}^2 \right].
$$

$$
(3) \qquad \left| \prod_{m(q)=m_0}^{m_1-1} \left[1 + \frac{i\hbar\lambda^{m(q)}}{n} \right]^{\frac{1}{2}} - 1 \right| < C\hbar\delta_n^\varepsilon.
$$

3.12. It is clear that $|\Delta_n|$ tends to 0 as n goes to ∞. $\sum_{j=0}^{m_0} x_j w_j(t)$ is very close to $\gamma_\Delta^{c\ell}(X)$ if $|\Delta_n|$ is very small. Starting from this fact, we can prove the following lemma.

Lemma. $\lim_{n\to\infty} I_n(\Delta, b | \hbar, T, x, y) - I(\Delta_n; \hbar, T, x, y) = 0$.

3.13. We can show from Lemma 3.12 and Theorem 3.10 the following

Theorem. Under (ASS-I) and if $|T| \leq T_1$ then

$$
\lim_{n\to\infty} I_n(\hbar, T, x, y) = K(\hbar, T, x, y).
$$

References

1. Albeverio, S. and Høegh-Krohn, R. J., *Mathematical theory of Feynman path integral*, Lecture Notes in Math., Springer, Berlin, **523** (1976).
2. Albeverio, S. and Høegh-Krohn, R. J., *Oscillatory integrals and the method of stationary phase in infinitely many dimensions, with application to the classical limit of quantum mechanics*, I. Inv. Math., **40** (1977), 59–106.
3. Albeverio, S., Høegh-Krohn, R. J. and others ed., *Feynman Path integrals, Preceedings Marseille 1978*, Lecture Notes in Physics, Springer, Berlin, **106** (1978).

4. Asada, K. and Fujiwara, D., *On some oscillatory integral transformations on $L^2(\mathbf{R}^n)$*, Japan. J. Math., **27** (1978), 299–361.

5. Cameron, R. H., *A family of integrals serving to connect the Wiener and Feynman integrals*, J. Mathematics and Physics, **39** (1960), 126–140.

6. Feynman, R. P., *Space time approach to non relativistic quantum mechanics*, Rev. of Modern Phys., **20** (1948), 367–387.

7. Feynman R. P. and Hibbs A. R., *Quantum mechanics and path integral*, McGraw-Hill, New York, (1965).

8. Fujiwara, D., *Fundamental solution of partial differential operator of Schrödinger's type*, I. Proc. Japan Acad., **50** (1974), 566–569.

9. Fujiwara, D., *A construction of the fundamental solution for the Schrödinger equations*, J. d'Analyse Math., **35** (1979), 41–96.

10. Fujiwara, D., *Remarks on convergence of some Feynman path integrals*, Duke Math. J. **47** (1980), 559–600.

11. Fujiwara, D., *Some Feynman path integrals as an improper integral over a Sobolev space*, Proc. conference of symposium on PDE, Saint Jean de Mont (1989).

12. Fujiwara, D., *The stationary phase method with an estimate of the remainder term on a space of large dimension*, to appear in Nagoya Math. J., **24** (1991), 61–97.

13. Gelfand, I. M. and Yaglom, A. M., *Integrals in functional spaces and its applications in quantum physics*, J. Math. Physics, **1**, 48–69.

14. Itô, K., *Generalized uniform complex measure in Hilbertian metric spaces and its application to the Feynman path integrals*, Proc. 5th Berkeley symposium on Math. Statistics and Probability, Univ. of California Press, Berkeley, **2** part 1, (1967), 145–161.

15. Nelson, E., *Feynman path integrals and Schrödinger equation*, J. Mathematical Physics, **5** (1964), 332–343.

16. Pauli, W., (edited by Enz, C. P.), *Pauli Lectures on Physics, vol.6, Selected topics in field quantization*, MIT Press (1973).

L^p estimates for the Stokes system

MARIKO GIGA, YOSHIKAZU GIGA
AND HERMANN SOHR

1. Introduction.

This paper investigates the fractional powers $(A + B)^\alpha$, $0 \leq \alpha \leq 1$ of the sum $A + B$ of two closed (resolvent commuting) operators A and B in a ζ-convex Banach space X. We compare the domain $D((A + B)^\alpha)$ of $(A + B)^\alpha$ with the domain $D(A^\alpha + B^\alpha) = D(A^\alpha) \cap D(B^\alpha)$ of the sum $A^\alpha + B^\alpha$ and show in particular the relation

$$(1.1) \qquad D((A + B)^\alpha) = D(A^\alpha) \cap D(B^\alpha)$$

with equivalent norms $\|(A + B)^\alpha u\|$ and $\|A^\alpha u\| + \|B^\alpha u\|$, under some assumptions on the pure imaginary powers of A and B.

Our results will be applied to L^p estimates for (generalized) solutions of the evolution equation

$$(1.2) \qquad \frac{du}{dt} + Au = f \quad \text{in} \quad (0,T), \quad 0 < T \leq \infty, \quad u(0) = 0.$$

Here we restrict ourselves to the Stokes operator $A = A_q$. Formally, we get such an equation if we apply the L^q Helmholtz projection P_q to the Stokes system

$$(1.3) \qquad \frac{\partial u}{\partial t} - \triangle u + \nabla p = f, \ \text{div } u = 0 \text{ in } \Omega \times (0,T)$$

$$u|_{\partial\Omega} = 0 \text{ on } \partial\Omega \times (0,T), \ u = 0 \text{ on } \Omega \text{ at } t = 0,$$

where Ω is a domain in $\mathbf{R}^n$ with smooth boundary $\partial\Omega$ and $1 < q < \infty$; see e.g. [GS1,2] for notations.

Our perturbation result is based on a theory recently developed by Dore and Venni [DV] which has been extended by Giga and Sohr [GS2] to the case that the inverse operators A^{-1} and B^{-1} need not be bounded. See also [PS] for another proof. The original theory of Dore and Venni is applicable to the evolution equation (1.2) only for a finite interval $[0,T]$; it yields a constant $C = C(\Omega, r, q, T) > 0$ such that

$$(1.4) \qquad \int_0^T \|\frac{du}{dt}\|_q^r dt + \int_0^T \|A_q u\|_q^r dt \leq C \int_0^T \|f\|_q^r dt,$$

where $1 < r, q < \infty$ and $\| \cdot \|_q$ denotes the $L^q(\Omega)$-norm. The extension by [GS2] strengthens the estimate (1.4) so that A_q^{-1} is allowed to be unbounded and that C may be chosen independently of T. Therefore, one may take $T = \infty$ in (1.4) which yields asymptotic properties of the solution u of (1.3) as $t \to \infty$ even when Ω need not be bounded [GS2]. In [GS2] the estimates applied to the nonlinear Navier-Stokes system. In [GGS] the estimate (1.4) has been extended to the case that $A = A(t)$ in (1.2) depends on t, and in [GS2] and [GGS] non zero initial values $u(0) = u_0$ are treated.

Recently Dore and Venni [DV2] applied their theory to get higher derivative estimates for solutions of (1.2).

The application of our abstract result (1.1) on fractional powers $(A + B)^\alpha$ to the evolution equation (1.2) yields now estimates of the form

$$(1.5) \qquad \int_0^T \|(\frac{d}{dt})^{1-\alpha} u\|_q^r dt + \int_0^T \|A_q^{1-\alpha} u\|_q^r dt \leq C \int_0^T \|A_q^{-\alpha} f\|_q^r dt$$

with C independent of f and T, and $0 < \alpha < 1$. Here u is a generalized solution of (1.2) and f may be a distribution which is regularized by $A_q^{-\alpha}$. The case $\alpha = 1/2$ is especially important because (1.5) yields an a priori estimate

$$(1.6) \qquad \int_0^T \|(\frac{d}{dt})^{1/2} u\|_q^r dt + \int_0^T \|\nabla u\|_q^r dt \leq C \int_0^T \|F\|_q^r dt$$

for solutions of (1.3) when $f = \text{div } F$; here we restrict $n \geq 3$ and $n/(n-1) < q < n$ when Ω is an exterior domain. This estimate is considered as a nonstationary version of Cattabriga's estimate (see e.g. [BM]).

The class $BIP(a, K)$ of operators we consider here consists of nonnegative closed operators A in X which satisfy the estimate $\|A^{is} u\|_X \leq K e^{a|s|} \|u\|_X$ for all $s \in \mathbf{R}$ where $K \geq 1$ and $0 \leq a < \pi$ (independent of u and s). The well known application of this estimate of the pure imaginary powers A^{is} is the identification

$$[X, D(A)]_\alpha = D(A^\alpha),$$

where $[X, D(A)]_\alpha$ is the complex interpolation space; see e.g. [Tr]. The Dore-Venni theory gives now another important application of the above estimate. This theory requires the ζ-convexity of the Banach space. For various properties of ζ-convex space we refer to the nice review article [B]. For the theory of complex powers A^z, $z \in \mathbf{C}$ we refer to the comprehensive article [Ko].

Our main abstract result is given in Section 3; Section 2 contains preliminary lemmas and Section 4 is devoted to the application to the Stokes system.

2. Sum of operators with bounded imaginary powers.

Let A be a closed linear operator with dense domain $D(A)$ in a Banach space X equipped with norm $\| \cdot \|$. We say A is *nonnegative* if its resolvent set contains all negative real numbers and

$$\sup_{t>0} t \|(t + A)^{-1}\| < \infty,$$

where $\|\cdot\|$ denotes the operator norm in $\mathcal{L}(X)$, the space of all bounded linear operators. If a nonnegative operator has a dense range $R(A)$ in X, one can define its complex power A^z for every $z \in \mathbf{C}$ as a densely defined closed operator in X. (cf. [Ko]). For $a \geq 0$ and $K \geq 1$ we say a nonnegative operator A belongs to $BIP(a; K)$ if $A^{is} \in \mathcal{L}(X)$ and is estimated as

$$\|A^{is}\| \leq K e^{a|s|}, \quad s \in \mathbf{R}$$

where $D(A)$ and $R(A)$ are assumed to be dense in X. Let $BIP(a)$ denote the union of $BIP(a, K)$ for $K \geq 1$.

2.1. FUNDAMENTAL LEMMA. (i) If $A \in BIP(a; K)$, then $A^\alpha \in BIP(a\alpha; K)$ for $0 < \alpha < 1$.
(ii) If $A \in BIP(a)$, $0 \le a < \pi$, then for each $\delta > 0$ with $\delta < \pi - a$ there is a constant M_δ independent of λ such that

$$\|(\lambda + A)^{-1}\| \le M_\delta/|\lambda|, \quad |\arg \lambda| \le \pi - a - \delta, \ 0 \ne \lambda \in \mathbf{C}.$$

In particular, if $a < \pi/2$, then $-A$ generates an analytic semigroup e^{-tA} in X.

PROOF: (i) As well known, if A is nonnegative so is A^α ($0 < \alpha < 1$); see e.g. [Kr, p.119, (5.25)] or [Ka]. If $A \in BIP(a)$, then $A^\alpha \in BIP(a\alpha)$ since

$$\|(A^\alpha)^{is}\| = \|A^{i\alpha s}\| \le Ke^{a\alpha|s|}.$$

Here we use the property $(A^\alpha)^{is} = A^{i\alpha s}$ which can be shown as follows. First we prove this property with A replaced by $(\varepsilon + A)^{-1}$, $\varepsilon > 0$; here we use the well known Dunford integral calculus. Then the assertion follows by letting $\varepsilon \to 0$ and using [PS, Theorem 3].
(ii) See [PS, Theorem 2].

2.2. SUMMATION LEMMA. Let X be a ζ-convex Banach space. Let A and B belong to $BIP(a, K)$ and $BIP(b, K)$, respectively. Suppose that A and B are resolvent commuting, i.e.,

$$(t + A)^{-1}(t + B)^{-1} = (t + B)^{-1}(t + A)^{-1} \quad \text{for all} \quad t > 0.$$

Then $A + B \in BIP(a \vee b, K')$ provided that $a \ne b$, where $a \vee b = \max(a, b)$ and $K' = K'(a, b, K, X)$.

This is Theorem 5 in [PS], where the dependence of constants is not explicitly stated. For various properties of ζ-convex spaces there is the nice review article by Burkholder [B] so we do not touch them here.

We next recall the Dore-Venni theory [DV] on the inverse of $A + B$. Let T be an injective closed linear operator in a Banach space X. Let $\hat{D}(T)$ be the completion of $D(T)$ in the norm $\|Tu\|$. Since T may not have a bounded inverse, $\hat{D}(T)$ may not be a subspace of X. The element $Tv \in X$ for $v \in \hat{D}(T)$ is defined by $Tv = \lim_{j\to\infty} Tv_j$, where $\{v_j\}$ is a Cauchy sequence converging to v in $\hat{D}(T)$. The norm of v in $\hat{D}(T)$ is defined by

$$\|v\|_{\hat{D}(T)} = \|Tv\| = \lim_{j\to\infty} \|Tv_j\|.$$

Let T' be another injective closed linear operator in X. Let $T + T'$ be the operator defined on $D(T + T') = D(T) \cap D(T')$. By $D(T + T')^\wedge$ we represent the completion of $D(T + T')$ in the norm $\|Tu\| + \|T'u\|$. Clearly, this space is continuously embedded in $\hat{D}(T)$ and $\hat{D}(T')$. However, the intersection $\hat{D}(T) \cap \hat{D}(T')$ is not meaningful unless the norms $\|Tv\|$ and $\|T'v\|$ are consistent in the sense of the interpolation theory [RS, p.35]. Note that $D(T + T')^\wedge$ need not be equal to $\hat{D}(T + T')$.

2.3. THEOREM ON INVERSES. *Let X be ζ-convex. Suppose that $A \in BIP(a; K)$ and $B \in BIP(b; K)$ are resolvent commuting and that $a + b < \pi$. Then the operator $A + B : D(A + B)^{\wedge} \to X$ is bijective and boundedly invertible. Moreover there is $C = C(a, b, K, X)$ such that*

$$\|A(A + B)^{-1}\| \leq C, \quad \|B(A + B)^{-1}\| \leq C.$$

REMARK: Observe as a consequence that $\|Au\| + \|Bu\|$ and $\|(A + B)u\|$ are equivalent norms on $D(A) \cap D(B)$ so that $D(A + B)^{\wedge} = \hat{D}(A + B)$.

This result was first proved by Dore and Venni [DV] under the assumption that both A and B have bounded inverses. The key observation is the following integral representation

$$(A + B)^{-1} = \frac{1}{2i} \int_{c-i\infty}^{c+i\infty} \frac{A^{-z} B^{z-1}}{\sin \pi z} dz, \quad 0 < c < 1.$$

It turns out that the assumptions A^{-1} and $B^{-1} \in \mathcal{L}(X)$ can be removed. The first proof is given by Y. Giga and Sohr [GS2] by introducing appropriate dense subspaces of X so that the argument in [DV] can be justified without $A^{-1}, B^{-1} \in \mathcal{L}(X)$. Another proof is given by Prüss and Sohr [PS]. They established a functional calculus generated by the group A^{is} and proved that $A \in BIP(a)$ implies $A_\epsilon = \epsilon I + A \in BIP(a; L)$ with L independent of $\epsilon > 0$. This is considered as a special case of the summation lemma. Since A_ϵ has a bounded inverse, they applied the Dore-Venni estimate to A_ϵ and sent $\epsilon \to 0$ to get the desired estimates in Theorem 2.3. The first proof is more direct because it does not use the approximated operator A_ϵ.

The injectivity of the operators A, B is not explicitly assumed. It follows from the fact that these operators are nonnegative and have dense ranges; see [Ko, Theorem 3.2 and 3.7]. Indeed $Au = 0$ implies $u = t(t + A)^{-1}u$, so letting $t \to 0$ yields $u = 0$.

It is convenient to consider appropriate dense subspaces in X as in [GS2]. For $\xi = (\zeta, \eta)$ and $\Lambda = (h, j, k, \ell)$ with nonnegative integers h, j, k, ℓ we set

$$g_\Lambda(\xi) = I_A^j(t) J_A^h(\tau) I_B^\ell(s) J_B^k(\sigma) g, \quad g \in X$$
$$\zeta = (t, \tau^{-1}), \quad \eta = (s, \sigma^{-1}), \quad t, \tau, s, \sigma > 0$$

with $I_A(t) = A(t + A)^{-1}$ and $J_A(\tau) = \tau(\tau + A)^{-1}$. We introduce the subspace

$$G_\Lambda = \text{linear hull of } \{g_\Lambda(\xi); g \in X, \xi = (t, \tau^{-1}, s, \sigma^{-1}), t, \tau, s, \sigma > 0\}.$$

2.4. DENSITY LEMMA. *Suppose that A and B are nonnegative and resolvent commuting with dense ranges and domains in X. Then G_Λ is dense in X. Moreover G_Λ is dense in $D(A) \cap D(B)$ under the norm $\|Av\| + \|Bv\|$.*

PROOF: By a standard argument [Ko] we see $g_\Lambda(\xi) \to g, Ag_\Lambda(\xi) \to Ag, Bg_\Lambda(\xi) \to Bg$ in X as $\xi \to 0$, which proves the lemma. We give a proof for completeness. Since A is nonnegative, one observes

$$t(t + A)^{-1}f = t(A(t + A)^{-1}u) \to 0 \quad \text{as} \quad t \to 0$$

for $f = Au \in R(A)$. Since $R(A)$ is dense in X and $\sup_t \|I_A(t)\| < \infty$, we conclude

$$I_A(t)f \to f \quad \text{in} \quad X \quad \text{as} \quad t \to 0.$$

A similar observation shows

$$J_A(\tau)f \to f \quad \text{in} \quad X \quad \text{as} \quad \tau \to \infty$$

and the same for B. Since all $I_A(t)$, $I_B(s)$, $J_A(t)$, $J_B(s)$ are bounded in $\mathcal{L}(X)$, these convergences for A and B imply that $g_\Lambda(\xi) \to g$ in X as $\xi \to 0$. The proofs of $Ag_\Lambda(\xi) \to Ag$ and $Bg_\Lambda(\xi) \to Bg$ under $g \in D(A) \cap D(B)$ are parallel, so they are omitted.

2.5. COMMUTATIVITY LEMMA. *Suppose that A and B are nonnegative and resolvent commuting with dense domains and ranges in X. Then*

$$A^z B^w A^u B^v f = B^w A^{z+u} B^v f \quad \text{for} \quad f \in G_\Lambda$$
$$\text{with } z, w, u, v \in \mathbf{C} \text{ and } \Lambda = (h, j, k, \ell)$$

provided that h, j, k, ℓ are sufficiently large and the largeness only depends on the modulus of the real parts of z, w, u, v.

For the proof we use an integral representation of the complex powers of A and B by their resolvents [Ko, (1.3) and (4.11)]. Since A and B are resolvent commuting, it is not difficult to prove

$$(t + A)^{-1}(s + B)^{-1} = (s + B)^{-1}(t + A)^{-1}, \quad t, s > 0.$$

Applying this commutativity to the integral representation yields the commutativity of complex powers on G_Λ. The proof is straightforward, so we omit the details.

2.6. COROLLARY TO THE THEOREM ON INVERSES. *Assume the hypotheses of the theorem on inverses. Let m be a positive integer. Then*

$$A^m(A + B)^{-m} = (A(A + B)^{-1})^m$$
$$B^m(A + B)^{-m} = (B(A + B)^{-1})^m$$

on an appropriate dense subspace of X. In particular, $A^m(A+B)^{-m}$ and $B^m(A+B)^{-m}$ can be extended to bounded linear operators on X with a bound depending only on a, b, K, m, X.

PROOF: We give a proof for $m = 2$; the proof for general $m \geq 3$ is parallel, so it is omitted. We use the Dore-Venni representation of $(A + B)^{-1}$. Formally for $z \in \mathbf{C}$, Re $z = c$ with $0 < c < 1$

$$AB^{z-1}(A + B)^{-1}f = AB^{z-1} \int_{c-i\infty}^{c+i\infty} \frac{A^{-w} B^{w-1} f}{2i \sin \pi w} dw$$
$$= B^{z-1}A(A + B)^{-1}f.$$

This calculation is justified by the commutativity lemma for $f \in G_\Lambda$, $\Lambda = (h, j, k, \ell)$ with h, j, k, ℓ sufficiently large. We thus observe

$$A^2(A+B)^{-2}f = A \int_{c-i\infty}^{c+i\infty} \frac{A^{-z}AB^{z-1}(A+B)^{-1}f}{2i \sin \pi z} dz$$
$$= A(A+B)^{-1}A(A+B)^{-1}f.$$

Since G_Λ is dense in X and $A(A+B)^{-1}$ is bounded by the theorem on inverses, $A^2(A+B)^{-2}$ can be extended to a bounded linear operator $(A(A+B)^{-1})^2$. The same argument applies to $B^2(A+B)^{-2}$.

3. Spaces of fractional powers

For $A \in BIP(a)$ let $\hat{D}(A^\alpha)$ be the completion of the domain $D(A^\alpha)$ in the norm $\|A^\alpha u\|$, where $0 < \alpha < 1$. The space $\hat{D}(A^\alpha)$ can be characterized by a complex interpolation space, namely

$$\hat{D}(A^\alpha) = [X, \hat{D}(A)]_\alpha.$$

This follows from the general interpolation theory (see e.g. [Tr], [BB]). For the proof see e.g. [GS1, Proposition 6.1] or [BM]. In this section we compare various norms on $D(A) \cap D(B)$.

3.1. MAIN THEOREM. *Suppose that X is ζ-convex. Suppose that $A \in BIP(a, K)$ and $B \in BIP(b, K)$ are resolvent commuting and that $a + b < \pi$. Then for $0 \leq \alpha \leq 1$*

$$D(A^\alpha) \cap D(B^\alpha) = D((A+B)^\alpha),$$
$$\hat{D}(A^\alpha + B^\alpha) = D(A^\alpha + B^\alpha)^\wedge = \hat{D}((A+B)^\alpha) = [X, \hat{D}(A+B)]_\alpha$$

and there are constants $C_j = C_j(a, b, \alpha, K, X) > 0$, $j = 1, 2, 3, 4$ such that

$$\|A^\alpha u\| + \|B^\alpha u\| \leq C_1 \|(A^\alpha + B^\alpha)u\| \leq C_2 \|(A+B)^\alpha u\| \leq$$
$$\leq C_3 \|u\|_{[X, \hat{D}(A+B)]_\alpha} \leq C_4(\|A^\alpha u\| + \|B^\alpha u\|)$$

for all $u \in D(A^\alpha) \cap D(B^\alpha)$.

PROOF: Since the summation lemma implies $A + B \in BIP(a \vee b + \delta, K')$, $\delta > 0$ with some $K' \geq 1$, it follows the identity

$$\hat{D}((A+B)^\alpha) = [X, \hat{D}(A+B)]_\alpha$$

with equivalent norms

$$\|(A+B)^\alpha u\| \quad \text{and} \quad \|u\|_{[X, \hat{D}(A+B)]_\alpha}.$$

Furthermore, since $A^\alpha \in BIP(a\alpha, K)$, $B^\alpha \in BIP(b\alpha, K)$ and $a\alpha + b\alpha < \pi$, by the theorem on inverses we observe that the norms

$$\|A^\alpha u\| + \|B^\alpha u\| \quad \text{and} \quad \|(A^\alpha + B^\alpha)u\|$$

are equivalent on $D(A^\alpha) \cap D(B^\alpha)$.

It remains to prove that $D(A^\alpha + B^\alpha) = D((A+B)^\alpha)$ and

$$(3.1) \qquad \|(A+B)^\alpha u\| \le C\|(A^\alpha + B^\alpha)u\|,$$

$$(3.2) \qquad \|(A^\alpha + B^\alpha)u\| \le C'\|(A+B)^\alpha u\|$$

for all $u \in D(A^\alpha + B^\alpha) = D(A^\alpha) \cap D(B^\alpha)$. Let us show the first inequality (3.1). To prove (3.1) it suffices to show that

$$(3.3) \qquad \|(A+B)^\alpha (A^\alpha + B^\alpha)^{-1}v\| \le C\|v\|$$

for all v belonging to an appropriate dense subspace of X. Let G_Λ be as in the density lemma with $\Lambda = (h, j, k, \ell)$. For sufficiently large h, j, k, ℓ the function

$$F(z) = e^{z^2}(A+B)^z(A^\alpha + B^\alpha)^{-z/\alpha}v, \quad v \in G_\Lambda$$

is holomorphic in a neighborhood of $0 \le \mathrm{Re}\ z \le 1$. Since $A + B \in BIP(a \vee b + \delta, K')$ and $A^\alpha + B^\alpha \in BIP((a \vee b + \delta)\alpha, K'')$ for all $\delta > 0$ with K', K'' depending on K, a, b, δ, α, X, estimating F on the imaginary axis yields

$$\|F(is)\| \le e^{-s^2} K' K'' e^{\rho|s|} e^{\rho|s|/\alpha}\|v\|, \quad \rho = a \vee b + \delta$$

$$\le M_0\|v\| \quad \text{with} \quad M_0 = \sup_{s \in \mathbf{R}} K K'' \exp(\rho|s|(1 + 1/\alpha) - s^2) < \infty,$$

where δ is now a fixed sufficiently small number. Similarly,

$$\|F(1 + is)\| = e^{1-s^2}\|(A+B)^{is}(A+B)(A^\alpha + B^\alpha)^{-1/\alpha}(A^\alpha + B^\alpha)^{-is/\alpha}v\|$$

$$\le e^{1-s^2} K' K'' e^{\rho|s|} e^{\rho|s|/\alpha}\|(A+B)(A^\alpha + B^\alpha)^{-1/\alpha}\|\ \|v\|$$

$$\le e M_0 \|(A+B)(A^\alpha + B^\alpha)^{-1/\alpha}\|\ \|v\|.$$

If $A(A^\alpha + B^\alpha)^{-1/\alpha}$ and $B(A^\alpha + B^\alpha)^{-1/\alpha}$ can be extended to bounded operators in X with

$$(3.4) \qquad \|A(A^\alpha + B^\alpha)^{-1/\alpha}\| \le c, \quad \|B(A^\alpha + B^\alpha)^{-1/\alpha}\| \le c,$$

then

$$\|F(1 + is)\| \le M_1\|v\|, \quad M_1 = 2eM_0 c.$$

Applying the three line theorem [RS, p.33] yields

$$\|F(\alpha)\| \le M_0^{1-\alpha} M_1^\alpha \|v\|, \quad v \in G_\Lambda.$$

This deduces (3.3), $D(A^\alpha + B^\alpha) \subset D((A+B)^\alpha)$ and (3.1) with $C = e^{-\alpha^2} M_0^{1-\alpha} M_1^\alpha$ since G_Λ is dense in X. The inequalities (3.4) are proved in the next lemma.

To prove the converse direction (3.2) we need that $A^\alpha(A+B)^{-\alpha}$ and $B^\alpha(A+B)^{-\alpha}$ extend to bounded operators in X, this is also proved in the next lemma. Similarly as above we then obtain $D((A+B)^\alpha) \subset D(A^\alpha + B^\alpha)$,

$$\|A^\alpha u\| + \|B^\alpha u\| \le C\|(A+B)^\alpha u\|, \quad u \in D((A+B)^\alpha);$$

this implies (3.2) and the proof is complete.

3.2. LEMMA. *Assume the hypotheses of the theorem on inverses.*
(i) *For $\sigma > 0$ the operators $A^\sigma(A+B)^{-\sigma}$ and $B^\sigma(A+B)^{-\sigma}$ can be extended to bounded linear operators in X with a bound depending only on a, b, K, σ, X.*
(ii) *For $0 < \alpha < 1$ the operators $A(A^\alpha + B^\alpha)^{-1/\alpha}$ and $B(A^\alpha + B^\alpha)^{-1/\alpha}$ can be extended to bounded linear operators in X with a bound depending only on a, b, K, α, X.*

PROOF: Part (ii) follows from (i) by setting $A = A^\alpha$, $B = B^\alpha$, $\sigma = 1/\alpha$ so it remains to prove (i). In the corollary to the theorem on inverses, we have proved (i) when σ is a positive integer. For general σ we again appeal to the three line theorem. Let m be a nonnegative integer. If we take an appropriate dense subspace G_Λ of X, the function

$$H(z) = e^{z^2} A^{m+z}(A+B)^{-(m+z)}v, \quad v \in G_\Lambda$$

is holomorphic in a neighborhood of $0 \le \mathrm{Re}\ z \le 1$. Since $A + B \in BIP(a \vee b + \delta, K')$ for all $\delta > 0$ with some $K' = K'(K, a, b, \delta, \alpha, X)$, estimating on the imaginary axis yields

$$\|H(is)\| \le e^{-s^2} K e^{a|s|} \|A^m(A+B)^{-m}\| K' e^{\rho|s|} \|v\|$$

with $\rho = a \vee b + \delta$, where δ is a fixed sufficiently small number. By the corollary to the theorem on inverses, $\|A^m(A + B)^{-m}\|$ is bounded by c_m; we now observe

$$\|H(is)\| \le c_m L \|v\|, \quad L = \sup_{s \in \mathbf{R}} K K' \exp(-s^2 + (a + \rho)|s|) < \infty.$$

Similarly, on $\mathrm{Re}\, z = 1$ we have

$$\|H(1 + is)\| \le c_{m+1} L e \|v\|.$$

Applying the three line theorem yields

$$\|H(\tau)\| \le M \|v\|, \quad M = c_m^{1-\tau} c_{m+1}^\tau e^\tau L < \infty, \quad v \in G_\Lambda.$$

Since G_Λ is dense in X, we now obtain

$$\|A^{m+\tau}(A + B)^{-(m+\tau)}\| \le e^{-\tau^2} M, \quad 0 < \tau < 1.$$

The proof for $B^\sigma(A + B)^{-\sigma}$ is parallel, so is omitted.

4. Application to the Stokes system.

Although our abstract result applies to a very general class of evolution equations (1.2), we consider here as an example only the Stokes system (1.3) on some domain Ω in $\mathbf{R}^n$.

Assumptions on the domain Ω.

In the following let $\Omega \subset \mathbf{R}^n$ ($n \ge 2$) be either the whole space $\mathbf{R}^n$, a bounded domain, a halfspace or an exterior domain. The boundary $\partial\Omega$ is always assumed at least of class $C^{2+\mu}$ with $0 < \mu < 1$. If Ω is an exterior domain we suppose $n \ge 3$.

Stokes operator.

For $1 < q < \infty$ let L^q_σ denote the L^q closure of the space $C^\infty_{0,\sigma}$ of all smooth divergence-free vector fields with compact support in Ω. Let $P = P_q$ denote the projection operator from $L^q = (L^q(\Omega))^n$ to L^q_σ associated with the Helmholtz decomposition. The *Stokes operator* A_q is defined in L^q_σ by $A_q = -P\Delta$ with the dense domain

$$D(A_q) = \{u \in L^q_\sigma; \nabla^2 u \in L^q, \ u|_{\partial\Omega} = 0\},$$

where Δ denotes the Laplacian and $\nabla^2 u$ denotes the tensor of all second order derivatives. In [G] and [GS1] it is shown that for all $0 < a < \pi/2$, $A_q \in BIP(a, K)$ with K depending on a. For more information on the Stokes operator and the Helmholtz decomposition we refer to [GS1, 2] and [BM] and the references cited there.

Evolution equation.

Applying the projection P_q to the Stokes system (1.3), one formally obtains its abstract form

$$(4.1) \qquad \frac{du}{dt} + A_q u = f \quad \text{in} \quad (0, T), \quad u(0) = 0.$$

For $1 < r < \infty$, $0 < T \leq \infty$ let B denote the derivative operator on $X = L^r(0, T; L^q_\sigma)$ defined by $B = d/dt$ (weak derivative) with

$$D(B) = \{u \in X; du/dt \in X, u(0) = 0\}.$$

The operator A in X is defined by $(Au)(t) = A_q u(t)$ for a.e. $t \in (0, T)$ where

$$u \in D(A) = \{u \in X; u(t) \in D(A_q) \ \text{for a.e.} \ t \in (0, T)$$

$$\text{and} \ \int_0^T \|A_q u(t)\|^r_q dt < \infty\}.$$

Using A and B we may rewrite (4.1) as

$$(4.2) \qquad Bu + Au = f.$$

The space X is ζ-convex because L^q_σ is ζ-convex; see [GS2] and the references cited there. As shown in [DV] for each $\delta > 0$ the operator $B \in BIP(\pi/2 + \delta, K)$ with K depending on δ but independent of T, $0 < T \leq \infty$. The property $A_q \in BIP(a, K)$ yields $A \in BIP(a, K)$, where a is arbitrary $0 < a < \pi/2$ and K depends on a but is independent of T. Clearly, A and B are resolvent commuting. Applying the extended Dore-Venni theorem in [GS2] one observes that there is a unique solution $u \in D(A+B)^\wedge$ of (4.2) for each $f \in X$. If $T < \infty$, B^{-1} exists as a bounded operator so that

$$D(A + B)^\wedge = D(A) \cap D(B).$$

For $0 < T < \infty$ we call $u : (0, T) \to L^q_\sigma$ a *strong* solution of (4.2) if it satisfies (4.2) with $u \in D(A) \cap D(B)$. In case $T = \infty$ we call $u : (0, \infty) \to L^q_\sigma$ a *strong* solution if so is u

on each finite time interval $(0, T)$.

Generalized solutions.

In order to apply our abstract Theorem 3.1 to (4.2) we have to consider generalized solutions u of (4.1) for a class of distributions f. This is caused by the fractional powers $(B + A)^\alpha$. For simplicity we will avoid here the definition via test functions and prefer the definition via regularization. Roughly speaking, u is a generalized solution of (4.1) if the "regularization" $A_q^{-\alpha} u$ is a strong solution of (4.1) with f replaced by $A_q^{-\alpha} f$.

Let us give a precise definition. For $0 < \alpha < 1$ the space $D(A_q^{-\alpha}) = R(A_q^\alpha)$ is equipped with the norm $\|A_q^{-\alpha} u\|_q$ and $\hat{D}(A_q^{-\alpha})$ denotes the completion of $D(A_q^{-\alpha})$ under this norm. For $v = (v_j)_{j=1}^\infty \in \hat{D}(A_q^{-\alpha})$ we define $A_q^{-\alpha} v = (A_q^{-\alpha} v_j)$ and get $A_q^{-\alpha} v \in L_\sigma^q$ for each $v \in \hat{D}(A_q^{-\alpha})$; $A_q^{-\alpha} v$ is called the regularization of $v \in \hat{D}(A_q^{-\alpha})$. In the case $T < \infty$ we say $u \in L^r(0, T; D(A_q^{1-\alpha}))$ is a *generalized solution* of (4.1) with $f \in L^r(0, T; \hat{D}(A_q^{-\alpha}))$ if $A_q^{-\alpha} u$ solves (4.2) as a strong solution with f replaced by $A_q^{-\alpha} f \in L^r(0, T; L_\sigma^q)$. If $u : (0, \infty) \to D(A_q^{1-\alpha})$ is a generalized solution of (4.1) on each finite time interval $(0, T)$, u is called a *generalized solution* in case $T = \infty$.

4.1. UNIQUE EXISTENCE OF GENERALIZED SOLUTIONS. *Let Ω be as above, $0 < T < \infty$, $1 < r < \infty$, $1 < q < \infty$, $0 < \alpha < 1$. Suppose $f \in L^r(0, T; \hat{D}(A_q^{-\alpha}))$. Then there exists a unique generalized solution $u \in L^r(0, T; D(A_q^{1-\alpha}))$ of (4.1). Moreover, $u \in D(B^{1-\alpha})$ and*

$$(4.3) \qquad \int_0^T \|(\tfrac{d}{dt})^{1-\alpha} u\|_q^r dt + \int_0^T \|A_q^{1-\alpha} u\|_q^r dt \leq C \int_0^T \|A_q^{-\alpha} f\|_q^r dt$$

with $C = C(\Omega, q, r, \alpha) > 0$ independent of T and f where $(d/dt)^{1-\alpha} = B^{1-\alpha}$.

REMARKS: a) The condition $u(0) = 0$ is implicitly contained in $u \in D(B^{1-\alpha})$ for small α (i.e. $0 < \alpha < 1 - 1/r$) while no condition is imposed on $u(0)$ for large α (i.e. $1 - 1/r < \alpha < 1$).

b) The case $T = \infty$ can be admitted in Theorem 4.1 if we replace $D(A_q^{1-\alpha})$ by $\hat{D}(A_q^{1-\alpha})$ and $D(B^{1-\alpha})$ by $\hat{D}(B^{1-\alpha})$. In this case (4.3) is

$$\int_0^\infty \|(\tfrac{d}{dt})^{1-\alpha} u\|_q^r dt + \int_0^\infty \|A_q^{1-\alpha} u\|_r^q dt \leq C \int_0^\infty \|A_q^{-\alpha} f\|_q^r dt$$

which yields asymptotic properties of u as $t \to \infty$.

c) Of course, this theorem extends to the class of all evolution equations for which Theorem 3.1 is applicable.

PROOF: We apply the extended Dore-Venni theorem in [GS2] to $A_q^{-\alpha} f \in X$ and obtain a unique solution $v \in D(B) \cap D(A)$ of $Bv + Av = A_q^{-\alpha} f$. The function $u = A_q^\alpha v$ is a generalized solution of (4.1) since $A_q^{-\alpha} u$ is a strong solution; the uniqueness of u is obvious.

To prove (4.3) we use the Yosida approximation $J_m = J_A(m) = m(m + A)^{-1}$ in Section 2 and obtain

$$BA^{-\alpha}J_m u + AA^{-\alpha}J_m u = A^{-\alpha}(BJ_m u + AJ_m u) = A^{-\alpha}J_m f$$
$$BJ_m u + AJ_m u = J_m f.$$

Here $J_m f$ is defined in the same way as $A_q^{-\alpha}f$. We know that $\lim_{m \to \infty} J_m u = u$ in $X = L^r(0, T; L_\sigma^q)$. Setting $u_m = J_m u$, $w = u_m - u_l$ and applying Theorem 3.1 yields

$$\|B^{1-\alpha}w\|_X + \|A^{1-\alpha}w\|_X \le C\|(B + A)^{1-\alpha}w\|_X$$
$$= C\|(B + A)^{-\alpha}(B + A)w\|_X = C\|A^\alpha(B + A)^{-\alpha}(J_m - J_l)A^{-\alpha}f\|_X$$
$$\le C'\|(J_m - J_l)A^{-\alpha}f\|_X;$$

here we used the fact that $A^\alpha(B + A)^{-\alpha}$ is bounded by Lemma 3.2. From this estimate we conclude $u \in D(B^{1-\alpha}) \cap D(A^{1-\alpha})$ since $B^{1-\alpha}$ and $A^{1-\alpha}$ are closed and $u \in X$. The same estimate with w replaced by u_m yields (4.3) by letting $m \to \infty$. This proves 4.1.

We next consider some concrete cases of distributions f in Theorem 4.1. In case a) of the following Corollary we consider a distribution of the form $f = \sum_{\nu=1}^n \partial_\nu f_\nu$ with $f_\nu \in X$ and $\partial_\nu = \partial/\partial x_\nu$ and in b) we let $f \in L^r(0, T; L_\sigma^\gamma)$ with some exponent γ different from q.

4.2. COROLLARY. *Suppose Ω as above and $0 < T < \infty$, $1 < q < \infty$, $1 < r < \infty$.*
a) Let $f = \sum_{\nu=1}^n \partial_\nu f_\nu$ with $f_\nu \in X = L^r(0, T; L_\sigma^q)$, $\nu = 1, \cdots, n$. If Ω is unbounded, suppose additionally $q > n/(n-1)$, $n \ge 3$. Then $A_q^{-1/2}f \in X$, $f \in L^q(0, T; \hat{D}(A_q^{-1/2}))$. There exists a unique generalized solution $u \in L^r(0, T; D(A_q^{1/2}))$ of (4.1) with $u \in D(B^{1/2})$ and

$$(4.4) \qquad \int_0^T \|(\tfrac{d}{dt})^{1/2}u\|_q^r dt + \int_0^T \|A_q^{1/2}u\|_q^r dt \le C \sum_{\nu=1}^n \int_0^T \|f_\nu\|_q^r dt$$

with $C = C(\Omega, q, r)$ independent of f and T.
b) For $1 < \alpha < 1$ let γ be defined by $2\alpha + n/q = n/\gamma$ and $f \in L^r(0, T; L_\sigma^\gamma)$. If Ω is an exterior domain, suppose additionally $1 < \gamma < n/2$, $n \ge 3$. Then $A_q^{-\alpha}f \in L^r(0, T; L_\sigma^q)$, $f \in L^r(0, T; \hat{D}(A_q^{-\alpha}))$. There exists a unique generalized solution $u \in L^r(0, T; D(A_q^{1-\alpha}))$ of (4.1) with $u \in D(B^{1-\alpha})$ and

$$(4.5) \qquad \int_0^T \|(\tfrac{d}{dt})^{1-\alpha}u\|_q^r dt + \int_0^T \|A_q^{1-\alpha}u\|_q^r dt \le C \int_0^T \|f\|_\gamma^r dt$$

with $C = C(\Omega, q, r, \alpha)$ independent of f and T.

REMARKS: (i) To prove a) and b) it suffices to prove that $f \in L^r(0, T; \hat{D}(A_q^{-1/2}))$ and

$$\|A_q^{-1/2}f\|_X \le C \sum_{\nu=1}^n \|f_\nu\|_X$$

in a) and that $f \in L^r(0,T;\hat{D}(A_q^{-\alpha}))$ and

$$\|A_q^{-\alpha}f\|_X \leq C(\int_0^T \|f\|_\gamma^r dt)^{1/r}$$

in b) respectively with C independent of f and T.

(ii) The estimate (4.4) yields (1.6) by applying of

$$\|\nabla u\|_q \leq C\|A_q^{1/2}u\|_q$$

which needs additionally the restriction $1 < q < n$, $n \geq 3$ when Ω is an exterior domain ([BM], [GS1]).

PROOF: a) In [GS1, p.123] it has been shown that $C_{0,\sigma}^\infty \subset R(A_q)$ if $q > n/(n-2)$ and Ω is the $\mathbf{R}^n$ or an exterior domain; the same proof works also for the half-space and the restriction becomes $q > n/(n-1)$ if A_q is replaced by $A_q^{1/2}$. If Ω is bounded, no restriction is needed.

So for each f_ν ($\nu = 1, 2, \cdots, n$) we find a sequence $(f_{\nu j})_{j=1}^\infty$ in $L^r(0,T;C_{0,\sigma}^\infty) \subset L^r(0,T;D(A_q^{-1/2}))$ with $f_\nu = \lim_{j\to\infty} f_{\nu j}$ in $L^r(0,T;L_\sigma^q)$. It follows that $(\tilde{f}_j) = (\sum_{\nu=1}^n \partial_\nu f_{\nu j})$ is a sequence in $L^r(0,T;D(A_q^{-1/2}))$.

We next use the estimate

$$\|A_q^{-1/2}\nabla u\|_q \leq C\|u\|_q$$

(see [BM], [GS1]) which is valid in all cases for Ω but in exterior domains under the restriction $q > n/(n-1)$; observe that this estimate is equivalent to $\|\nabla u\|_{q'} \leq C\|A_{q'}^{1/2}u\|_{q'}$, where by duality the restriction is now given by $1 < q' < n$. This leads to

$$\|A_q^{-1/2}(\tilde{f}_i - \tilde{f}_j)\|_X = \|\sum_{\nu=1}^n A_q^{-1/2}\partial_\nu(f_{\nu i} - f_{\nu j})\|_X \leq C\sum_{\nu=1}^n \|f_{\nu i} - f_{\nu j}\|_X$$

which yields $f \in L^r(0,T;\hat{D}(A_q^{-1/2}))$. This estimate also yields

$$\|A_q^{-1/2}f\|_X \leq C\sum_{\nu=1}^n \|f_\nu\|_X$$

so Theorem 4.1 is applicable.

b) Since $R(A_q^\alpha) \subset L_\sigma^\gamma$, is dense in L_σ^γ, one can choose $f_j \in L^r(0,T;D(A_q^{-\alpha}))$, $j = 1, 2, \cdots$ with $f = \lim_{j\to\infty} f_j$ in $L^r(0,T;L_\sigma^\gamma)$. Then we use the estimate

$$\|A_q^{-\alpha}u\|_q \leq C\|u\|_\gamma$$

in [GS1, p.104] which holds for $2\alpha + n/q = n/\gamma$; in exterior domains the restriction $1 < \gamma < n/2$, $n \geq 3$ is needed. This leads to

$$\|A_q^{-\alpha}(f_i - f_j)\|_X \leq C(\int_0^T \|f_i - f_j\|_\gamma^r dt)^{1/r}$$

which yields $f \in L^r(0,T; \hat{D}(A_q^{-\alpha}))$ and

$$\|A_q^{-\alpha}f\| \leq C(\int_0^T \|f\|_\gamma^r dt)^{1/r},$$

so Theorem 4.1 is applicable.

Further applications. The estimates above can be applied to weak solutions of the nonlinear Navier-Stokes equations if we take the nonlinear term to the right hand side in (4.1). The procedure is completely analogous to that in [GS2].

REFERENCES

[BM] W. Borchers and T. Miyakawa, *Algebraic L^2-decay for Navier-Stokes flows in exterior domains*, Acta Math. **165** (1990), 89-227.

[B] D.L. Burkholder, *Explorations in martingale theory and its applications*, Ecole d'Ete de Probabilités de Saint-Flour XIX-1989, Lecture Notes in Math. **1464** (1991), 1-66. (ed. P.L. Hennequin), Springer.

[BB] P. Butzer and H. Berens, *Semi-Groups of Operators and Approximations*, Berlin-Heidelberg-New York (1967).

[DV] G. Dore and A. Venni, *On the closedness of the sum of two closed operators*, Math. Z. **196** (1987), 189-201.

[DV2] ──────────, *Maximal regularity for parabolic initial-boundary value problems in Sobolev spaces*, Math. Z. **208** (1991), 297-308.

[GGS] M. Giga, Y. Giga and H. Sohr, *L^p estimate for abstract linear parabolic equations*, Proc. Japan Acad. **67** (1991), 197-202.

[G] Y. Giga, *Domains of fractional powers of the Stokes operators in L_r spaces*, Arch. Rational Mech. Anal. **89** (1985), 251-265.

[GS1] Y. Giga and H. Sohr, *On the Stokes operator in exterior domains*, J. Fac. Sci. Univ. Tokyo Sec. IA **36** (1989), 103-130.

[GS2] ──────────, *Abstract L^p estimates for the Cauchy problem with applications to the Navier-Stokes equations in exterior domains*, J. Func. Anal. **102** (1991), 72-94.

[Ka] T. Kato, *Note on fractional powers of linear operators*, Proc. Japan Acad. **36** (1960), 94-96.

[Ko] H. Komatsu, *Fractional powers of operators*, Pacific J. Math. **19** (1966), 285-346.

[Kr] S. Krein, *Linear Differential Equations in Banach Spaces*, Amer. Math. Soc., Providence, 1972.

[PS] J. Prüss and H. Sohr, *On operators with bounded imaginary powers in Banach spaces*, Math. Z. **203** (1990), 429-452.

[RS] M. Reed and B. Simon, *Methods of Mordern Mathematical Physics vol. II*, Academic Press, New York-San Francisco-London 1975.

[Tr] H. Tribel, *Interpolation Theory, Function Spaces, Differential Operators*. North-Holland-Amsterdam-New York-Oxford (1978).

Mariko Giga
School of General Education
Nippon Medical School
Kosugi 2-297
Kawasaki 211, JAPAN

Yoshikazu Giga
Department of Mathematics
Hokkaido University
Sapporo 060, JAPAN

Hermann Sohr
Department of Mathematics
University of Paderborn
D-4790 Paderborn, Germany

Semigroups in Probability Theory

Kiyosi Itô

RIMS, Kyoto University

1. Introduction

In his famous paper: Analytische Methoden in der Wahrscheinlichkeitsrechnung (Math. Ann. 104, 1931) A. Kolmogorov introduced stochastic dynamical systems, now called Markov processes. In explaining his idea, we restrict ourselves, for the sake of simplicity, to the time homogeneous case.

Let us consider a random motion x_t, $t \geq 0$ of a particle in the real state space $\mathbf{R}$ governed by a system of transition probabilities:

$$p_t(x, E), \quad t > 0, \quad x \in \mathbf{R}, \quad E \subset \mathbf{R}. \tag{1}$$

From the intuitive meaning of transition probabilities we can easily derive the Chapman-Kolmogorov equation:

$$p_{t+s}(x, E) = \int_s p_t(x, dy) p_s(y, E). \tag{2}$$

Assuming that the motion x_t is continuous, Kolmogorov showed that for every E, $U(t, x) \equiv p_t(x, E)$ is determined by the Kolmogorov backward equation:

$$\frac{\partial u}{\partial t} = Au, \quad A = a(x)\frac{\partial}{\partial x} + \frac{1}{2}b(x)\frac{\partial^2}{\partial x^2}, \tag{3}$$

where

$$a(x) = \lim_{t \to 0} \frac{1}{t} \int_{\mathbf{R}} (y - x) p_t(x, dy), \tag{4}$$

and

$$b(x) = \lim_{t \to 0} \frac{1}{t} \int_{\mathbf{R}} (y - x)^2 p_t(x, dy). \tag{5}$$

Kolmogorov also showed that for every x the transition probability density $v(t, y) \equiv p_t(x, dy)/dy$ is determined by the Kolmogorov forward equation, called the Fokker-Planck equation by physicists:

$$\frac{\partial v}{\partial t} = A^* v, \tag{6}$$

where A^* is the adjoint differential operator of A.

Kolmogorov did not discuss the existence and the uniqueness of the solutions of the equations above and the existence of the transition probability density, all of which were proved by W. Feller under some reasonable conditions (Math. Ann. 113, 1936). Feller also treated the general case where the motion may have jumps. Thus Kolmogorov and Feller initiated the analytic methods in the theory of Markov processes.

The analytic methods were extended by a number of mathematicians, including Kolmogorov and Feller, to cover the case where the state space is $\mathbf{R}^n$ or a differentiable manifold, and culminated in Yosida's semigroup-theoretical treatment applicable to the case where the state space is a compact (or locally compact) metric space, as we will explain in the next section.

From the probabilistic view-point we are also concerned with the path-theoretical treatment of Markov processes that enables us to study the rich structure of sample paths of Markov processes, as we will explain in section 3 with emphasis on the relation to analytic methods.

2. The semigroup-theoretical treatment of Markov processes

In this section we are mainly concerned with the transition probabilities of Markov processes.

2.1. Let S be a compact metric space. A system

$$\mathbf{T} = \{p_t(x, E), \ t \geq 0, \ x \in S, \ E \in \mathcal{B}(S)\}$$

is called a *transition probability system*, if the following five conditions are satisfied:

(T.1) $p_t(x, E)$ is a probability measure in E for every (t, x), and $p_0(x, E) = \delta_x(E)$ for every x,

(T.2) $p_t(x, E)$ is $\mathcal{B}(S)$-measurable in x for every (t, E), ($\mathcal{B}(S) = $ the Borel subsets of S),

(T.3) (*Chapman-Kolmogorov equation*)

$$p_{t+s}(x, E) = \int_S p_t(x, dy) p_s(y, E) \tag{1}$$

for every (t, s, x, E).

(T.4) (*weak continuity*) $\lim_{t \to 0} p_t(x, U) = 1$ for every x and every neighborhood $U = U(x)$.

(T.5) (*Feller property*)

$$\lim_{\xi \to x} \int_S p_t(\xi, dy) f(y) = \int_S p_t(x, dy) f(y)$$

for every (t, x) and every continuous function f.

Remark 1. (T.2) follows from (T.1) and (T.5).

Remark 2. In many concrete Markov processes S is locally compact but not compact. Even in this case it often happens that we can extend the transition probability system onto a certain compactification of S in a reasonable way.

Given a transition probability system $\mathbf{T}$ we define

$$T_t f(x) := \int_S p_t(x, dy) f(y), \quad t \geq 0, \quad x \in S. \tag{2}$$

Then $\mathbf{G} = \{T_t : t \geq 0\}$ satisfies the following six conditions:

(G.1) T_t is a linear map from $C(S)$ into itself, ($C(S) =$ the continuous real functions defined on S),

(G.2) (*semigroup property*) $T_{t+s} = T_t T_s$, $T_0 = I$,

(G.3) $\|T_t f\| \leq \|f\|$, $\forall t \geq 0$, $\forall f \in C(S)$, ($\| \ \| =$ the maximum norm on $C(S)$),

(G.4) $\lim\limits_{t \to 0} \|T_t f - f\| = 0$, $\forall f \in C(S)$,

(G.5) $T_t \geq 0$, i.e. $T_t f \geq 0$, $\forall f \geq 0$,

(G.6) $T_t 1 = 1$.

Remark 3. Denote the bounded linear operators on $C(S)$ by $BL(C(S))$ and the operator norm on $BL(C(S))$ by $\| \ \|$, i.e. $\|L\| = \sup\{\|Lf\| : \|f\| \leq 1\}$, $\forall L \in BL(C(S))$. Then (G.3) is written $\|T_t\| \leq 1$.

Remark 4. (G.3) follows from (G.1), (G.5) and (G.6).

A family of operators $\mathbf{G} = \{T_t\}$ satisfying (G.1), (G.2), $\cdots$, (G.6) is called a *transition semigroup* on $C(S)$. If we are given a transition semigroup on $C(S)$, then we can find a unique transition probability system $\mathbf{T} = \{p_t(x, E)\}$ satisfying (2). Hence the correspondence

$$\mathbf{T} \longleftrightarrow \mathbf{G} \quad \text{under (2)} \tag{3}$$

is one-to-one.

A transition semigroup $\mathbf{G}$ on $C(S)$ is obviously a strongly continuous semigroup of linear contractions on the Banach space $C(S) = (C(S), \| \ \|)$. Hence we can apply the Hille-Yosida theory to define the *generator* of $\mathbf{G} = \{T_t, t \geq 0\}$:

$$Au = \lim_{t \downarrow 0} \frac{1}{t}(T_t u - u), \quad u \in \mathcal{D}(A), \tag{4}$$

where the domain $\mathcal{D}(A)$ is the set of all u's for which the limit exists. The semigroup $\mathbf{G} = \{T_t\}$ is determined by its generator A through the so-called *evolution equation*:

$$\frac{d}{dt}T_t u = AT_t u, \quad u \in \mathcal{D}(A). \tag{5}$$

Note that if $u \in \mathcal{D}(A)$, then $T_t u \in \mathcal{D}(A)$ and

$$AT_t u = T_t Au.$$

Remark 5. In this note convergence, limit, differentiation and integration are defined with respect to the norm on $C(S)$ or on $BL(C(S))$.

For the Kolmogorov case observed in Section 1 the backward equation is nothing but the evolution equation (5), if we ignore the problem related to the determination of $\mathcal{D}(A)$ in terms of boundary conditions; the problem was fully discussed by W. Feller [1].

In any case the generator A determines the transition semigroup $\mathbf{G}$, which in turn determines the original transition probability system $\mathbf{T}$. Thus we have a one-to-one correspondence:

$$\mathbf{T} \longleftrightarrow \mathbf{G} \longleftrightarrow A. \tag{6}$$

This is Yosida's general idea of the semigroup-theoretical treatment of Markov processes. From this view point he investigated a number of Markov processes and obtained many interesting new results and new proofs of known facts; see Kôsaku Yosida Collected Papers [2], Sections VI, VII and IX and E. Hille and R. Phillips [10]. Yosida's idea was so attractive that a large number of papers on Markov processes were written along his line.

The following fact or a similar fact seems to have been known by Yosida, Feller and many others. However, as far as we know, there is no clear statement with a rigorous proof in spite of its importance in the semigroup-theoretical treatment of Markov processes.

Proposition 1. *Let A be a densely defined linear operator in $C(S)$. If A satisfies the following three conditions (A.1), (A.2) and (A.3), then A is the generator of a transition semigroup on $C(S)$:*

(A.1) There exists at least one $\alpha_0 > 0$ such that for every $f \in C(S)$, the equation

$$(\alpha_0 I - A)u = f \tag{7}$$

has at least one solution $u \in \mathcal{D}(A)$.
(A.2) $1 \in \mathcal{D}(A)$ and $A \cdot 1 = 0$,
(A.3) *(Feller's minimum principle)* If x_0 is a minimum point for $u \in \mathcal{D}(A)$, then

$$Au(x_0) \geq 0.$$

Remark 6. If A is the generator of a transition semigroup, then (A.2) and (A.3) are obvious and (A.1) holds for

$$u := \int_0^\infty e^{-\alpha_0 t} T_t f \, dt.$$

To prove Proposition 1 we will use the following fact that is part of the Hille-Yosida semigroup theory:

Proposition 2. *Let A be a densely defined linear operator on a Banach space $E = (E, \| \ \|)$. If for every $\alpha > 0$ and every $f \in E$ there exists one and only one $u \in \mathcal{D}(A)$ satisfying*

$$(\alpha I - A)u = f \quad \text{and} \quad \|u\| \leq \alpha^{-1}\|f\|, \tag{8}$$

then A is the generator of a strongly continuous semigroup of linear contractions on E. (The converse is obvious.)

Proof. We follow Yosida's idea and use the nice technique in P. Meyer's book [3], pp 210-215.

Denote the u satisfying (8) by $R_\alpha f$. Then R_α is obviously a bounded linear operator on E, i.e. $R_\alpha \in BL(E)$, and

$$\mathcal{D}(A) = R_\alpha(E), \quad \forall \alpha > 0, \tag{9}$$

$$\|R_\alpha\| \le \alpha^{-1}, \tag{10}$$

$$A(R_\alpha f) = \alpha R_\alpha f - f. \tag{11}$$

Observing that

$$\beta I - A = (\beta - \alpha)I + (\alpha I - A),$$

we obtain

$$(\beta I - A)R_\alpha f = (\beta - \alpha)R_\alpha f + f, \quad \forall \in E,$$

so

$$R_\alpha f = R_\beta((\beta - \alpha)R_\alpha f + f),$$

which implies *the resolvent equation*:

$$R_\alpha - R_\beta + (\alpha - \beta)R_\alpha R_\beta = 0, \tag{12}$$

and so

$$\|R_\alpha - R_\beta\| \le |\frac{1}{\alpha} - \frac{1}{\beta}|, \tag{13}$$

$$\frac{d}{d\alpha}R_\alpha = -R_\alpha^2, \tag{14}$$

$$\alpha R_\alpha R_1 - R_1 = \frac{1}{\alpha - 1}(R_1 - \alpha R_\alpha). \tag{15}$$

We will prove that

$$\|\alpha R_\alpha f - f\| \to 0 \ (\alpha \to \infty), \quad \forall f \in E. \tag{16}$$

If $f \in \mathcal{D}(A)$, then $f = R_1 g$ for some $g \in E$ by (9). Hence

$$\|\alpha R_\alpha f - f\| = \|(\alpha R_\alpha R_1 - R_1)g\|$$

$$= \frac{1}{\alpha - 1}\|(R_1 - \alpha R_\alpha)g\| \qquad (by \ (15))$$

$$\le \frac{2}{\alpha - 1}\|g\| \qquad (by \ (10))$$

for $\alpha > 1$. This proves (16) for $f \in \mathcal{D}(A)$. Since $\mathcal{D}(A)$ is dense in E and $\|\alpha R_\alpha\| \le 1$, (16) holds for every $f \in E$.

In view of (16) and (11) the operator

$$A^\alpha := A(\alpha R_\alpha) = \alpha(\alpha R_\alpha - I) \in BL(E)$$

approximates A as $\alpha \to \infty$, so

$$T_t^\alpha = e^{tA^\alpha} = e^{-\alpha t}e^{t\alpha^2 R_\alpha} = e^{-\alpha t}\sum_{n=0}^{\infty}\frac{1}{n!}t^n\alpha^{2n}R_\alpha^n$$

approximates the semigroup $\{T_t\}$ whose generator is A. This is Yosida's idea to construct $\{T_t\}$. Note that $\{T_t^\alpha, t \geq 0\}$ is a strongly continuous semigroup of linear contractions with *bounded* generator A^α.

To be rigorous, we should first prove that for every $f \in E$, $T_t^\alpha f$ $(\alpha \to \infty)$ is convergent uniformly on any bounded t-interval. For this purpose we use the technique in Meyer's book. Observing that

$$\frac{d}{d\alpha} A^\alpha = 2\alpha R_\alpha - \alpha^2 R_\alpha^2 - I = -(\alpha R_\alpha - I)^2,$$

$$\frac{d}{d\alpha} T_t^\alpha = t T_t^\alpha \frac{d}{d\alpha} A^\alpha = -t T_t^\alpha (\alpha R_\alpha - I)^2,$$

$$\frac{d}{d\alpha} T_t^\alpha R_1^2 = -t T_t^\alpha (\alpha R_\alpha R_1 - R_1)^2$$

$$= -\frac{t}{(\alpha - 1)^2} T_t^\alpha (R_1 - \alpha R_\alpha)^2 \quad (\alpha > 1)$$

by (15), we obtain

$$\|\frac{d}{d\alpha} T_t^\alpha R_1^2\| \leq \frac{4t}{(\alpha - 1)^2} \quad (\alpha > 1)$$

by (10). Hence

$$\|T_t^\beta R_1^2 - T_t^\alpha R_1^2\| \leq \int_\alpha^\beta \|\frac{d}{d\gamma} T_t^\gamma R_1^2\| d\gamma \leq \frac{4t}{\alpha - 1} - \frac{4t}{\beta - 1}$$

for $\beta > \alpha > 1$.

Suppose that $f = R_1^2 g$ for some $g \in E$. Then

$$\|T_t^\beta f - T_t^\alpha f\| = \|T_t^\beta R_1^2 g - T_t^\alpha R_1^2 g\| \leq (\frac{4t}{\alpha - 1} - \frac{4t}{\beta - 1})\|g\|$$

$$\to 0 \quad (\beta > \alpha \to \infty)$$

uniformly on every bounded t-interval. Hence this holds for every $f \in F$, because $R_1^2(E)$ is dense in E and $\|T_t^\alpha\| \leq 1$ for every (α, t). We will prove that $R_1^2(E)$ is dense in E. Since $R_1(E) = \mathcal{D}(A)$, $R_1(E)$ is dense in E. If $f \in E$, then $\|R_1 f_n - f\| \to 0$ for some $\{f_n\} \subset E$ and $\|R_1 g_n - f_n\| \to 0$ for some $\{g_n\} \subset E$. Hence

$$\|R_1^2 g_n - f\| \leq \|R_1(R_1 g_n - f_n)\| + \|R_1 f_n - f\|$$

$$\leq \|R_1 g_n - f_n\| + \|R_1 f_n - f\| \to 0.$$

Denote $\lim_{\alpha \to \infty} T_t^\alpha f$ by $T_t f$. As $\{T_t^\alpha\}$ is a strongly continuous semigroup of linear contractions on E, so is $\{T_t\}$. It remains to prove that the generator of $\{T_t\}$ is A. Let $u \in \mathcal{D}(A)$. Then $u = R_1 f$ for some $f \in E$. But

$$T_t u - u = \lim_{\alpha \to \infty} (T_t^\alpha u - u) = \lim_{\alpha \to \infty} \int_0^t T_s^\alpha A^\alpha u \, ds$$

$$= \lim_{\alpha \to \infty} \int_0^t T_s^\alpha \alpha (\alpha R_\alpha - I) R_1 f \, ds$$

$$= \lim_{\alpha \to \infty} \frac{\alpha}{\alpha - 1} \int_0^t T_s^\alpha (R_1 - \alpha R_\alpha) f ds. \qquad (by\ (15))$$

Since

$$\|\alpha T_s^\alpha R_\alpha f - T_s^\alpha f\| \le \|\alpha R_\alpha f - f\| \to 0, \qquad (by\ (16))$$

we obtain

$$T_t u - u = \lim_{\alpha \to \infty} \int_0^t T_s^\alpha (R_1 f - f) ds$$

$$= \lim_{\alpha \to \infty} \int_0^t T_s^\alpha A u ds \qquad (by\ (11)\ and\ u = R_1 f)$$

$$= \int_0^t T_s A u ds,$$

which implies that

$$\lim_{t \to 0} \frac{1}{t} (T_t u - u) = A u, \quad u \in \mathcal{D}(A).$$

This completes the proof of Proposition 2.

Proof of Proposition 1. We apply Proposition 2 to $E = C(S)$ in order to prove that A is the generator of a strongly continuous semigroup of linear contractions $\{T_t\}$ on $C(S)$. For this purpose we will check the assumption of Proposition 2.

First we will note that if $(\alpha I - A)u = f$ has a solution u, then

$$\|u\| \le \alpha^{-1} \|f\|. \qquad (17)$$

Let x_0 be a minimum point for u. Then $Au(x_0) \ge 0$ by Feller's minimum principle (A.3). Hence, for every x we obtain

$$\alpha u(x) \ge \alpha u(x_0) \ge (\alpha I - A)u(x_0) = f(x_0) \ge -\|f\|.$$

Since $(\alpha I - A)(-u) = -f$, we obtain

$$-\alpha u(x) \ge -\| - f\|, \quad \text{i.e.} \quad \alpha u(x) \le \|f\|, \ \forall x$$

similarly. This proves (17).

From (17) we see that if $(\alpha I - A)u_i = f, i = 1, 2$, then $(\alpha I - A)(u_1 - u_2) = 0$, so $\|u_1 - u_2\| = 0$; i.e. $u_1 = u_2$. This implies that $(\alpha I - A)u = f$ has at most one solution.

Now we will prove that $(\alpha I - A)u = f$ has at least one solution u for every (α, f). By assumption (A.1) this holds for $\alpha = \alpha_0$. Hence $(\alpha_0 I - A)u = f$ has a unique solution satisfying $\|u\| \le \alpha_0^{-1} \|f\|$ by the observation above. Denote the solution u by Rf. Then $R : C(S) \to \mathcal{D}(A)$ is linear and

$$\|Rf\| \le \alpha_0^{-1} \|f\|, \quad \text{so} \quad R \in BL(C(S)). \qquad (18)$$

Define, for $\alpha \in (0, 2\alpha_0)$,

$$R_\alpha := \sum_{n=0}^{\infty} (\alpha_0 - \alpha)^n R^{n+1}. \tag{19}$$

Since $|\alpha_0 - \alpha| < \alpha_0$ and $\|R^{n+1}\| \leq \|R\|^{n+1} \leq \alpha_0^{-n-1}$, the infinite series above is convergent. It is obvious by (19) that

$$R_\alpha f = R(I + (\alpha_0 - \alpha)R_\alpha)f,$$

so

$$(\alpha_0 - A)R_\alpha f = f + (\alpha_0 - \alpha)R_\alpha f,$$
$$\text{i.e.} \quad (\alpha - A)R_\alpha f = f.$$

As we observed above, $u = R_\alpha f$ is the unique solution of $(\alpha - A)u = f, R_\alpha \in BL(C(S))$ and

$$\|R_\alpha\| \leq \alpha^{-1}.$$

Let $\alpha_1 = 3\alpha_0/2 \in (0, 2\alpha_0)$. Replacing R by $R' := R_{\alpha_1}$ in the discussion above, we obtain $R'_\alpha \in BL(C(S))$ for $\alpha \in (0, 2\alpha_1 \equiv 3\alpha_0)$ such that $(\alpha I - A)R'_\alpha f = f$ and $\|R'_\alpha\| \leq \alpha^{-1}$. Since $(\alpha I - A)R_\alpha f = f$ for $\alpha \in (0, 2\alpha_0)$,

$$R'_\alpha = R_\alpha \quad \text{for} \quad \alpha \in (0, 2\alpha_0)$$

by the observation above. Hence we denote R'_α by R_α. Repeating this procedure, we obtain $\{R_\alpha, \alpha > 0\} \subset BL(C(S))$ such that $u = R_\alpha f$ is the unique solution of

$$(\alpha - A)u = f \quad \text{and} \quad \|u\| \leq \alpha^{-1}\|f\|.$$

Thus we have checked that A sastisfies the assumption of Proposition 1. Hence A is the generator of a strongly continuous semigroup of linear contractions on $C(S), \{T_t, t > 0\}$.

From the proof of Proposition 1 we see that

$$T_t f = \lim_{\alpha \to \infty} T_t^\alpha f, \quad \text{where} \quad T_t^\alpha = e^{-\alpha t} \sum_{n=0}^{\infty} \frac{1}{n!} t^n \alpha^{2n} R_\alpha^n.$$

Since $A.1 = 0$ by (A.2), $(\alpha I - A) \cdot 1 = \alpha$ and so $R_\alpha \cdot \alpha = 1$, i.e. $R_\alpha 1 = \alpha^{-1}$. Hence $R_\alpha^n 1 = \alpha^{-n}$, so $T_t^\alpha 1 = 1$. This implies that $T_t 1 = 1$.

Suppose that $f \geq 0$. Let x_0 be a minimum point of $R_\alpha f$. Then $AR_\alpha f(x_0) \geq 0$ by (A.3). Hence

$$\alpha R_\alpha f(x_0) \geq (\alpha I - A)R_\alpha f(x_0) = f(x_0) \geq 0,$$

so $\alpha R_\alpha f(x) \geq 0$ for every x. Hence $T_t^\alpha f \geq 0$, so $T_t f \geq 0$ for $f \geq 0$.

Since $T_t 1 = 1$ and $T_t \geq 0, \{T_t\}$ is a transition semigroup. This completes the proof of Proposition 1.

2.2. Pregenerators of transition semigroups

Let $\mathbf{T} = \{T_t, t \geq 0\}$ be a transition semigroup on $C(S)$ and A its generator. A densely defined linear operator A_0 in $C(S)$ is called a *pregenerator* of $\{T_t\}$ or a *core* of A, if A is the closed extension of A_0. Every pregenerator determines the generator A, which in turn determines the transition semigroup $\{T_t, t > 0\}$. In practical problems it is usual to construct a transition semigroup from a pregenerator A_0 rather than from the generator A.

Proposition 3. *Let A_0 be a densely defined operator in $C(S)$. If A_0 satisfies the following four conditions, then A_0 is a pregenerator of a transition semigroup on $C(S)$:*
$(A_0.1)$ *There exists $\alpha_0 > 0$ such that $\mathcal{R}_0 := (\alpha_0 I - A_0)(\mathcal{D}(A_0))$ is dense in $C(S)$.*
$(A_0.2)$ $\quad 1 \in \mathcal{D}(A_0)$ *and $A.1 = 0$.*
$(A_0.3)$ *(Feller's minimum principle) If x_0 is a minimum point for $u \in \mathcal{D}(A_0)$, then*

$$A_0 u(x_0) \geq 0.$$

$(A_0.4)$ *For every $x_0 \in S$ there exists $\rho = \rho_{x_0} \in \mathcal{D}(A_0)$ such that $\rho(x_0) = A_0\rho(x_0) = 0$ and $\rho(x) > 0$ for every $x \neq x_0$.*
(The converse is not always true.)

Proof. Let A be the closed extension of A_0. It suffices to prove that A satisfies the conditions (A.1),(A.2) and (A.3) of Proposition 1. (A.2) is obvious by $(A_0.2)$. We will prove (A.3). Let $u \in \mathcal{D}(A)$ and x_0 any minimum point for u. Take a function ρ satisfying the conditions of $(A_0.4)$. Then x_0 is the only one minimum point for $u + \rho$. Since $u \in \mathcal{D}(A)$ and $\rho \in \mathcal{D}(A_0)$, $u + \rho \in \mathcal{D}(A)$. Hence we can find $u_n \in \mathcal{D}(A_0), n = 1, 2, \cdots$ such that

$$\|u_n - (u + \rho)\| \to 0 \quad \text{and} \quad \|A_0 u_n - A(u + \rho)\| \to 0. \tag{20}$$

Let x_n be a minimum point for u_n. By taking a subsequence of $\{u_n\}$ we can assume that $x_n \to x_0'$ for some $x_0' \in S$. Hence

$$(u + \rho)(x_0') = \lim_n u_n(x_n), \tag{21}$$

$$A(u + \rho)(x_0') = \lim_n A_0 u_n(x_n) \geq 0 \tag{22}$$

by (20). But $u_n(x_n) \leq u_n(x), \forall x$, so

$$(u + \rho)(x_0') \leq (u + \rho)(x), \ \forall x.$$

Hence x_0' is a minimum point for $u + \rho$, so $x_0' = x_0$. Hence

$$A(u + \rho)(x_0) \geq 0, \quad \text{i.e.} \quad Au(x_0) \geq 0$$

by (22) and $A\rho(x_0) = 0$. This proves (A.3).

Using the argument at the beginning of the proof of Proposition 1 we can deduce from $(A_0, 2)$ and $(A_0.3)$ that $(\alpha_0 I - A_0) : \mathcal{D}(A_0) \to \mathcal{R}_0 := (\alpha_0 I - A_0)(\mathcal{D}(A_0))$ is bijective and that $\|(\alpha_0 I - A_0)^{-1} f\| \le \alpha_0^{-1} \|f\|, \forall f \in \mathcal{R}_0$. Let $f \in C(S)$. By $(A_0.1)$ we can find $\{f_n\} \subset \mathcal{R}_0$ such that $\|f_n - f\| \to 0$. Let $u_n := (\alpha_0 I - A)^{-1} f_n \in \mathcal{D}(A_0)$. Then

$$(\alpha_0 I - A_0) u_n = f_n$$

and

$$\|u_m - u_n\| = \|(\alpha_0 I - A)^{-1}(f_m - f_n)\| \le \alpha_0^{-1} \|f_m - f_n\|$$
$$\to 0 \quad (m, n \to \infty).$$

Hence we can find $u \in C(S)$ such that

$$\|u_n - u\| \to 0 \quad (n \to \infty).$$

But

$$\|(\alpha_0 I - A_0) u_n - f\| = \|f_n - f\| \to 0 \quad (n \to \infty).$$

Therefore u belongs to $\mathcal{D}(\alpha_0 I - A) \equiv \mathcal{D}(A)$ and

$$(\alpha_0 I - A) u = f.$$

Hence $(\alpha_0 I - A)(\mathcal{D}(A)) = C(S)$. This proves that A satisfies (A.1) of Proposition 1. This completes the proof of Proposition 3.

Example. Let S be a compact C^∞-manifold with no boundary and A_0 a strictly elliptic operator:

$$A_0 = \sum_i a^i(x) \frac{\partial}{\partial x^2} + \sum_{i,j} a^{ij}(x) \frac{\partial^2}{\partial x^i \partial x^j}, \quad \mathcal{D}(A_0) = C^\infty(S)$$

with C^∞-coefficients. Then A_0 satisfies $(A_0.1), (A_0.2), (A_0.3)$ and $(A_0.4)$. Hence A_0 is a pregenerator of a transition semigroup.

2.3. The following semigroups are also investigated by K. Yosida, E. Hille, W. Feller and many others.

(a) Let $\mathcal{M}$ be the space of all signed measures with the weak* topology and consider the semigroup:

$$(T_t^* \mu)(E) = \int_S \mu(dx) p_t(x, E), \quad \mu \in \mathcal{M}.$$

The Kolmogorov forward equation is interpreted in terms of the generator of $\{T_t^*\}$.

(b) Let m be an invariant σ-finite measure on S, i.e.

$$m(E) = \int_S m(dx) p_t(x, E), \quad \forall t \ge 0, \quad \forall E.$$

Then

$$T_t f(x) = \int_S p_t(x, dy) f(g), \quad f \in L_p(S, m), \quad 1 \le p \le \infty$$

carries $L_p(S, m)$ into itself. This gives a semigroup of linear contractions on the Banach space $L_p(S, m)$.

(c) Let m be a σ-finite measure on S and suppose that the transition probabilities are given in the form

$$p_t(x, E) = \int_S p_t(x, y) m(dy)$$

where $p_t(x, y) = p_t(y, x)$. Then m becomes an invariant measure. Hence we obtain a semigroup of linear contractions on $L_p(S, m)$. The case where $p = 2$ has recently been extensively investigated in connection with Dirichlet forms [4].

(d) Taking a transition subprobability system $(p_t(x, S) \le 1)$ instead of a transition probability system, we can define a transition semigroup. In this case an operator of the form:

$$A_0 u(x) = L u(x) - c(x) u(x)$$

appears as a pregenerator, where L is an elliptic operator and $c(x) \ge 0$.

3. The path-theoretical approach to Markov processes

First of all we define D-paths (or càdelàg paths) in a compact metric space S. An S-valued function defined on $[0, \infty)$ is called a *path* in S. A path in S is called a D-path if it is right continuous and has left limit. We denote the set of all D-paths by Ω. The Kolmogorov σ-algebra of subsets of Ω is denoted by $\mathcal{B}$. For every t we define $X_t : \Omega \to S$ by

$$X_t(\omega) = \omega(t), \quad \forall \omega \in \Omega. \tag{1}$$

Let $\mathcal{B}_t$ denote the σ-algebra $\sigma[X_s, s \le t]$, the least σ-algebra with respect to which $X_s : \Omega \to S$ is measurable for every $s \le t$. Then $\mathcal{B} = \vee_{t \in [0, \infty]} \mathcal{B}_t$.

Let Ω_C denote the set of all continuous paths. It is obvious that

$$\Omega_C \subset \Omega \quad \text{and} \quad \Omega_C \in \mathcal{B}. \tag{2}$$

Let Ω_J denote the set of all D-paths varying only by jumps. Then

$$\Omega_J \subset \Omega \quad \text{and} \quad \Omega_J \in \mathcal{B}. \tag{3}$$

A family of probability measures on $(\Omega, \mathcal{B})$, $\{P_x, x \in S\}$ is called a *Markov process* if $P_x(A)$ is $\mathcal{B}(S)$-measurable in x for every $A \in \mathcal{B}$,

$$P_x(X_0 = x) = 1, \tag{4}$$

and

$$P_x(X_{t+s} \in E \mid \mathcal{B}_t) = P_{X(t)}(X_s \in E) \quad P_x\text{-a.e.}, \tag{5}$$

where the left hand side is the conditional probability and the right hand side is $P_\xi(x_s \in E)|_{\xi = X_t}$, so both hand sides depend on $\omega \in \Omega$. The second condition is called *Markov property*, because it implies that the stochastic process $X_t(\omega)$, $t \ge 0$, $\omega \in (\Omega, \mathcal{B}, P_x)$ is a Markov process in the usual sense for every x.

Let $\tau = \tau(\omega) \in [0, \infty]$ be a *stopping time* relative to the filtration $\mathcal{B}_t$, $t \ge 0$. If a Markov process $M = \{P_x, x \in S\}$ satisfies the *strong Markov property*:

$$P_x(X_{t+s} \in E \mid \mathcal{B}_\tau) = P_{X(\tau)}(X_s \in E) \quad P_x\text{-a.e.}, \tag{6}$$

for every stopping time τ with $P_x(\tau < \infty) = 1$ P_x-a.e., then M is called a *strong Markov process*.

The rigorous treatment of general Markov processes was initiated by J.L. Doob [5], but the systematic set-up in the style above was first made by E.B. Dynkin [6].

A strong Markov process $M = \{P_x, x \in S\}$ is called a *diffusion process* if $P_x(\Omega_C) = 1$, $\forall x$, and a *pure jump* (strong Markov) *process* if $P_x(\Omega_J) = 1$.

Let $M = \{P_x, x \in S\}$ be a Markov process. Then the system

$$\mathbf{T} = \{p_t(x, E) := P_x\{X_t \in E\}, \quad t \geq 0, \quad x \in S, \quad E \in \mathcal{B}(S)\} \tag{7}$$

is called the *transition probability system* of M. Using Markov property (5), we obtain

$$P_x\{X_{t_1} \in E_1, X_{t_2} \in E_2, \ldots, X_{t_n} \in E_n\}$$

$$= \int_{E_1} \int_{E_2} \cdots \int_{E_n} p_{t_1}(x, dx_1) p_{t_2 - t_1}(x, dx_2) \cdots p_{t_n - t_{n-1}}(x_{n-1}, dx_n). \tag{8}$$

This implies that P_x is determined by $\mathbf{T} = \{p_t(x, E)\}$.

It is easy to check (T.1), (T.2) and (T.3) of Section 2. But we are not sure of (T.4) and (T.5) in general. If the transition probability system of M satisfies all conditions (T.1),(T.2),$\cdots$ (T.5), then M is called a *Feller process*.

Proposition 4.
(i) *Every Feller process is strong Markov.*
(ii) *If $\mathbf{T} = \{p_t(x, E)\}$ satisfies all conditions (T.1), (T.2), $\cdots$, (T.5), then there exists one and only one Feller process whose transition probability system is $\mathbf{T} = \{p_t(x, E)\}$.*

J.L. Doob [5] proved (ii) of this proposition in a slightly different form by using his famous theory of martingales. (i) can be verified by appealing to Feller's property (T.4).

For the sake of simplicity we restrict ourselves to the Feller processes in this note. Then Proposition 4 ensures that there is one-to-one correspondence:

$$M \longleftrightarrow \mathbf{T} \longleftrightarrow \mathbf{G} \longleftrightarrow A$$

where $\mathbf{T}$, $\mathbf{G}$ and A were defined in Section 2. In view of this correspondence A is called the generator of the Feller process M.

We denote $\int_\Omega f(\omega) P(d\omega)$ by $E_x(f)$ as usual. We can represent the transition semigroup $\{T_t\}$ and the resolvent operators $\{R_\alpha; \alpha > 0\}$ in terms of the corresponding Markov process $M = \{P_x, x \in E\}$ as follows:

$$T_t f(x) = E_x(f(X_t)) \tag{9}$$

$$R_\alpha f(x) = \int_0^\infty e^{-\alpha t} T_t f(x) dt = E_x\left(\int_0^\infty e^{-\alpha t} f(X_t) dt\right). \tag{10}$$

Suppose that $\{\tau_t, t > 0\}$ be a decreasing family of stopping times and that

$$P_x(\lim_{t \to 0} \tau_t = 0) = 1 \quad \text{and} \quad E_x(\tau_1) < \infty, \quad \forall x \in S.$$

Then the generator A of $\{T_t\}$ defined in Section 1 is represented by Dynkin [6] as follows:

$$Af(x) = \lim_{t \to 0} E_x(f(X_{\tau_t}) - f(X_0))/E_x(\tau_t). \tag{11}$$

If $\tau_t \equiv t$, then this formula is nothing but the definition of A given in Section 2.

If we choose

$$\tau_t = \inf\{s \in [0,1] : d(X_s, X_0) > t\} \quad (d = \text{metric in } S)$$

(set $\tau_t = 1$ by convention if $d(X_s, X_0) < t$, $\forall s \in [0,1]$), then the formula (10) gives a new representation of A. Usually the formula (10) for this τ_t is called the *Dynkin representation* of A.

If $M = \{P_x, x \in S\}$ is a diffusion process, then $d(X_{\tau_t}, X_0) \leq t$ P_x-a.e. and so (11) shows that $Af(x)$ depends on the behavior of f near x. Thus we obtain

Proposition 5. *(W. Feller) The generator of any diffusion process has local property.*

Feller obtained this fact in the one-dimensional case and used it to establish his famous theory of one-dimensional diffusions [8]. See K. Itô and H.P. McKean [9] for the purely path-theoretical treatment of this theory.

Proposition 6. *Let $f \in \mathcal{D}(A)$. Then for every $x \in S$ the stochastic process*

$$Y_t := f(X_t) - \int_0^t Af(X_u)du, \ t \geq 0, \ \omega \in (\Omega, P_x), \ t \geq 0$$

is a martingale relative to the filtration $\{\mathcal{B}_t, t \geq 0\}$.

Proof. First note that

$$E_x(g(X_{t+s})|\mathcal{B}_t) = E_{X_t}(g(X_s)) = (T_s g)(X_t). \tag{12}$$

But

$$Y_{t+s} - Y_t = f(X_{t+s}) - f(X_t) - \int_t^{t+s} Af(X_u)du$$

$$= f(X_{t+s}) - f(X_t) - \int_0^s Af(X_{t+u})du.$$

Applying $E_x(\cdot|\mathcal{B}_t)$ to both hand sides and using (12), we obtain

$$E_x(Y_{t+s}|\mathcal{B}_t) - Y_t = (T_s f)(X_t) - f(X_t) - \int_0^s T_u Af(X_t)du$$

$$= 0 \quad P_x\text{-a.e.}$$

by the evolution equation (5) of Section 2. This completes the proof.

Proposition 6 holds even if we replace A by a pregenerator A_0. In view of this fact D.W. Stroock and S.R.S. Varadhan [7] proposed to construct a Markov process $M = \{P_x, x \in A\}$ whose pregenerator is A_0 by solving the *martingale problem*:

"Find $X_t, t \geq 0$ such that $X_0 = x$ and

$$f(X_t) - \int_0^t A_0 f(X_u)du, \quad t \geq 0$$

is a martingale".

This method is sometimes more powerful than the analytic method of determining the corresponding transition semigroup. For example, consider the strictly elliptic operator A_0 on a compact manifold observed in Section 2, where we assumed that the coefficients are C^∞. If we assume that the coefficients are only continuous, then the theory of partial differential equations does not work at present, whereas the Stroock-Varadhan method works well.

Let us consider the probabilistic meaning of Yosida's approximating semigroups $\{T_t^\alpha, t \geq 0\}(\alpha \to \infty)$ (See Section 2). Remember that

$$T_t^\alpha = e^{tA^\alpha}, \quad A^\alpha = \alpha(\alpha R_\alpha - I).$$

Since $\alpha R_\alpha \geq 0$ and $\alpha R_\alpha 1 = 1, \alpha R_\alpha$ is represented in the form:

$$\alpha R_\alpha f(x) = \int_S \mu_\alpha(x, dy) f(y),$$

where $\mu_\alpha(x, \cdot)$ is a probability measure for every x, when α is fixed. Let $M^\alpha = \{P_x^\alpha, \alpha \in S\}$ be a Markov process corresponding to $\{T_t^\alpha, t \geq 0\}$ or a Markov process whose generator is

$$A^\alpha f(x) = \alpha \int_S \mu_\alpha(x, dy)(f(y) - f(x)), \quad \mathcal{D}(A^\alpha) = C(S).$$

Then the Markov process M^α can be constructed as follows. Starting at x, the process stays there for a exponential holding time τ_1, jumps to a new random point that is independent of τ_1 and μ_x-distributed, and repeat a similar motion from the new point over and over again. The sample path of this motion belongs to Ω_J. Let P_x^α be the probability law of the sample path. Then $M^\alpha = \{P_x^\alpha, x \in S\}$ is a pure jump Markov process whose generator is A^α. Since $T_t^\alpha \to T_t$ in the sense as was mentioned in Section 2, it is conceivable to believe that $M^\alpha \to M$ in a certain sense. In fact M^α is a pure jump Markov process but the exponential holding time at each point tends to 0 in the average sense. Hence the limit process M can be a diffusion process, i.e. a sample continuous process. It may be an interesting exercise to find, in the space of all Feller processes, a nice topology with respect to which $M^\alpha \to M$.

If A is the generator of a Markov process M, then the Markov process with generator $A - c(x) \cdot (c(x) \geq 0)$ is constructed by killing M with rate $c(x)$ and the Markov process with generator $-(-A)^\alpha$ (fractional power) is constructed by time change by a subordinator with index α independent of the process M.

Finally we remark that determining the domain $\mathcal{D}(A)$ is equivalent to assigning the boundary conditions. In concrete problems this gives rise to very interesting problems. See W. Feller [1] for one-dimensional diffusions and A. D. Wentzell [11] for several dimensional diffusions. See K. Itô and H.P. McKean [12], K. Itô [13] and S. Watanabe [14] for the probabilistic approaches to these problems.

References

[1] W. Feller: The parabolic differential equations and the associated semigroups, Ann. Math. 55 (1952), 468-519.

[2] Kôsaku Yosida Collected Papers, Springer-Verlag, to be published.

[3] P. A. Meyer: Probability and potentials, Ginn (Blaisdell), Boston, 1966.

[4] M. Fukushima: Dirichlet forms and Markov processes, Kodansha and North Holland, 1980.

[5] J. L. Doob: Stochastic processes, John Wiley, 1953.

[6] E. B. Dynkin: Markov processes I,II, Springer-Verlag, 1965.

[7] D. W. Stroock-S. R. S. Varadhan: Multidimensional diffusion processes, Springer-Verlag, 1979.

[8] W. Feller: The general diffusion operator and positivity preserving semi-groups in one-dimension, Ann. Math. 60 (1954), 417-436.

[9] K. Itô and H. P. Mckean: Diffusion processes and their sample paths, Springer-Verlag, 1965.

[10] E. Hille and R. Phillips: Functional analysis and semi-groups, Colloq. Publ. Amer. Math. Soc. 1948 (1st. ed.), 1957 (2nd ed.).

[11] A. D. Wentzell: On boundary conditions for multi-dimensional diffusion processes, Th. Prob. and Appl. 4 (1959), 164-177.

[12] K. Itô and H. P. McKean: Brownian motions on a half-line, Illinois Journ. Math. 7 (1963), 181-231.

[13] K. Itô: Poisson point processes attached to Markov processes, Proc. Sixth Berkeley Symp. Math. Statist. Prob. III (1970), 225-239.

[14] S. Watanabe: Construction of diffusion processes with Wentzell's boundary conditions by means of Poisson point processes of Brownian excursions, Prob. Th., Banach Center Publications Vol. 5, 255-271, Polish Scientific Publishers, Warsaw, 1979.

Characterization of Nonlinearly Perturbed Semigroups

Toshiyuki Iwamiya   Tadayasu Takahashi

National Aerospace Laboratory
Chofu, Tokyo 182

Shinnosuke Oharu

Department of Mathematics
Hiroshima University
Higashi-Hiroshima 724, Japan

Introduction.

This paper is concerned with nonlinear semigroups which provide mild solutions of semilinear evolution equations in Banach spaces of the form

(SE) $$u'(t) = (A + B)u(t), \quad t > 0.$$

In (SE) the operator A is assumed to be the infinitesimal generator of a linear (C_0)-semigroup $\mathcal{T} \equiv \{T(t) : t \geq 0\}$ in a Banach space X and B is a nonlinear operator from a subset D of X into X.

Equation (SE) does not necessarily admit strong solutions and the variation of parameters formula

$$u(t) = T(t)x + \int_0^t T(t-s)Bu(s)\, ds, \quad t \geq 0,$$

is employed to define the generalized solutions that are called mild solutions of (SE). Our objective here is to discuss necessary and sufficient conditions on the semilinear operator $A + B$ for the unique mild solutions of (SE) to exist in a global sense. We interpret this problem as a characterization problem of a nonlinear semigroup which provides unique mild solutions of (SE) and discuss the characterization of such a nonlinear semigroup in terms of the "semilinear generator" $A + B$.

In this paper we mainly deal with nonlinear continuous perturbations of $\mathcal{T}$ under the assumption that D is closed in X and B is continuous on D. Although this is the simplest case, we describe the main points of our nonlinear perturbation theory. Naturally, it is possible to extend the results presented here to much more general cases and in various directions.

Our results contain three features. First, the perturbation results are formulated in the following framework: The nonlinear perturbing operator B is not necessarily quasidissipative, but the semilinear operator $A + B$ is quasidissipative and generates a quasicontractive semigroup $\mathcal{S} \equiv \{S(t) : t \geq 0\}$ on D such that for each $x \in D$ the X-valued function $S(\cdot)x$ is a unique mild solution of (SE). In our previous works [9] and [10], both A and B are assumed to be quasidissipative. However, as pointed out in [8], there are many important cases in which B is not quasidissipative but $A + B$ is. Secondly, we treat the case where the domain D of B is not necessarily convex. As discussed in the papers [8], [9] and [10], the convexity of the domains of semigroups is essential to investigate their generators. If the domains are convex, a Hille-Yosida type theorem is obtained. Thirdly, the generation of a nonlinear semigroup associated with (SE) is treated under the subtangential condition

$$\liminf_{h\downarrow 0} h^{-1}d(T(h)x + hBx, D) = 0 \quad \text{for } x \in D.$$

We first show that the subtangential condition holds uniformly in a local sense. The local uniformity enables us to construct approximate solutions to (SE) and the mild solutions are obtained as limits of the approximate solutions. The convergence of the approximate solutions is obtained by imposing the semilinear stability condition

$$\liminf_{h\downarrow 0} h^{-1}[|T(h)(x - y) + h(Bx - By)| - |x - y|] \leq \omega|x - y| \quad \text{for } x, y \in D.$$

The semilinear stability condition corresponds to the quasidissipativity condition for $A+B$ in the case where the domain D is convex, and this approach to (SE) makes it possible to extend the results given in the papers cited above.

Our main results are stated in Section 1 along with comments. Section 2 deals with the local uniformity of the subtangential condition and Section 3 discusses the uniqueness of mild solutions of (SE). The approximate solutions are constructed in Section 4. Section 5 is devoted to the proof of our main result in the case where D is not necessarily convex. Finally, in Section 6, the convex case is treated and a semilinear Hille-Yosida theorem is established.

1. Characterization theorems

Let X be a Banach space with norm $|\cdot|$ and D a fixed closed subset of X. Let $T \equiv \{T(t) : t \geq 0\}$ be a semigroup of class (C_0) on X, A the infinitesimal generator of T with domain $D(A)$, and let B be a possibly nonlinear operator from D into X. We then define an operator $A+B$ with domain $D(A)\cap D$ by $(A+B)x = Ax+Bx$ for $x \in D(A)\cap D$. If the linear part A is considered the principal part of $A + B$, we call the operator $A + B$ a semilinear operator in X.

Given a semilinear operator $A + B$ in X and an initial-value $x \in D$ we formulate the initial-value problem:

$$(SP) \qquad u'(t) = (A + B)u(t), \qquad t > 0; \qquad u(0) = x \in D.$$

The semilinear problem (SP) does not necessarily admit strong solutions and the variation of constants formula is employed to define solutions in a generalized sense. Namely, a function $u(\cdot) \in \mathcal{C}([0, \infty); X)$ is said to be a *mild solution* of (SP), if $u(t) \in D$ for $t \geq 0$, $Bu(\cdot) \in \mathcal{C}([0, \infty); X)$, and $u(\cdot)$ satisfies the integral equation

$$(IE) \qquad u(t) = T(t)x + \int_0^t T(t - s)Bu(s)ds \qquad t \geq 0.$$

In this sense $A + B$ is sometimes called a nonlinear perturbation of T. For basic properties of the mild solutions and related notions of generalized solutions to (SP), we refer to [1], [8], [9] and [10].

By a continuous semigroup on D is meant a one-parameter family $\mathcal{S} \equiv \{S(t) : t \geq 0\}$ of continuous operators from D into itself. In this paper we are concerned with a continuous semigroup $\mathcal{S}$ on D such that for each $x \in D$ the function $u(\cdot) \equiv S(\cdot)x$ gives a mild solution of (SP). If a semigroup $\mathcal{S}$ provides unique mild solutions of (SP) for all initial data in D, we call the operators $S(t)$ the solution oparators and say that $\mathcal{S}$ is associated with (SP).

The dual space of X is denoted by X^*. For $x \in X$ and $f \in X^*$ the value of f at x is written as $\langle x, f \rangle$. The duality mapping of X is the mapping F from X into the power set of X^* which assigns to each $y \in X$ the weak-star compact convex set $F(y) \equiv \{f \in X^* : \langle y, f \rangle = |y|^2 = |f|^2\}$. For $x, y \in X$ the symbols $\langle x, y \rangle_i$ and $\langle x, y \rangle_s$ stand for the infimum and the supremum of $\{\langle x, f \rangle : f \in F(y)\}$, respectively.

We write $d(x, D)$ for the distance from $x \in X$ to D, namely, $d(x, D) = \inf\{|x - y| : y \in D\}$. For $r > 0$ and $x \in X$ we write $U_r(x) = \{y \in X : |y - x| < r\}$. The identity operator on X is denoted by I and for $\lambda \in \mathbb{R}$ the range of the operator $I - \lambda(A + B)$ is represented by $R(I - \lambda(A + B))$. If the linear operator A satisfies $\langle Ax, x \rangle_i \leq \omega_A |x|^2$ for $x \in D(A)$ and some $\omega_A \in \mathbb{R}$, we say that $A - \omega_A I$ is dissipative, or that A is quasidissipative in X. If the nonlinear operator B satisfies $\langle Bx - By, x - y \rangle_i \leq \omega_B |x - y|^2$ for $x, y \in D$ and some $\omega_B \in \mathbb{R}$, we say that $B - \omega_B I$ is dissipative, or that B is quasidissipative in X. Likewise, the semilinear operator $A + B$ is said to be quasidissipative in X, if

$$\langle (A + B)x - (A + B)y, x - y \rangle_i \leq \omega |x - y|^2 \quad \text{for } x, y \in D(A) \cap D \text{ and some } \omega \in \mathbb{R}.$$

In what follows, the following conditions are assumed to be valid for A and B:

(A) The operator A generates a C_0-semigroup $\mathcal{T} \equiv \{T(t) : t \geq 0\}$ on X.

(B) The operator B is continuous from D into X.

We now state our main results. Theorem 1 below gives a characterization theorem for semigroups associated with (SP) in the case where D is not necessarily convex.

THEOREM 1. *Let $\omega \in \mathbb{R}$. Suppose that the operators A and B satisfy* (A) *and* (B), *respectively. Then the following two conditions are equivalent:*

(I) *There is a nonlinear semigroup $\mathcal{S} \equiv \{S(t) : t \geq 0\}$ on D such that*

(a)
$$S(t)x = T(t)x + \int_0^t T(t - s)BS(s)x\,ds \quad \text{for } t \geq 0 \text{ and } x \in D,$$

(b)
$$|S(t)x - S(t)y| \leq e^{\omega t}|x - y| \text{ for } t \geq 0 \text{ and } x, y \in D.$$

(II) *The operators A and B satisfy the subtangential condition and the semilinear stability condition stated below:*

(a)
$$\liminf_{h \downarrow 0} h^{-1} d(T(h)x + hBx, D) = 0 \quad \text{for } x \in D,$$

(b)
$$\liminf_{h \downarrow 0} h^{-1}[|T(h)(x - y) + h(Bx - By)| - |x - y|] \leq \omega |x - y|$$

$$\text{for } x, y \in D.$$

As will be seen in Section 5, condition (II) gives a necessary and sufficient condition for the existence of unique mild solutions of (SP). The implication (II) $\Rightarrow$ (I) in the above theorem may be called a generation theorem for semigroups associated with (SP) on nonconvex domains, and the main point here is that the new condition (II-b) is employed. The proof is based on the ideas and methods which have been developed in [11], [6], [7], [4] and [5].

The next result gives a characterization theorem for smigroups associated with (SP) in the convex case.

THEOREM 2. *Let $\omega \in \mathbb{R}$. Suppose that the operators A and B satisfy (A) and (B), respectively. Assume in addition that D is convex. Then conditions (I), (II) and the following condition are equivalent:*

(III) *The semilinear operator $A + B$ satisfies the following density condition, quasidissipativity condition and range condition*

(a)
$$\overline{D(A) \cap D} = D,$$

(b)
$$\langle (A+B)x - (A+B)y, x-y \rangle_i \leq \omega |x-y|^2 \quad for\ x,\ y \in D(A) \cap D,$$

(c)
$$R(I - \lambda(A+B)) \supset D \quad for\ \lambda \in (0, \infty)\ with\ \lambda\omega < 1.$$

The implication (III) $\Rightarrow$ (I) is a special case of the Crandall-Liggett theorem established in [2], and the main point in the convex case is that the reverse implication (I) $\Rightarrow$ (III) is obtained by applying Theorem 1. The equivalence between (I) and (III) may be called the semilinear Hille-Yosida theorem. If in particular $D = X$ and $B = 0$, Theorem 2 is nothing but the Hille-Yosida theorem. In contrast with the linear perturbation theory contained in [3] and [12], purely nonlinear methods are required.

In the subsequent discussions $\mathcal{T}$ can be an arbitrary (C_0)-semigroup. However, for notational simplicity, we assume throughout this paper that $\mathcal{T}$ is a contraction semigroup of class (C_0).

2. Local uniformity of the subtangential condition

In this section we demonstrate that the subtangential condition (II-a) holds locally uniformly, and that condition (II-a) is eventually equivalent to the stronger form below:

(II-a')
$$\limsup_{h \downarrow 0} h^{-1} d(T(h)x + hBx, D) = 0 \quad for\ x \in D.$$

The local uniformity of (II-a) plays an essential role to construct approximate solutions converging to the mild solutions of (SP) and makes it possible to connect the subtangential condition, semilinear stability condition and quasidissipativity condition. In order to verify the local uniformity, we need the following two lemmas:

LEMMA 2.1. *Let $(s_n)_{n\geq 0}$ and $(x_n)_{n\geq 0}$ be sequences such that $s_n \leq s_{n+1}$ and $x_n \in D$. Then we have*

$$x_n - T(s_n - s_0)x_0 - \sum_{k=0}^{n-1}(s_{k+1} - s_k)T(s_n - s_{k+1})Bx_k$$

$$= \sum_{k=0}^{n-1} T(s_n - s_{k+1})[x_{k+1} - T(s_{k+1} - s_k)x_k - (s_{k+1} - s_k)Bx_k].$$

The identities stated in the above lemma are obtained straightforwardly.

LEMMA 2.2. *Let $\epsilon > 0$ and $M > 0$. Let $(s_n)_{n\geq 0}$ and $(x_n)_{n\geq 0}$ be sequences such that $s_n \leq s_{n+1}$, $x_n \in D$, $|Bx_n| \leq M$ and*

$$|x_{n+1} - T(s_{n+1} - s_n)x_n - (s_{n+1} - s_n)Bx_n| \leq (s_{n+1} - s_n)\epsilon$$

for $n \geq 0$. If $s_n \uparrow s$ as $n \to \infty$, then the sequence $(x_n)_{n\geq 0}$ is a Cauchy sequence in X.

PROOF. Let k be a nonnegative integer. The application of Lemma 2.1 with x_0 replaced by x_k implies

$$|x_n - T(s_n - s_k)x_k| \leq (s_n - s_k)(M + \epsilon) \qquad \text{for} \quad n \geq k.$$

Hence we have

$$|x_m - x_n| \leq |T(s_m - s_k)x_k - T(s_n - s_k)x_k| + (s_m + s_n - 2s_k)(M + \epsilon)$$

for m, $n \geq k$. Letting m, $n \to \infty$ yields

$$\limsup_{m,n\to\infty} |x_m - x_n| \leq 2(s - s_k)(M + \epsilon).$$

Since $s_k \uparrow s$ as $k \to \infty$, it is concluded that $(x_n)_{n\geq 0}$ is Cauchy in X. ∎

The following proposition asserts that the subtangential condition (II-a) holds uniformly in a local sense. See [5, Section 5]. As mentioned in the Introduction, this result enables us to construct approximate solutions to the mild solutions of (SP). The construction is treated in Section 4.

PROPOSITION 2.3. *Suppose that condition (II-a) is satisfied. Let $x \in D$ and $\epsilon \in (0, 1)$. Let $r \equiv r(x)$ be a positive number such that*

$$(2.1) \qquad |By - Bx| \leq 4^{-1}\epsilon \qquad and \qquad \sup_{0\leq\sigma\leq r}|T(\sigma)Bx - Bx| \leq 4^{-1}\epsilon$$

for any $y \in D \cap U_r(x)$. Choose a number M so that $|By| \leq M$ for $y \in D \cap U_r(x)$, and set

$$(2.2) \qquad h(x) = \sup\{h > 0; h(M + 1) + \sup_{0\leq\sigma\leq h}|T(\sigma)x - x| \leq r\}.$$

Let $h \in [0, h(x))$ and let y be any element of D satisfying

$$(2.3) \qquad |y - T(h)x| \leq h(M + 1).$$

Then for any $\eta > 0$ with $h + \eta \leq h(x)$, there exists an element $z \in D \cap U_r(x)$ satisfying

(2.4)
$$|z - T(\eta)y - \eta By| \leq \eta\epsilon.$$

PROOF. Let $h \in [0, h(x))$, $y \in D$, let (2.3) hold. Let $\eta \in (0, h(x) - h]$. In order to find an element $z \in D \cap U_r(x)$ satisfying (2.4), we apply the subtangential condition (II-a), Lemma 2.2 and use (2.1), (2.2), (2.3) to construct a sequence $(s_n)_{n \geq 0}$ in $[0, \eta]$ and a sequence $(x_n)_{n \geq 0}$ in $D \cap U_r(x)$ such that

(i) $\qquad s_0 = 0$, $x_0 = y$, $0 \leq s_n \leq s_{n+1} \leq \eta \qquad$ for $n \geq 0$,

(ii) $\qquad \lim_{n \to \infty} s_n = \eta$, $\lim_{n \to \infty} x_n = z$, $\qquad$ and

(iii) $\qquad |x_{n+1} - T(s_{n+1} - s_n)x_n - (s_{n+1} - s_n)Bx_n| \leq 4^{-1}(s_{n+1} - s_n)\epsilon \qquad$ for $n \geq 0$.

Set $s_0 = 0$ and $x_0 = y$. Suppose that s_k and x_k, $k = 1, \cdots, n$, have been defined in such a way that (i) and (iii) are valid for $k = 1, \cdots, n-1$, and $|x_n - T(s_n)x| \leq s_n(M+1)$ holds. We define $\overline{\sigma}_n$ to be the supremum of those $\sigma \geq 0$ satisfying $s_n + \sigma \leq \eta$ and

(2.5)
$$d(T(\sigma)x_n + \sigma Bx_n, D) \leq 5^{-1}\sigma\epsilon.$$

If $s_n < \eta$, the number $\overline{\sigma}_n$ is positive by the subtangential condition (II-a). Choosing a number $\sigma_n \in [2^{-1}\overline{\sigma}_n, \overline{\sigma}_n]$, we put $s_{n+1} = s_n + \sigma_n$ and take an element $x_{n+1} \in D$ such that (iii) holds. This is possible by (2.5). Since $|x_{n+1} - T(s_{n+1} - s_n)x_n| < (s_{n+1} - s_n)(M+1)$ by (iii), we have

$$\begin{aligned} |x_{n+1} - T(s_{n+1})x| &\leq |x_{n+1} - T(s_{n+1} - s_n)x_n| + |T(s_{n+1} - s_n)(x_n - T(s_n)x)| \\ &< (s_{n+1} - s_n)(M+1) + s_n(M+1) \leq s_{n+1}(M+1) \end{aligned}$$

and

$$\begin{aligned} |x_{n+1} - x| &\leq |x_{n+1} - T(s_{n+1})x| + |T(s_{n+1})x - x| \\ &< s_{n+1}(M+1) + \sup_{0 \leq \sigma \leq h(x)} |T(\sigma)x - x| \leq r \end{aligned}$$

by (2.2). This shows that $x_{n+1} \in D \cap U_r(x)$. In this way we obtain a sequence $(s_n)_{n \geq 0}$ in $[0, \eta]$ and a sequence (x_n) in $D \cap U_r(x)$ satisfying (i) and (iii). It now remains to show (ii): $s = \lim_{n \to \infty} s_n = \eta$. Suppose to the contrary that $s < \eta$. Then Lemma 2.2 implies that $(x_n)_{n \geq 0}$ is Cauchy in X and $z = \lim_{n \to \infty} x_n$ exists in X. Moreover, $|z - T(h+s)x| \leq |z - T(s)y| + |T(s)y - T(h+s)x| \leq s(M+1) + |y - T(h)x| < (s+h)(M+1)$ and $|z - y| \leq |z - T(h+s)x| + |T(h+s)x - x| < (s+h)(M+1) + \sup_{\sigma \leq s+h} |T(\sigma)x - x| \leq r$, which means that $z \in D \cap U_r(x)$. On the other hand, by (II-a), one finds $\delta \in (0, \eta - s)$ such that

(2.6)
$$d(T(\sigma)z + \delta Bz, D) \leq 6^{-1}\delta\epsilon.$$

Since $s < \eta$, it follows that $s_n < s_{n+1} < s$ for all n. Choose an integer N so that $s - s_n \leq 2^{-1}\delta$ for $n \geq N$. Then $s + \delta \leq \eta$ and $\delta \geq 2(s - s_n) > 2\sigma_n \geq \overline{\sigma}_n$ for $n \geq N$.

Hence, by the definition of $\overline{\sigma}_n$, we have $d(T(\delta)x_n + \delta Bx_n, D) > 5^{-1}\delta\epsilon$ for $n \geq N$. Since B is continuous at z, we would finally obtain $d(T(\delta)z + \delta Bz, D) \geq 5^{-1}\delta\epsilon$. This contradicts (2.6) and hence $\lim_{n\to\infty} s_n = s = \eta$. Thus the sequences $(s_n)_{n\geq 0}$ and $(x_n)_{n\geq 0}$ have the properties (i)-(iii). Applying Lemma 2.1, condition (iii) and condition (2.1), we have

$$(2.7) \qquad |x_n - T(s_n)y - s_n By|$$

$$\leq |x_n - T(s_n)y - \sum_{k=0}^{n-1}(s_{k+1} - s_k)T(s_n - s_{k+1})Bx_k|$$

$$+ \sum_{k=0}^{n-1}(s_{k+1} - s_k)|T(s_n - s_{k+1})Bx_k - By|$$

$$\leq 4^{-1}s_n\epsilon + \sum_{k=0}^{n-1}(s_{k+1} - s_k)[|T(s_n - s_{k+1})Bx_k - Bx)|$$

$$+ |T(s_n - s_{k+1})Bx - Bx| + |Bx - By|]$$

$$\leq 4^{-1}s_n\epsilon + s_n(\sup_k |Bx_k - Bx| + \sup_{0\leq\sigma\leq\eta} |T(\sigma)Bx - Bx| + |Bx - By|) \leq s_n\epsilon.$$

Passing to the limit as $n \to \infty$, we obtain $|z - T(\eta)y - \eta By| \leq \eta\epsilon$. Since $z \in D \cap U_r(x)$, (2.4) is valid for the limit point z in (ii). This completes the proof. $\blacksquare$

The following is an immediate consequence of Proposition 2.3.

COROLLARY 2.4.　*Condition* (II-a) *is equivalent to the following stronger form:*

$$(\text{II} - \text{a}') \qquad \limsup_{h\downarrow 0} h^{-1}d(T(h)x + hBx, D) = 0 \quad \text{for } x \in D.$$

3.　Semilinear stability condition and uniqueness of mild solutions

In this section we discuss the relations between the semilinear stability condition (II-b), subtangential condition and quasidissipativity condition. It is not necessarily simple to check the semilinear stability condition. With regard to this the first result shows that if both A and B are quasidissipative then the semilinear stability condition follows from the subtangential condition.

PROPOSITION 3.1.　*Assume that* $A - \omega_A I$ *and* $B - \omega_B I$ *are dissipative in* X *for some* ω_A *and* ω_B *in* $\mathbb{R}$. *Then the subtangential condition* (II-a) *implies the semilinear stability condition* (II-b) *for* $\omega = \omega_A + \omega_B$.

PROOF.　By Corollary 2.4 we may assume that the stronger form (II-a') holds. Let $\epsilon > 0$ and $x, y \in D$. Then there exist $h > 0$ and $x_h, y_h \in D$ such that

$$|x_h - T(h)x - hBx| \leq h\epsilon, \qquad |y_h - T(h)y - hBy| \leq h\epsilon,$$

$$|Bx_h - Bx| \leq \epsilon, \qquad |By_h - By| \leq \epsilon.$$

Hence
$$|T(h)(x-y)+h(Bx-By)| \le |x_h - y_h| + 2h\epsilon.$$

Since $B - \omega_B I$ is dissipative, we have

$$(1 - h\omega_B)|x_h - y_h| \le |x_h - y_h - h(Bx_h - By_h)|$$
$$\le |T(h)(x-y)| + |x_h - T(h)x - hBx| + |y_h - T(h)y - hBy|$$
$$+ h|Bx_h - Bx| + h|By_h - By|$$
$$\le |T(h)(x-y)| + 4h\epsilon \le e^{\omega_A h}|x-y| + 4h\epsilon.$$

Therefore we obtain

$$h^{-1}[|T(h)(x-y)+h(Bx-By)| - |x-y|] \le \omega_B|x_h - y_h| + h^{-1}(e^{\omega_A h} - 1)|x-y| + 6\epsilon.$$

This shows that A and B satisfies the semilinear stability condition (II-b), and the proof is complete. ∎

The next result shows that the semilinear stability condition implies the so-called strong quasidissipativity condition.

PROPOSITION 3.2. *For x, $y \in D(A) \cap D$ we have the identity*

$$\lim_{h \downarrow 0} h^{-1}[|T(h)(x-y)+h(Bx-By)| - |x-y|]|x-y|$$
$$= \langle (A+B)x - (A+B)y, x-y \rangle_s.$$

PROOF. Let $f_h \in F(T(h)(x-y)+h(Bx-By))$. Then there is a weak-star cluster point f of the net $(f_h : h \downarrow 0)$ in X^* such that $f \in F(x-y)$. Since

$$\langle h^{-1}(T(h)x - x) + Bx - h^{-1}(T(h)y - y) - By, f_h \rangle$$
$$\ge h^{-1}[|T(h)(x-y)+h(Bx-By)| - |x-y|] \cdot |T(h)(x-y)+h(Bx-By)|,$$

we have

$$\langle (A+B)x - (A+B)y, f \rangle$$
$$\ge \limsup_{h \downarrow 0} h^{-1}[|T(h)(x-y)+h(Bx-By)| - |x-y|]|x-y|.$$

Conversely, for any $f \in F(x-y)$, we have

$$\langle h^{-1}(T(h)x - x) + Bx - h^{-1}(T(h)y - y) - By, f \rangle$$
$$\le h^{-1}[|T(h)(x-y)+h(Bx-By)| - |x-y|]|x-y|.$$

Letting $h \downarrow 0$ yields

$$\langle (A+B)x - (A+B)y, f \rangle$$
$$\le \liminf_{h \downarrow 0} h^{-1}[|T(h)(x-y)+h(Bx-By)| - |x-y|]|x-y|.$$

Combining these two estimates, we obtain the desired identity. ∎

Condition (II-b) guarantees the uniqueness of mild solutions to the semilinear problem

(SP) $$u'(t) = (A + B)u(t), \ t > 0; \quad u(0) = x \in D.$$

PROPOSITION 3.3. *Suppose that condition* (II-b) *is satisfied. If* $u(\cdot)$ *and* $v(\cdot)$ *are mild solutions of* (SP), *then*

(3.1) $$|u(t) - v(t)| \le e^{\omega t}|u(0) - v(0)| \qquad \text{for } t \ge 0.$$

In particular, for each $x \in D$ *there exists at most one mild solution* $u(\cdot)$ *of* (SP).

PROOF. Assume that

(3.2) $$\liminf_{h \downarrow 0} h^{-1}[|T(h)(x-y) + h(Bx - By)| - |x - y|] \le \omega|x - y|$$

for x, $y \in D$. Let $u(\cdot)$ and $v(\cdot)$ be a pair of mild solutions to (SP) with $x = u(0)$ and $y = v(0)$, respectively. Then we have

$$h^{-1}[|u(t + h) - v(t + h)| - |u(t) - v(t)|]$$

$$\le \ h^{-1}[|T(h)(u(t) - v(t)) + h(Bu(t) - Bv(t))| - |u(t) - v(t)|]$$

$$+h^{-1} \int_t^{t+h} |T(t + h - s)Bu(s) - Bu(t)|ds$$

$$+h^{-1} \int_t^{t+h} |T(t + h - s)Bv(s) - Bv(t)|ds$$

for $t \ge 0$ and $h > 0$. From this and (3.2) it follows that

$$D_+|u(t) - v(t)| \le \omega|u(t) - v(t)| \qquad \text{for } t \ge 0,$$

where $D_+\varphi(t)$ stands for the Dini lower right derivative of a real-valued continuous function φ at $t \ge 0$. Solving this differential inequality, we obtain the desired estimate (3.1). This shows that mild solutions depend continuously upon initial data, and in particular that a mild solution is uniquely determined by the initial-value. ∎

Theorem 1 asserts that if a nonlinear semigroup $S \equiv \{S(t)\}$ on D satisfying (I) exists, then (II-b) holds. Therefore, for each $x \in D$, $S(\cdot)x$ is a unique mild solution to (SP) in the sense that every mild solution $u(t; x)$ with initial-value $u(0; x) = x$ is represented as $S(t)x$. That is, the existence of semigroup solutions on D implies the unique existence of mild solutions of (SP) for initial-data in D.

4. Approximate solutions

The main point of the proof of Theorem 1 is to construct approximate solutions which converge to the mild solutions of (SP). Since D is not convex in general and does not necessarily have interior points, it is not trivial to show the existence of such approximate solutions under condition (II). We here construct the approximate solutions by applying Proposition 2.3. See Lemma 4.1 and (4.3) below.

In view of conditions (II-a) and (2.1), it is natural to think of approximate solutions which are piecewise continuous and may have a finite number of jumps. In fact, as defined in (4.3), our approximate solution $u_\epsilon : [0, \tau] \to X$ does not necessarily take its values in D, the jumps may occur at points t_i^ϵ, $i = 0, \cdots, N(\epsilon) - 1$, satisfying $0 = t_0^\epsilon < t_1^\epsilon < \cdots < t_{N(\epsilon)}^\epsilon = \tau$ and $t_{i+1}^\epsilon - t_i^\epsilon \leq \epsilon$, and these points do not form an equipartition of $[0, \tau]$. Therefore, it is not straightforward to discuss the convergence of the approximate solutions by means of standard methods. In order to overcome this difficulty we prepare Proposition 4.2 below which plays a crucial role to discuss the convergence of the approximate solutions. Applying this result, one finds auxiliary functions to estimate the difference between two approximate solutions u_ϵ and u_η.

LEMMA 4.1. *Suppose that condition (II-a) is satisfied. Let $x \in D$. Let $R > 0$ and $M > 0$ be such that $|By| \leq M$ for $y \in D \cap U_R(x)$. Let $\tau > 0$ be small enough to satisfy $\tau(M + 1) + \sup_{0 \leq \sigma \leq \tau} |T(\sigma)x - x| \leq R$. Then for each $\epsilon \in (0, 1)$ there exists a sequence $((t_i, x_i))_{0 \leq i \leq N}$ such that*

(i) $t_0 = 0$, $x_0 = x$ and $t_N = \tau$,

(ii) $0 < t_{i+1} - t_i \leq \epsilon$ for $0 \leq i \leq N - 1$,

(iii) $x_i \in D \cap U_R(x)$ for $0 \leq i \leq N - 1$,

(iv) $|x_{i+1} - T(t_{i+1} - t_i)x_i - (t_{i+1} - t_i)Bx_i| \leq (t_{i+1} - t_i)\epsilon$,

(v)   *for each i with $0 \leq i \leq N - 1$ there exists a number $r_i \in (0, \epsilon]$ such that $|By - Bx_i| \leq 4^{-1}\epsilon$ for $y \in D \cap U_{r_i}(x_i)$, $\sup_{0 \leq \sigma \leq r_i} |T(\sigma)Bx_i - Bx_i| \leq 4^{-1}\epsilon$ and*

$$(t_{i+1} - t_i)(M + 1) + \sup\{|T(\sigma)x_i - x_i|; 0 \leq \sigma \leq t_{i+1} - t_i\} < r_i.$$

PROOF.   Set $t_0 = 0$ and $x_0 = x$. Let $k \geq 1$ and assume that t_i and x_i, $i = 0, \cdots, k$, have been constructed in such a way that conditions (ii) through (v) hold for $0 \leq i \leq k - 1$. If $t_k < \tau$, we define r_k to be the supremum of the numbers $r \in (0, \epsilon]$ such that

(4.1) $|By - Bx_k| \leq 4^{-1}\epsilon$ for $y \in D \cap U_r(x_k)$ and $\sup_{0 \leq \sigma \leq r} |T(\sigma)Bx_k - Bx_k| \leq 4^{-1}\epsilon.$

We then define $h_k = \min\{\tau - t_k, \eta_k\}$ and $t_{k+1} = t_k + h_k$, where η_k is a positive number satisfying

(4.2) $$\eta_k(M + 1) + \sup_{0 \leq \sigma \leq \eta_k} |T(\sigma)x_k - x_k| \leq r_k.$$

It should be noted here that $h_k > 0$. In view of (4.1) and (4.2), we may apply Proposition 2.3 to find an element $x_{k+1} \in D \cap U_R(x)$ such that

$$|x_{k+1} - T(h_k)x_k - h_k Bx_k| \leq h_k \epsilon.$$

One can continue constructing t_k and x_k in this manner so far as $t_k < \tau$. Moreover, it is clear that the sequence $(t_0, x_0), \cdots, (t_{k+1}, x_{k+1})$ satisfies conditions (ii) through (v). Therefore, in order to complete the proof of this proposition, it suffices to show that $t_N = \tau$ for some $N \geq 1$. Suppose to the contrary that $t_i < \tau$ for all $i \geq 1$. Then $\eta_i \to 0$ as $i \to \infty$ and $(x_i)_{i \geq 0}$ is Cauchy in X by Lemma 2.2. But B is continuous on D, and so $\inf_{i \geq 1} r_i > 0$. Therefore $\inf_{i \geq 1} \eta_i > 0$ by the choice of η_i. This is a contradiction. Thus the proof of the proposition is obtained.   ∎

Let $x \in D$. Then by Lemma 4.1 one finds positive numbers R, M and τ such that for each $\epsilon \in (0,1)$ there exists a partition $\{0 = t_0 < \cdots < t_N = \tau\}$ and a finite sequence $(x_k)_{0 \le k \le N}$ satisfying (i)-(iii). Hence a piecewise continuous function $u_\epsilon : [0, \tau) \to X$ is defined by

$$(4.3) \qquad u_\epsilon(0) = x \quad \text{and} \quad u_\epsilon(t) = T(t - t_i)x_i + (t - t_i)Bx_i$$

for $t \in (t_i, t_{i+1}]$ and $0 \le i \le N - 1$. If $t \in (t_i, t_{i+1}]$, then

$$(4.4) \qquad \begin{aligned} |x_{i+1} - u_\epsilon(t)| \quad &\le \quad |x_{i+1} - T(t_{i+1} - t_i)x_i - (t_{i+1} - t_i)Bx_i| \\ &+ \quad |T(t_{i+1} - t)x_i - x_i| + (t_{i+1} - t)|Bx_i| \\ &\le \quad \epsilon + r_i \le 2\epsilon \end{aligned}$$

by Lemma 4.1 (iv) and (4.2). In view of this, Lemmas 2.1 and 4.1, it is expected that the piecewise continuous function u_ϵ gives an approximate solution to (SP) on the interval $[0, \tau]$. In the next section we show that for any null sequence $\epsilon_n \downarrow 0$ the corresponding approximate solution u_n converges uniformly on $[0, \tau]$. To this end the following proposition is crucial.

PROPOSITION 4.2. *Suppose that conditions* (II-a) *and* (II-b) *are satisfied. Let* $x, \hat{x} \in D$ *and* $\epsilon, \hat{\epsilon} \in (0,1)$. *Let* $r \equiv r(x)$ *and* $\hat{r} \equiv r(\hat{x})$ *be positive numbers such that*

$$|By - Bx| \le 4^{-1}\epsilon \quad \text{and} \quad \sup_{0 \le \sigma \le r} |T(\sigma)Bx - Bx| \le 4^{-1}\epsilon,$$

$$|B\hat{y} - B\hat{x}| \le 4^{-1}\hat{\epsilon} \quad \text{and} \quad \sup_{0 \le \sigma \le \hat{r}} |T(\sigma)B\hat{x} - B\hat{x}| \le 4^{-1}\hat{\epsilon}$$

for $y \in D \cap U_r(x)$ *and* $\hat{y} \in D \cap U_{\hat{r}}(\hat{x})$. *Choose a number* M *so that* $|By| \le M$ *for* $y \in D \cap U_r(x)$ *and* $|B\hat{y}| \le M$ *for* $\hat{y} \in D \cap U_{\hat{r}}(\hat{x})$. *Set*

$$h(x) = \sup\{h > 0; h(M + 1) + \sup_{0 \le \sigma \le h} |T(\sigma)x - x| \le r\},$$

$$h(\hat{x}) = \sup\{h > 0; h(M + 1) + \sup_{0 \le \sigma \le h} |T(\sigma)\hat{x} - \hat{x}| \le \hat{r}\}.$$

Let $h \in [0, h(x))$, $\hat{h} \in [0, h(\hat{x}))$, $y, \hat{y} \in D$ *and assume that*

$$(4.5) \qquad |y - T(h)x| \le h(M + 1) \quad \text{and} \quad |\hat{y} - T(\hat{h})\hat{x}| \le \hat{h}(M + 1).$$

Then for each $\delta > 0$ *and for each* $\eta > 0$ *with* $h + \eta \le h(x)$ *and* $\hat{h} + \eta \le h(\hat{x})$, *there exist elements* $z \in D \cap U_r(x)$ *and* $\hat{z} \in D \cap U_{\hat{r}}(\hat{x})$ *such that*

$$(4.6) \qquad |z - T(\eta)y - \eta By| < 2\eta\epsilon, \quad |\hat{z} - T(\eta)\hat{y} - \eta B\hat{y}| < 2\eta\hat{\epsilon}$$

and

$$(4.7) \qquad |z - \hat{z}| \le \exp(\omega\eta)|y - \hat{y}| + \eta \exp(-\omega_+\eta)(\epsilon + \hat{\epsilon} + \delta)$$

where $\omega_+ = \max\{\omega, 0\}$.

PROOF. Let $\eta > 0$ be such that $h + \eta \le h(x)$ and $\hat{h} + \eta \le h(\hat{x})$. By induction we define a sequence $(s_n)_{n \ge 0}$ of nonnegative numbers, and sequences $(x_n)_{n \ge 0}$ in $D \cap U_r(x)$ and $(\hat{x}_n)_{n \ge 0}$ in $D \cap U_r(\hat{x})$ such that

(i) $s_0 = 0$, $x_0 = y$, $\hat{x}_0 = \hat{y}$ and $0 \leq s_n \leq s_{n+1}$ for $n \geq 0$,

(ii) $\lim_{n \to \infty} s_n = \eta$,

(iii) $|x_{n+1} - T(s_{n+1} - s_n)x_n - (s_{n+1} - s_n)Bx_n| \leq (s_{n+1} - s_n)\epsilon$,

(iv) $|\hat{x}_{n+1} - T(s_{n+1} - s_n)\hat{x}_n - (s_{n+1} - s_n)B\hat{x}_n| \leq (s_{n+1} - s_n)\hat{\epsilon}$,

(v) $e^{-\omega(s_{n+1} - s_n)}|T(s_{n+1} - s_n)(x_n - \hat{x}_n) + (s_{n+1} - s_n)(Bx_n - B\hat{x}_n)|$
      $\leq |x_n - \hat{x}_n| + (s_{n+1} - s_n)\delta$.

Set $s_0 = 0$, $x_0 = y$ and $\hat{x}_0 = \hat{y}$. Suppose that s_k, x_k and $\hat{x}_k$, $k = 1, \cdots, n$, have been defined in such a way that (i), (iii), (iv) and (v) are valid. We define $\bar{\sigma}_n$ to be the supremum of those numbers $\sigma \geq 0$ satisfying $s_n + \sigma \leq \eta$ and

(4.8) $$e^{-\omega\sigma}|T(\sigma)(x_n - \hat{x}_n) + \sigma(Bx_n - B\hat{x}_n)| \leq |x_n - \hat{x}_n| + \sigma\delta.$$

The number $\bar{\sigma}_n$ is positive by the semilinear stability condition (II-b), whenever $s_n < \eta$. Then, choosing a number $\sigma_n \in [2^{-1}\bar{\sigma}_n, \bar{\sigma}_n]$, we put $s_{n+1} = s_n + \sigma_n$ and choose elements x_{n+1} and $\hat{x}_{n+1} \in D$ such that

$$|x_{n+1} - T(s_{n+1} - s_n)x_n - (s_{n+1} - s_n)Bx_n| \leq (s_{n+1} - s_n)\epsilon,$$

$$|\hat{x}_{n+1} - T(s_{n+1} - s_n)\hat{x}_n - (s_{n+1} - s_n)B\hat{x}_n| \leq (s_{n+1} - s_n)\hat{\epsilon},$$

and (4.8) holds for $\sigma = \sigma_n$. We note that $s_n < s_{n+1}$ so far as $s_n < \eta$, and that the three conditions (iii), (iv) and (v) hold for $k = n$. Furthermore, $x_{n+1} \in U_r(x)$ and $\hat{x}_{n+1} \in U_{\hat{r}}(\hat{x})$. In fact, Lemma 2.1 implies that

$$|x_{n+1} - T(s_{n+1})x_0| < s_{n+1}(M + 1)$$

and hence

$$|x_{n+1} - T(h + s_{n+1})x| < (h + s_{n+1})(M + 1)$$

by condition (4.5) on $x_0 = y$. Therefore

$$|x_{n+1} - x| < (h + s_{n+1})(M + 1) + \sup\{|T(\sigma)x - x|; 0 \leq \sigma \leq h(x)\} \leq r(x).$$

Similarly, we have

$$|\hat{x}_{n+1} - \hat{x}| < (\hat{h} + s_{n+1})(M + 1) + \sup\{|T(\sigma)\hat{x} - \hat{x}|; 0 < \sigma \leq h(\hat{x})\} \leq r(\hat{x})$$

Thus we can continue constructing three sequences (s_n) in $(0, \eta)$, (x_n) in $D \cap U_r(x)$ and $(\hat{x}_n)$ in $D \cap U_{\hat{r}}(\hat{x})$.

Now it remains to show that $s \equiv \lim_{n \to \infty} s_n = \eta$. Suppose to the contrary that $s < \eta$. Lemma 2.2 implies that the sequences $(x_n)_{n \geq 0}$ and $(\hat{x}_n)_{n \geq 0}$ are Cauchy in X, $z = \lim_{n \to \infty} x_n$ exists in D, and that $\hat{z} = \lim_{n \to \infty} \hat{x}_n$ exists in D. Moreover, the semilinear stability condition (II-b) implies that there exist $\zeta > 0$ such that $\zeta \leq \eta - s$ and

$$e^{-\omega\zeta}|T(\zeta)(z - \hat{z}) + \zeta(Bz - B\hat{z})| \leq |z - \hat{z}| + 2^{-1}\zeta\delta.$$

Choose an integer $N \geq 1$ such that $s - s_n \leq 2^{-1}\zeta$ for $n \geq N$ and set $\zeta_n = s - s_n + \zeta$ for each $n \geq N$. Then $s_n + \zeta_n = s + \zeta \leq \eta$ and $\zeta_n > \zeta \geq 2(s - s_n) > 2\sigma_n \geq \bar\sigma_n$ for all $n \geq N$. Hence it follows from the definition of $\bar\sigma_n$ that

$$\exp(-\omega\zeta_n)|T(\zeta_n)(x_n - \hat{x}_n) + \zeta_n(Bx_n - B\hat{x}_n)| > |x_n - \hat{x}_n| + \zeta_n\delta \quad \text{for } n \geq 0.$$

Using the continuity of $T(\cdot)$ and B, we obtain

$$e^{-\omega\zeta}|T(\zeta)(z - \hat{z}) + \zeta(Bz - B\hat{z})| \geq |z - \hat{z}| + \zeta\delta.$$

This is a contradiction. Therefore, $s = \lim_{n \to \infty} s_n = \eta$.

We now demonstrate that (4.6) and (4.7) hold for the limit points z and $\hat{z}$ obtained above. In the same way as in (2.3), the application of (iii) and (iv) implies

$$\begin{aligned}
|x_n - T(s_n)y - s_n By| &\leq s_n\epsilon + \sum_{k=0}^{n-1}(s_{k+1} - s_k)[|T(s_n - s_{k+1})Bx_k - Bx| \\
&\quad + |T(s_n - s_{k+1})Bx - Bx| + |Bx - By|] \\
&\leq (1 + 3/4)s_n\epsilon < 2s_n\epsilon
\end{aligned}$$

and

$$|\hat{x}_n - T(s_n)\hat{y} - s_n B\hat{y}| \leq (1 + 3/4)s_n\hat\epsilon < 2s_n\hat\epsilon$$

for $n \geq 1$. Passing to the limit as $n \to \infty$, we obtain $|z - T(\eta)y - \eta By| < 2\eta\epsilon$. Similarly, we have $|\hat{z} - T(\eta)\hat{y} - \eta B\hat{y}| < 2\eta\hat\epsilon$ and

$$|z - \hat{z}| \leq \exp(\omega\eta)|y - \hat{y}| + \eta\exp(\omega_+\eta)(\epsilon + \hat\epsilon + \delta).$$

Thus (4.6) and (4.7) are obtained, and this completes the proof. ∎

REMARK . Proposition 4.2 implies that under the subtangential condition (II-a), the semilinear stability condition (II-b) is equivalent to anyone of the following:

(II-b′) $\quad \limsup_{h\downarrow 0} h^{-1}[|T(h)(x - \hat{x}) + h(Bx - B\hat{x})| - |x - \hat{x}|] \leq \omega|x - \hat{x}|$ for x, $\hat{x} \in D$;

(II-b″) $\quad \limsup_{h\downarrow 0} h^{-1}[e^{-\omega h}|T(h)(x - \hat{x}) + h(Bx - B\hat{x})| - |x - \hat{x}|] \leq 0$ for x, $\hat{x} \in D$.

5. General case : Proof of Theorem 1

This section is devoted to the proof of our main theorem. First it is easily seen that (I) implies (II). In fact, assume that a semigroup S on D exists and satisfies (I-a) and (I-b). Let x, $y \in D$ and put $x_h = S(h)x$ and $y_h = S(h)y$ for $h > 0$. Then

$$\lim_{h\downarrow 0} h^{-1}|T(h)x + hBx - x_h| = \lim_{h\downarrow 0} |h^{-1}(S(h)x - T(h)x) - Bx| = 0.$$

Since $x_h \in D$ for $h > 0$, we obtain the subtangential condition (II-a). Furthermore, we have

$$\limsup_{h \downarrow 0} h^{-1}[|T(h)(x-y) + h(Bx - By)| - |x-y|]$$

$$\leq \limsup_{h \downarrow 0} h^{-1}[|S(h)x - S(h)y| - |x-y|]$$

$$\leq \lim_{h \downarrow 0} h^{-1}(e^{\omega h} - 1)|x-y| = \omega|x-y|.$$

This means that (II-b) is valid. Thus (I) implies (II).

In order to prove the reverse implication (II) $\Rightarrow$ (I), we prepare a local existence theorem for mild solutions to (SP).

PROPOSITION 5.1. *Suppose that conditions (II-a) and (II-b) hold. Let $x \in D$. Let $R > 0$, $M > 0$ and $\tau > 0$ be the numnbers appearing in Lemma 4.1. Then there exists a unique mild solution of (SP) on $[0, \tau]$ satisfying the initial condition $u(0) = x$.*

PROOF.   For simplicity, we assume that $\omega \geq 0$. Let $x \in D$ and let $\{\epsilon_n\}$ be a null sequence in $(0, 1)$. For each $n \geq 1$, let $((t_i^n, x_i^n))_{0 \leq i \leq N_n}$ be a sequence with $\epsilon = \epsilon_n$ obtained through Lemma 4.1. By Proposition 2.3, we can construct a sequence of partitions $\{0 = t_0^n < t_1^n < \cdots < t_{N_n}^n = \tau\}$ of $[0, \tau]$ in such a way that

$$\{t_i^n \ ; \ 0 \leq i \leq N_n\} \subset \{t_i^{n+1} \ ; \ 0 \leq i \leq N_{n+1}\} \quad \text{for} \quad n \geq 1.$$

For each $n \geq 1$, we define a function $u_n \ : \ [0, \tau] \to X$ by (4.3), namely,

$$(5.1) \qquad u_n(0) = x \quad \text{and} \quad u_n(t) = T(t - t_i^n)x_i^n + (t - t_i^n)Bx_i^n$$

for $t \in (t_i^n, t_{i+1}^n]$ and $0 \leq i \leq N_n - 1$. As seen from (4.4), $d(u_n(t), D) \leq 2\epsilon_n$ for $t \in [0, \tau]$ and $n \geq 1$. We then establish the estimate

$$(5.2) \qquad |u_n(t) - u_m(t)| < 5te^{\omega t}(\epsilon_n + \epsilon_m)$$

for $t \in [0, \tau]$ and $m, n \geq 1$. For this purpose, let $m > n \geq 1$, $t \in (0, \tau]$, $0 \leq i \leq N_n - 1$, $0 \leq j \leq N_m - 1$, and assume that $t_{j+1}^m \leq t_{i+1}^n$ and $t \in (t_i^n, t_{i;1}^n] \cap (t_j^m, t_{j+1}^m]$. We then define a new subdivision $\{0 = s_1 < \cdots < s_{j+1} = t\}$ of $[0, t]$ by

$$(5.3) \qquad s_l = t_l^m \quad \text{for} \quad 0 \leq l \leq j \text{ and } s_l = t \text{ for } l = j + 1.$$

Applying Proposition 4.2 with $\delta = \epsilon_m$, we can inductively construct two auxiliary sequences $(z_l)_{0 \leq l \leq j+1}$ and $(\hat{z}_l)_{0 \leq l \leq j+1}$ in D with the following properties: (a) $z_0 = \hat{z}_0 = x$. (b) For each $k \in \{0, 1, \cdots, i\}$,

$$(5.4) \qquad |z_{l+1} - T(s_{l+1} - s_l)x_k^n - (s_{l+1} - s_l)Bx_k^n| < 2(s_{l+1} - s_l)\epsilon_n,$$

$$(5.5) \qquad |\hat{z}_{l+1} - T(s_{l+1} - s_l)x_l^m - (s_{l+1} - s_l)Bx_l^m| < 2(s_{l+1} - s_l)\epsilon_m,$$

$$(5.6) \qquad |z_{l+1} - \hat{z}_{l+1}| \leq e^{\omega(s_{l+1} - s_l)}[|x_k^n - x_l^m| + (s_{l+1} - s_l)(\epsilon_n + 2\epsilon_m)]$$

for $\ell \in \{0, 1, \cdots, j\}$ with $s_\ell = t_k^n$ and

(5.4)' $\qquad |z_{\ell+2} - T(s_{\ell+2} - s_{\ell+1})z_{\ell+1} - (s_{\ell+2} - s_{\ell+1})Bz_{\ell+1}| < 2(s_{\ell+2} - s_{\ell+1})\epsilon_n,$

(5.5)' $\qquad |\hat{z}_{\ell+2} - T(s_{\ell+2} - s_{\ell+1})\hat{z}_{\ell+1} - (s_{\ell+2} - s_{\ell+1})B\hat{z}_{\ell+1}| < 2(s_{\ell+2} - s_{\ell+1})\epsilon_m,$

(5.6)' $\qquad |z_{\ell+2} - \hat{z}_{\ell+2}| \le e^{\omega(s_{\ell+2} - s_{\ell+1})}\left[\left|z_{\ell+1} - x_{\ell+1}^m\right| + (s_{\ell+2} - s_{\ell+1})(\epsilon_n + 2\epsilon_m)\right]$

for $\ell \in \{0, 1, \cdots, j - 1\}$ satisfying $t_k^n < s_\ell < t_{k+1}^n$. Now, in the same way as in (2.7), the application of Lemma 4.1 and (5.4) implies

(5.7) $\qquad |z_{\ell+1} - T(s_{\ell+1} - t_k^n)x_k^n - (s_{\ell+1} - t_k^n)Bx_k^n| \le (2 + 3/4)s_n\epsilon_n < 3s_n\epsilon_n.$

Moreover (5.7) and Lemma 4.1 together imply

$$
\begin{aligned}
\text{(5.8)} \qquad |z_{\ell+1} - x_{k+1}^n| \;\; &\le \;\; |z_{\ell+1} - T(s_{\ell+1} - t_k^n)x_k^n - (s_{\ell+1} - t_k^n)Bx_k^n| \\
&+ |x_{k+1}^n - T(s_{\ell+1} - t_k^n)x_k^n - (s_{\ell+1} - t_n^k)Bx_k^n| \\
&\le \;\; 3(s_{\ell+1} - t_k^n)\epsilon_n + (s_{\ell+1} - t_k^n)\epsilon_n = 4(t_{k+1}^n - t_k^n)\epsilon_n
\end{aligned}
$$

for $\ell \in \{0, \cdots, j - 1\}$ and $k \in \{0, \cdots, i - 1\}$ satisfying $s_\ell = t_k^n$, and we see from (5.5)', (5.3) and Lemma 4.1 that

(5.8)' $\qquad |\hat{z}_{\ell+1} - x_{\ell+1}^m| \le 2(s_{\ell+1} - s_\ell)\epsilon_m + (s_{\ell+1} - s_\ell)\epsilon_m = 3(s_{\ell+1} - s_\ell)\epsilon_m.$

The estimate (5.6) with $\ell = j$ can be written as

$$|z_{j+1} - \hat{z}_{j+1}| \le e^{\omega(s_{j+1} - s_j)}\left[\left|z_j - x_j^m\right| + (s_{j+1} - s_j)(\epsilon_n + 2\epsilon_m)\right].$$

Since $|z_j - x_j^m| \le |z_j - \hat{z}_j| + |\hat{z}_j - x_j^m| \le |z_j - \hat{z}_j| + 3(s_j - s_{j-1})\epsilon_m$ by (5.8)', $|z_{j+1} - \hat{z}_{j+1}|$ is bounded above by

$$e^{\omega(s_{j+1} - s_j)}[|z_j - \hat{z}_j| + 3(s_j - s_{j-1})\epsilon_m + (s_{j+1} - s_j)(\epsilon_n + 2\epsilon_m)].$$

We then use (5.6) or (5.6)' and then (5.8) or (5.8)' inductively and obtain the estimate

(5.9) $\qquad |z_{j+1} - \hat{z}_{j+1}| \le e^{\omega t}[5t_i^n\epsilon_n + 5t_j^m\epsilon_m + (t - t_i^n)\epsilon_n + 2(t - t_j^m)\epsilon_m].$

Since $t = s_{j+1}$, $|u_n(t) - z_{j+1}| \le 3(t - t_i^n)\epsilon_n$ by (5.1), (5.3) and (5.4) and $|u_m(t) - \hat{z}_{j+1}| \le 2(t - t_j^m)\epsilon_m$ by (5.1), (5.3) and (5.5), combining these estimates with (5.9) implies the desired estimate (5.2). From (5.2) it follows that there exists a function $u : [0, \tau] \to X$ such that

$$u(t) = \lim_{n\to\infty} u_n(t) \qquad \text{uniformly for } t \in [0, \tau].$$

As mentioned after Lemma 4.1, it is easy to check that $u(t) \in D$ for $t \in [0, \tau]$, $u(\cdot) \in C([0, \tau]; X)$ and $u(\cdot)$ satisfies

$$u(t) = T(t)x + \int_0^t T(t - s)Bu(s)ds$$

for $t \in [0, \tau]$. This completes the proof. $\blacksquare$

The implication (II) $\Rightarrow$ (I) is a restatement of the following result.

PROPOSITION 5.2. *Suppose that conditions* (II-a) *and* (II-b) *hold. Then there exists a semigroup* $\mathcal{S} \equiv \{S(t) : t \geq 0\}$ *on* D *satisfying* (I-a) *and* (I-b).

PROOF. Let $x \in D$ and let $u(\cdot) \equiv u(\cdot; x)$ be a nonextendable mild solution of (SP) obtained by Proposition 5.1. We write $[0, \tau^*)$ for the maximal interval of existence of $u(\cdot)$. Suppose that $\tau^* < \infty$. Then by Proposition 3.3 we have

$$(5.10) \qquad |u(t + h) - u(t)| \leq e^{\omega t} |u(h) - x|$$

for $t \in [0, \tau^*)$ and $h > 0$ with $t + h \in [0, \tau^*)$. Since $u(h) \to x$ as $h \downarrow 0$, it follows that $u(\tau^*) = \lim_{t \uparrow \tau^*} u(t)$ exists in D since (5.10) implies that $(u(t_n))_{n \geq 1}$ is Cauchy in X for any sequence $(t_n)_{n \geq 1}$ tending to τ^* from below. Therefore Proposition 5.1 implies that the mild solution $u(\cdot)$ can be extended over some interval $[0, \tau^{**})$ with $\tau^{**} > \tau^*$. This is a contradiction, and $u(\cdot)$ gives a unique mild solution of (SP). We then define a one-parameter family of operators $S(t)$, $t \geq 0$, by

$$S(t)x = u(t; x) \quad \text{for } t \geq 0 \text{ and } x \in D.$$

From Proposition 3.3 it follows that $\mathcal{S} \equiv \{S(t) : t \geq 0\}$ is a semigroup on D satisfying (I-a) and (I-b). The proof of the proposition is now complete. ∎

6. Convex case : Proof of Theorem 2

In this section we treat the characterization in the convex case. As stated in Theorem 2, it is possible to obtain a semilinear version of the so-called Hille-Yosida theorem in this case.

In view of Theorem 1, it suffices to show that conditions (I) and (III) are equivalent. The Crandall-Liggett theorem implies that if (III) holds, then the desired nonlinear semigroup $\{S(t); t \geq 0\}$ is generated via the exponential formula

$$S(t)x = \lim_{n \to \infty} \left(I - \frac{t}{n}(A + B) \right)^{-n} x \qquad \text{for } t \geq 0 \text{ and } x \in D.$$

For the related results, we refer to [2], [7], [9] and [11]. It should be noted that the convexity assumption on D is not necessary for the proof of the implication (III) $\Rightarrow$ (I).

Now suppose that (I) holds. First we note that $A + B$ satisfies the density condition (III-a);

$$\overline{D(A) \cap D} = D.$$

In fact, let $x \in D$ and set $x_h = h^{-1} \int_0^h S(s)x \, ds$ for $h > 0$. Since D is closed and convex by assumption, we see from a result of [9] that $x_h \in D(A) \cap D$ and x_h satisfies

$$Ax_h = \frac{1}{h}(S(h)x - x) - \frac{1}{h} \int_0^h BS(s)x \, ds.$$

Since $x_h \to x$ as $h \to 0$ and $x \in D$ is arbitrary, it follows that (III-a) holds. The quasidissipativity condition (III-b) follows from Proposition 3.2, condition (I-b) and the fact that

$$\lim_{h \downarrow 0} h^{-1} |T(h)x + hBx - S(h)x| = 0, \qquad x \in D.$$

It now remains to show that $A + B$ satisfies the range condition (III-c), that is,

$$R(I - \lambda(A + B)) \supset D \qquad \text{for } \lambda > 0 \text{ with } \lambda\omega < 1.$$

Let $x \in D$ and $\lambda > 0$ be such that $\lambda\omega < 1$. Let $\hat{x}, z \in D$ and set

$$\hat{x}_h = hx + (1 - h)S(\lambda h)\hat{x},$$
$$z_h = hx + (1 - h)S(\lambda h)z$$

for $h \in (0, 1)$. Noting that $S(\lambda h)\hat{x} \to \hat{x}$ and $h^{-1} \int_0^{\lambda h} T(\lambda h - s)BS(s)\hat{x}\,ds \to \lambda B\hat{x}$ as $h \to 0$, we have

$$\lim_{h \downarrow 0} h^{-1}|T(\lambda h)\hat{x} + h(\lambda B\hat{x} - \hat{x} + x) - \hat{x}_h|$$

$$= \lim_{h \downarrow 0} \left| S(\lambda h)\hat{x} - \hat{x} - h^{-1} \int_0^{\lambda h} T(\lambda h - s)BS(s)\hat{x}\,ds + \lambda B\hat{x} \right| = 0.$$

Since D is convex, $\hat{x}_h \in D$ for $h \in (0, 1)$. This implies

$$\limsup_{h \downarrow 0} h^{-1}d(T(\lambda h)\hat{x} + h(\lambda B\hat{x} - \hat{x} + x), D) = 0.$$

Moreover, by condition (I-b), we have

$$\limsup_{h \downarrow 0} h^{-1}[|T(\lambda h)(\hat{x} - z) + h(\lambda B\hat{x} - \lambda Bz - \hat{x} + z)| - |\hat{x} - z|]$$

$$\leq \limsup_{h \downarrow 0} h^{-1}[|\hat{x}_h - z_h| - |\hat{x} - z|]$$

$$\leq -|\hat{x} - z| + \limsup_{h \downarrow 0} h^{-1}[|S(\lambda h)\hat{x} - S(\lambda h)z| - |\hat{x} - z|]$$

$$\leq (\lambda\omega - 1)|\hat{x} - x|.$$

These relations together imply that condition (II) holds with A, B and ω replaced by λA, $\lambda B - I + x$ and $\lambda\omega - 1$, respectively. Hence we can apply Theorem 1 to conclude that there exists a nonlinear semigroup $\{S_\lambda(t);\ t \geq 0\}$ on D such that

$$(6.1) \quad S_\lambda(t)\hat{x} = T(\lambda t)\hat{x} + \int_0^t T(\lambda(t-s))[\lambda BS_\lambda(s)\hat{x} - S_\lambda(s)\hat{x} + x]ds \qquad \text{for } t \geq 0 \text{ and } \hat{x} \in D,$$

$$(6.2) \qquad |S_\lambda(t)\hat{x} - S_\lambda(t)z| \leq e^{(\lambda\omega - 1)}|\hat{x} - z| \qquad \text{for } t \geq 0 \text{ and } \hat{x}, z \in D.$$

Since $\lambda\omega < 1$, the estimate (6.2) ensures that there exists a unique common fixed point $x_\lambda \in D$ of the commutative strict contraction operators $S_\lambda(t)$, $t \geq 0$, namely,

$$(6.3) \qquad S_\lambda(t)x_\lambda = x_\lambda \qquad \text{for all } t \geq 0.$$

Since $\lambda A + \lambda B - I + x$ is the infinitesimal generator of $\{S_\lambda(t);\ t \geq 0\}$, it follows from (6.3) that $x_\lambda \in D(A) \cap D$ and

$$\lambda(A + B)x_\lambda - x_\lambda + x = 0.$$

Since $x \in D$ and $\lambda > 0$ with $\lambda\omega < 1$ are arbitrary, it is proved that the range condition (III-c) holds. Thus it is concluded that (III) holds under (I), and that conditions (I)-(III) are equivalent.

References

[1] J. Ball, Strongly continuous semigroups, weak solutions, and the variation of constants formula, Proc. Amer. Math. Soc., 63(1977), 370-373.

[2] M. Crandall and T. Liggett, Generation of semigroups of nonlinear transformations on general Banach spaces, Amer. J. Math., 93(1971), 265-298.

[3] E. Hille and R. Phillips, Functional Analysis and Semi-Groups, Amer. Math. Soc., Providense, R. I., 1957.

[4] N. Kenmochi and T. Takahashi, Nonautonomous differential equations in a Banach space, Nonlinear Analysis, TMA, 5(1980), 1109-1121.

[5] T. Iwamiya, Global existence of mild solutions to semilinear differential equations in Banach spaces, Hiroshima Math. J. 16(1986), 499-530.

[6] R. H. Martin, Jr., Invariant sets for perturbed semigroups of linear operators, Ann. Mat. Pura Appl. 150(1975), 221-239.

[7] R. H. Martin, Jr., Nonlinear Operators and Differential Equations in Banach Spaces, Wiley-Interscience, New York, 1976.

[8] R. H. Martin, Jr., S. Oharu, T. Takahashi, Uniqueness of weak solutions of semilinear evolution equations, to appear.

[9] S. Oharu and T. Takahashi, Characterization of nonlinear semigroups associated with semilienar evolution equations, Trans. Amer. Math. Soc., 311(2)(1989), 593-619.

[10] S. Oharu, Nonlinear perturbations of analytic semigroups, Semigroup Forum, 42(1991), 127-146.

[11] G. Webb, Continuous nonlinear perturbations of linear accretive operators in Banach spaces, J. Func. Anal., 10(1972), 191-203.

[12] K. Yosida, A perturbation theorem for semi-groups of linear operators, Proc. Japan Acad., 41(1965), 645-647.

Abstract evolution equations, linear and quasilinear, revisited

TOSIO KATO

Department of Mathematics, University of California, Berkeley,
CA 94720, USA

Introduction

In this paper we continue the study of abstract evolution equations. Specifically, we intend to strengthen the theory of quasilinear equations of "hyperbolic" type

$$(Q) \qquad \partial_t u + A(t, u)u = f(t, u),$$

which has been developed in [6,12,13,14,15]. (For the underlying linear theory, we refer also to [2,4,7,9,10,11,16,18,21,22].)

In these papers, use is made of (at least) two Banach spaces X, Y, assumed to be reflexive (except in [14]), with Y densely and continuously embedded in X. The following are typical assumptions made. The linear operator $A(t,u)$, depending on t and the unknown u, has the dual property of being the negative generator of a C_0-semigroup on X and, at the same time, a bounded operator on Y into X. Moreover, there is an isomorphism S of Y onto X with the property that

$$SA(t, u)S^{-1} = A(t, u) + B(t, u), \qquad B(t, u) \in L(X). \qquad (0.1)$$

The theory proved to be useful in applications, but it has often been felt that the assumptions are still rather restrictive.

Recently, Sanekata [19,20] and Kobayasi-Sanekata [17] have successfully eliminated the reflexivity condition, opening the way to dealing with problems in pairs of spaces such as $Y = C^1(\mathbf{R}^m)$ and $X = C(\mathbf{R}^m)$, etc. But they still use the isomorphism S, which does not seem to exist in such a pair except when $m = 1$.

In the present paper we shall not only eliminate the reflexivity but also relax the conditions imposed on S. Instead of an isomorphism of Y onto X, we use a closed linear operator S from X to a third Banach space Z such that $D(S) = Y$, and replace (0.1) with the "intertwining condition"

$$Se^{-sA(t, u)} \supset e^{-s\hat{A}(t, u)}S, \qquad s \geqslant 0. \qquad (0.2)$$

Here $\hat{A}(t, u)$, an operator in Z, is to be related to $A(t, u)$ via $\hat{A}(t, u) = \overline{A}(t, u) + B(t, u)$, where $\overline{A}(t, u)$ is a "shadow" of $A(t, u)$ and $B(t, u)$ is a bounded perturbation. By "shadow" we mean that $\overline{A}(t, u)$ is the image of $A(t, u)$ under a certain homomorphism Ξ of the operator algebra $L(X)$ into $L(Z)$, extended to generators of semigroups in a natural way. (In a typical case this will be achieved by direct sum construction). Thus $\overline{A}(t, u)$ and $\hat{A}(t, u)$ will also be negative generators of C_0-semigroups on Z.

The plan of the paper is as follows. As in previous papers, the quasilinear theory is based on the linear theory, which has to be constructed first. In section 1 we study the intertwining condition (0.2) in detail, deducing some criteria for its validity. Also we deduce some simple lemmas for the homomorphism Ξ. In section 2 an operator calculus involving "weights" and "kernels", which was used in [10,11], is further developed. In section 3 the evolution operator for a given family $\{A(t)\}$ of generators is constructed, which solves the associated linear evolution equation. In section 4 we solve the quasilinear equation (Q); here a basic tool is a fixed point theorem, which generalizes the contraction map theorem and which is implicit in [17] in a simpler situation. Section 5 proves the continuous dependence of the solution u on A, f and the initial condition; the proof uses the ingenious method given in [19]. The fixed point theorem is proved in Appendix A.

As an application in which nonreflexivity of the spaces is essential, we give in section 6 a C^1-theory of hyperbolic systems of quasilinear partial differential equations in $\mathbf{R} \times \mathbf{R}^m$ with C^1-coefficients, thereby recovering the classical results of Douglis [3], Hartman and Wintner [5] (both for $m = 1$), and Cinquini Cibrario [1] (for $m \geqslant 1$). Technical details on semigroups generated by first order systems are given in Appendix B.

Here we summarize the notations used in the sequel. Given Banach spaces $X, Y, \cdots$, the associated norms are denoted by $\| \ \|_X$, $\| \ \|_Y$, etc. $L(Y; X)$ denotes the Banach space of all bounded linear operators on Y to X. The norm in $L(Y; X)$ is denoted by $\| \ \|_{Y; X}$. We write $L(X)$ for $L(X; X)$; $L(X)$ is a Banach algebra. If Q is an unbounded operator, $D(Q)$ denotes its domain, $R(Q)$ its range, and $\rho(Q)$ its resolvent set. $G(X)$ denotes the set of the negative generators of C_0-semigroups on X. If $A \in G(X)$, there are constants M, β such that $\|e^{-sA}\|_X \leqslant Me^{\beta s}$, $s \geqslant 0$; then we write $A \in G(X, M, \beta)$. A family $\{A(t) \in G(X); 0 \leqslant t \leqslant T\}$ is said to be *stable*, in symbols $A(\cdot) \in \mathrm{sta}(X, M, \beta)$, if

$$\|e^{-s_k A(t_k)} \cdots e^{-s_1 A(t_1)}\|_{X; X} \leqslant Me^{\beta(s_k + \cdots + s_1)}$$
$$\text{for} \qquad s_j \geqslant 0, \quad 0 \leqslant t_1 \leqslant \cdots \leqslant t_k \leqslant T, \qquad (0.3)$$

with some constants M, β. This implies in particular that $A(t) \in G(X, M, \beta)$ for each t. Obviously $A(\cdot) \in \mathrm{sta}(X, 1, \beta)$ if $A(t) \in G(X, 1, \beta)$ for all t.

C $[BC, BUC, PC]$ denotes the class of continuous [bounded continuous, bounded uniformly continuous, piecewise continuous] functions, and C_* the class of strongly continuous operator functions. For such functions we use the sup norm $\|f\|_X = \sup_t \|f(t)\|_X$, $\|A\|_{Y; X} = \sup_t \|A(t)\|_{Y; X}$, etc.

1. Intertwining relations, operator homomorphisms

In this section we introduce two new notions which we need in this paper.

1. Let X, Z be Banach spaces, let $A \in G(X, M, \beta), \hat{A} \in G(Z, \hat{M}, \hat{\beta})$. Let S be a densely defined, closed linear operator from X into Z. We are interested in the *intertwining relation*

$$Se^{-tA} \supset e^{-t\hat{A}}S \qquad \text{for} \quad t \geqslant 0. \qquad (1.1)$$

Due to the basic properties of C_0-semigroups and the closedness of S, (1.1) implies that

$$S(A + \lambda)^{-1} \supset (\hat{A} + \lambda)^{-1} S \qquad (1.2)$$

for all $\lambda > \beta' \equiv \max\{\beta, \hat{\beta}\}$. Conversely, (1.1) is true if (1.2) holds for a single $\lambda > \beta'$. Indeed, (1.2) implies that

$$S(A + \lambda)^{-n} \supset (\hat{A} + \lambda)^{-n} S \qquad n = 1, 2, \cdots, \qquad (1.3)$$

and the analyticity of the resolvents shows that (1.3) is true for all $\lambda > \beta'$. Then (1.1) follows by the exponential formula for semigroups.

Lemma 1.1. Assume (1.1). If Y denotes the Banach space $D(S)$ with the graph norm, Y is A-admissible (see [10]), i.e. e^{-tA} maps Y into itself and forms a C_0-semigroup on Y. Thus the part A_Y of A in Y belongs to $G(Y, M', \beta')$ with $M' = \max\{M, \hat{M}\}, \beta' = \max\{\beta, \hat{\beta}\}$. Moreover, $Y \cap D(A)$ is a core for A. (Note that Y need not be a subset of $D(A)$.)

Proof. (1.1) shows that $y \in Y$ implies $e^{-tA}y \in Y$. The continuity of $t \mapsto e^{-tA}y$ in Y-norm follows from (1.1) by the closedness of S. Moreover,

$$\|Se^{-tA}y\|_Z = \|e^{-t\hat{A}}Sy\|_Z \leqslant \hat{M}e^{\hat{\beta}t}\|Sy\|_Z,$$
$$\|e^{-tA}y\|_X \leqslant Me^{t\beta}\|y\|_X.$$

Hence $\|e^{-tA}\|_{Y;Y} \leqslant M'e^{t\beta'}$. This proves the first part of the lemma.

Again, since (1.1) implies (1.2), we have $(A + \lambda)^{-1}Y \subset D(A) \cap Y$, hence $Y \subset (A + \lambda)(D(A) \cap Y)$. Since Y is dense in X, it follows that $D(A) \cap Y$ is a core for A.

Remark. Given any Banach spaces X, Z, (1.1) is satisfied by any $A \in G(X)$, $\hat{A} \in G(Z)$, and $S = 0$ (i.e. $D(S) = X$ with $Sx = 0 \in Z$). Of course nothing interesting results from this; we have simply $Y = X$ and $A_Y = A$. But this example shows that there is in general no relationship between the constants β, $\hat{\beta}$, etc.

2. We now turn to sufficient conditions for (1.1) or (1.2). It is not difficult to show that (1.2) is equivalent to $S(A + \lambda) \subset (\hat{A} + \lambda)S$, but the latter relation is unmanageable since it involves the products of unbounded operators. It is more convenient to consider a condition of the form

$$SA\varphi = \hat{A}S\varphi \qquad \text{for} \qquad \varphi \in \tilde{D} \subset D(SA) \cap D(\hat{A}S). \qquad (1.4)$$

(Recall that $D(SA)$ consists of all $x \in D(A)$ such that $Ax \in D(S)$.) If $\tilde{D}$ is not too small, it is expected that (1.4) implies (1.2).

Lemma 1.2. (1.4) implies that $\tilde{D} \subset D(A_Y)$, with

$$\|(A_Y + \lambda)\varphi\|_Y \geqslant \delta_\lambda\|\varphi\|_Y, \qquad \varphi \in \tilde{D}, \quad \lambda > \beta', \quad \text{where} \quad \delta_\lambda > 0.$$

Proof. Assume (1.4). Then $\tilde{D} \subset D(S) = Y$, $\tilde{D} \subset D(A)$, and $A\tilde{D} \subset D(S) = Y$. Hence $\tilde{D} \subset D(A_Y)$. Moreover, we have $S(A + \lambda)\varphi = (\hat{A} + \lambda)S\varphi$ for $\varphi \in \tilde{D}$. Hence $\|S(A+\lambda)\varphi\|_Z = \|(\hat{A}+\lambda)S\varphi\|_Z \geqslant c'\|S\varphi\|_Z$ with a constant $c' > 0$, since $(\hat{A}+\lambda)^{-1} \in L(Z)$ exists. Since $\|(A + \lambda)\varphi\|_X \geqslant c\|\varphi\|_X$, $c > 0$, similarly, the required result follows.

Lemma 1.3. (1.4) implies (1.1) if $(A + \lambda)\tilde{D}$ is dense in Y for some $\lambda > \beta'$.

Proof. Let $y \in Y = D(S)$. By hypothesis there is a sequence $\varphi_n \in \tilde{D}$ such that $\lim(A + \lambda)\varphi_n = y$ in Y. In other words,

$$\lim(A + \lambda)\varphi_n = y \text{ in } X \quad \text{and} \quad \lim S(A + \lambda)\varphi_n = Sy \text{ in } Z. \tag{1.5}$$

Since $(A+\lambda)^{-1}$ and $(\hat{A}+\lambda)^{-1}$ are bounded, (1.5) implies that $\lim \varphi_n = (A+\lambda)^{-1}y$ and $\lim(\hat{A}+\lambda)^{-1}S(A+\lambda)\varphi_n = (\hat{A}+\lambda)^{-1}Sy$. But (1.4) implies that $S\varphi_n = (\hat{A}+\lambda)^{-1}S(A+\lambda)\varphi_n$. Since S is closed, it follows that $S(A + \lambda)^{-1}y$ exists and equals $(\hat{A} + \lambda)^{-1}Sy$. This proves (1.2) for the particular λ considered. As remarked above, this implies (1.1).

Lemma 1.4. Assume that $S^{-1} \in L(Z; X)$ exists. If $(\hat{A}+\lambda)S\tilde{D}$ is dense in Z for some $\lambda > \beta'$, then (1.4) implies (1.1). Moreover, we have $\hat{A} = SAS^{-1}$.

Proof. $S^{-1}(\hat{A}+\lambda)^{-1}$ and $(A+\lambda)^{-1}S^{-1}$ are both in $L(Z; X)$ and coincide on $S(A+\lambda)\tilde{D} = (\hat{A}+\lambda)S\tilde{D}$, which is dense in Z. Therefore $S^{-1}(\hat{A}+\lambda)^{-1} = (A+\lambda)^{-1}S^{-1}$. If $y \in D(S)$, then $(A + \lambda)^{-1}y = (A + \lambda)^{-1}S^{-1}Sy = S^{-1}(\hat{A} + \lambda)^{-1}Sy \in D(S)$, hence $S(A + \lambda)^{-1}y$ exists and equals $(\hat{A} + \lambda)^{-1}Sy$. This leads to (1.1), as in the proof of Lemma 1.3. Also we have $(\hat{A} + \lambda)^{-1} = S(A + \lambda)^{-1}S^{-1}$ because $SS^{-1} = 1$ on Z. Taking the inverse then gives $\hat{A} + \lambda = S(A + \lambda)S^{-1}$, hence $\hat{A} = SAS^{-1}$.

Remark. Relation (1.2) makes sense for more general operators A, $\hat{A}$ which need not generate C_0-semigroups, provided that $\lambda \in \rho(A) \cap \rho(\hat{A})$. Some of the above results relating to (1.2) are true in this case, so long as the semigroup properties are not involved.

3. Let X, Z be Banach spaces. Let Ξ be a *strongly continuous, nonexpansive homomorphism* of the algebra $L(X)$ into $L(Z)$, in other words, an algebraic homomorphism such that $\Xi 1_X = 1_Z$, $\|\Xi B\|_{Z; Z} \leqslant \|B\|_{X; X}$, and that $s\text{-}\lim B_n = B$ implies $s\text{-}\lim \Xi B_n = \Xi B$ for a sequence $B_n \in L(X)$. The nonexpansive property implies norm continuity.

A typical example of such a triplet X, Z, Ξ is given by

$$Z = X \oplus \cdots \oplus X, \quad \Xi B = B \oplus \cdots \oplus B \text{ (direct sums)}, \quad B \in L(X). \tag{1.6}$$

If $\{U(t)\}$ is a C_0-semigroup on X, $\{\Xi U(t)\}$ is a C_0-semigroup on Z. If $U(t) = e^{-tA}$, $\Xi U(t) = e^{-t\overline{A}}$, then we set $\overline{A} = \Xi A$, thereby extending Ξ to a map of $G(X)$ into $G(Z)$. If $A \in G(X, M, \beta)$, then $\|\Xi U(t)\|_{Z; Z} \leqslant \|U(t)\|_{X; X} \leqslant Me^{\beta t}$, so that Ξ maps $G(X, M, \beta)$ into $G(Z, M, \beta)$. Moreover we have $\Xi(A+\lambda)^{-1} = (\overline{A}+\lambda)^{-1}$, $\lambda > \beta$, since the resolvent is recovered from $U(t)$ by integration.

As is easily seen, this extension is consistent with the original definition of $\overline{A} = \Xi A$ if $A \in L(X)$. We also note that $\Xi(A + \lambda) = \Xi A + \lambda$ for any scalar λ, since $\Xi(e^{-t\lambda}U(t)) = e^{-t\lambda}\Xi U(t)$.

2. Kernels, chains, and evolution operators

This section deals with a formal calculus with operator valued functions, extending some results given in [10 ,11].

1. **A convolution calculus with operator valued kernels.** Let $I = [a, b]$ be a finite interval and let $\Delta(I)$ denote the triangle $b \geqslant t \geqslant s \geqslant a$ in the (t, s)-plane. Let X be a Banach space. We consider two kinds of operator valued functions.

A function $B \in PC_*(I; L(X))$ will be called an $L(X)$-*weight* (on I), and a function $U \in C_*(\Delta(I); L(X))$ an $L(X)$-*kernel* (on $\Delta(I)$). We may omit the affix $L(X)$, I, or $\Delta(I)$, if there is no possibility of confusion.

If U, V are kernels and B is a weight, we define the product UBV as the kernel such that

$$UBV(t, r) = \int_r^t U(t, s)B(s)V(s, r)ds.$$

It is easy to see that the associative law: $(UBV)CW = UB(VCW)$ holds, which we simply write $UBVCW$.

Given a kernel U and a weight B, we construct a Volterra type series

$$V = \text{vol}(U, B) \equiv U + UBU + UBUBU + \cdots; \tag{2.1}$$

this converges in norm and defines a kernel V. Indeed, suppose that

$$\|U(t, s)\| \leqslant Me^{\beta(t-s)}, \quad t, s \in \Delta(I); \qquad \|B(t)\| \leqslant K, \quad t \in I, \tag{2.2}$$

where $\| \ \| = \| \ \|_{X; X}$. (Such M, β can be found in many ways.) Then it is easily seen that $\|U(BU)^n\| \leqslant M^{n+1}K^n(t - s)^n e^{\beta(t-s)}/n!$. Hence

$$\|V(t, s)\| \leqslant Me^{(\beta+MK)(t-s)}. \tag{2.3}$$

(2.1) implies that

$$V = U + UBV = U + VBU, \tag{2.4}$$

and either equality in (2.4) implies (2.1). Hence $V = \text{vol}(U, B)$ if and only if $U = \text{vol}(V, -B)$.

More generally we have

$$\text{vol}(\text{vol}(U, B), C) = \text{vol}(U, B + C). \tag{2.5}$$

Indeed, let $V = \text{vol}(U, B)$, $W = \text{vol}(V, C)$. It suffices to show that $U(B+C)W = W-U$. But $U(B+C)W = UB(V+VCW)+(V-UBV)CW = UBV+VCW = V-U+W-V = W - U$.

2. **Chains.** We say that a kernel U obeys the *chain rule* (or U is a chain), if

$$U(t, r) = U(t, s)U(s, r), \qquad U(s, s) = 1, \qquad t \geqslant s \geqslant r. \tag{2.6}$$

Lemma 2.1. Let U be a chain and B a weight. Then $V = \mathrm{vol}(U, B)$ is also a chain.

Proof. Using (2.4) twice we obtain, for $t \geqslant s \geqslant r$,

$$W(t, s, r) \equiv V(t, s)V(s, r) = (U + UBV)(t, s)V(s, r)$$
$$= U(t, s)(U + UBV)(s, r) + UBV(t, s)V(s, r)$$
$$= U(t, r) + \int_r^s U(t, \tau)B(\tau)V(\tau, r)d\tau + \left(\int_s^t U(t, \tau)B(\tau)V(\tau, s)d\tau \right) V(s, r)$$
$$= (U + UBV)(t, r) + \int_s^t U(t, \tau)B(\tau)(W(\tau, s, r) - V(\tau, r))d\tau.$$

The first term on the right equals $V(t, r)$. If we fix s and r, and write $Z(t) = W(t, s, r) - V(t, r)$, it follows that

$$Z(t) = \int_s^t U(t, \tau)B(\tau)Z(\tau)d\tau.$$

Obviously this implies that $Z(t) = 0$ for $t \geqslant s$. Hence $V(t, s)V(s, r) = W(t, s, r) = V(t, r)$. Since it is obvious that $V(s, s) = 1$, this proves the lemma.

3. **Juxtaposition of kernels.**

Lemma 2.2. Let $I' = [a', b']$, $I'' = [a'', b'']$ be two adjoining intervals, with $b' = a''$. Let $I = I' \cup I'' = [a', b'']$. Let U', U'' be chains on $\Delta(I')$, $\Delta(I'')$, respectively. Then there is a unique chain U on $\Delta(I)$ that coincides with U', U'' on $\Delta(I')$, $\Delta(I'')$, respectively. (We shall write $U = U' \cup U''$.) A similar result holds for several adjoining intervals.

Proof. Set $U = U'$ on $\Delta(I')$, $U = U''$ on $\Delta(I'')$. If $t \in I''$ and $s \in I'$, set $U(t, s) = U''(t, a'')U'(b', s)$. It is easy to see that U is a unique chain with the required property.

Lemma 2.3. In the situation of Lemma 2.2, let B be a weight on I, and $B'[B'']$ its restriction on $I'[I'']$. Then $\mathrm{vol}(U, B) = \mathrm{vol}(U', B') \cup \mathrm{vol}(U'', B'')$. Similarly for several intervals.

Proof. By Lemmas 2.1 and 2.2, the two kernels $\mathrm{vol}(U, B)$ and $\mathrm{vol}(U', B') \cup \mathrm{vol}(U'', B'')$ both obey the chain rule. Since they coincide on $\Delta(I')$ and on $\Delta(I'')$, they are identical by the uniqueness in Lemma 2.2.

4. **Finitely generated kernels.** If $A \in G(X)$, it is obvious that $U(t, s) = e^{-(t-s)A}$ defines an $L(X)$-chain on $\Delta(I)$ for any interval I. If $A(t) \in G(X)$ is piecewise constant in $t \in I$ (a step function), with values $A_j \in G(X)$ on subintervals $I_j \subset I$, $j = 1, ..., n$, it follows from Lemma 2.2 that there is a unique chain U that equals $e^{-(t-s)A_j}$ on $\Delta(I_j)$. We shall say that U is *generated* by the family $A(\cdot)$.

Lemma 2.4. Let U be generated by a piecewise constant family $A(t) \in G(X)$. Let $B(t) \in L(X)$ be piecewise constant. Then vol$(U, -B)$ is generated by $A(t) + B(t) \in G(X)$, which is likewise piecewise constant.

Proof. In view of Lemma 2.3, it suffices to prove this on each subinterval on which A, B are constant. Thus the assertion reduces to a perturbation theorem for semigroups (see e.g. [8,Theorem IX-2.1]).

5. Evolution operators. Let Y, X be Banach spaces such that $Y \subset X$ densely and continuously. Let $A \in PC_*(I; L(Y; X))$. A kernel U on $\Delta(I)$ is called a Y/X-evolution operator for A if the following conditions are satisfied.

(a) U is an $L(X)$-chain on $\Delta(I)$.
(b) $U(t, s)Y \subset Y$ for all t, $s \in \Delta(I)$, and the restriction of $U(t, s)$ on Y forms an $L(Y)$-chain.
(c) $\partial_t U(t, s)\phi = -A(t)U(t, s)\phi$, $\partial_s U(t, s)\phi = U(t, s)A(s)\phi$, $\phi \in Y$.

Since A may be discontinuous, (c) should be interpreted in the integrated sense:

$$\int_s^t A(\tau)U(\tau, s)\phi d\tau = \phi - U(t, s)\phi = \int_s^t U(t, \tau)A(\tau)\phi d\tau, \quad \phi \in Y. \qquad (2.7)$$

The evolution operator is uniquely determined by A. This is known when A is continuous (see [10]). The case of piecewise continuity can be dealt with by using Lemma 2.2.

Existence of the evolution operator for a given family A is a major problem, to be discussed in next section. But the special case of a piecewise constant A is simple. The following lemma is easily proved by juxtaposition of semigroup operators (cf.[10]).

Lemma 2.5. Let U be an $L(X)$-chain generated by a piecewise constant family $A(t) \in G(X)$. Let $Y \subset X$ be as above, with $Y \subset D(A(t))$, and let Y be $A(t)$-admissible for $t \in I$ (i.e. the part of $A(t)$ in Y belongs to $G(Y)$). Then U is a Y/X-evolution operator for A.

Remark. By generalization we may define $L(X; Y)$-kernels and $L(X; Y)$-weights in an obvious way. Thus we can construct an $L(X; W)$-kernel UBV from an $L(X; Y)$-kernel V, an $L(Y; Z)$-weight B, and an $L(Z; W)$-kernel U, etc.

3. The linear evolution equation

1. The purpose of this section is to solve the linear evolution equation

(L) $$\partial_t u + A(t)u = f(t), \qquad t \in I = [0, T].$$

This will be done by constructing the Y/X-evolution operator for the family $\{A(t)\}$, where Y, X are Banach spaces such that $Y \subset X$ densely and continuously. To this end we need an auxiliary Banach space Z, and an operator $S \in L(Y; Z)$, such that

$$\|u\|_Y = \|u\|_X + \|Su\|_Z, \qquad u \in Y. \tag{3.1}$$

Condition (3.1) is equivalent to saying that S is a closed (unbounded) operator from X to Z with $D(S) = Y$. ((3.1) is a standard graph norm, but other norms are possible.) We assume for simplicity that S does not depend on t. Generarization to the t-dependent case will be discussed in Appendix C.

Moreover, we assume that there is a strongly continuous and nonexpansive homomorphism Ξ of the algebra $L(X)$ into $L(Z)$, which is extended to a map of $G(X, M, \beta)$ into $G(Z, M, \beta)$ (see section 1.3).

In this setting, we introduce the following conditions (i) to (iii) for A.

(i) (stability) $A(\cdot) \in \mathrm{sta}(X, M, \beta)$ (see (0.3)).

Remark. This is a rather implicit condition; criteria for its validity are given in [10]. It is obvious that (i) holds with $M = 1$ if $A(t) \in G(X, 1, \beta)$ for all t.

(ii) (continuity) $Y \subset D(A(t))$, $t \in I$, and $A(\cdot) \in PC(I; L(Y; X))$.

Remark. This means that the restriction of $A(t)$ on Y is in $L(Y; X)$ and is piecewise norm-continuous in t. We admit piecewise continuous families simply for consistency, since we have to use step function approximations in the proof.

(iii) (intertwining condition, see section 1)

$$Se^{-sA(t)} \supset e^{-s\hat{A}(t)}S, \qquad s \geqslant 0, \ t \in I, \tag{3.2}$$

where

$$\hat{A}(t) = \overline{A}(t) + B(t) \in G(Z), \quad \overline{A}(t) = \Xi A(t) \in G(Z), \quad B \in PC_*(I; L(Z)). \tag{3.2a}$$

Remark. Obviously $\overline{A} = \Xi A \in \mathrm{sta}(Z, M, \beta)$ by (i). Since $\||B\||_{Z;Z} \equiv K < \infty$, (3.2a) implies that $\hat{A} \in \mathrm{sta}(Z, M, \beta')$ with $\beta' = \beta + MK$, by a perturbation theorem (see [10,Prop.3.5]). Moreover, (3.2) implies that $e^{-sA(t)}$ maps Y into Y, with

$$Se^{-s_k A(t_k)} \cdots e^{-s_1 A(t_1)} \supset e^{-s_k \hat{A}(t_k)} \cdots e^{-s_1 \hat{A}(t_1)} S. \tag{3.3}$$

Theorem I. Under assumptions (i), (ii), (iii), there is a unique Y/X-evolution operator U for the family A. We have $\||U\||_{X;X} \leqslant Me^{\beta T}$, $\||U\||_{Y;Y} \leqslant Me^{\beta' T}$.

Corollary. If $\phi \in Y$ and $f \in C(I; Y), (L)$ has a unique solution $u \in C(I; Y) \cap PC^1(I; X)$ with $u(0) = \phi$; it is given by

$$u(t) = U(t, 0)\phi + \int_0^t U(t, s)f(s)ds.$$

The proof of Theorem I is parallel to that of a similar theorem in [10], and is even simpler since S is independent of t. We may assume without loss of generality that $A \in C(I; L(Y; X))$.

We begin with constructing step function approximations A_n to A, $n = 1, 2, ...$, on setting $A_n(t) = A((T/n)[nt/T])$ (where [ ] is the Gaussian symbol). Similarly we define B_n.

Proposition 3.1. The Y/X-evolution operator U_n exists for A_n. Moreover, we have

$$SU_n(t, s) \supset \hat{U}_n(t, s)S, \quad \hat{U}_n = \text{vol}(\overline{U}_n, -B_n), \quad \overline{U}_n = \Xi U_n. \tag{3.4}$$

Proof. U_n exists by Lemma 2.5 (see a remark after (iii)). (3.4) follows from (3.3), since U_n is the $L(X)$-chain generated by A_n while $\hat{A}_n = \Xi A_n + B_n$ generates an $L(Z)$-chain $\hat{U}_n$; see Lemma 2.4.

Proposition 3.2. $U(t, s) = s\text{-}\lim U_n(t, s) \in L(X)$ exists, uniformly in t, s, and forms an $L(X)$-chain. Moreover, $s\text{-}\lim \overline{U}_n = \Xi U \equiv \overline{U}$ and $s\text{-}\lim \hat{U}_n = \text{vol}(\overline{U}, -B) \equiv \hat{U}$ in $L(Z)$ in a similar sense ($\Delta = \Delta(I)$). We have, alternatively,

$$\hat{U} = \overline{U} - \overline{U}B\hat{U} = \overline{U} - \hat{U}B\overline{U}. \tag{3.5}$$

Proof. (i) implies that $\|U_n(t, s)\|_{X; X} \leqslant Me^{\beta(t-s)}$, hence $\|\overline{U}_n(t, s)\|_{Z; Z} \leqslant Me^{\beta(t-s)}$ by the property of Ξ, hence $\|\hat{U}_n(t, s)\|_{Z; Z} \leqslant Me^{\beta'(t-s)}$ by (2.3) (recall that $\beta' = \beta + MK$). Thus both U_n and $\hat{U}_n$ are uniformly bounded. But U_n is also an $L(Y)$-chain, with $\|U_n(t, s)\|_{Y; Y} \leqslant Me^{\beta'(t-s)}$ (cf. proof of Lemma 1.1). Thus the existence of $U(t, s) = s\text{-}\lim U_n(t, s)$ follows by an argument used in [10]. Obviously U is an $L(X)$-chain. The rest of the proposition follows from (3.4). (3.5) follows from (2.4).

Proposition 3.3. $s\text{-}\lim U_n(t, s) = U(t, s)$ holds also in $L(Y)$. Moreover, we have $U \in C_*(\Delta; L(X)) \cap C_*(\Delta; L(Y))$, with

$$SU(t, s) \supset \hat{U}(t, s)S. \tag{3.6}$$

Proof. Propositions 3.1~2 imply that $\lim SU_n(t, s)\phi = \lim \hat{U}_n(t, s)S\phi = \hat{U}(t, s)S\phi$ exists in X if $\phi \in Y$. Since S is closed, this implies the first assertion of the proposition. The remainder follows from Propositions 3.1~2 by passing to the limit.

Proposition 3.4. U satisfies the differentiability condition, in the sense of (2.7).

Proof. We recall that (2.7) holds with U, A replaced by U_n, A_n, respectively (see Lemma 2.5). Since $s\text{-}\lim U_n(t, s) = U(t, s)$ in $L(X)$ as well as in $L(Y)$, and since $\lim A_n(t) = A(t)$ in $L(Y; X)$, (2.7) follows in the limit $n \to \infty$.

Propositions 3.1~4 prove that U is a Y/X-evolution operator for A. The second part of Theorem I follows easily from the estimates given in the proof of Proposition 3.2. The corollary can be proved as in [10].

2. For later use, we deduce a refined estimate for $U(t, s) - 1$. Here we have to make an additional assumption.

(iv) There is a Banach space $Z_0 \subset Z$ (densely and continuously) such that $\overline{A} \in PC_*(I; L(Z_0; Z))$. (This condition is automatically satisfied if Z and Ξ are given by (1.6), with $Z_0 = Y \oplus \cdots \oplus Y$.)

Lemma 3.5. Assume conditions (i), (ii), (iii), and (iv). For any $\phi \in Y$ and $\psi \in Z_0$, we have

$$\|(U(t, s) - 1)\phi\|_Y \leqslant (1 + C')\|S\phi - \psi\|_Z + (t - s)(C\|\|A\|\|_{Y; X}\|\phi\|_Y + C'\|\|\hat{A}\psi\|\|_Z), \quad (3.7)$$

where $C = Me^{\beta T} \geqslant \|\|U\|\|_{X; X}$, $C' = Me^{\beta' T} \geqslant \|\|\hat{U}\|\|_{Z; Z}$.

Proof. We estimate $p = \|(U(t, s) - 1)\phi\|_X$ and $q = \|S(U(t, s) - 1)\phi\|_Z$ and apply (3.1). Using (2.7) we obtain

$$p = \left\|\int_s^t U(t, \tau)A(\tau)\phi d\tau\right\|_X \leqslant (t - s)C\|\|A\|\|_{Y; X}\|\phi\|_Y. \quad (3.8)$$

To estimate q, we use (3.6), obtaining

$$q = \|(\hat{U}(t, s) - 1)S\phi\|_Z \leqslant \|(\hat{U}(t, s) - 1)(S\phi - \psi)\|_Z + \|(\hat{U}(t, s) - 1)\psi\|_Z.$$

The first term on the right is majorized by $(1 + C')\|S\phi - \psi\|_Z$. To estimate the second term, we use $\psi - \hat{U}(t, s)\psi = \int_s^t \hat{U}(t, \tau)\hat{A}(\tau)\psi d\tau$, which can be proved as in (2.7) by using the approximations $\hat{U}_n$ for $\hat{U}$; note that $\hat{A}$ also satisfies condition (iv) due to (3.2a). An estimates as in (3.8) then gives $\|(\hat{U}(t, s) - 1)\psi\|_Z \leqslant (t - s)C'\|\|\hat{A}\psi\|\|_Z$. Collecting the terms, we obtain the desired result.

Lemma 3.6. Assume conditions (i), (ii), (iii), and (iv). For each $\phi \in Y$ and $\varepsilon > 0$, there is a constant R, depending on ε and ϕ (in addition to such constants as M, β, β', $\|\|A\|\|_{Y; X}$) but not on t or s, such that

$$\|(U(t, s) - 1)\phi\|_Y \leqslant \varepsilon + (t - s)R, \qquad (t, s) \in \Delta.$$

Moreover, R can be chosen common to all ϕ in a compact subset of Y.

Proof. This follows from (3.7) on choosing ψ inside an Z-ball of diameter ε containing $S\phi$. The last assertion follows from the fact that a compact set can be covered by a finite number of such balls.

4. The quasilinear equation

We now consider the quasilinear equation

$$\text{(Q)} \qquad \partial_t u + A(t, u)u = f(t, u).$$

We use the same setting as in the previous section, with the spaces Y, X, Z, the operator S with the property (3.1), and the homomorphism Ξ of $L(X)$ into $L(Z)$. In addition, we need another Banach space $\tilde{X}$ such that $Y \subset \tilde{X} \subset X$, with $\| \ \|_X \leqslant \| \ \|_{\tilde{X}} \leqslant \| \ \|_Y$. It is assumed that X and $\tilde{X}$ have the same (uniform) topology on a bounded set in Y. In other words, given any bounded subset Σ of Y and $\varepsilon > 0$, there is $\delta > 0$ such that $x, y \in \Sigma$ with $\|x - y\|_X < \delta$ implies $\|x - y\|_{\tilde{X}} < \varepsilon$. (This is the case, for example, if $\| \ \|_{\tilde{X}} \leqslant c\| \ \|_X^\alpha \ \| \ \|_Y^{1-\alpha}$ with $0 < \alpha < 1$.).

For the maps A and f in (Q), we introduce the following conditions, in which $I = [0, T]$ is fixed with $0 < T < \infty$, W is a bounded open set in Y, $\tilde{W}$ is its closure in $\tilde{X}$ (or, equivalently, in X), and λ_A, μ_A, etc. are fixed constants. We note that $\tilde{W} \subset Y$ if Y is reflexive.

(I) A maps $I \times W$ into $G(X)$. If $v \in C(I; W) \cap \mathrm{Lip}(I; \tilde{X})$, then $A(\cdot, v(\cdot)) \in \mathrm{sta}(X, M, \beta)$, with M, β depending only on $\mathrm{Lip}_{\tilde{X}}(v)$.

(II) $A \in C(I \times W; L(Y; \tilde{X}))$ (which implies that $Y \subset D(A(t, w))$). Moreover,

$$\|A(t, w)\|_{Y; \tilde{X}} \leqslant \lambda_A, \qquad \|A(t, w) - A(t, w')\|_{Y; X} \leqslant \mu_A \|w - w'\|_X,$$
$$t \in I, \quad w, w' \in W.$$

(III) The intertwining relation holds:

$$Se^{-sA(t,w)} \supset s^{-s\hat{A}(t,w)}S, \ s \geqslant 0, \ \hat{A}(t, w) = \Xi A(t, w) + B(t, w, Sw) \in G(Z),$$

where B is a strongly continuous map of $I \times \tilde{W} \times Z$ into $L(Z)$. Moreover,

$$\|B(t, w, Sw)\|_{Z; Z} \leqslant \lambda_B, \quad \|B(t, w, z) - B(t, w, z')\|_{Z; Z} \leqslant \mu_B \|z - z'\|_Z,$$
$$t \in I, \quad w \in \tilde{W}, \quad z, z' \in Z.$$

(IV) There is a Banach space $Z_0 \subset Z$ (densely and continuously) such that $\Xi A \in C_*(I \times W; L(Z_0; Z))$.

(V) $f \in C(I \times W; Y)$, with

$$\|f(t, w)\|_Y \leqslant \lambda_f, \qquad \|f(t, w) - f(t, w')\|_X \leqslant \mu_f \|w - w'\|_X.$$

Moreover, there is a continuous map $g : I \times \tilde{W} \times Z \to Z$ such that $Sf(t, w) = g(t, w, Sw)$, with

$$\|g(t, w, z) - g(t, w, z')\|_Z \leqslant \mu_g \|z - z'\|_Z, \quad t \in I,\ w \in \tilde{W},\ z, z' \in Z.$$

Remarks. (a) In some problems we may set $\tilde{X} = X$, but in general we need different spaces, see [6]. (The notation in [6] is slightly different from the present one; in [6] our X and Z are identified and written Z, while our $\tilde{X}$ is written X.)

(b) Condition (I) is rather implicit. A convenient criterion for it is given in [6,10], which uses variable norms in X.

(c) In view of (II), condition (IV) is automatically satisfied if Z and Ξ are given by (1.6) (see the remark on condition (iv), section 3).

Theorem II. Assume conditions (I) to (V). Given any $\phi \in W$, there is $T' > 0$ such that (Q) has a unique solution $u \in C([0, T']; W) \cap C^1([0, T']; \tilde{X})$ with $u(0) = \phi$. T' can be chosen common to all ϕ in a compact subset of W.

The proof will be reduced to a fixed point theorem to be proved in Appendix A. We choose a closed ball in W, with center ϕ and radius ρ, and introduce the set E of functions v such that

$$v \in C(I'; Y) \cap \text{Lip}(I'; \tilde{X}), \quad \|v(t) - \phi\|_Y \leqslant \rho, \quad \|v(t) - v(s)\|_{\tilde{X}} \leqslant L|t - s|, \quad (4.1)$$

where $I' = [0, T']$, $T' \leqslant T$, with T', L yet to be determined.

For each $v \in E$, we write $A^v(t) = A(t, v(t))$, $f^v(t) = f(t, v(t))$, and solve the linear initial value problem

$$\partial_t u + A^v(t)u = f^v(t), \qquad u(0) = \phi. \tag{4.2}$$

Proposition 4.1. The Y/X-evolution operator $U^v(t, s)$ for A^v exists, with the estimates $\|U^v\|_{X; X} \leqslant C$, $\|U^v\|_{Y; Y} \leqslant C'$, where C, C' are determined by the stability constants for A^v and by $\|B\|_{Z; Z}$, which are in turn determined by L and λ_B. U^v also satisfies Lemmas 3.5 and 3.6.

Proof. We have only to verify the assumptions of Theorem I. The stability condition (i) follows from (I). (ii) is satisfied by virtue of (II) and (4.1). By (III), A^v satisfies the interwining condition (iii). Regarding Lemmas 3.5~6, note that (IV) implies (iv).

Proposition 4.2. For any $\phi \in Y$, (4.2) has a unique solution $u \in C(I'; Y) \cap C^1(I'; \tilde{X})$ with $u(0) = \phi$. The constants ρ, T' and L can be chosen so that $u \in E$. This defines a map $\Phi : E \to E$ by $u = \Phi(v)$.

Proof. According to Corollary to Theorem I, u is given by

$$u(t) = U^v(t, 0)\phi + \int_0^t U^v(t, s)f^v(s)ds. \tag{4.3}$$

With ρ fixed, we shall determine L and T' in such a way that $u \in E$. To this end we use Lemma 3.6 and (4.3), obtaining

$$\|u(t) - \phi\|_Y \leqslant \|(U^v(t, 0) - 1)\phi\|_Y + \int_0^t \|U^v(t, s)f^v(s)\|_Y ds$$

$$\leqslant \varepsilon + (R + C'\lambda_f)T', \tag{4.4}$$

$$\|\partial_t u(t)\|_{\tilde{X}} \leqslant \lambda_A \|u(t)\|_Y + \|f^v(t)\|_{\tilde{X}} \leqslant \lambda_A \|u(t)\|_Y + \|f^v(t)\|_Y, \tag{4.5}$$

where $\varepsilon > 0$ is arbitrary and R depends on ε, λ_A, T' and L. We require that they satisfy

$$\varepsilon + (R + C'\lambda_f)T' \leqslant \rho, \qquad \lambda_A(\rho + \|\phi\|_Y) + \lambda_f = L. \tag{4.6}$$

This is possible by setting L as shown and $\varepsilon = \rho/2$ and then choosing T' small enough; note that although R and C' depend on T', the dependence is monotone. Then (4.4) and (4.5) imply that $\|u(t) - \phi\|_Y \leqslant \rho$ and $\|\partial_t u(t)\|_{\tilde{X}} \leqslant L$, hence $u \in E$.

Proposition 4.3. Φ is a contraction in the sup-norm of $\mathcal{X} \equiv C(I'; X)$ if $T' > 0$ is sufficiently small.

Proof. Conditions (II) and (V) imply that $\||A^{v'} - A^v\||_{Y;X} \leqslant \mu_A \||v - v'\||_X$, $\||f^{v'} - f^v\||_X \leqslant \mu_f \||v' - v\||_X$. Since (see [10])

$$U^{v'} - U^v = -U^{v'}(A^{v'} - A^v)U^v \in C(\Delta; L(Y; X)) \tag{4.7}$$

(for notation see the remark at the end of section 2), we obtain

$$\||\Phi(v') - \Phi(v)\||_X \leqslant T'(CC'\mu_A(\|\phi\|_Y + T'\lambda_f) + C\mu_f)\||v - v'\||_X.$$

by (4.3) and Proposition 4.1. Thus Φ is a contracation if T' is sufficiently small.

Proposition 4.4. For $u = \Phi(v)$, we have the identity

$$Su(t) = \overline{U}^v(t, 0)S\phi + \int_0^t \overline{U}^v(t, s)[g(s, v(s), Sv(s))$$

$$- B(s, v(s), Sv(s))Su(s)]ds, \tag{4.8}$$

where $\overline{U}^v(t, s) = \Xi U^v(t, s)$.

Proof. (4.3) may be written $u = U^v \phi + U^v f^v$, in an obvious symbolic notation. The interwining relation (3.6) then gives $Su = \hat{U}^v S\phi + \hat{U}^v Sf^v$. Since $\hat{U}^v = \overline{U}^v - \overline{U}^v B^v \hat{U}^v$ by (3.5), we obtain

$$Su = \overline{U}^v S\phi + \overline{U}^v Sf^v - \overline{U}^v B^v(\hat{U}^v S\phi + \hat{U}^v Sf^v)$$

$$= \overline{U}^v S\phi + \overline{U}^v Sf^v - \overline{U}^v B^v Su.$$

This gives (4.8) when written explicitly; note that $Sf^v(s) = g(s, v(s), Sv(s))$ by (V).

Proposition 4.5. Let $\tilde{E}$ be the closure of E in $\mathcal{X} = C(I'; X)$ (or, equivalently, in $\tilde{\mathcal{X}} = C(I'; \tilde{X})$). Then the map $v \mapsto U^v$ $[v \mapsto \overline{U}^v]$ can be extended to a continuous map from $\tilde{E}$ into $C_*(\Delta; L(X))$ $[C_*(\Delta; L(Z))]$.

Proof. In view of (II), we see from (4.7) that $\|(U^v - U^{v'})(t, s)y\|_X \leqslant CC'\mu_A \|y\|_Y \|\|v - v'\|\|_X$. The desired result for U^v then follows by the density of Y in X. The result for $\overline{U}^v = \Xi U^v$ follows from the strong continuity of Ξ.

Proposition 4.6. For $v \in \tilde{E}$ (so that $v(t) \in \tilde{W}$) and w, $z \in \mathcal{B} \subset \mathcal{Z} \equiv C(I'; Z)$, where $\mathcal{B}$ is a ball about 0 of radius $2C\|S\phi\|_Z$ (which contains $\overline{U}^v(t, 0)S\phi$), set

$$
\Psi(v, w, z)(t) = \overline{U}^v(t, 0)S\phi + \int_0^t \overline{U}^v(t, s)[g(s, v(s), w(s)) \\
- B(s, v(s), w(s))z(s)]ds. \tag{4.9}
$$

If T' is sufficiently small, Ψ maps $\tilde{E} \times \mathcal{B} \times \mathcal{B}$ into $\mathcal{B}$ continuously. Moreover, it is Lipschitz continuous in the last two variables, with Lipschitz constant arbitraly small if T' is small.

Proof. This follows conditions (III), (V) and Proposition 4.5.

Proposition 4.7. Let $v \in E$, $u = \Phi(v)$. Then

$$
Su = \Psi(v, Sv, Su). \tag{4.10}
$$

Proof. This follows directly from (4.8) and (4.9).

Proposition 4.8. Φ has a unique fixed point $u \in E$. u is the unique solution in E of (Q) with $u(0) = \phi$.

Proof. This follows from Theorem A in Appendix A, in which the spaces X, Y, Z should be replaced by $\mathcal{X}, \mathcal{Y} = C(I'; Y)$, $\mathcal{Z}$, respectively.

Completion of the proof of Theorem II. It remains to prove the last statement of Theorem II. From the proof given above it is easy to see that T' depends, apart from various constants λ_A, μ_A, etc. (which are all fixed), on the size of R in (4.6), which in turn depends on ϕ. This dependence cannot be described by a single or several real parameters (such as $\|\phi\|_Y$). According to Lemma 3.6, however, R can be chosen uniformly in ϕ in a compact subset of W. Hence the same is true of T'.

5. Continuous dependence

We shall show that the solution u of (Q) depends continuously on A, f, and $\phi = u(0)$. To formulate it precisely, we introduce a sequence (Q_n), $n = 1, 2, ..., \infty$, of equations of the form (Q), in which A, f are replaced by A_n, f_n, respectively. We assume that A_n and f_n satisfy conditions (I) to (V) *uniformly*, with the operators B_n defined in an obvious manner. This means that the constants λ_A, μ_A, etc. and the dependence of M, β on $\||\partial_t v\||_{\tilde{X}}$ are common to all n, and that $A_n \in C(I \times W; L(Y; \tilde{X}))$ holds uniformly in n (equicontinuity), etc. The spaces $W \subset Y \subset \tilde{X} \subset X$, $Z_0 \subset Z$, the operator S, and the homomorphism Ξ are assumed to be common to all these equations.

In addition, we assume that

$$s\text{-}\lim A_n(t, w) = A_\infty(t, w) \text{ in } L(Y; X), \quad t \in I, \ w \in W, \tag{5.1}$$

$$s\text{-}\lim \Xi A_n(t, w) = \Xi A_\infty(t, w) \text{ in } L(Z_0; Z), \quad t \in I, \ w \in W, \tag{5.2}$$

$$s\text{-}\lim B_n(t, w, z) = B_\infty(t, w, z) \text{ in } L(Z), \quad t \in I, \ w \in \tilde{W}, \ z \in Z. \tag{5.3}$$

$$\lim f_n(t, w) = f_\infty(t, w) \text{ in } Y, \quad t \in I, \ w \in W. \tag{5.4}$$

Under these assumptions, we prove

Theorem III. Let $u_\infty \in C(\overline{I}; W) \cap C^1(\overline{I}; \tilde{X})$, $\overline{I} = [0, \overline{T}]$, be a solution of (Q_∞). Let $\phi_n \in W$, $n = 1, 2, \ldots$, such that $\lim \phi_n = \phi_\infty = u_\infty(0)$ in Y. Then there are unique solutions $u_n \in C(\overline{I}; W) \cap C^1(\overline{I}; \tilde{X})$ to (Q_n) with $u_n(0) = \phi_n$ for sufficiently large n, and $\lim u_n = u_\infty$ in $C(\overline{I}; W) \cap C^1(\overline{I}; \tilde{X})$.

For the proof we follow the method of [19], which reduces the theorem to an existence problem in the spaces $\underline{Y} = c(Y)$, $\underline{X} = c(X)$, the Banach spaces of convergent sequences in Y, X, respectively. To this end we first summarize basic results on operators in such spaces.

Given any Banach space X, let $\oplus X$ denote the algebraic sequence space consisting of vectors $(x_n) \equiv (x_1, x_2, \cdots, x_\infty)$, $x_n \in X$. $\underline{X} = c(X)$ is the subspace of $\oplus X$ consisting of (x_n) such that $x_\infty = \lim x_n$. $\underline{X}$ becomes a Banach space with norm $\|(x_n)\|_{\underline{X}} = \sup \|x_n\|_X$. (Unless otherwise stated, the subscript n always takes values $1, 2, \cdots, \infty$.)

Given Banach spaces X, Y, let $\{A_n\}$ be a sequence of linear operators from Y to X. We construct a "diagonal operator" (A_n) from $\oplus Y$ to $\oplus X$ by $(A_n)(y_n) = (A_n y_n)$, where $y_n \in D(A_n)$. In general (A_n) need not define an operator from $\underline{Y}$ to $\underline{X}$.

The following lemmas are easy to prove, if not trivial.

Lemma 5.1. If $Y \subset X$ continuously, then the same is true for $\underline{Y} \subset \underline{X}$.

Lemma 5.2. For each n, let D_n be a dense subset of X. Then the set $\underline{D}$ consisting of all $(x_n) \in \underline{X}$ with $x_n \in D_n$ is dense in $\underline{X}$.

Lemma 5.3. Let $B_n \in L(Y; X)$. In order that $(B_n) \in L(\underline{Y}; \underline{X})$, it is necessary and sufficient that $s\text{-}\lim B_n = B_\infty$ in $L(Y; X)$. If this condition is satisfied, then $\|(B_n)\|_{\underline{Y}; \underline{X}} = \sup \|B_n\|_{Y; X}$.

Lemma 5.4. Let $\underline{B}^j = (B_n^j) \in L(\underline{Y}; \underline{X})$, $j = 1, 2, \cdots$ (so that $s\text{-}\lim_n B_n^j = B_\infty^j$). In order that $s\text{-}\lim \underline{B}^j = \underline{B}$ exist in $L(\underline{Y}; \underline{X})$, it is necessary and sufficient that $s\text{-}\lim_j B_n^j$ exist in $L(Y; X)$ uniformly in n.

Lemma 5.5. Let $A_n \in G(X, M, \beta)$. In order that $(A_n) \in G(\underline{X}, M, \beta)$, it is necessary and sufficient that $s\text{-}\lim U_n(t) = U_\infty(t)$ uniformly for $t \in [0, 1]$, where U_n is the semigroup generated by $-A_n$. If this condition is satisfied, $-(A_n)$ generates the C_0-semigroup $(U_n(\cdot))$.

Remark. The condition in Lemma 5.5 is equivalent to the *strong resolvent convergence* of A_n to A_∞ (cf. e.g. [8,Theorem IX-2.16]). As is easily seen, it also implies that $[0, 1] \ni t \mapsto U_n(t) \in L(X)$ is strongly *equicontinuous*, so that $(U_n(t))$ forms a C_0-semigroup on $\underline{X}$ by Lemma 5.4. Finally we note that for the resolvent convergence, it is sufficient (but not necessary) that there is a core D for A_∞ such that $D \subset D(A_n)$ for all n and $\lim A_n \varphi = A_\infty \varphi$ for $\varphi \in D$ (cf.[8,Theorem VIII-1.5].)

Proof of Theorem III. Introducing the spaces $\underline{Y} = c(Y)$, $\underline{X} = c(X)$, $\underline{Z} = c(Z)$, etc. as above, we can write the system $\{(Q_n)\}$ as a single equation

$$(\underline{Q}) \qquad\qquad \partial_t \underline{u} + \underline{A}(t, \underline{u})\underline{u} = \underline{f}(t, \underline{u}),$$

with the unknown $\underline{u} = (u_n)$ and with $\underline{f}(t, \underline{u}) = (f_n(t, u_n))$, $\underline{A}(t, \underline{u}) = (A_n(t, u_n))$. We denote by $\underline{W}$ the set of $\underline{w} = (w_n) \in \underline{Y}$ such that $w_n \in W$; note that $\underline{W}$ is a bounded open set in $\underline{Y}$. It is easy to see that S induces a diagonal operator $\underline{S} = \oplus S \in L(\underline{Y}, \underline{X})$, and that the homomorphism Ξ of $L(X)$ into $L(Z)$ induces a homomorphism of the subalgebra $L(\underline{X})$, consisting of diagonal operators, into the corresponding subalgebra of $L(\underline{Z})$, which is nonexpansive and strongly continuous.

For this equation $(\underline{Q})$, we shall verify conditions (I) to (V). (5.1) implies that $A_n(t, u_n) \to A(t, u_\infty)$ in the sense of resolvent convergence; see the remark above and note that Y is a core for $A(t, u_\infty)$ by Lemma 1.1. It follows from Lemma 5.5 that $\underline{A}(t, \underline{u}) \in G(\underline{X}, M, \beta)$. The stability condition (I) then follows easily from the uniformity assumption. This verifies (I).

(II) follows from the corresponding conditions for A_n, again by the uniformity. Similarly (IV) follows from (5.2), (III) from (5.3), and (V) from (5.4).

In view of Theorem II, we have thus proved

Theorem III′. Given any $\phi = (\phi_n) \in \underline{W}$, $(\underline{Q})$ is solvable for $\underline{u} \in C([0, T']; \underline{Y}) \cap C^1([0, T']; \underline{X})$ for some $T' > 0$.

As is easily seen, Theorem III′ implies a *local* version of Theorem III. To prove the latter fully, we have only to use the standard argument of repeating the process a finite number of times; here it is essential to note that, according to Theorem II, T' can be chosen uniformly when the initial value ϕ_∞ varies over the set $\{u_\infty(t); t \in \bar{I}\}$, which is compact in Y.

6. A C^1-theory of first-order hyperbolic systems

As an application of the abstract theory, we consider the system of quasilinear partial differential equtions

$$\text{(HS)} \qquad \partial_t u + a_j(t, u)\partial_j u = f(t, u).$$

Here the unknown $u = u(t, x)$ is a function on $\mathbf{R} \times \mathbf{R}^m$ into $\mathbf{R}^N$; $\partial_t = \partial/\partial t$, $\partial_j = \partial x_j$; a_j are $N \times N$-matrix valued functions, and f is an $\mathbf{R}^N$-valued function, all defined on a ball $\Omega = B_\rho(\mathbf{R}^N)$. ($B_\rho(Z)$ denotes the open ball in a Banach space Z with center 0 and radius ρ.) Summation convention is used for $j = 1, \cdots, m$. For simplicity we assume that the a_j and f do not depend on x explicitly, but the general case can be handled in essentially the same way.

It is assumed that (HS) has the *Schauder canonical form*, which means that the a_j are simultaneously diagonalizable by a common nonsingular matrix function q, so that

$$a_j(t, u) = q(t, u)^{-1} a_j^0(t, u) q(t, u), \qquad j = 1, \cdots, m, \tag{6.1}$$

where a_j^0 are diagonal matrices. Moreover, it is assumed that

$$q, q^{-1}, a_j^0 \in C^1(\mathbf{R} \times \Omega; \mathbf{R}^{N \times N}), \quad j = 1, \cdots, m, \quad f \in C^1(\mathbf{R} \times \Omega; \mathbf{R}^N). \tag{6.2}$$

We want to solve (HS) for u in the class

$$u \in C([-T, T]; Y) \cap C^1([-T, T]; X), \qquad u(t, x) \in \Omega, \tag{6.3}$$
$$Y = BUC^1(\mathbf{R}^m; \mathbf{R}^N), \quad X = BUC(\mathbf{R}^m; \mathbf{R}^N),$$

with some $T > 0$, under a given initial condition

$$u(0) = \phi \in B_\rho(Y). \tag{6.4}$$

Remark. (HS) has automatically the Schauder canonical form if $N = 1$, or if $m = 1$ and $a_1(t, u)$ has distinct eigenvalues for all $t, u \in \mathbf{R} \times \Omega$.

To define the norms in X, Y, we first set $|\xi| = \max_j |\xi_j|$ for $\xi = (\xi_1, \cdots, \xi_N) \in \mathbf{R}^N$. If $g : \mathbf{R}^m \to \mathbf{R}^N$, we define its X-norm by

$$\|g\|_X = \sup\{|g(x)|; \, x \in \mathbf{R}^m\},$$

and Y-norm by

$$\|g\|_Y = \|g\|_X \vee \max\{\|\partial_k g\|_X; \, 1 \leqslant k \leqslant m\} \quad (\vee \text{ means max}).$$

The system (HS) can be written in the form (Q), in which A is formally given by

$$A(t, w) = a_j(t, w(t, x))\partial_j.$$

We shall apply to (HS) the abstract theory, with the Banach spaces Y, X given above. We set

$$\tilde{X} = X, \quad Z = X \oplus \cdots \oplus X, \quad \Xi Q = Q \oplus \cdots \oplus Q \quad (\text{m-fold direct sums}), \quad (6.5)$$
$$Sy = \partial y = (\partial_1 y, \cdots, \partial_m y) \in Z, \quad y \in Y;$$

S is a closed operator from X to Z with domain Y. We set $W = B_\sigma(Y)$, where $\|\phi\|_Y < \sigma < \rho$. It is obvious that $\tilde{W}$, the closure of W in X, is a subset of $\text{cl}(B_\sigma(X))$ (cl denotes the closure). We shall verify conditions (I) to (V) of section 4, assuming that $t \geqslant 0$ without loss of generality. For this purpose we need various properties of the operator $A(t, w)$ with $w(t) \in W$. The relevant results are proved in Appendix B, which will be used in the sequel.

If $w(t) \in W$, then Lemma B.3 shows that $q(t, w)A(t, w)q(t, w)^{-1} \in G(X, 1, \beta)$ with a constant $\beta < \infty$ (which depends only on σ). If $v \in C^1([0, T]; X)$, then the same is true of the matrix elements of $q(\cdot, v)$ and $q(\cdot, v)^{-1}$. Thus the family $A(\cdot, v(\cdot))$ is quasi-m-accretive with respect to a variable norm that depends on t Lipschitz continuously. Hence the family is stable with stability constants depending only on $\|v\|_Y$ (which is smaller than σ) and $\||\partial_t v\||_X$ (see [10,Prop.3.4]). This verifies condition (I).

Condition (II) is satisfied since $(A(t, w) - A(t, w'))y = (a_j(t, w) - a_j(t, w'))\partial_j y$ is dominated in X-norm by $c\|w - w'\|_X \|y\|_Y$.

The intertwining relation (III) is proved in Lemma B.4. Here $B(t, w, \partial w)$ is a "supermatrix operator" with matrix elements $a_{j,k} = \partial_k a_j(t, w) = (\partial a_j(t, w)/\partial w_i)\partial_k w_i$. B involves $\partial_k w$ only linearly, with coefficients depending continuously on $w \in \tilde{W} \subset \text{cl}(B_\sigma(X))$. Hence we may write $B = B(t, w, Sw)$, where $B(t, w, z) \in L(Z)$ is strongly continuous in t and $w \in \tilde{W}$, and is Lipschitz continuous in $z \in Z$. This verifies condition (III).

Condition (IV) is automatically true for our Z and Ξ (see remark for (IV), section 4).

(V) can be verified in the same way as the similar condition for B in (III) is handled.

Thus we have proved

Theorem 6.1. Given ϕ as in (6.4), there is $T > 0$ and a unique solution u to (HS) in the class (6.3). u depends continuously (in this class) on the data a_j, f, and ϕ (varying in their classes).

Remark. For $m = 1$, Theorem 6.1 was proved by Douglis [3] and Hartmen-Wintner [5]. Their proofs, depending on the estimates for moduli of continuity of ∂u, were rather complicated. For $m > 1$, it was solved by Cinquini-Cibrario [1], with no simpler proof. All these previous papers are local theories dealing with a bounded domain in space time, while we consider the space domain $\mathbf{R}^m$. But there is no essential difference, since local data can be extended to all $\mathbf{R}^m$ and, due to the finite propagation property, the local results can be recovered by restriction. Sanekata [20] solved (HS) for $m = 1$ as an application of his abstract theory, but unfortunately he had to assume that the $a_j(t, w)$ are in C^2 in w. His proof, like ours, avoids the use of moduli of continuity. The most delicate part in our method is the verification of the interwining relation (Lemma B.4).

Appendix A. A fixed point theorem

We prove a fixed point theorem in a Banach space Y (which is essentially hidden in [17]). A feature of this theorem is the use of two auxiliary Banach spaces X and Z. Assume that $Y \subset X$ algebraically and topologically, and that there is an operator $S \in L(Y; Z)$ such that

$$\|u\|_Y \leqslant c(\|u\|_X + \|Su\|_Z), \qquad u \in Y, \ c > 0. \tag{A.1}$$

Theorem A. Let E be a (nonempty) bounded closed subset of Y, $\tilde{E}$ its closure in X. Let Φ be a map of E into itself with the following properties.

(i) Φ is a contraction in X-norm, i.e. $\|\Phi(u) - \Phi(v)\|_X \leqslant \alpha \|u - v\|_X$, $u, v \in E$, where $\alpha < 1$ is a constant.

(ii) There is a closed set $F \subset Z$ containing SE, and a continuous function $\Psi : \tilde{E} \times F \times F \to F$ such that

$$S\Phi(u) = \Psi(u, Su, S\Phi(u)), \quad u \in E. \tag{A.2}$$
$$\|\Psi(u, v, w) - \Psi(u, v', w')\|_Z \leqslant \beta \|v - v'\|_Z + \gamma \|w - w'\|_Z, \tag{A.3}$$
$$u \in \tilde{E}, \quad v, v', w, w' \in F,$$

where $\beta + \gamma < 1$. Then Φ has a unique fixed point $\overline{u}$. $\overline{u}$ can be computed by standard iteration as the limit in Y-norm of a sequence $\{u^n\}$ defined by $u^{n+1} = \Phi(u^n)$, $n = 0, 1, \cdots$, starting with any $u^0 \in E$.

Remark. Condition (A.2) is implicit, inasmuch as $S\Phi(u)$ appears on the right too. It is not difficult to solve (A.2) for $S\Phi(u)$ to give it an explicit form, but it is convenient for applications to keep the implicit form.

Proof. Define a sequence $u^n \in E$ as described above. Since Φ is a contraction in X-norm, $\{u^n\}$ is a Cauchy sequence in X and has a limit $\overline{u} \in \tilde{E}$.

With this $\overline{u}$, we solve the equation

$$v = \Psi(\overline{u}, v, v). \tag{A.4}$$

This has a unique solution $v = \overline{v} \in Z$, since $\Psi(\overline{u}, v, v)$ is a contraction in v by (A.3).

We now show that $\lim Su^n = \overline{v}$ in Z as $n \to \infty$. Indeed,

$$\|Su^{n+1} - \overline{v}\|_Z = \|S\Phi(u^n) - \overline{v}\|_Z \leqslant \|\Psi(u^n, Su^n, Su^{n+1}) - \Psi(\overline{u}, \overline{v}, \overline{v})\|_Z$$
$$\leqslant \|\Psi(u^n, Su^n, Su^{n+1}) - \Psi(u^n, \overline{v}, \overline{v})\|_Z + \|\Psi(u^n, \overline{v}, \overline{v}) - \Psi(\overline{u}, \overline{v}, \overline{v})\|_Z$$
$$\leqslant \beta \|Su^n - \overline{v}\|_Z + \gamma \|Su^{n+1} - \overline{v}\|_Z + \varepsilon_n, \tag{A.5}$$

where $\lim \varepsilon_n = 0$ (recall that $\Psi(u, \overline{v}, \overline{v}) \in Z$ is continuous in $u \in \tilde{E}$ and $\lim u^n = \overline{u}$ in $\tilde{E}$). As is easily seen from the fact that $\beta + \gamma < 1$, (A.5) implies that $\lim \|Su^n - \overline{v}\|_Z = 0$.

Thus $\{Su^n\}$ forms a Cauchy sequence in Z, and (A.1) shows that $\{u^n\}$ is Cauchy in Y. Therefore $\overline{u} \in E$, and it is routine to show that $\overline{u}$ is a unique fixed point of Φ.

Appendix B. Semigroups generated by first-order linear systems

1. This section contains basic results on the semigroups generated by the operators $A(t, u)$ that appear in (HS). Here t and $u = u(t)$ are frozen, so that it suffices to consider a formal linear differential operator $\mathcal{A}$ in the Schauder canonical form:

$$\mathcal{A} = a_j(x)\partial_j, \quad a_j(x) = q(x)^{-1}a_j^0(x)q(x), \quad j = 1, \cdots, m, \tag{B.1}$$
$$a_j^0, q, q^{-1} \in BUC^1(\mathbf{R}^m; \mathbf{R}^{N \times N}),$$

where $a_j^0(x)$ are diagonal matrices. We denote by A the realization of $\mathcal{A}$ in $X = BUC(\mathbf{R}^m; \mathbf{R}^N)$. A is defined by $Au = \mathcal{A}u$ whenever $u \in X$ and $\mathcal{A}u \in X$; note that $\mathcal{A}u$ makes sense as a distribution for every $u \in X$, due to the property (B.1). Similarly we define the realization A_Y of $\mathcal{A}$ in $Y = BUC^1$.

In what follows a constant depending only on the BUC^1-norms of q and a_j^0 will be called a *constant of order* 1.

Lemma B.1. $A[A_Y]$ is a densely defined closed linear operator in $X[Y]$, with $\tilde{D} = BUC^\infty$ as a core (i.e. $A[A_Y]$ is the closure of its restriction with domain $\tilde{D}$).

Proof. This is essentially the well known theorem "weak = strong,"due to Friedrichs.

2. As a preliminary, we begin with the special case with $N = 1$; thus the a_j are scalar valued, and we may set $a_j^0 = a_j$, $q = 1$. In this case the semigroup $\{e^{-sA}\}$ can be constructed by the classical method. Define the characteristic function $x(s) = \kappa(s, \xi)$ as the solution of the system of ordinary differential equations

$$dx/ds = a(x), \quad x = (x_1, \cdots, x_m), \quad a = (a_1, \cdots, a_m). \tag{B.2}$$

with the initial condition $x(0) = \xi \in \mathbf{R}^m$. Due to condition (B.1), $\kappa(s, \xi)$ exists for all s, ξ, is C^1 in the two variables, and is BUC^1 in ξ. In particular we have $\kappa(0, \xi) = \xi$. Moreover,

$$|\partial\kappa(s, \xi)| \leqslant e^{\beta s}, \quad \beta = \sup_j \sum_k \|a_{j,k}\|, \quad a_{j,k} = \partial_k a_j \tag{B.3}$$

($|\ |$ denotes the matrix norm associated with the vector norm $|x| = \sup|x_j|$). β is a constant of order 1.

Lemma B.2. $A \in G(X, 1, 0)$ (m-accretive), $A_Y \in G(Y, 1, \beta)$ (quasi-m-accretive). e^{-sA} is explicitly given by

$$e^{-sA}\phi(x) = \phi(\kappa(-s, x)). \tag{B.4}$$

Proof. (B.4) is the classical formula for the solution of the initial value problem $\partial_s u + \mathcal{A}u = 0$ with $u(0) = \phi$. It is easy to see that if $\phi \in BCU^1$, then (B.4) belongs to $C(BCU^1) \cap C^1(BCU)$ and solves the problem. Since the right member of (B.4) has the same X-norm as ϕ, it follows that $A \in G(X, 1, 0)$. Moreover we have $A_Y \in G(Y)$, since $\phi \in Y$ implies $e^{-sA}\phi \in Y$ as remarked above. On applying ∂ to (B.4) and using (B.3), we see that $A_Y \in G(Y, 1, \beta)$.

3. Now we turn to the general case $N \geqslant 1$.

Lemma B.3. $A \in G(X, M, \beta_1)$, and $qAq^{-1} \in G(X, 1, \beta_1)$, where M and β_1 are constants of order 1.

Proof. Since $a_j = q^{-1}a_j^0 q$, we have

$$A = q^{-1}(A^0 + A')q, \quad A^0 = a_j^0 \partial_j, \quad A' = -a_j^0 q_{,j}q^{-1}, \quad q_{,j} = \partial_j q. \quad \text{(B.5)}$$

Here $A^0 \in G(X, 1, 0)$, being the direct sum of N scalar operators of the form considered above. A' is an operator of multiplication with a matrix function in BUC. Hence $A' \in L(X)$, and $qAq^{-1} = A^0 + A' \in G(X, 1, \beta_1)$ by a perturbation theorem for semigroups. Since A is similar to qAq^{-1}, we conclude that $A \in G(X, M, \beta_1)$.

Lemma B.4. Let $Z = X \oplus \cdots \oplus X$ (m-fold), $S = \partial \in L(Y; Z)$, $\Xi A = A \oplus \cdots \oplus A$ (see (1.6)). We have the intertwining relation

$$S(A + \lambda)^{-1} \supset (\hat{A} + \lambda)^{-1}S, \quad \hat{A} = \Xi A + B \in G(Z, M', \beta_2), \quad \lambda > \beta_2, \quad \text{(B.6)}$$

where M' and $\beta_2 \geqslant \beta_1$ are constants of order 1, and where $B \in L(Z)$ is a "supermatrix operator" given by

$$B(v_1 \oplus \cdots \oplus v_m) = w_1 \oplus \cdots \oplus w_m, \quad w_k = a_{j,k}v_j, \quad a_{j,k} = \partial_k a_j. \quad \text{(B.7)}$$

Proof. A straightforward calculation shows that condition (1.4) is satisfied with $\hat{A}$ given above and with $\tilde{D} = BUC^\infty \subset D(SA) \cap D(\hat{A}S)$. Note that $\hat{A} \in G(Z, M', \beta_2)$ for some $\beta_2 \geqslant \beta_1$, since it is obvious that $\Xi A \in G(Z, M, \beta_1)$ by Lemma B.3, and since $B \in L(Z)$. β_2 is a constant of order 1.

We first prove (B.6) under the additional assumption that $q \in BUC^2$. Then $q_{,j} \in BUC^1$, so that the proof of Lemma B.3 can be adapted to show that $A_Y \in G(Y)$. This implies that sufficiently large numbers belong to $\rho(-A_Y)$. But Lemma 1.2 shows that if $\lambda > \beta_2$, then $A_Y + \lambda$ is semi-Fredholm with nullity zero. Therefore, it follows by the standard argument (based on the stability of index) that $\lambda \in \rho(-A_Y)$ for $\lambda > \beta_2$. This implies that $(A_Y + \lambda)\tilde{D}$ is dense in Y for $\lambda > \beta_2$, since $\tilde{D}$ is a core for A_Y (Lemma B.1). According to Lemma 1.3, this proves (B.6) under the additional assumption.

In the general case, we fix $a_j^0 \in BUC^1$ and approximate $q \in BUC^1$ by a sequence $q_n \in BUC^2$ of invertible matrices in such a way that $q_n \to q$ in BUC^1, denoting the resulting operator by A_n. According to what was just proved, we have the intertwining relation

$$S(A_n + \lambda)^{-1} \supset (\hat{A}_n + \lambda)^{-1}S, \quad \lambda > \beta_3, \quad \text{(B.8)}$$

where $\hat{A}_n = \Xi A_n + B_n$ and B_n is obtained from (B.7) by replacing a_j by $q_n^{-1}a_j^0 q_n$. Note that β_3 can be chosen as a constant of order 1 independent of n, since the q_n are uniformly bounded in BUC^1. To prove (B.6), therefore, it suffices to prove the following lemma (recall that S is closed); here it is important that strong convergence suffices.

Lemma B.5. For $\lambda > \beta_3$ we have, as $n \to \infty$,

$$s\text{-}\lim(A_n + \lambda)^{-1} = (A + \lambda)^{-1} \quad \text{in} \quad L(X), \tag{B.9}$$

$$s\text{-}\lim(\hat{A}_n + \lambda)^{-1} = (\hat{A} + \lambda)^{-1} \quad \text{in} \quad L(Z). \tag{B.10}$$

Proof. We may forget that $q_n \in BUC^2$; we use only the relation $\lim q_n = q$ in BUC. According to a criterion for strong resolvent convergence (see [8,Theorem IIIV-1.5]), (B.9) follows from the fact that $A_n\varphi \to A\varphi$ in X for $\varphi \in \tilde{D}$, which is a core for A (Lemma B.1), since the resolvents in (B.9) are bounded uniformly in n. Similarly, for (B.10) it suffices to note that $\hat{A}_n\hat{\varphi} \to \hat{A}\hat{\varphi}$ in Z for $\hat{\varphi} \in \tilde{D} \oplus \cdots \oplus \tilde{D}$, which is obviously a core for ΞA, hence of $\hat{A}$ too.

Appendix C. A remark on variable S

In the text we have assumed that the operator $S \in L(Y; Z)$ is fixed. It is possible to remove this restriction, allowing S to depend on t in the linear theory, and on t and u in the quasilinear theory. Here we shall restrict ourselves to the linear case and indicate, without proof, how this can be done.

We replace S in (3.1) and (3.2) with a time dependent operator $S(t) \in L(Y; Z)$. A major problem is how $S(t)$ may depend on t. A reasonable assumption appears to be that $S \in C(I; L(Y; Z))$ with

$$\partial_t S(t) = C(t)S(t) + D(t), \quad C \in PC_*(I; L(Z)), \quad D \in PC_*(I; L(X; Z)), \tag{C.1}$$

which generalizes condition (6.4) of [10]. Under this assumption, a formal computation leads to the following relation, which generalizes (3.6).

$$S(t)U(t, s) \supset \hat{V}(t, s)S(s) + \hat{V}DU(t, s), \tag{C.2}$$

where $\hat{V} = \text{vol}(\hat{U}, C) = \text{vol}(\hat{U}, -B + C)$ is an $L(Z)$-chain and $\hat{V}DU$ is an $L(X; Z)$-kernel (see the remark at the end of section 2). As in Proposition 3.3, (C.2) shows that $U(t, s)$ maps Y into Y and in fact gives an evolution operator for the family $A(t)$.

A rigorous proof of (C.2) is rather tedious. A computation using the approximations U_n and $\hat{U}_n$, such as given in [9], appears to be necessary.

References

[1]. M. Cinquini Cibrario, *Ulteriori resultati per sistemi di equazioni quasi lineari a derivate párziali in piu' variabili independenti*, Rend. Istituto Lombardo, A, **103** (1969), 373-407.

[2]. J. R. Dorroh and R. A. Graff, *Integral equations in Banach spaces, a general approach to the linear Cauchy problem, and applications to nonlinear problem*, J. Integral Equations 1 (1979), 309-359.

[3]. A. Douglis, *Some existence theorems for hyperbolic systems of partial differential equations in two independent variables*, Comm. Pure Appl. Math. **5** (1952), 119-154.

[4]. J. A. Goldstein, *Semigroups of linear operators and applications*, Oxford University Press, 1985.

[5]. P. Hartman and A. Wintner, *On the hyperbolic partial differential equations*, Amer. J. Math. **74** (1952), 834-864.

[6]. T. J. R. Hughes, T. Kato, and J. E. Marsden, *Well-posed quasi-linear second-order hyperbolic systems with applications to nonlinear elastodynamics and general relativity*, Arch. Rational Mech. Anal. **63** (1977), 273-294.

[7]. S. Ishii, *Linear evolution equations $\partial u/\partial t + A(t)u = 0$: a case where $A(t)$ is strongly uniform-measurable*, J. Math. Soc. Japan **34** (1982), 413-424.

[8]. T. Kato, *Perturbation theory for linear operators*, Springer, 1966, 1984.

[9]. _____, *Integration of the equation of evolution in a Banach space*, J. Math. Soc. Japan **5** (1953), 208-234.

[10]. _____, *Linear evolution equations of "hyperbolic" type*, J. Fac. Sci. Univ. Tokyo **17** (1970), 241-258.

[11]. _____, _____ II, J. Math. Soc. Japan **25** (1973), 648-666.

[12]. _____, *Quasi-linear equations of evolution, with applications to partial differential equations*, Lecture Notes in Mathematics **448**, pp. 25-70, Springer, 1975.

[13]. _____, *Linear and quasi-linear equations of evolution of hyperbolic type, Hyperbolicity*, pp. 125-191, CIME, II Ciclo, Cortona, 1976.

[14]. _____, *Quasi-linear equations of evolution in nonreflexive Banach spaces, Nonlinear Partial Differential Equations in Applied Sciences*, Proc. U.S.-Japan Seminar, Tokyo, 1982, pp. 61-76, North-Holland, 1983.

[15]. _____, *Abstract differential equations and nonlinear mixed problems*, Lezioni Fermiane, Accademia Nazionale dei Lincei, Scuola Normale Superiore, 1985.

[16]. K. Kobayasi, *On a theorem for linear evolution equations of hyperbolic type*, J. Math. Soc. Japan **31** (1979), 647-654.

[17]. K. Kobayasi and N. Sanekata, *A method of iterations for quasi-linear evolution equations in nonreflexive Banach spaces*, Hiroshima Math. J. **19** (1989), 521-540.

[18]. A. Pazy, *Semigroups of linear operators and applications to partial differential equations*, Springer, 1983.

[19]. N. Sanekata, *Abstract quasi-linear equations of evolution in nonreflexive Banach spaces*, Hiroshima Math. J. **19** (1989), 109-139.

[20]. _____, *Abstract quasi-linear equations of evolution*, Thesis, Waseda University, 1989.

[21]. H. Tanabe, *Equations of evolution*, Pitman, 1971.

[22]. K. Yosida, *Functional analysis*, Springer, 1965, 1980.

Exactly solvable orbifold models and subfactors

Yasuyuki Kawahigashi*
Department of Mathematics
University of California, Berkeley, CA 94720

§1 Introduction

Since the pioneering work of V. F. R. Jones [J], the theory of subfactors has had deep and unexpected relations to 3-dimensional topology, conformal field theory, quantum groups, etc. Here we present a relation between paragroup theory of Ocneanu on subfactors and exactly solvable lattice models in statistical mechanics, and in particular, show usefulness of the notion of orbifold lattice models in subfactor theory.

Classification of subfactors of the approximately finite dimensional (AFD) factor of type II$_1$ is one of the most important and challenging problems in the theory of operator algebras. This is a study of inclusions of certain infinite dimensional simple algebras of bounded linear operators on a Hilbert space. The AFD type II$_1$ factor, the operator algebra we work on, is the most natural infinite dimensional analogue of finite dimensional matrix algebras $M_n(\mathbf{C})$ in a sense. (As a general reference on operator algebras, see [T], and as a basic reference on subfactor theory, we cite [GHJ]. All the basic notions are found there.) Here, by "a subfactor $N \subset M$" we mean that N and M are both AFD type II$_1$ factor and N is a subalgebra of M. V. Jones studied the Jones index [J], a real-valued invariant for the inclusion $N \subset M$, which measures the relative size of M with respect to N, roughly speaking.

A classification approach based on higher relative commutants were studied by several people such as Jones, Ocneanu, Pimsner and Popa. In this approach, the classification problem can be divided into the following 3 steps.

(1) Prove that the subfactor can be approximated by certain increasing sequence of finite dimensional algebras called higher relative commutants.

(2) Characterize the higher relative commutants in an axiomatic way.

(3) Work on the axioms to classify higher relative commutants.

Part (1) is quite functional analytic, but parts (2) and (3) are rather algebraic and combinatorial. We are concerned mainly with (2) and (3) in this paper.

V. Jones noticed that some graph called "principal graph" appears naturally from the higher relative commutants. A. Ocneanu [O1] first claimed that Step (1) is possible in the case with certain finiteness condition called "finite depth," which means that the principal graph is finite. As to Step (2), he further obtained a complete combinatorial characterization of higher relative commutants for the finite depth case, found a new algebraic structure, and named it *paragroup*. It was known that if the Jones

---

*Miller Research Fellow
Permanent address:Department of Mathematics, Faculty of Science
University of Tokyo, Hongo, Tokyo, 113, JAPAN

index is less than 4, then the principal graph must be one of the Coxeter-Dynkin diagrams A_n, D_n, E_6, E_7, E_8. Then Ocneanu announced a complete classification of these subfactors through his paragroup approach in [O1].

But unfortunately, details of Ocneanu's proofs have not appeared. As to Step (1) of the above, S. Popa published a complete proof [P2] with the finite depth condition and without assuming the so-called trivial relative commutant property $N' \cap M = \mathbf{C}$. Furthermore, he announced an ultimate result along this line [P3] and full details are soon appearing. Thus, we now have a satisfactory theory about Step (1), and in this paper, we work on steps (2) and (3).

On step (2), A. Ocneanu has the following four different viewpoint for paragroups.

(a) "Quantized" Galois groups
(b) "Discrete" version of compact manifolds
(c) Exactly solvable lattice models without a spectral parameter
(d) Finite tensor categories giving topological invariants of 3-manifolds

Though we mainly work on the third viewpoint (c), we make a quick explanation on each aspect. The classical Galois theory deals with inclusions of a subfield in another field. Here we work on inclusions of an (infinite dimensional non-commutative) algebra in another. Passing to non-commutative settings is often referred (vaguely) as quantization.

In paragroup theory, we work on finite graphs and some structure on it called "connections." This can be regarded as a discrete version of connections in differential geometry, and an analogue of flatness of connections will play a key role.

In solvable lattice model theory, they have a notion of IRF (interaction around faces) models. They have a graph and certain structure called Boltzmann weights on it. This Boltzmann weight depends on a parameter called a spectral parameter, and satisfies certain axioms. Axioms for paragroup are very similar to those of IRF models without a spectral parameter. Indeed, we have the following correspondence table.

Paragroups	IRF models
Connections	Boltzmann weights
Unitarity	First inversion relations
Commuting square conditions	Second inversion relations
Commuting square conditions for higher relative commutants	Crossing symmetry
Flatness of connections	Yang-Baxter equation *plus something* (?)

We show more details on this later. (See [Ba, DJMO, Ji] for more on lattice models.)

A. Ocneanu recently announced in [O4] that paragroups are in bijective correspondence to certain type of complex-valued topological invariants of 3-dimensional manifolds. This construction is a generalization of the Turaev-Viro invariant [TV].

In this paper, we first make a review of Ocneanu's paragroup theory, and then go into our idea of using orbifold models in paragroup settings. Roughly speaking, this idea is making a quotient of a paragroup for certain symmetry having a fixed point. In this way, we can solve some problems and give new constructions in subfactor theory. The notion of orbifold subfactors were studied through the joint work with D. E. Evans. The author thanks Professors D. Bisch, M. Choda, U. Haagerup, M. Izumi, V. F. R. Jones, T. Miwa, A. Ocneanu, M. Okado, S. Okamoto, S. Popa, H. Wenzl, and S. Yamagami for several useful comments and conversations during this work.

§2 Bijective correspondence between paragroups and finite depth subfactors

Although A. Ocneanu announced a combinatorial axiomatization of higher relative commutants as paragroups in [O1], full details of his proof have not appeared. But this characterization is now fully understood from his various lectures [O2, O3]. Because necessary arguments are scattered among the several literatures like [O1, O2, O3, P2, Ka] some of which are not widely available, here we give an exposition on how to obtain the bijective correspondence between paragroups and finite depth subfactors.

We follow a formal string algebra approach of [O2] rather than II_1 factor bimodule approach [O1, O3]. This has a disadvantage that meaning is less clear, but has an advantage that the method is entirely elementary and more general than the bimodule approach. (For example, formal approach based on connection also easily works for subfactors of Goodman-de la Harpe-Jones [GHJ]. See Remark 2.2.) For details of the II_1 factor bimodule approach, also see [Y].

Let $N \subset M$ be a subfactor of the AFD factor of type II_1 with finite index and finite depth. (In the rest of this paper, we mean by a subfactor only the above type of subfactors.) Repeating the downward basic construction, we choose a tunnel

$$\cdots N_3 \subset N_2 \subset N_1 \subset N \subset M$$

(See [J, GHJ] for basic definitions.) Consider the following canonical commuting squares in the sense of Popa.

$$
\begin{array}{ccccccc}
N \cap N' & \subset & N \cap N_1' & \subset & N \cap N_2' & \subset & N \cap N_3' \cdots \\
\cap & & \cap & & \cap & & \cap \\
M \cap N' & \subset & M \cap N_1' & \subset & M \cap N_2' & \subset & M \cap N_3' \cdots
\end{array}
$$

By Theorem of S. Popa [P2] (announced by A. Ocneanu), we may assume that each line above converges to N and M respectively by a certain choice of a tunnel. When we look at the Bratteli diagram of the sequence above, we find that there are four bipartite graphs $\mathcal{G}_1, \mathcal{G}_2, \mathcal{H}_1, \mathcal{H}_2$ connected as follows.

$$
\begin{array}{ccc}
\cdot & \xrightarrow{\;\mathcal{G}_1\;} & \cdot \\
{\scriptstyle \mathcal{G}_2}\downarrow & & \downarrow{\scriptstyle \mathcal{H}_1} \\
\cdot & \xrightarrow[\;\mathcal{H}_2\;]{} & \cdot
\end{array}
$$

That is, we consider the following sequence of graphs and there is a distinguished vertex at the upper left corner of the diagram above so that all the paths starting from $*$ in the following diagram gives the above-mentioned Bratteli diagram.

Here the notation ˜ means the reflection of the graph. Furthermore, $\mathcal{G}_1$ and $\mathcal{G}_2$ are the same (the principal graph), and $\mathcal{H}_1$ and $\mathcal{H}_2$ are the same (the dual principal graph). The ways $\mathcal{G}_1$ and $\mathcal{H}_1$ are connected and $\mathcal{G}_2$ and $\mathcal{H}_2$ are connected are the same. But the ways $\mathcal{G}_1$ and $\mathcal{G}_2$ are connected and $\mathcal{H}_1$ and $\mathcal{H}_2$ are connected may be non-trivial. This non-triviality is handled by the contragredient map in the sense of Ocneanu [O1]. (For these, see [P2, §6], for example.)

Because we assume the finite depth condition, the Bratteli diagram above starts to have period 2 after finite stages. Let k be the minimal even integer at which the commuting square

$$
\begin{array}{ccc}
N \cap N_k' & \subset & N \cap N_{k+1}' \\
\cap & & \cap \\
M \cap N_k' & \subset & M \cap N_{k+1}'
\end{array}
$$

starts to be periodic. If we have inclusions of four finite dimensional C^*-algebras, this diagram is described by string algebras and connections as in [O1, O2, O3]. (String algebra description of an increasing system of finite dimensional C^*-algebra was introduced in [E1, E2] with the name of path algebra.) A full proof of this was given in [O2], and only elementary linear algebra is needed for the proof. Here by "connection", we mean an assignment of complex numbers to each square of the following form.

$$
\begin{array}{ccc}
a & \xrightarrow{\ \xi_1\ } & b \\
\xi_2 \downarrow & & \downarrow \xi_3 \\
c & \xrightarrow[\ \xi_4\ ]{} & d
\end{array}
$$

where $\xi_1, \xi_2, \xi_3, \xi_4$ are edges of $\mathcal{G}_1, \mathcal{G}_2, \mathcal{H}_1, \mathcal{H}_2$ respectively, and a, b, c, d are vertices of the appropriate graphs. (Remember that each graph is bipartite and we identify even vertices of $\mathcal{G}_1$ and $\mathcal{G}_2$, and so on.) The connection is not uniquely determined from the four finite dimensional algebras, but it is unique up to certain equivalence relation discussed in [O1, O2]. Equivalent connections differ by so-called gauge choice freedom. This setting is quite similar to that of exactly solvable IRF (interaction around faces) lattice models [Ba, DJMO]. Usually in their setting, our four graphs are the same single one, which is sometimes oriented. They call above type of squares arising from the graph admissible, and their notion of Boltzmann weight corresponds to our notion of connection. (Further similarity between IRF models and paragroups will be discussed later, but we do not have a parameter corresponding to the spectral parameter in IRF models.) Rougly speaking, the connection gives identification of two basis on the algebra $M \cap N_{k+1}'$ arising from the two inclusions $N \cap N_k' \subset N \cap N_{k+1}' \subset M \cap N_{k+1}'$ and $N \cap N_k' \subset M \cap N_k' \subset M \cap N_{k+1}'$. (Each inclusion gives an expression of a string algebra.) Because this identification must be a $*$-isomorphism, we have the following unitarity axiom.

$$\sum_{b,\xi_2,\xi_3} \xi_1 \downarrow \begin{array}{ccc} a \xrightarrow{\xi_2} b & b \xleftarrow{\xi_2} a \\ & \downarrow \xi_3 \cdot \xi_3 & \downarrow \eta_1 \\ c \xrightarrow{\xi_4} d & d \xleftarrow{\eta_4} c' \end{array} = \delta_{\xi_1,\eta_1}\delta_{\xi_4,\eta_4}\delta_{c,c'},$$

$$\sum_{c,\xi_1,\xi_4} \xi_1 \downarrow \begin{array}{ccc} a \xrightarrow{\xi_2} b & b' \xleftarrow{\eta_2} a \\ & \downarrow \xi_3 \cdot \eta_3 & \downarrow \xi_1 \\ c \xrightarrow{\xi_4} d & d \xleftarrow{\xi_4} c \end{array} = \delta_{\xi_2,\eta_2}\delta_{\xi_3,\eta_3}\delta_{b,b'},$$

Here we used the following conventions.

$$\begin{array}{cc} a \xrightarrow{\xi_2} b \\ \xi_1 \downarrow \quad \downarrow \xi_3 \\ c \xrightarrow{\xi_4} d \end{array} = \begin{array}{cc} \overline{b \xleftarrow{\xi_2} a} \\ \xi_3 \downarrow \quad \downarrow \xi_1 \\ d \xleftarrow{\xi_4} c \end{array} \quad \begin{array}{cc} \overline{c \xrightarrow{\xi_4} d} \\ \xi_1 \uparrow \quad \uparrow \xi_3 \\ a \xrightarrow{\xi_2} b \end{array} \quad \begin{array}{cc} d \xleftarrow{\xi_4} c \\ \xi_3 \uparrow \quad \uparrow \xi_1 \\ b \xleftarrow{\xi_2} a \end{array}$$

In IRF model theory, they call the corresponding axiom unitarity or the first inversion relations [Ba].

Now our four algebras give not just inclusions but a commuting square. This gives a certain condition on the connection. Because we can write down the trace and the conditional expectation explicitly as in [O1, O2, O3], we can write down the commuting square condition in terms of the connection, and by comparing the coefficients in the both hand sides, we get the following.

$$\sum_{d,\xi_3,\xi_4} \sqrt{\frac{\mu(a)\mu(d)}{\mu(b)\mu(c)}} \xi_2 \downarrow \begin{array}{cc} \overline{a \xrightarrow{\xi_1} b} \\ \\ c \xrightarrow{\xi_4} d \end{array} \sqrt{\frac{\mu(a')\mu(d)}{\mu(b)\mu(c)}} \eta_2 \downarrow \begin{array}{cc} a \xrightarrow{\eta_1} b \\ \downarrow \xi_3 \\ c \xrightarrow{\xi_4} d \end{array} = \delta_{a,a'}\delta_{\xi_1,\eta_1}\delta_{\xi_2,\eta_2}.$$

Here the notation $\mu(\cdot)$ means the entry of the Perron-Frobenius eigenvector. A proof for this is given in [O2, Sc] and quite elementary. This formula corresponds to the second inversion relations in the IRF models [Ba], so we also call this the second inversion relations. Then for our admissible squares, we *define* the following values for other types of squares.

$$
\begin{array}{ccc}
d \xrightarrow{\;\tilde{\xi}_4\;} c & \qquad & a \xrightarrow{\;\xi_1\;} b \\
\tilde{\xi}_3 \downarrow \qquad \downarrow \tilde{\xi}_2 = \xi_2 \downarrow & & \downarrow \xi_3 \\
b \xrightarrow[\;\tilde{\xi}_1\;]{} a & \qquad & c \xrightarrow[\;\xi_4\;]{} d
\end{array}
$$

$$
\begin{array}{cccc}
c \xrightarrow{\;\xi_4\;} d & b \xrightarrow{\;\tilde{\xi}_1\;} a & & \overline{a \xrightarrow{\;\xi_1\;} b} \\
\tilde{\xi}_2 \downarrow \qquad \downarrow \tilde{\xi}_3 = \xi_3 \downarrow & \downarrow \xi_2 = \sqrt{\dfrac{\mu(a)\mu(d)}{\mu(b)\mu(c)}}\, \xi_2 \downarrow & & \downarrow \xi_3 \\
a \xrightarrow[\;\xi_1\;]{} b \qquad d \xrightarrow[\;\tilde{\xi}_4\;]{} c & & & c \xrightarrow[\;\tilde{\xi}_4\;]{} d
\end{array}
$$

where the notation $\tilde{\xi}_j$ means the edge with its orientation reversed. Then by the second inversion relation, the above also satisfy unitarity. The definitions above are an analogue of crossing symmetry in IRF model theory [Ba]. So we also call these crossing symmetry. We call the above assignment of complex numbers to all the admissible squares (biunitary) connection. (The name "biunitary" connection of Ocneanu comes from that there are two kinds of unitarity: first and second inversion relations.) In this way, we obtain a connection from a subfactor. In short, commuting squares arising as the higher relative commutants correspond to crossing symmetry and more general commuting squares correspond to the second inversion relations in IRF model theory. Note that the renormalization convention for crossing symmetry here is slightly different from that in [O1] and the same as in [O3, EK, IK, Ka].

A connection arising from a subfactor satisfies another important axiom called flatness. In order to explain it, we first show how to construct a subfactor from a biunitary connection on two graphs $\mathcal{G}, \mathcal{H}$ with the same Perron-Frobenius eigenvalues.

We have the distinguished point $*$ among the even vertices of $\mathcal{G}$, which corresponds to the starting vertex of the Bratteli diagram for $N_k' \cap N$. Then, we can construct a double sequence of string algebras starting from $*$:

$$
\begin{array}{ccccccc}
A_{0,0} & \subset & A_{0,1} & \subset & \cdots & \to & A_{0,\infty} \\
\cap & & \cap & & & & \cap \\
A_{1,0} & \subset & A_{1,1} & \subset & \cdots & \to & A_{1,\infty} \\
\cap & & \cap & & & & \cap \\
A_{2,0} & \subset & A_{2,1} & \subset & \cdots & \to & A_{2,\infty} \\
\vdots & & \vdots & & & & \vdots
\end{array}
$$

Identification for different expressions of strings are again given by the connection. See [O1, page 128] or [O3, II.1–2] for more details. A trace compatible with the embeddings above can be defined with the Perron-Frobenius eigenvector entries, and $A_{0,\infty}$, etc., are the GNS-completions with respect to this trace. (See [O1, page 129] or [O3, II.1] for this trace.) Then the inclusion $A_{0,\infty} \subset A_{1,\infty}$ is the string model subfactor of Ocneanu. We claim that if the connection arises from a subfactor $N \subset M$ in the above way, this subfactor $A_{0,\infty} \subset A_{1,\infty}$ is conjugate to $N \subset M$.

This is proved as follows. First note that we have a sequence of the Jones projections $\{e_{-n}\}$ with $N_n = N_{n-1} \cap \{e_{-n+1}\}'$. We identify the increasing sequence of the string algebra

$$A_{0,0} \subset A_{0,1} \subset \cdots \subset A_{0,k} \subset A_{0,k+1} \subset A_{1,k+1}$$

with the sequence

$$N \cap N' \subset N \cap N_1' \subset \cdots \subset N \cap N_k' \subset N \cap N_{k+1}' \subset M \cap N_{k+1}',$$

where k is as in the definition of the connection. Next we make identification of the following two sequences

$$A_{0,k+1} \subset A_{0,k+2} \subset A_{0,k+3} \subset \cdots$$

$$N \cap N_{k+1}' \subset N \cap N_{k+2}' \subset N \cap N_{k+3}' \subset \cdots$$

so that the Jones projections $\{e_{-n}\}$, $n \geq k+1$, have the expressions

$$e_{-n} = \sum_{\substack{|\alpha|=n-1 \\ |v|=|w|=1}} \frac{\mu(r(v))^{1/2} \mu(r(w))^{1/2}}{\beta \mu(r(\alpha))} (\alpha \cdot v \cdot \tilde{v}, \alpha \cdot w \cdot \tilde{w}),$$

where β is the Perron-Frobenius eigenvalue of the graph (which is equal to the square root of $[M:N]$), α is any horizontal path from $*$ of $\mathcal{G}$, and v, w are chosen so that the compositions are possible, and $|\cdot|$ denote the length of a path. This is possible because the expressions above satisfy all the required properties for the Jones projections. The expressions above for the Jones projections are due to Ocneanu [O3, II.3] and Sunder [Su]. We also identify two sequences

$$A_{1,k+1} \subset A_{1,k+2} \subset A_{1,k+3} \subset \cdots$$

$$M \cap N_{k+1}' \subset M \cap N_{k+2}' \subset M \cap N_{k+3}' \subset \cdots$$

so that the Jones projections have the same expressions for the other graph $\mathcal{H}$. Then we claim that these two identifications are compatible with the connection. To see this, it is enough to check the Jones projections on the first horizontal line are transformed to the Jones projections on the second horizontal line by the connection, but this is valid as in [O3], which is the same as flatness of the Jones projections implied by the crossing symmetry. In this way, we get an isomorphism of $A_{1,\infty}$ onto $\vee_n(M \cap N_n')$ which carries $A_{0,\infty}$ onto $\vee_n(N \cap N_n')$. Because we assumed that the tunnel has the generating property, we are done.

Now we work on the double sequence of string algebras arising from a biunitary connection which may not come from a subfactor. We define the vertical Jones projections $\{e_n\}$ with $e_n \in A_{n-1,0}$ with the same expression as above for the vertical string algebra of $\mathcal{G}$ from $*$. Then the crossing symmetry again implies the flatness of the vertical Jones projections and we can conclude that

$$A_{0,\infty} \subset A_{1,\infty} \subset A_{2,\infty} \subset A_{3,\infty} \subset \cdots$$

is the Jones tower of the inclusion $A_{0,\infty} \in A_{1,\infty}$ by [PP, Proposition 1.2].

One of the main problems in subfactor theory is a computation of higher relative commutants $N' \cap M_n$ for a given construction of a subfactor $N \subset M$. Suppose we have a subfactor $N = A_{0,\infty} \subset M = A_{1,\infty}$ as above constructed from a double sequence of string algebras with a connection on finite graphs which may not come from a subfactor. Then Ocneanu found that $N' \cap M_n$ is always a subalgebra of $A_{n+1,0}$ and gave a nice combinatorial characterization for strings in $A_{n+1,0}$ to be in $N' \cap M_n$ with his compactness argument [O3, II.6]. In many natural cases, we have equality $N' \cap M_n = A_{n+1,0}$. Because we have two graphs $\mathcal{G}$ and $\mathcal{H}$, we can repeat the same construction with $\mathcal{G}$ and $\mathcal{H}$ interchanged. If we have equalities $N' \cap M_n = A_{n+1,0}$ in the both cases, we say that the connection is flat. We now claim that a connection arising from a subfactor $N \subset M$ is flat.

We already know that $A_{0,\infty} = N$, $A_{1,\infty} = M$, and $N' \cap M_n \subset A_{n+1,0}$. But both $N' \cap M_n$ and $A_{n+1,0}$ are identified with the string algebra with length $n+1$ of $\mathcal{G}$ starting from $*$, so these two algebras have the same dimensions. With the inclusion above, we get $N' \cap M_n = A_{n+1,0}$. Next we define $A_{1,-1} = \mathbf{C}$ and regard the sequence

$$A_{1,-1} \subset A_{1,0} \subset A_{1,1} \subset A_{1,2} \subset \cdots$$

as an increasing sequence of string algebras of $\mathcal{H}$ starting from $*$ of $\mathcal{H}$. This is possible because we have the same number of edges from $*$ of $\mathcal{G}$ and $*$ of $\mathcal{H}$ and each pair of vertices of $\mathcal{G}$ and $\mathcal{H}$ is identified. Furthermore, we construct a vertical increasing sequence of string algebras

$$A_{1,-1} \subset A_{2,-1} \subset A_{3,-1} \subset A_{4,-1} \subset \cdots$$

as an increasing sequence of string algebras of $\mathcal{H}$ starting from $*$ of $\mathcal{H}$. Then we embed $A_{n,-1}$ into $A_{n,0}$ using the graph $\mathcal{G}$ or $\mathcal{H}$ according to parity of n. In this way, we extend the double sequence to the following form.

$$
\begin{array}{ccccccc}
A_{0,0} & \subset & A_{0,1} & \subset & \cdots & \to & A_{0,\infty} \\
\cap & & \cap & & & & \cap \\
A_{1,-1} & \subset & A_{1,0} & \subset & A_{1,1} & \subset \cdots \to & A_{1,\infty} \\
\cap & & \cap & & \cap & & \cap \\
A_{2,-1} & \subset & A_{2,0} & \subset & A_{2,1} & \subset \cdots \to & A_{2,\infty} \\
\vdots & & \vdots & & \vdots & & \vdots
\end{array}
$$

Now the same kind of argument as above gives $M' \cap M_n = A_{n+1,-1}$, which means flatness as desired.

By the above, we know that if we make a flat connection from a subfactor and then make a subfactor from the flat connection, we get the same subfactor back. Conversely, we will prove that if we make a subfactor from a flat connection and then make a flat connection from the subfator, we get the same flat connection back. So suppose we have a double sequence of string algebras arising from a flat connection. By similar method to the above, we can extend the double sequence to the following form.

$$
\begin{array}{ccccccc}
 & & & & \cdots & & \vdots \\
 & & & & \cdots & & \vdots \\
 & & & A_{-2,2} & \cdots & \rightarrow & A_{-2,\infty} \\
 & & & \cap & & & \cap \\
 & & A_{-1,1} & \subset\; A_{-1,2} & \cdots & \rightarrow & A_{-1,\infty} \\
 & & \cap & \cap & & & \cap \\
A_{0,0} & \subset & A_{0,1} & \subset\; A_{0,2} & \cdots & \rightarrow & A_{0,\infty} \\
\cap & & \cap & \cap & & & \cap \\
A_{1,0} & \subset & A_{1,1} & \subset\; A_{1,2} & \cdots & \rightarrow & A_{1,\infty} \\
\cap & & \cap & \cap & & & \cap \\
A_{2,0} & \subset & A_{2,1} & \subset\; A_{2,2} & \cdots & \rightarrow & A_{2,\infty} \\
\vdots & & \vdots & \vdots & & & \vdots
\end{array}
$$

That is, we have $A_{m,n}$, for all m,n with $-m \le n, 0 \le n$. By the same kind of argument as above, we can prove that

$$
A_{0,\infty} \supset A_{-1,\infty} \supset A_{-2,\infty} \supset \cdots
$$

is a tunnel. Flatness now implies that $A_{0,n} = A'_{-n,\infty} \cap A_{0,\infty}$ and $A_{1,n} = A'_{-n,\infty} \cap A_{1,\infty}$. In this way, we get the conclusion. (By this arguments, we also proved that the tunnel above has the generating property.)

Note that identification of flat connections are given by gauge choice freedom and graph isomorphism as in [O1, O2, O3]. Thus, we have established the bijective correspondence between conjugacy classes of subfactors and equivalence classes of flat connections. Flatness does not have a direct analogue in IRF model theory unlike the first and second inversion relations, but in some "good" cases, the flatness is obtained from the Yang-Baxter equation in the IRF model theory. This will be discussed in §4 in connection with Hecke algebra subfactors of Wenzl.

The next topic is how to check flatness for a given biunitary connection. We work only on the equality $N' \cap M_n = A_{n+1,0}$ because the other is handled in the same way. First note that this can be stated as $xy = yx$ for $x \in A_{n,0}$, $y \in A_{0,m}$. This form is conceptually simple. But the difficulty is that we have infinitely many strings for which we have to check commutativity. Because we assume that the both graphs $\mathcal{G}, \mathcal{H}$ are finite, the increasing sequence of the string algebra is generated by the Jones projections and some finitely many elements. Flatness is automatically satisfied for the Jones projections, so it is enough to check commutativity only for finitely many elements. In this way, checking flatness is reduced to finitely many computations. Furthermore, the following theorem gives computational methods to check flatness.

Theorem 2.1. *The following conditions are equivalent.*

(1) *In the double sequence of string algebras, any two elements* $x \in A_{\infty,0}$, *the vertical string algebra, and* $y \in A_{0,\infty}$, *the horizontal string algebra, commute.*

(2) For each vertical string $\rho = (\rho_+, \rho_-) \in A_{k,0}$, we get

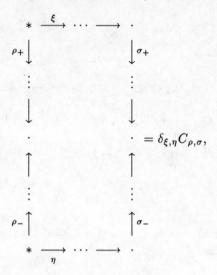

$$= \delta_{\xi, \eta} C_{\rho, \sigma},$$

where $C_{\rho, \sigma} \in \mathbf{C}$ depends only on $\rho, \sigma = (\sigma_+, \sigma_-)$.

(3) For any horizontal paths ξ_+, ξ_- and vertical paths η_+, η_- with all the sources and ranges equal to $*$, we get

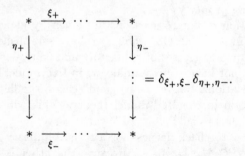

$$= \delta_{\xi_+, \xi_-} \delta_{\eta_+, \eta_-}.$$

In the theorem above, the meaning of large diagrams are as follows. We consider all the possible ways of filling the large diagram with admissible squares. Each such a choice is called a configuration. We multiply the connection values of all the admissible squares in each configuration and sum them over all the configurations. This is the value assigned to the large diagram above, and we mean this value by the diagram. This is an analogue of a partition function in the IRF model theory.

Condition (3) above was used as definition of flatness by Ocneanu [O1], and equivalence to (2) was claimed in [O1] without proof. The name "flat connection" comes from a fact that these conditions are analogues of conditions for flat connections in differential geometry. A proof for the theorem above was given in [Ka]. Again, in order to check flatness for a given connection, we need to verify (2) or (3) only for finitely many diagrams. Biunitarity is a local axiom in the sense that we can check it just by looking at a part of the graph for each equality, but this flatness axiom is a global axiom, and in general, very hard to verify. For example, in the case of index less than 4, the principal graph must be one of the Dynkin diagrams of type A, D, E and the dual principal graph must be the same. Then it is not so difficult to get all the connections on them [O1, O3, Ka]. That is, there is a unique connection on each of A_n, D_n, and there are two connections on each of E_6, E_7, E_8. So classification of subfactors with index less that 4 is reduced to flatness problem of these connections, which requires deeper considerations.

Remark 2.2. A series of subfactors was constructed by [GHJ] for the Ćoxeter-Dynkin diagrams, and S. Okamoto [Ok] computed the principal graphs for them. This result follows very easily from the above general settings of a double sequence of string algebras with a connection which does not come from a subfactor. By [GHJ], we know that the subfactor arises from the double sequence of string algebras with a biunitary connection on the following type of diagram.

$$
\begin{array}{ccc}
* & \xrightarrow{\quad A_n \quad} & \cdot \\
G_1 \downarrow & & \downarrow G_2 \\
\cdot & \xrightarrow{\text{one of } A,D,E} & \cdot
\end{array}
$$

where G_1 and G_2 are some graphs with the same Perron-Frobenius eigenvalues, which can be computed easily. Then it is easy to see the method above of computation of the principal graph also works for this connection. The horizontal string algebra is now generated by the Jones projections, so flatness of this connection immediately follows from the flatness of the Jones projections. (That is, each vertical string commutes with each horizontal string.) Thus, the vertical graph G_1 is the principal graph of the subfactor. (Both Ocneanu and Okamoto said to the author that they had known this simple method.)

It is also possible to compute the connection explicitly. For example, [R] has a table of connection values for the subfactor with index $3 + \sqrt{3}$ arising from E_6 and [DZ] has more. (Roche calls it a cell system and used it for a different purpose.)

The graph G_2 in the diagram above is not the "dual" principal graph in general, and it is very difficult to compute the dual principal graph from Ocneanu's general machinery. But in some cases, purely combinatorial arguments determine the dual principal graph. For example, in the case with index $3 + \sqrt{3}$, the dual principal graph must be the same as G_1, Okamoto's graph [Ok], as conjectured at the end of [Ok]. (But Haagerup [H] proved a much stronger result. That is, if a finite depth subfactor has index $3 + \sqrt{3}$, then its principal graph must be Okamoto's.)

If we choose the graph A_n from A-D-E's in the construction above, then we get the index values $(\sin^2 k\pi/N)/(\sin^2 \pi/N)$, same as Wenzl's index values for his Hecke algebra subfactors [We]. But Ocneanu's machinery tells that the construction here gives a subfactor arising from basic construction of subfactors of type A. This corresponds to Wenzl's remark in [We, page 360] and is different from his Hecke algebra subfactors. Principal graphs of these subfactors were computed by Izumi [I1, Figures 5, 7] and principal graphs of the Hecke algebra subfactors were computed by unpublished work of Wenzl and [EK].

For the diagrams D_{2n}, E_6, E_8, for which flat connections exist, each diagram has the distinguished vertex $*$ which is characterized as having the smallest entry of the Perron-Frobenius eigenvector. If we choose a vertex different from $*$ in the construction above, then the resulting subfactor contains a non-trivial intermediate subfactor corresponding to the string algebra starting from $*$. In this sense, it is not a "simple" object.

§3 Orbifold construction I: $*$ is fixed.

Now we come to the main topic of this paper: orbifold subfactors. This is an analogue of orbifold models in IRF model theory [Ko, FG] and the idea is quite simple. That is, if we have a certain symmetry of a paragroup, we can make a quotient of the paragroup by the symmetry and produce a new paragroup. In operator algebraic setting, this means that we take a fixed point algebra of the string algebra by the automorphism with finite order arising from the paragroup symmetry. Recall that in C^*-algebra theory, an idea of non-commutative orbifold has been also successful [BEEK1–3, BEK, BK].

First we point out that we have to consider two type of symmetries of a paragroup separately. The first case is symmetries fixing the distinguished vertex $*$ of the paragroup, and the second is symmetries moving $*$. Though the second case is more interesting, the first case is easier, so we work on the first case in this section. This type of symmetry is also related to Loi's work [L1], [L2, Lemma 4.2].

In order to keep arguments simple, we assume that the two graphs of the paragroup have no multiple edges and let σ be a non-trivial automorphism of the graphs which keeps the connection invariant:

$$
\begin{array}{ccc}
a \longrightarrow b & \sigma(a) \longrightarrow \sigma(b) \\
\downarrow \qquad \downarrow = \downarrow & \qquad \downarrow \\
c \longrightarrow d & \sigma(c) \longrightarrow \sigma(d)
\end{array}.
$$

(Because we have no multiple edges, we drop labeling of edges.)

Then this σ defines an automorphism of the string algebra, and this is well-defined on the double sequence of string algebras because of the invariance of the connection. We denote this automorphism by σ again and take fixed point algebra $A_{m,n}^\sigma$ of each $A_{m,n}$ in the double sequence. Then $A_{m,n}^\sigma$ gives a double sequence of commuting squares, and the Jones projections are invariant under σ, so we have another double sequence with each horizontal line defining the Jones tower. Next we would like to check flatness of this system, but this is automatically satisfied because each algebra in the double sequence in a subalgebra of an algebra in the original double sequence where we have flatness and then commutativity is trivial. In this way, we get another subfactor with the same index and a different principal graph. (It is easy to see that the index is value is kept in this procedure because the Jones projections are invariant under the symmetry. We can also appeal to [GHJ] or [Wa, page 227].) The new principal graph is an "orbifold" graph of the original principal graph. (M. Choda worked on this kind of graph problem in more abstract settings in [Ch].)

A simple example is as follows. An abelian group $\mathbf{Z}_n$ is realized as a paragroup as follows. The two graphs $\mathcal{G}$ and $\mathcal{H}$ are the same and it has a single odd vertex x and n single edges with length 1 from x. Each even vertex of $\mathcal{G}$ corresponds to an element of

Z_n and each of $\mathcal{H}$ to an element of $\hat{Z}_n$. The distinguished vertex $*$ corresponds to 0 in Z_n and $\hat{Z}_n$. The connection is given by

$$
\begin{array}{ccc}
i & \longrightarrow & x \\
\downarrow & & \downarrow \\
x & \longrightarrow & j
\end{array} = \exp(2\pi\sqrt{-1}ij/n).
$$

(The other types of admissible squares have the values determined by the crossing symmetry.) Choose a positive integer p which is relatively prime to n and another positive integer q such that $pq \equiv 1 \bmod n$. We define a paragroup symmetry σ by $\sigma(i) = pi \bmod n$ for $i \in Z_n$ and $\sigma(j) = qj \bmod n$ for $j \in \hat{Z}_n$. Then the connection above is invariant under this σ. We then have flatness automatically, so we get many examples of new principal graphs with integer indices very easily.

Next example of a symmetry fixing $*$ is a $\mathbf{Z}_2$-symmetry of $A_{\text{odd}}^{(1)}$ considered in [IK]. This is related to classification of subfactors with index 4. We label vertices of $A_{2n-5}^{(1)}$ as follows.

$$
A_{2n-5}^{(1)} : \qquad a_0 \Big\langle \begin{array}{l} a_1 - a_2 \cdots a_{n-4} - a_{n-3} \\ a_1' - a_2' \cdots a_{n-4}' - a_{n-3}' \end{array} \Big\rangle a_{n-2}
$$

The symmetry σ fixed a_0 and a_{n-2}, and switches a_i and a_i', $1 \le i \le n-3$. There are $n-2$ mutually non-equivalent flat connections on this diagram, and they correspond to locally trivial subfactors with index 4. (So this classification also follows from Connes' classification of automorphisms [Co].) Then we can prove that the connection is chosen so that it is invariant under σ, then we can get orbifold subfactors from them. Our new principal graph is the following $D_n^{(1)}$.

$$
D_n^{(1)} : \qquad \begin{array}{c} b_0 \\ \\ b_0' \end{array} \Big\rangle \begin{array}{c} \\ b_1 - b_2 \cdots b_{n-4} - b_{n-3} \\ \\ \end{array} \Big\langle \begin{array}{c} b_{n-2} \\ \\ b_{n-2}' \end{array}
$$

For classification of subfactors with the principal graph $D_n^{(1)}$, we also have to prove that all of these $n-2$ flat connections are really non-equivalent, and there are nothing more. But these can be proved with the use of the intertwining Yang-Baxter equation for the two graphs as in [IK].

Remark 3.1. S. Popa [P3] announced a complete classification of subfactors with index 4. He gave numbers of conjugacy classes for each possible principal graphs, including infinite graphs $A_\infty, A_{\infty,\infty}, D_\infty$, but the number was left open only for $D_n^{(1)}$. (His classification of subfactors with principal graph $D_n^{(1)}$ was given in terms of certain 3rd cohomology group elements and it was hard to compute the number of conjugacy classes.) On the other hand, Ocneanu showed a classification table of subfactors with index 4 in his invited talk at the ICM-90, and in it, he claimed that there is a unique subfactor with a principal graph $D_n^{(1)}$. But our method as above easily shows that the right number is $n-2$ and Ocneanu's announced uniqueness is invalid, and gives the last missing number in Popa's list.

§4 Orbifold construction II: $*$ is moved.

Now we work on more interesting orbifold constructions. These are in duality to the orbifold constructions in §3. First we consider subfactors with principal graph D_n. Ocneanu announced in 1987 that there is a unique subfactor for each D_{2n} and there are no subfactors with the principal graph D_{2n+1} among his announced classification of subfactors with index less than 4 [O1]. But he has not published his proof. (See Remark 4.2 below.) Here we show that our orbifold method produces a proof of the claim above on D_n as in [Ka].

First note that the graph A_{2n-3} has a $\mathbf{Z}_2$-symmetry σ. It can be proved that a connection on A_{2n-3} can be chosen so that it is invariant under σ. (It is possible to choose a connection to be real, but then this is not a right choice because it is not invariant under σ.) Now we want to apply the same procedure as in §3, but we cannot do so immediately, because this σ does move the distinguished vertex $*$. That is, we cannot apply σ even at the starting algebra $A_{0,0} = \mathbf{C}$ of the double sequence of string algebras. To overcome this difficulty, we modify the construction of the double sequence of string algebras as follows. We allow strings to start one of $*$ and $\sigma(*)$. Multiplication rules and identification based on connections are kept same. In this way, we get a double sequence $\{A_{n,m}\}$ of string algebras starting from $A_{0,0} = \mathbf{C} \oplus \mathbf{C}$ as in [Ka]. Then, we can apply σ to each $A_{m,n}$ and we get another double sequence $A^{\sigma}_{m,n}$. It is not hard to see that this is indeed the double sequence of string algebras of D_n. Thus the flatness problem of D_n is reduced to some computational problem of the connection of A_{2n-3}. That is, as in Theorem 2.1 (3), we can prove that flatness of the connection on D_n is equivalent to the equality that some large partion function has value 1. Now induction shows that this partition function value is $(-1)^n$, which proves Ocneanu's announcement on D_n. This can be understood as follows in our general settings. If the symmetry fixes the $*$, then the orbifold construction automatically has flatness as in §3, but if the symmetry does move the $*$, there arises an obstruction for flatness in the orbifold procedure. This obstruction eliminates D_{2n+1} but such an obstruction does not exist for D_{2n}.

Furthermore, one can prove that the modification above of the string algebra double sequence construction still gives the same subfactor of type A_{2n-3}. One way to see this is that we can compute the principal graph of this subfactor by Ocneanu's compactness argument [O3]. Then it is easy to see that the principal graph is A_{2n-3}. Another way is that we prove the modified construction gives a subfactor with relative McDuff property. This is done by seeing that the horizontal Jones projections make central sequences in the subfactor for the ambient factor. (See [Bi].) Then it is easy to see that we get a conjugate subfactor by cutting the factors by a projection in the subfactor. (Also see [P1, page 200].) This means that a subfactor with the principal graph D_{2n} is realized as $N^{\sigma} \subset M^{\sigma}$, where $N \subset M$ has a principal graph A_{4n-3} and σ is an automorphism of M with order 2 fixing N globally. (Note that the orbifold construction as in §3 for D_n and Takesaki duality easily produce the following: If D_n principal graph is realized, the subfactor is of the above form of the simultaneous fixed point algebra of the A_{2n-3} subfactor. Thus the difference between D_{2n} and D_{2n+1} comes from the difference between A_{4n-3} and A_{4n-1}.)

Remark 4.1. Impossibility of D_{2n+1} as a principal graph has also been independently proved by [I1], [SV]. Their method is to prove inconsistency of decomposition rules of multiplication of endomorphisms or bimodules for D_{2n+1}. This method was also claimed by Ocneanu without a proof. For impossibility of E_7, the author verified non-flatness by a computer [Ka], and inconsistency of decomposition rules was verified by [SV], as Ocneanu announced. Izumi [I1] gave a much simpler proof just by looking at the Perron-Frobenius eigenvector entry. (This was essentially in [P2, Theorem 3.8].) For E_6, Bion-Nadal [BN] gave a construction, and then Ocneanu's general machinery easily produces that there are two and only two subfactors for the E_6 principal graph [O1, O3, Ka]. For E_8, Izumi [I2] checked flatness and thus proved that there are two and only two subfactors for the E_8 principal graph. In this way, we now have a complete proof of classification of subfactors with index less than 4 announced by Ocneanu.

U. Haagerup said to the author that he has also verified this classification by his method based on bimodules.

Remark 4.2. A. Ocneanu sketched his original proof of flatness of D_{2n} in his lectures at Tokyo in July of 1990 [O3]. But the details were not clear and the author could not complete the proof along this line, so the author found a different proof based on orbifold method in December of 1990 [Ka] as described above. Then in October of 1991, A. Ocneanu showed further details to the author on his original method and the author understood his complete proof. We include his arguments in Appendix here because his method is quite different from ours, and is of another interest. His proof is somewhat shorter than ours, but it does not give a realization of D_{2n} as a simultaneous fixed point algebra of A_{4n-3} and it seems difficult to extend this method to the Hecke algebra settings while ours does as below. The author thanks Prof. A. Ocneanu for showing this proof and permitting us to include it here.

H. Wenzl constructed his Hecke algebra subfactors with index values $\dfrac{\sin^2(N\pi/k)}{\sin^2(\pi/k)}$ as a natural generalization of Jones' subfactors of type A_n with index values $4\cos^2\dfrac{\pi}{n+1}$ in [We]. It turned out that these subfactors correspond to 2-variable link invariant (HOMFLY) polynomial [FYHLMO], quantum groups $U_q(sl_N)$, and the Jimbo-Miwa-Okado solutions to the Yang-Baxter equation [JMO1, JMO2]. Now we show that we can also apply the orbifold construction to Wenzl's Hecke algebra subfactors [EK].

In order to apply the orbifold method, we have to know the paragroups of the Hecke algebra subfactors. Wenzl computed the principal graphs of his subfactors in an unpublished work but could not obtain paragroups. We compute the paragroups with the Yang-Baxter equation in [EK]. Because Wenzl's representation is a certain trigonometric limit of Boltzmann weights of Jimbo-Miwa-Okado [JMO1, JMO2] in the form of elliptic functions, we still have the Yang-Baxter equation. Roche's result [R] implies that the face operators are flat when we have the Yang-Baxter equation, and now the Hecke algebra generators are essentially face operators. This is a typical example that the Yang-Baxter equation implies flatness. (But in general, the subfactor does not have good series of generators and thus the Yang-Baxter equation does not produce full flatness.) Furthermore, Wenzl's construction of subfactor uses commuting squares of period N, but the canonical form arising as higher relative commutants should have a period 2. This means that the construction of Wenzl's Hecke algebra subfactor is not in the canonical form in the sesne of Popa and Ocneanu. But by the use of the Yang-Baxter equation as above and the crossing symmetry, we can compute the paragroup and modify the construction so that it has a period 2.

Here we show how to proceed in the case $N = 3$ for the orbifold construction. Then first we have to modify the double sequence again so that it now starts with $A_{0,0} = \mathbf{C} \oplus \mathbf{C} \oplus \mathbf{C}$. Then we can define a symmetry σ of order 3 and make a fixed point algebra $A_{m,n}^{\sigma}$. Again, flatness for orbifolds can be reduced to computations of certain partition functions of the original connection. By the Yang-Baxter equation, we can compute the values. In the case D_n, there was a difference between D_{2n} and D_{2n+1}, but now it turns out that an obstruction for flatness vanishes in all the cases. It appears that this difference between $N = 2$ and $N = 3$ comes from parity of N. Because Wenzl's subfactors are regarded as a generalization of the A_n-sequence, our construction gives a "generalized D_n" sequence. (Graphs $\mathcal{D}$ were considered in [FG] as a generalization of D_n-sequence, but our principal graph is only a part of their graphs.) For some examples of our principal graphs, see [EK].

Appendix: Ocneanu's original proof of flatness of D_{2n}

Here we include an outline of Ocneanu's original proof of flatness of D_{2n} along the line suggested in [O3]. The author thanks him for showing the arguments and permitting him to include this here. We did not try to supply all the computational details, but readers familiar with [O3, Ka] should have no difficulty in understanding.

We identify $\mathcal{G}$ and $\mathcal{H}$ and label some vertices of it as follows.

$$D_{2n}: \quad \begin{matrix} a \\ \diagdown \\ \diagup \\ b \end{matrix} x\!-\!\cdot\ \cdots *,$$

Define a connection on this diagram as follows. We start from $*$, and until the vertex x, we choose a real connection as in the case of diagrams A_n. (See [O3], for example.) At the vertex x, we have 3×3-unitary matrix as follows.

$$\begin{pmatrix} c_{11} & c_{12} & c_{13} \\ c_{21} & c_{22} & c_{23} \\ c_{31} & c_{32} & c_{33} \end{pmatrix},$$

where columns are labeled by $\cdot, a, b$ from the left to the right and rows are labeled by $\cdot, a, b$ from the top to the bottom. We choose the connection so that $c_{11}, c_{12}, c_{13}, c_{21}, c_{31} \in \mathbf{R}$ and $c_{22} = c_{33} = \bar{c}_{23} = \bar{c}_{32}$. (This is possible. See [O3, IV.2] and [Ka, §3]. Note that this choice of gauges is different from the standard one in [O1, O3, Ka].) We fix this connection. Note that this connection is invariant under the graph automorphism σ flipping a and b.

We make a double sequence of string algebras starting from $*$. Denote by q, q' horizontal strings $(*\!-\!\cdots\!-\!x\!-\!a, *\!-\!\cdots\!-\!x\!-\!a)$ and $(*\!-\!\cdots\!-\!x\!-\!b, *\!-\!\cdots\!-\!x\!-\!b)$ respectively. (Each string has length $2n - 2$.) We also denote by p, p' the vertical strings with the same expressions. Note that $p + p'$ is a flat projection orthogonal to the vertical Jones projections $e_1, e_2, \cdots, e_{2n-3}$. Then $q(p + p')q$ can be identified with a string with length $2n - 2$ starting from a. We can express this as a sume of mutually orthogonal projections by decomposing it according to the endpoints of strings. Then we have the following.

Claim. *For each n, one of the following holds under the situation above.*

(1) If the endpoint for $q(p + p')q$ is a, then the corresponding projection is 0 and if the endpoints for $q(p + p')q$ is not a, then the corresponding projection is of rank 1.

(2) If the endpoint for $q(p + p')q$ is b, then the corresponding projection is 0 and if the endpoints for $q(p + p')q$ is not b, then the corresponding projection is of rank 1.

Furthermore, we have the same claim for $q'(p+p')q'$, and if we have (1) [resp. (2)] for $q(p + p')q$, then we have (2) [resp. (1)] for $q'(p + p')q'$. This claim is proved by looking at the embeddings of the Jones projections into the string algebra of D_{2n} starting from a or b. (See [GHJ], [Ok, page 99].)

Now it is enough to prove

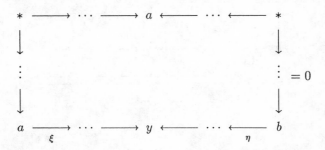

for all the ξ, η, y, where the diagram above is of size $(2n - 2) \times 2(2n - 2)$. We prove this as follows.

If $y = a$ or $y = b$, then the claim above proves the desired equality. Suppose $y \neq a, b$. Again by the claim above, it is enough to prove the equality above for a single σ and $\eta = \sigma(\xi)$ after a certain change of basis in the path Hilbert space. Because $q + q'$ is flat, and the connection is σ-invariant, we get the following identities.

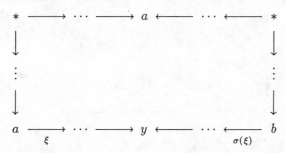

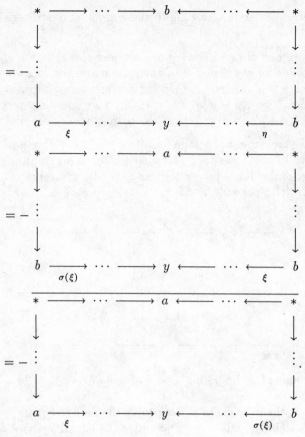

This implies that

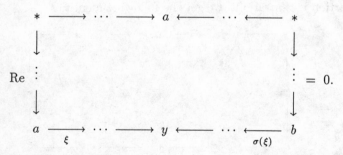

Now the following claim finishes the proof.

Claim. *Each 2×2-diagram with even vertices of D_{2n} on the boundary has a real value.*

This claim is checked directly.

REFERENCES

[Ba] R. J. Baxter, "Exactly solved models in statistical mechanics", Academic Press, New York, 1982.

[BN] J. Bion-Nadal, *An example of a subfactor of the hyperfinite II_1 factor whose principal graph invariant is the Coxeter graph E_6*, in "Current Topics in Operator Algebras", World Scientific Publishing, 1991, pp. 104–113.

[Bi] D. Bisch, *On the existence of central sequences in subfactors*, Trans. Amer. Math. Soc. **321** (1990), 117–128.

[BEEK1] O. Bratteli, G. A. Elliott, D. E. Evans, & A. Kishimoto, *Non-commutative spheres*, Inter. J. Math. **2** (1991), 139–166.

[BEEK2] O. Bratteli, G. A. Elliott, D. E. Evans, & A. Kishimoto, *Finite group actions on AF algebras obtained by folding the interval*, (to appear in K-theory).

[BEEK3] O. Bratteli, G. A. Elliott, D. E. Evans, & A. Kishimoto, *Non-commutative spheres II: rational rotations*, (to appear in J. Operator Theory).

[BEK] O. Bratteli, D. E. Evans, & A. Kishimoto, *Crossed products of totally disconnected spaces by $Z_2 * Z_2$*, preprint, Swansea, 1991.

[BK] O. Bratteli & A. Kishimoto, *Non-commutative spheres III: irrational rotations*, preprint, Trondheim, 1991.

[Ch] M. Choda, *Duality for finite bipartite graphs*, to appear in Pac. J. Math.

[Co] A. Connes, *Periodic automorphisms of the hyperfinite factor of type II_1*, Acta Sci. Math. **39**, 39–66 (1977).

[DJMO] E. Date, M. Jimbo, T. Miwa, & M. Okado, *Solvable lattice models*, in "Theta functions — Bowdoin 1987, Part 1," Proc. Sympos. Pure Math. Vol. 49, Amer. Math. Soc., Providence, R.I., pp. 295–332.

[DZ] P. Di Francesco & J.-B. Zuber, *$SU(N)$ lattice integrable models associated with graphs*, Nucl. Phys. **B338** (1990), 602–646.

[E1] D. E. Evans, *The C^*-algebras of topological Markov chains*, Tokyo Metropolitan University Lecture Notes, 1984.

[E2] D. E. Evans, *Quasi-product states on C^*-algebras*, in "Operator algebras and their connections with topology and ergodic theory", Springer Lecture Notes in Math., 1132 (1985), 129–151.

[EG] D. E. Evans & J. D. Gould, *Embeddings and dimension groups of non-commutative AF algebras associated to models in classical statistical mechanics*, preprint, Swansea.

[EK] D. Evans & Y. Kawahigashi, *Orbifold subfactors from Hecke algebras*, preprint.

[FG] P. Fendley & P. Ginsparg, *Non-critical orbifolds*, Nucl. Phys. **B324** (1989), 549–580.

[FYHLMO] P. Freyd, D. Yetter; J. Hoste; W. Lickorish, K. Millet; & A. Ocneanu, *A new polynomial invariant of knots and links*, Bull. Amer. Math. Soc. **12** (1985), 239–246.

[GHJ] F. Goodman, P. de la Harpe, & V. F. R. Jones, "Coxeter graphs and towers of algebras", MSRI publications 14, Springer, 1989.

[HS] U. Haagerup & J. Schou, in preparation.

[H] U. Haagerup, in preparation.

[I1] M. Izumi, *Application of fusion rules to classification of subfactors*, to appear in Publ. RIMS Kyoto Univ.

[I2] M. Izumi, *On flatness of the Coxeter graph E_8*, preprint.

[IK] M. Izumi & Y. Kawahigashi, *Classification of subfactors with the principal graph* $D_n^{(1)}$, to appear in J. Funct. Anal.

[Ji] M. Jimbo (editor), "Yang-Baxter equation in integrable systems", Advanced Series in Mathematical Physics, Vol. 10, World Scientific, 1989.

[JMO1] M. Jimbo, T, Miwa, & M. Okado, *Solvable lattice models whose states are dominant integral weights of* $A_{n-1}^{(1)}$, Lett. Math. Phys. **14** (1987), 123–131.

[JMO2] M. Jimbo, T, Miwa, & M. Okado, *Solvable lattice models related to the vector representation of classical simple Lie algebras*, Comm. Math. Phys. **116** (1988), 507–525.

[J] V. F. R. Jones, *Index for subfactors*, Invent. Math. **72** (1983), 1–15.

[Ka] Y. Kawahigashi, *On flatness of Ocneanu's connections on the Dynkin diagrams and classification of subfactors*, University of Tokyo, preprint, 1990.

[Ko] I. Kostov, *Free field presentation of the* A_n *coset models on the torus*, Nucl. Phys. **B300** (1988), 559–587.

[L1] P. Loi, *On automorphisms of subfactors*, preprint.

[L2] P. Loi, *On the derived towers of certain inclusions of type* III_λ *factors of index 4*, preprint.

[O1] A. Ocneanu, *Quantized group string algebras and Galois theory for algebras*, in "Operator algebras and applications, Vol. 2 (Warwick, 1987)," London Math. Soc. Lect. Note Series Vol. 136, Cambridge University Press, 1988, pp. 119–172.

[O2] A. Ocneanu, *Graph geometry, quantized groups and nonamenable subfactors*, Lake Tahoe Lectures, June–July, 1989.

[O3] A. Ocneanu, "Quantum symmetry, differential geometry of finite graphs and classification of subfactors", University of Tokyo Seminary Notes 45, (Notes recorded by Y. Kawahigashi), 1991.

[O4] A. Ocneanu, in preparation.

[Ok] S. Okamoto, *Invariants for subfactors arising from Coxeter graphs*, in "Current Topics in Operator Algebras", World Scientific Publishing, 1991, pp. 84–103.

[PP] M. Pimsner & S. Popa, *Iterating the basic constructions*, Trans. Amer. Math. Soc. **310** (1988), 127–134.

[P1] S. Popa, *Relative dimension, towers of projections and commuting squares of subfactors*, Pac. J. Math. **137** (1989), 181-207.

[P2] S. Popa, *Classification of subfactors: reduction to commuting squares*, Invent. Math. **101** (1990), 19–43.

[P3] S. Popa, *Sur la classification des sousfacteurs d'indice fini du facteur hyperfini*, C. R. Acad. Sc. Paris. **311** (1990), 95–100.

[R] Ph. Roche, *Ocneanu cell calculus and integrable lattice models*, Comm. Math. Phys. **127** (1990), 395–424.

[Sc] J. Schou, *Commuting squares and index for subfactors*, Ph.D. thesis, Odense University, 1990.

[Su] V. S. Sunder, *A model for AF-algebras and a representation of the Jones projections*, J. Operator Theory **18** (1987), 289–301.

[SV] V. S. Sunder & A. K. Vijayarajan, *On the non-occurrence of the Coxeter graphs* E_7, D_{2n+1} *as principal graphs of an inclusion of* II_1 *factors*, preprint.

[T] M. Takesaki, "Theory of Operator Algebras I", Springer, New York, 1979.

[TV] V. G. Turaev & O. Y. Viro, *State sum invariants of 3-manifolds and quantum 6j-symbols*, preprint.

[Wa] A. J. Wassermann, *Coactions and Yang-Baxter equations for ergodic actions and subfactors*, in "Operator algebras and applications, Vol. 2 (Warwick, 1987)," London Math. Soc. Lect. Note Series Vol. 136, Cambridge University Press, 1988, pp. 203–236.

[We] H. Wenzl, *Hecke algebras of type A and subfactors*, Invent. Math. **92** (1988), 345–383.

[Y] S. Yamagami, *A report on Ocneanu's lecture*, preprint.

Asymptotic Completeness of N-body Wave Operators II. A New Proof for the Short-Range Case and the Asymptotic Clustering for Long-Range Systems

HITOSHI KITADA
DEPARTMENT OF MATHEMATICAL SCIENCES
UNIVERSITY OF TOKYO
KOMABA, MEGURO, TOKYO 153, JAPAN

Introduction and Main Result

It is well-known that the asymptotic completeness holds for short-range quantum systems. The first proof was given by Sigal-Soffer in reference [13] of Part I [9], which we refer to as reference [I.13], and the alternative proofs were obtained by Graf [I.8], Kitada [9], Dereziński [3], Tamura [I.17], and Yafaev [18].

In the present paper we shall prove the asymptotic clustering for N-body long-range quantum systems

$$(1) \qquad H = H_0 + V, \qquad V = \sum_\alpha V_\alpha(x_\alpha),$$

in $\mathcal{H} = L^2(R^n)$, $n = (N-1)\nu$, with pair potentials $V_\alpha(x_\alpha)$ of general decay: $V_\alpha(x_\alpha) = O(|x_\alpha|^{-\epsilon})$ as $|x_\alpha| \to \infty$ with $\epsilon > 0$, where $\alpha = (i,j)$, $1 \le i < j \le N$, $x_\alpha = r_i - r_j \in R^\nu$ ($\nu \ge 1$). We also give a new proof of the asymptotic completeness for the short-range case: $\epsilon > 1$. The notion "asymptotic clustering" was first introduced in Sigal-Soffer [15] and was proved for Coulomb-type pair potentials. The result has been extended to general long-range pair potentials with decay rate $\epsilon > 1/2$ recently by Gérard-Dereziński [7]. This notion is equivalent to the asymptotic completeness, if the potentials are short-range. However when there are the long-range parts of the pair potentials, it does not necessarily imply the asymptotic completeness. We shall prove below a sharper "asymptotic clustering" result than Sigal-Soffer's in the sense that the orthogonal decomposition

$$(1.32) \qquad \psi(H)f = \bigoplus_{2 \le |b| \le N} \Omega_b^\psi f, \qquad (f \in \mathcal{H}, \psi \in C_0^\infty(R^1 - \mathcal{T}))$$

of $\mathcal{H}^c(H)$ in our Theorem 1.1 below has the property that $\mathcal{R}(\Omega_b^\psi) \supset \mathcal{R}(\psi(H)W_b^{D+})$ for each $2 \le |b| \le N$. Here $\mathcal{T}$ is the set of thresholds and point spectra of the total Hamiltonian H and W_b^{D+} ($2 \le |b| \le N$) are the modified wave operators defined by (7) below. Thus (1.32) gives a necessary and sufficient condition for the asymptotic completeness to hold. The asymptotic completeness for the short-range case will be shown by seeing this condition valid. The asymptotic clustering in the usual form for the long-range case will be proved by using decomposition (1.32).

The decomposition (1.32) is a stronger version of the one given in my unpublished paper [12, Proposition 4.3], where we have tried to prove the asymptotic completeness utilizing that decomposition. The present decomposition (1.32) is stronger than it in the sense that $\Omega_b^\psi f$'s are mutually orthogonal. The underlying ideas are similar, but the technical developments in these seven or eight years made it possible to get the present necessary and sufficient condition.

We follow the notation and conventions in Part I [9], and the formulae, Theorems, references, etc. in Part I will be quoted as formula (I.2.12), Theorem I.1.2, reference [I.10], etc. We review some of our notation which will be used in the following. We denote by a, b, $\cdots$ the cluster decompositions of the set $\{1, \cdots, N\}$. $b < a$ means that b is a refinement of a. $\alpha = (i, j) \leq a$ means that the particles i and j lie in one and the same cluster of a. $\sim (\alpha \leq a)$ is its negation. $I_a = \sum_{\sim(\alpha \leq a)} V_\alpha$ is the intercluster interaction associated with cluster decomposition a. $H_a = H - \bar{I}_a = H^a + T_a$ is the truncated Hamiltonian, where H^a is the internal (subsystem) Hamiltonian and $T_a = T_a(D_a)$ $(D_a = D_{x_a} = -i\partial/\partial x_a)$ is the intercluster free Hamiltonian. P_a denotes the eigenprojection for H^a, and for any integer $M \geq 1$, P_a^M denotes the M-dimensional partial projection of P_a such that $P_a = s - \lim_{M \to \infty} P_a^M$. As our coordinate systems, we adopt clustered Jacobi coordinates: $x = (x_a, x^a) \in R^n$, $n = (N-1)\nu$, $x_a \in R^{n_a}$, $n_a = (|a|-1)\nu$, $x^a \in R^{n^a}$, $n^a = (N-|a|)\nu$. x_a is the outer or intercluster Jacobi coordinates, and x^a is the inner Jacobi coordinates, namely coordinates inside the clusters of a.

We can now explain our idea to get (1.32). Let supp ψ be close to $\Lambda \in R^1 - \mathcal{T}$, and take $\psi_{j_0} \in C_0^\infty(R^1 - \mathcal{T})$ such that $\sum_{j_0}^{\text{finite}} \psi_{j_0}\psi = \psi$ and dist $((\text{supp } \psi_{j_0})^c, \Lambda_{j_0}) < \delta_0$ for some $\Lambda_{j_0} \in R^1 - \mathcal{T}$. We decompose: $e^{-itH}\psi(H)f = \sum_{2 \leq |d| \leq N} \tilde{P}_d^{M_d} e^{-itH}\psi(H)f$, where $\tilde{P}_d^{M_d}$ is an approximate projection defined by (I.1.13) on the eigenspace of the subsystem H^d. Introducing a decomposition of identity $1 = \Phi_{\delta_0 j_0}(x/t)^{-1} \cdot \Phi_{\delta_0 j_0}(x/t)$ (Lemma I.2.1), and utilizing Enss' asymptotic micro-local relation (Proposition I.1.1, see [I.4, I.6]) which says that $x_d/t \sim q_d$ asymptotically on $\tilde{P}_d^{M_d} e^{-itH}\psi(H)f$, we will translate $\Phi_{\delta_0 j_0}(x/t)\psi_{j_0}(H)e^{-itH}\psi(H)f$ asymptotically to $\sum_{2 \leq |b_1| \leq N} \sum_{\lambda_1 \in \mathcal{T}_{\Lambda_{j_0}}^{r_0}} L_{b_1, \lambda_1}^{\delta_0, j_0}(t)\psi_{j_0}(H)$ $\times e^{-itH}\psi(H)f$ along some sequence $t = t_n \to \infty$ determined by $\psi(H)f$. Here $\mathcal{T}_{\Lambda_{j_0}}^{r_0}$ will be defined by (1.4), and $L_{b_1, \lambda_1}^{\delta_0, j_0}(t)$ is a phase space operator defined by (1.8) below that localizes the state $\psi_{j_0}(H)e^{-itH}\psi(H)f$ to the phase space region

$$\frac{|x|^2}{t^2} \approx 2M \equiv 2(\Lambda_{j_0} - \lambda_1)$$

$$\frac{x_{b_1}}{t} \approx q_{b_1} \qquad (q_{b_1} = \text{the momentum conjugate to } x_{b_1}),$$

$$|q_k|^2 > 2M\rho_{|b_1|} \qquad (q_k = \text{the momentum between any 2-clusters in } b_1),$$

$$2T_{b_1} > 2M(1 - \theta_{|b_1|}),$$

where $1 > \rho_{|b_1|} >> \theta_{|b_1|} > 0$ are some constants. Note that these relations imply that $|x^{b_1}|^2/t^2 < 2M\theta_{|b_1|}$, where we can let $\theta_{|b_1|} \to 0$ as $\delta_0 \downarrow 0$. If we use some positivity arguments of Heisenberg derivatives, we have the existence of the limit

$$s - \lim_{t \to \infty} e^{itH} L_{b_1, \lambda_1}^{\delta_0, j_0}(t) e^{-itH}\psi(H)f.$$

This suffices to prove the asymptotic completeness for the short-range case, if one assumes the completeness of the subsystems as induction hypotheses. We can go further to a finer decomposition however, which makes the induction unnecessary. On the left of $L_{b_1,\lambda_1}^{\delta_0,j_0}(t)$, let us insert $I = (I - \psi^{\delta_1}(H^{b_1})) + \psi^{\delta_1}(H^{b_1})$, where $\psi^{\delta_1} \in C_0^\infty(R^1)$ equals 1 near $\lambda \in T_{b_1} = \{$thresholds of $H^{b_1}\} \cup \sigma_p(H^{b_1})$ and satisfies dist $((\text{supp } \psi^{\delta_1})^c, T_{b_1}) < 2\delta_1$. $I - \psi^{\delta_1}(H^{b_1})$ restricts H^{b_1} to the continuous subspace $\mathcal{H}^c(H^{b_1})$, hence one can decompose the state $(I - \psi^{\delta_1}(H^{b_1}))L_{b_1,\lambda_1}^{\delta_0,j_0}(t)e^{-itH}\psi(H)f$ further, asymptotically as $t \to \infty$ in a way similar to the above. Continuing this procedure, one finally gets to the existence of the following limit and the equality (1.32):

$$(1.31) \qquad \Omega_b^\psi f = s - \lim_{\delta_N \downarrow 0} \cdots s - \lim_{\delta_0 \downarrow 0} s - \lim_{t \to \infty} \sum e^{itH} \psi_{k+1}^{\delta_{k+1}}(H^{b_{k+1}})$$
$$\times \Psi_{\delta_0,\cdots,\delta_k}^{b_1,\cdots,b_k}(q_{b_{k+1}})\tilde{L}_{\delta_0,\cdots,\delta_k}^{b_1,\cdots,b_{k+1}}(t)e^{-itH}\psi(H)f,$$

where $b = b_{k+1}$, $\sum$ is a certain sum, and $\Psi_{\cdots}^{\cdots}(q_b)$ and $\tilde{L}_{\cdots}^{\cdots}(t)$ are defined by (1.30) below. From the construction, $e^{-itH}\Omega_b^\psi f$ is expected to be close to $P_b e^{-itH}\Omega_b^\psi f$ asymptotically as $t \to \infty$, where P_b is the eigenprojection for H^b. We will show that this expectation is true for the short-range case. Using this information and some decay assumption of eigenvectors, we shall prove the asymptotic completeness for short-range systems. In the long-range case the limit (1.31) also exists, and according to Dereziński [4], it satisfies the above expectation as well, if $\epsilon > \sqrt{3} - 1$ (Remark 2.6 below). We shall also give some partial results on the asymptotic completeness for the long-range case in Sec.4.

We mention some history of N-body long-range problem. As far as we know, the first result for long-range case was obtained by Merkuriev [13], where three body systems with pure Coulomb potentials were discussed. Enss [5, I.6] subsequently discussed the three body long-range systems. His essential assumption is the decay rate $\epsilon > \sqrt{3}-1 = 0.732\cdots$ for pair potentials $V_\alpha(x_\alpha)$. The third work is Sigal-Soffer [15, 16], where they considered the asymptotic completeness of 3 and 4 body long-range quantum systems with pair potentials $V_\alpha(x_\alpha) = O(|x_\alpha|^{-1})$ being Coulomb-like. Recently Wang [17] proved the asymptotic completeness for 3-body systems with general long-range pair potentials with $\epsilon > 1/2$ under some conditions on the negative parts of pair potentials. When the same conditions are assumed for the positive parts of the pair potentials, his results reproduce Enss' results. For general N-body long-range case, quite recently Dereziński [4] has proved the asymptotic completeness under the same condition $\epsilon > \sqrt{3} - 1$ as Enss'. His proof relies on a detailed dynamical analysis of 0-momentum particles.

We formulate our results. Our first assumptions are concerned with the decay rates of V_α.

Assumption 1. $V_\alpha(x_\alpha)$ is split into $V_\alpha(x_\alpha) = V_\alpha^S(x_\alpha)+V_\alpha^L(x_\alpha)$, both summands being real-valued, and C^1 and C^∞ respectively, such that there exist constants $1 > \epsilon > 0$, $C_0 > 0$, $C_\beta > 0$ satisfying

$$(S) \qquad |V_\alpha^S(x_\alpha)| + |\nabla_{x_\alpha} V_\alpha^S(x_\alpha)| \leq C_0(1 + |x_\alpha|)^{-1-\epsilon},$$

(L) $$|\partial_{x_\alpha}^\beta V_\alpha^L(x_\alpha)| \le C_\beta(1+|x_\alpha|)^{-|\beta|-\epsilon}$$

for all ν-dimensional multi-indices β.

Our second assumption is the following

Assumption 2. For any integer $M \ge 1$, and cluster decomposition a with $2 \le |a| \le N-1$,

(2) $$\||x^a|^2 P_a^M\| < \infty.$$

Since it is known ([1], [6]) that the non-threshold eigenvectors decay exponentially in the configuration space, Assumption 2 is concerned only with the decay property of threshold eigenvectors of each subsystem. In Appendix B we will show that general polynomial decay of eigenvectors holds at least for 2- and 3-body Hamiltonians with homogeneous pair potentials $V_\alpha(x_\alpha) = c_\alpha|x_\alpha|^{-\epsilon}$, $c_\alpha \in R^1$, $2 > \epsilon > 0$. This result is useful for the scattering problem up to 4-body cases.

We introduce a (time-dependent) modifier $S_b(\xi_b, t)$ as a solution of Hamilton-Jacobi equation:

(3) $$\partial_t S_b(\xi_b, t) = T_b(\xi_b) + I_b^L(t, \nabla_{\xi_b} S_b(\xi_b, t), 0)$$

with initial conditions

(4) $$S_b(\xi_b, 0) = 0$$

$$|\partial_{\xi_b}^\alpha \{S_b(\xi_b, t) - tT_b(\xi_b)\}| \le C_{\alpha\epsilon'} < t >^{1-\epsilon'}$$

for $t \in R^1$ and all α and ϵ' with $1 > \epsilon > \epsilon' > 0$. The "long-range tail", that is, a time dependent intercluster potential $I_b^L(t, x_b, x^b)$ in (3) is defined as follows:

(5) $$I_b^L(t, x_b, x^b) = \sum_{\sim(\alpha \le b)} V_\alpha^L(t, x_\alpha),$$

where $V_\alpha^L(t, x_\alpha) = V_\alpha^L(x_\alpha)\phi(\frac{<\log<t>>}{<t>}|x_\alpha|)$, $\phi(\tau) \in C^\infty(R^1)$ with $0 \le \phi \le 1$, $\phi(\tau) = 1$ $(\tau \ge 2)$, $= 0$ $(\tau \le 1)$, and $< t >$ is a smooth function of t such that $< t >= |t|$ $(|t| \ge 1)$ and $\ge 1/2$ $(|t| \le 1)$. The definition (5) of $I_b^L(t, x_b, x^b)$ is an extension of the time dependent cutting off of the 2-body long-range potentials introduced in [10] and [11]. Note that $I_b^L(t, x_b, x^b)$ thus defined satisfies the decay property of time-dependent 2-body potentials of long-range type (see [10] and [11]):

(6) $$|\partial_x^\alpha I_b^L(t, x)| \le C_{\alpha\epsilon'} < t >^{-|\alpha|-\epsilon'}$$

for all α and ϵ' with $\epsilon > \epsilon' > 0$. For the existence of a solution $S_b(\xi_b, t)$ of (3) and (4), see Appendix A.

We denote by $U_b^D(t)$ the unitary propagator generated by $H_b^D(t) = \partial_t S_b(D_b, t) + H^b$: $(\partial_t + iH_b^D(t))U_b^D(t) = 0$, $U_b^D(0) = I$. Then it is not difficult (see Sec.3) to show that the modified channel wave operators

(7) $$W_b^{D\pm} = s - \lim_{t \to \pm\infty} e^{itH} U_b^D(t) P_b$$

exist under Assumptions 1 and 2. For general long-range many body systems, our proof of the existence of wave operators requires Assumption 2.

We also use the unitary propagator $U_b(t)$ generated by $H_b(t) = H_b + I_b^L(t, x)$:

(8) $$(\partial_t + iH_b(t))U_b(t) = 0, \qquad U_b(0) = I.$$

Now our main result is the following

Theorem 3. *Let Assumptions 1 and 2 be satisfied. Then the modified channel wave operators exist and the system is asymptotically clustering: Namely let $f \in \mathcal{H}$ and $\psi \in C_0^\infty(R^1 - \mathcal{T})$ with $\mathcal{T}$ being the set of thresholds and point spectra of H. Then $e^{-itH}\psi(H)f$ can be decomposed asymptotically as*

$$(9) \qquad \lim_{t \to \pm\infty} \|e^{-itH}\psi(H)f - \sum_{2 \le |b| \le N} U_b(t)f_b^\pm\| = 0$$

with $f_b^\pm \in \mathcal{H}$.

Further when $V_\alpha^L(x_\alpha) \equiv 0$ for all α in Assumption 1, the system is asymptotically complete:

$$(10) \qquad \bigoplus_{2 \le |b| \le N} \mathcal{R}(W_b^{D\pm}) = \mathcal{H}^c(H),$$

where $\mathcal{H}^c(H)$ is the continuous spectral subspace for H.

We consider in the sequel the case $t \to +\infty$ only omitting the superscript $+$. The other case can be treated similarly. The inclusion of local singularities into short-range pair potentials $V_\alpha^S(x_\alpha)$ in Assumption 1 is purely of technical matter, and is not discussed here.

In the following, we will state a proof of the theorem, where all results in Part I will be used freely. Some formulae will be quoted explicitly from Part I with tags like (I.2.6) when necessary.

1. Asymptotic decomposition of the scattering state

As in Part I we let $\mathcal{T}$ be the set of thresholds and point spectra of H: $\mathcal{T} = \bigcup_{1 \le |a| \le N} \sigma_p(H^a) \cup \{0\}$. Choose a real-valued $\psi \in C_0^\infty(R^1)$ so that $0 \le \psi \le 1$ and supp ψ is contained in an interval $(\Lambda-\delta, \Lambda+\delta)$ around $\Lambda \in R^1 - \mathcal{T}$ which is bounded away from $\mathcal{T}$. We take $\tau_0 > 0$ as $8\tau_0 << \inf\{|\Lambda - \lambda| | \lambda \in \mathcal{T}\}$ as in Sec.I.2, and choose $\delta_0 > \delta_0^0 > 0$ as $\delta_0^0 << 8\tau_0 < \delta_0 << \inf\{|\Lambda - \lambda| | \lambda \in \mathcal{T}\}$. Take a finite number of $\psi_{j_0} \in C_0^\infty(R^1)$, $0 \le \psi_{j_0} \le 1$, and $\Lambda_{j_0} \in R^1 - \mathcal{T}$ so that $\sum_{j_0}^{\text{finite}} \psi_{j_0}(H)\psi(H) = \psi(H)$, the width of supp ψ_{j_0} is at most δ_0^0 around $\Lambda_{j_0} \in R^1 - \mathcal{T}$, and $8\tau_0 << \inf(\{|\Lambda_{j_0} - \lambda| | \lambda \in \mathcal{T}, j_0\})$.

We take a finite set $\mathcal{T}^F = \{\lambda_\ell\}_{\ell=1}^K$ such that

$$(1.1) \qquad \mathcal{T} \subset \bigcup_{\ell=1}^K [\lambda_\ell - \tau_0, \lambda_\ell + \tau_0],$$

and choose constants $\sigma_0 > 0$ and $\tau_0 \ge \tau_0(\lambda) > 0$ for each $\lambda \in \mathcal{T}^F$ such that $0 < \sigma_0 < \tau_0$ and

$$(1.2) \qquad 0 < \tau_0(\lambda) \le \tau_0, \qquad \mathcal{T} \subset \bigcup_{\lambda \in \mathcal{T}^F} [\lambda - \tau_0(\lambda), \lambda + \tau_0(\lambda)],$$

$$\min_{\lambda, \mu \in \mathcal{T}^F, \lambda \ne \mu} \text{dist}\left([\lambda - \tau_0(\lambda), \lambda + \tau_0(\lambda)], [\mu - \tau_0(\mu), \mu + \tau_0(\mu)]\right) \ge 4\sigma_0 > 0$$

hold. We switch to another smaller $\delta_0^0 > 0$ such that

(1.3) $$0 < \delta_0^0 << \min(\{\sigma_0\} \cup \{\tau_0(\lambda)|\lambda \in \mathcal{T}^F\}),$$

if necessary. Then we define the sets $\mathcal{T}^{\tau_0} = \mathcal{T}_{a_1}^{\tau_0}$ and $\mathcal{T}_{\Lambda_{j_0}}^{\tau_0} = \mathcal{T}_{a_1}^{\tau_0} \cap (-\infty, \Lambda_{j_0})$ with $|a_1| = 1$ and $\mathcal{T}_a \equiv \{0\} \cup \bigcup_{b \le a} \sigma_p(H^b)$ as follows:

(1.4) $$\mathcal{T}_a^{\tau_0} = \{\lambda \in \mathcal{T}^F | \exists \tilde\lambda \in \mathcal{T}_a \text{ such that } |\lambda - \tilde\lambda| \le \tau_0(\lambda)\},$$
$$\mathcal{T}_{\Lambda^a}^{\tau_0} = \mathcal{T}_a^{\tau_0} \cap (-\infty, \Lambda^a).$$

Here we recall Lemma I.2.1. Define for $b < a$

(I.2.13) $$T_b^a(\rho, \theta) = \left(\bigcap_{k=1}^{k_b^a} \{x^a | |z_k^a|^2 > \rho\} \right) \cap \{x^a | |x^b|^2 \le \theta\}.$$

Lemma I.2.1. Let $1 \le |a| \le N$ and $0 < \delta_0 << 1$. Then for any constants $1 - \delta_0 = \theta_{|a|}^a > \rho_j^a >> \theta_j^a \ge \rho_N^a > 0$ $(j = |a| + 1, \cdots, N - 1)$ with $\theta_{j-1}^a \ge \theta_j^a + \rho_j^a$, there hold

(I.2.11) $$S_{\delta_0}^a \equiv \{x^a \in R^{n^a} | 1 - \delta_0 < |x^a|^2 < 1 + \delta_0\} \subset \bigcup_{b<a} T_b^a(\rho_{|b|}^a, \theta_{|b|}^a),$$

(I.2.12) $$T_b^a(\rho_{|b|}^a, 8\theta_{|b|}^a) \subset \{x^a | |x_\alpha|^2 \ge \rho_{|b|}^a/2 \text{ for all } \alpha \text{ with } \sim (\alpha \le b) \text{ and } \alpha \le a\}.$$

Corollary. Let $0 < 8\tau_0 << \inf(\{|\Lambda^a - \lambda| | \lambda \in \mathcal{T}_a\})$ for $\Lambda^a \notin \mathcal{T}_a$ as in Sec.I.2. Let $M_0 = \inf\{(\Lambda^a - \lambda) - \tau_0|\Lambda^a - \lambda > 0, \lambda \in \mathcal{T}_a\}(> 0)$, $M_1 = \sup\{(\Lambda^a - \lambda) + \tau_0|\Lambda^a - \lambda > 0, \lambda \in \mathcal{T}_a\}$, $0 < \sigma < \tau_0 << M_0\rho_N^a \le M_0\theta_j^a$, and $M_0 < M' < M_1$. Then there exists a constant $c_0 > 0$ such that, on $S_d^a(M') \equiv \{x^a \in R^{n^a} | 2M' - d < |x^a|^2 < 2M' + d\}$ with $M' >> \delta_0 > d > 0$, we have the following inequality associated with the decomposition (I.2.11) of the ring $S_d^a(M')$

(I.2.14)
$$\Phi_{M'}^a(x^a) \equiv \sum_{b<a} \varphi_b^{M'}(x_b^a)^2 h_b^{M'}(x_b^a)$$

$$\equiv \sum_{b<a} \{\prod_{k=1}^{k_b^a} \phi_\sigma(|z_k^a|^2 > 2M'\rho_{|b|}^a)\}^2 \phi_\sigma(|x_b^a|^2 > 2M'(1 - \theta_{|b|}^a) - d)^2$$

$$\ge c_0,$$

where '$\equiv$' means that the LHS is defined by the RHS.

Thus one can construct a real-valued C^∞ function $\Phi^a(x^a)$, which, on $R = \{x^a \in R^{n^a} | 2M_0 - \tilde\tau_0 < |x^a|^2 < 2M_1 + \tilde\tau_0\}$ $(\tilde\tau_0 = 2\tau_0 + 4\sigma_0 + 2\delta_0^0)$, has the form

(I.2.15) $$\Phi^a(x^a) = \sum_{\ell=1}^{L} \Phi_{\Lambda_\ell}^a(x^a) g_\ell(x^a).$$

Here $\Lambda_\ell \in [M_0, M_1]$, $g_\ell(x^a) \in C_0^\infty(S_{2\tilde\tau_0}^a(\Lambda_\ell))$, $\sum_{\ell=1}^{L} g_\ell = 1$ on R, and $\Phi_{\Lambda_\ell}^a$ is defined by (I.2.14) with $d = \tilde\tau_0(\lambda) \equiv 2\tau_0(\lambda) + 4\sigma_0 + 2\delta_0^0$ when $\Lambda_\ell = \Lambda^a - \lambda$ for some $\lambda \in \mathcal{T}_{\Lambda^a}^{\tau_0}$, and with $d = \tilde\tau_0$ otherwise. Further $\Phi^a(x^a)$ can be constructed to satisfy the following conditions (a)-(c):

(a) On $S^a_{\tilde{\tau}_0(\lambda)}(\Lambda^a - \lambda)$ with $\lambda \in \mathcal{T}^{\tau_0}_{\Lambda^a}$

$$(I.2.16) \qquad \Phi^a(x^a) = \Phi^a_{\Lambda^a - \lambda}(x^a),$$

(b) On R^{n^a}

$$(I.2.17) \qquad \Phi^a(x^a) \geq c_0/2,$$

and

(c) On each $S^a_{\tilde{\tau}_0}(M')$ with $M_0 - \tilde{\tau}_0 < M' < M_1 + \tilde{\tau}_0$, the condition that the summand $\varphi^{\Lambda^\ell}_b(x^a_b)^2 h^{\Lambda^\ell}_b(x^a_b) g_\ell(x^a)$ in $\Phi^a(x^a)$ does not vanish implies

$$(I.2.18) \qquad |x^b|^2 < 2M'\theta^a_{|b|} + 4\tilde{\tau}_0.$$

So far we studied a geometry of the configuration space associated with the N-body Hamiltonian H in (1). We turn to a study of the dynamics of the system defined by H.

Set $S_\tau(M,t) \equiv \{x \in R^n | 2M - \tau < |x|^2/t^2 < 2M + \tau\}$ for $t > 0$ and $0 < \tau < 2M$. Then Lemma I.2.1 and its Corollary with $|a| = 1$ and $\Lambda^a = \Lambda_{j_0}$ yield the existence of a sum of smooth functions $\Phi_{\delta_0 j_0}(x/t)$ $(t > 0)$ which decomposes the ring $\{x \in R^n | 2M_{0j_0} - \tilde{\tau}_0 < |x|^2/t^2 < 2M_{1j_0} + \tilde{\tau}_0\}$ $(M_{0j_0} = \inf\{\Lambda_{j_0} - \lambda - \tau_0 | \Lambda_{j_0} - \lambda > 0, \lambda \in \mathcal{T}\}(> 0)$, $M_{1j_0} = \sup\{\Lambda_{j_0} - \lambda + \tau_0 | \Lambda_{j_0} - \lambda > 0, \lambda \in \mathcal{T}\}$, $\tilde{\tau}_0 = 2\tau_0 + 4\sigma_0 + 2\delta^0_0 < 8\tau_0 < \delta_0)$ such that

$$\Phi_{\delta_0 j_0}(x/t) = \sum_{\ell=1}^{L} \Phi_{M_\ell}(x/t) g_\ell(x/t)$$

with $M_\ell \in [M_{0j_0}, M_{1j_0}]$, $g_\ell(x/t) \in C^\infty_0(S_{2\tilde{\tau}_0}(M_\ell, t))$, and Φ_{M_ℓ} being defined by (I.2.14). Further $\Phi_{\delta_0 j_0}(x/t)$ satisfies
(a) On $S_{\tilde{\tau}_0(\lambda)}(\Lambda_{j_0} - \lambda, t)$ with $\lambda \in \mathcal{T}^{\tau_0}_{\Lambda_{j_0}}$

$$(1.5) \qquad \Phi_{\delta_0 j_0}(x/t) = \Phi_{\Lambda_{j_0} - \lambda}(x/t),$$

where $\tilde{\tau}_0(\lambda) = 2\tau_0(\lambda) + 4\sigma_0 + 2\delta^0_0$;
(b) On R^n

$$(1.6) \qquad (0 <)c_0 \leq \Phi_{\delta_0 j_0}(x/t) \leq C_0(< \infty);$$

and
(c) On each $S_{\tilde{\tau}_0}(M', t)$ with $M_{0j_0} - \tilde{\tau}_0 < M' < M_{1j_0} + \tilde{\tau}_0$, the condition that the summand $\varphi^{M_\ell}_b(x_b/t)^2 h^{M_\ell}_b(x_b/t) g_\ell(x/t)$ in the definition of $\Phi_{\delta_0 j_0}(x/t)$ (see (I.2.14)) does not vanish implies

$$(1.7) \qquad |x^b|^2/t^2 < 2M'\theta_{|b|} + 4\tilde{\tau}_0,$$

where $\theta_{|b|} = \theta^a_{|b|}$ and $\rho_{|b|} = \rho^a_{|b|}$ for $|a| = 1$.

In addition to these properties the function $\Phi_{\delta_0 j_0}(x/t)$ can be constructed to satisfy the following condition.

(d) On R^n

$$|(\nabla_x \Phi_{\delta_0 j_0})(x/t)| \leq C$$

for some constant $C > 0$ independent of $\tau_0, \sigma_0, \delta_0^0, \tau_0(\lambda)$, and θ_j.

This is possible by adding other suitable smooth functions of x/t with the supports near the boundary of the sets $T_b(2M'\rho_{|b|}, 2M'\theta_{|b|})$ defined by (I.2.13) with $b < a_1$ for $|a_1| = 1$, to the terms of the definition of $\Phi_{\delta_0 j_0}$ in the above (see Corollary). These added terms are some functions of x_b/t near the boundary of the set $T_b(2M'\rho_{|b|}, 2M'\theta_{|b|})$ and are treated similarly to other original terms in considering the existence of the limits like (1.31) and (2.2) below due to the fact that their supports are near the boundary of the sets $T_b(2M'\rho_{|b|}, 2M'\theta_{|b|})$ (see the argument after (1.27) below). (Actually we can take $\Phi_{\delta_0 j_0}(x)$ identically equal to one on R^n. Note that the function $1 - \Phi_{\delta_0 j_0}(x)$ with $\Phi_{\delta_0 j_0}(x)$ as in the above does not vanish only near the boundary of $T_b(2M'\rho_{|b|}, 2M'\theta_{|b|})$ for some $b < a_1$. In particular the constant c_0 in (1.6) can be taken independently of $\tau_0, \sigma_0, \delta_0^0, \tau_0(\lambda)$, and θ_j.)

The decomposition of the identity $1 = \Phi_{\delta_0 j_0}(x/t)^{-1} \cdot \Phi_{\delta_0 j_0}(x/t)$ will be used in the following to get an asymptotic decomposition (1.10) below of the state $e^{-itH}\psi(H)f$ for $f \in \mathcal{H}$.

We define a selfadjoint operator γ by

$$(I.2.19) \qquad \gamma = \frac{1}{2}\left(\frac{x}{<x>} \cdot D_x + D_x \cdot \frac{x}{<x>}\right),$$

a ψdo $P_b^\delta(t) = p_b^\delta(x_b/t, q_b)$ as a ψdo with symbol

$$(I.2.21) \qquad p_b^\delta(x_b/t, \xi_b) = \phi(|x_b/t - \xi_b|^2 < \delta),$$

and functions $\varphi_\sigma(|\tau| < \delta)$, $\varphi_b^M(q_b) \equiv \varphi_{b a_1}^M(q_b^{a_1})$ $(M > 0, |a_1| = 1)$, and $\tilde{\varphi}_\theta^M(\tau)$ $(\theta > 0, M = \Lambda_{j_0} - \lambda > 0, \lambda \in \mathcal{T}_{\Lambda_{j_0}}^{\tau_0})$ by

$$(I.2.20) \qquad \varphi_\sigma(|\tau| < \delta) = \phi_\sigma(\tau < \delta + \sigma)\phi_\sigma(\tau > -\delta),$$

$$(I.2.22) \qquad \varphi_{ba}^M(q_b^a) = \prod_{k=1}^{k_b^a} \phi_\sigma(|q_k^a|^2 > 2M\rho_{|b|}^a),$$

$$(I.2.23) \qquad \tilde{\varphi}_\theta^M(\tau) = \phi_\sigma(\tau > 2M(1 - \theta) - \tilde{\tau}_0(\lambda)),$$

We also use the notation $\phi_\sigma(2\sigma < \tau < 2M) = \phi_\sigma(\tau > 2\sigma)\phi_\sigma(\tau < 2M + \sigma)$.

Then as in (I.2.24) we define for $2 \leq |b_1| \leq N$, $\lambda_1 \in \mathcal{T}_{\Lambda_{j_0}}^{\tau_0}$

$$(1.8) \qquad L_{b_1, \lambda_1}^{\delta_0, j_0}(t) = G_{b_1}(t)G_{b_1}(t)^*,$$

where

$$G_{b_1}(t) = P_{b_1}^{\delta_0^0}(t)\varphi_{b_1}^{\Lambda_{j_0}-\lambda_1}(q_{b_1})(g \cdot \tilde{\varphi}_{\theta_{|b_1|}}^{\Lambda_{j_0}-\lambda_1})(2T_{b_1})$$

$$\times \varphi_{\sigma_0}(|\,|x|^2/t^2 - 2M| < \tau_2(\lambda_1))\varphi_{\sigma_1}(|\gamma^2 - 2M| < \tau_1(\lambda_1))\phi_{\sigma_1}(2\sigma_1 < \gamma < 2\tilde{\tilde{M}}_1).$$

Here $M = \Lambda_{j_0} - \lambda_1$ with $\lambda_1 \in \mathcal{T}_{\Lambda_{j_0}}^{\tau_0}$, $\tau_1(\lambda_1) = 2\tau_0(\lambda_1) + 4\delta_0^0$, $\tau_2(\lambda_1) = 2\tau_0(\lambda_1) + 2\sigma_0$, $0 < 2\sigma_1 < \min\{\sigma_0, \sqrt{\tau_0}\}$, $\tilde{M}_1 = \sup\{\Lambda_{j_0} - \lambda_1|\Lambda_{j_0} - \lambda_1 > 0, \lambda_1 \in \mathcal{T}\}$, $\tilde{\tilde{M}}_1 = \sup\{\sqrt{2(\Lambda_{j_0} - \lambda_1)}|\Lambda_{j_0} \geq \lambda_1, \lambda_1 \in \mathcal{T}\}$, and $g \in C_0^\infty(R^1)$ with $0 \leq g \leq 1$, $g(\tau) = 1$ $(|\tau| \leq 6\tilde{M}_1)$, and $g(\tau) = 0$ $(|\tau| \geq 8\tilde{M}_1)$. These constants are chosen so that γ-supp φ'_{σ_1} is disjoint with $\Sigma(\Lambda_{j_0}) \equiv \{\pm\sqrt{2(\Lambda_{j_0} - \lambda)} \mid \Lambda_{j_0} \geq \lambda, \lambda \in \mathcal{T}\}$ and one can argue as in Sec.I.4 in proving the existence of the limits (1.13), (1.31) and (2.2) below.

Then as in Theorem I.2.3, using Proposition I.1.1 and choosing $\tau_0 > 0$ and $\delta_0^0 > 0$ so small that $8\tau_0 << \inf(\{|\Lambda_{j_0} - \lambda_1||\lambda_1 \in \mathcal{T}\} \cup \{M_{0j_0}\rho_N\})(> 0)$ and that (1.3) and $\delta_0^0\tilde{M}_1 << \tau_0^2$ hold, we have the following asymptotic relation along some sequence $t_n \to \infty$ $(n \to \infty)$ determined by $\psi(H)f$ $(f \in \mathcal{H})$:

$$(1.9) \qquad \psi_{j_0}(H)e^{-it_nH}\psi(H)f$$

$$\sim \Phi_{\delta_0 j_0}(x/t_n)^{-1} \sum_{2 \leq |b_1| \leq N} \sum_{\lambda_1 \in \mathcal{T}_{\Lambda_{j_0}}^{\tau_0}} L_{b_1,\lambda_1}^{\delta_0,j_0}(t_n)\psi_{j_0}(H)e^{-it_nH}\psi(H)f.$$

Here we have used the identity $\Phi_{\delta_0 j_0}(x/t_n)^{-1} \cdot \Phi_{\delta_0 j_0}(x/t_n) = 1$ and translated $\Phi_{\delta_0 j_0}(x/t_n)$ to $\sum_{2 \leq |b_1| \leq N} \sum_{\lambda_1 \in \mathcal{T}_{\Lambda_{j_0}}^{\tau_0}} L_{b_1,\lambda_1}^{\delta_0,j_0}(t_n)$ asymptotically as $n \to \infty$ using Proposition I.1.1. (See Sect.I.3 or the proof of Proposition I.2.3′ below for the detailed argument.) By $\sum_{j_0}^{\text{finite}} \psi_{j_0}\psi = \psi$, (1.9) yields

$$(1.10)$$

$$e^{-it_nH}\psi(H)f = \sum_{j_0}^{\text{finite}} \psi_{j_0}(H)e^{-it_nH}\psi(H)f$$

$$\sim \sum_{j_0}^{\text{finite}} \Phi_{\delta_0 j_0}(x/t_n)^{-1} \sum_{2 \leq |b_1| \leq N} \sum_{\lambda_1 \in \mathcal{T}_{\Lambda_{j_0}}^{\tau_0}} L_{b_1,\lambda_1}^{\delta_0,j_0}(t_n)\psi_{j_0}(H)e^{-it_nH}\psi(H)f$$

as $n \to \infty$. Here note that the last asymptotic equality holds termwise.

We here state some reasons which make the decomposition (1.10) possible. The main one is that $|x|^2/t_n^2 \geq \delta$ for some $\delta > 0$ on $e^{-it_nH}\psi(H)f$ asymptotically as $n \to \infty$. Actually using (I.1.15), we decompose $\psi_{j_0}(H)e^{-it_nH}\psi(H)f$:

$$F(t_n) = \psi_{j_0}(H)e^{-it_nH}\psi(H)f$$

$$= \sum_{2 \leq |d| \leq N} P_d^{M_{|d|}^n}\tilde{P}_{a_1d}^{M_{a_1d}^n}\psi_{j_0}(H)e^{-it_nH}\psi(H)f = \sum_{2 \leq |d| \leq N} f_d(t_n),$$

where $|a_1| = 1$. Then on each state $f_d(t_n)$, $H = T_d + I_d + H^d$ is asymptotically equal to $T_d + H^d$ by Proposition I.1.1. By the factor $P_d^{M_{|d|}^n}$, H^d equals some $\lambda_1 \in \tilde{T} = \{0\} \cup \bigcup_{d < a_1} \sigma_p(H^d) \subset T$. Thus $H \sim T_d + \lambda_1$ asymptotically. On the other hand, by $\psi_{j_0}(H)$, we have $H \approx \Lambda_{j_0}$. Hence $T_d \approx \Lambda_{j_0} - \lambda_1$ with $\Lambda_{j_0} \in R^1 - T$ and $\lambda_1 \in \tilde{T} \subset T$, where T is a closed set. Since $T_d \geq 0$, these yield $T_d \approx \Lambda_{j_0} - \lambda_1 > 0$. But $2T_d \sim |x_d|^2/t_n^2$ on $f_d(t_n)$ by Proposition I.1.1. Hence

$$\frac{|x|^2}{t_n^2} \geq \frac{|x_d|^2}{t_n^2} \sim 2T_d \approx 2(\Lambda_{j_0} - \lambda_1) \geq \delta > 0$$

for some $\delta > 0$. Similarly we have $|x|^2/t_n^2 \leq R$ for some $R > 0$. Thus on $F(t_n)$, x belongs to a ring $\{x | (0 <) \delta/2 \leq |x|^2/t_n^2 \leq R(< \infty)\}$ asymptotically as $n \to \infty$. Then we can decompose this ring as in Lemma I.2.1 above, and apply the decomposition of identity $\Phi_{\delta_0 j_0}(x/t_n)^{-1} \cdot \Phi_{\delta_0 j_0}(x/t_n) = 1$ and Proposition I.1.1 to get (1.9) or (1.10).

From the definition (1.8) of $L_{b_1,\lambda_1}^{\delta_0,j_0}(t)$ and Assumption 1, it is easily seen that

$$(1.11) \qquad \|[I_{b_1}, L_{b_1,\lambda_1}^{\delta_0,j_0}(t)]\| \leq C_{\delta_0} < t >^{-1-\epsilon},$$

where we have used some commutator estimates in [I.3] including functions of γ, the factor $\varphi_{\sigma_0}(||x|^2/t^2 - 2M| < \tau_2(\lambda))$, and (I.3.9). (Note that $L_{b_1,\lambda_1}^{\delta_0,j_0}(t)$ restricts R^n to the region $|x_\alpha| \geq d_{\delta_0}|t|$ for any α with $\sim (\alpha \leq b_1)$ and some $d_{\delta_0} > 0$ by (I.3.9).)

We next apply Theorem I.4.3 to the terms $L_{b_1,\lambda_1}^{\delta_0,j_0}(t)e^{-itH}\psi(H)f$ on the RHS of (1.10). Then, setting $D_t^{b_1} L_{b_1,\lambda_1}^{\delta_0,j_0}(t) = i[H_{b_1}, L_{b_1,\lambda_1}^{\delta_0,j_0}(t)] + \partial_t L_{b_1,\lambda_1}^{\delta_0,j_0}(t)$, we have

$$(1.12) \qquad \int_{-\infty}^{\infty} |(e^{itH} D_t^{b_1} L_{b_1,\lambda_1}^{\delta_0,j_0}(t)e^{-itH}\psi(H)f, \psi(H)g)| dt \leq C_{\delta_0}\|f\|\|g\|$$

for any $f, g \in \mathcal{H}$ and some constant $C_{\delta_0} > 0$.

Thus from (1.11) and (1.12) follows the existence of the strong limit for $\delta_0 > 0$:

$$(1.13) \qquad s - \lim_{t \to \infty} e^{itH}\psi_1(H^{b_1})\Psi(q_{b_1})L_{b_1,\lambda_1}^{\delta_0,j_0}(t)e^{-itH}\psi(H)f$$

for any $\Psi \in C^\infty(R^{n b_1})$ and $\psi_1 \in C_0^\infty(R^1)$ with bounded derivatives, because $\|[I_{b_1}, \Psi(q_{b_1})]L_{b_1,\lambda_1}^{\delta_0,j_0}(t)\| + \|[I_{b_1}, \psi_1(H^{b_1})]L_{b_1,\lambda_1}^{\delta_0,j_0}(t)\| \leq C_{\delta_0} t^{-1-\epsilon}$ by Assumption 1 and the property of $L_{b_1,\lambda_1}^{\delta_0,j_0}(t)$ stated in the parentheses after (1.11).

As was seen in Part I, the asymptotic decomposition (1.10) and the existence of the limit (1.13) suffice to prove the asymptotic completeness for the short-range quantum systems, if one assumes the induction hypotheses on the completeness of subsystems. We here make an asymptotic decomposition finer than (1.10), which enables one to prove, in the short-range case, the asymptotic completeness directly without assuming the induction hypotheses. Namely we introduce a decomposition of the identity

$$(1.14) \qquad I = \psi_1^{\delta_1}(H^{b_1}) + (I - \psi_1^{\delta_1}(H^{b_1}))$$

on the left of $L^{\delta_0,j_0}_{b_1,\lambda_1}(t_n)$ on the RHS of (1.10). Here $\psi_1^{\delta_1} \in C_0^\infty(R^1)$, $0 \le \psi_1^{\delta_1} \le 1$, $\psi_1^{\delta_1}(\lambda) = 1$ for $\lambda \in R^1$ with $|\lambda - \lambda_2| \le \delta_1$ for all $\lambda_2 \in T^{\tau_0}_{b_1}$ and $\psi_1^{\delta_1}(\lambda) = 0$ away from those neighborhoods of $\lambda_2 \in T^{\tau_0}_{b_1}$, where $8\tau_0 \le \delta_1$. Here $T^{\tau_0}_{b_1}$ is defined from $T_{b_1} \equiv \{0\} \cup \bigcup_{a \le b_1} \sigma_p(H^a)$ as in (1.4) using the same finite set T^F, constants $\sigma_0 > 0$ and $\tau_0 \ge \tau_0(\lambda) > 0$ $(\lambda \in T^F)$ as in the above. Further we take $\psi_1^{\delta_1}$ such that $0 \le \psi_1^{\delta_1}(\lambda) \le \psi_1^{\delta_2}(\lambda)$ when $\delta_1 \le \delta_2$. Thus $I - \psi_1^{\delta_1}(H^{b_1})$ restricts H^{b_1} to the continuous subspace for H^{b_1}. By Proposition I.1.1 and an argument similar to the proof of Theorem I.2.3 (see Sec.I.3), we can easily see that H^{b_1} is asymptotically bounded on the state $L^{\delta_0,j_0}_{b_1,\lambda_1}(t_n)e^{-it_n H}\psi(H)f$ on the RHS of (1.10) as $n \to \infty$. Thus asymptotically, $I - \psi_1^{\delta_1}(H^{b_1})$ can be decomposed as a finite sum of $\psi_{j_1}(H^{b_1})$'s $\in C_0^\infty((-\infty,\rho))$ for some $\rho > 0$, where ψ_{j_1}'s can be taken so that $0 \le \psi_{j_1} \le 1$ and supp $\psi_{j_1} \subset (\Lambda^{b_1}_{j_1} - \delta_0^0, \Lambda^{b_1}_{j_1} + \delta_0^0)$ with $\Lambda^{b_1}_{j_1} \in R^1 - T_{b_1}$ and $\delta_0^0 > 0$ being the constant in the above with $\delta_0^0 << \delta_0$. (Note that $\Lambda^{b_1}_{j_1}$ and ψ_{j_1} depend on $\delta_1 > 0$.) In the following we take $\delta_0 > 0$ and $\delta_1 > 0$ so small that $0 < \delta_0, \delta_1 << \inf(\{|\Lambda - \lambda| | \lambda \in T\})$, and we will let $\delta_0 \downarrow 0$ and $\delta_1 \downarrow 0$. This will cause no problem in the following, since we will let $t \to \infty$ first and δ's $\downarrow 0$ later (see (1.31) below). The order of the limits $\delta_k \downarrow 0$ is irrelevant. It suffices to leave these limits mutually independent. We set $T^{\tau_0}_{\Lambda^{b_1}_{j_1}} = T^{\tau_0}_{b_1} \cap (-\infty, \Lambda^{b_1}_{j_1})$ as in (1.4).

Lemma I.2.1 with $a = b_1$ and $\Lambda^a = \Lambda^{b_1}_{j_1}$ yields the existence of a smooth function $\Phi^{b_1}_{\delta_1 j_1}(x^{b_1}/t)$ such that

$$(1.15) \qquad \Phi^{b_1}_{\delta_1 j_1}(x^{b_1}/t) = \sum_{\ell=1}^{L} \Phi^{b_1}_{M_\ell}(x^{b_1}/t)g_\ell(x^{b_1}/t)$$

with $M_\ell \in [M_{0j_1}, M_{1j_1}]$ $(M_{0j_1} = \inf\{\Lambda^{b_1}_{j_1} - \lambda - \tau_0|\Lambda^{b_1}_{j_1} - \lambda > 0, \lambda \in T_{b_1}\}(>0)$, $M_{1j_1} = \sup\{\Lambda^{b_1}_{j_1} - \lambda + \tau_0|\Lambda^{b_1}_{j_1} - \lambda > 0, \lambda \in T_{b_1}\})$, $g_\ell(x^{b_1}/t) \in C_0^\infty(S^{b_1}_{2\tilde\tau_0}(M_\ell,t))$, $(\tilde\tau_0 = 2\tau_0 + 4\sigma_0 + 2\delta_0^0 < \delta_0)$, and $\Phi^{b_1}_{M_\ell}$ being defined by (I.2.14). Further $\Phi^{b_1}_{\delta_1 j_1}(x^{b_1}/t)$ satisfies the following properties with $S^{b_1}_\tau(M,t) \equiv \{x^{b_1} \in R^{n^{b_1}}|2M - \tau < |x^{b_1}|^2/t^2 < 2M + \tau\}$.

(a) On $S^{b_1}_{\tilde\tau_0(\lambda)}(\Lambda^{b_1}_{j_1} - \lambda, t)$ with $\lambda \in T^{\tau_0}_{\Lambda^{b_1}_{j_1}}$

$$(1.16) \qquad \Phi^{b_1}_{\delta_1 j_1}(x^{b_1}/t) = \Phi^{b_1}_{\Lambda^{b_1}_{j_1} - \lambda}(x^{b_1}/t),$$

where $\tilde\tau_0(\lambda) = 2\tau_0(\lambda) + 4\sigma_0 + 2\delta_0^0$;

(b) On $R^{n^{b_1}}$

$$(1.17) \qquad (0 <)c_0 \le \Phi^{b_1}_{\delta_1 j_1}(x^{b_1}/t) \le C_0(< \infty);$$

and

(c) On each $S^{b_1}_{\tilde\tau_0}(M',t)$ with $M_{0j_1} - \tilde\tau_0 < M' < M_{1j_1} + \tilde\tau_0$, the condition that the summand $\varphi^{M_\ell}_{b_2}(x^{b_2}_{b_2}/t)^2 h^{M_\ell}_{b_2}(x^{b_2}_{b_2}/t)g_\ell(x^{b_1}/t)$ in $\Phi^{b_1}_{\delta_1 j_1}(x^{b_1}/t)$ (see (I.2.14)) does not vanish implies

$$(1.18) \qquad |x^{b_2}|^2/t^2 < 2M'\theta^{b_1}_{|b_2|} + 4\tilde\tau_0.$$

In addition to these properties, similarly to the property (d) for $\Phi_{\delta_0 j_0}(x/t)$ in the above, we can construct the function $\Phi^{b_1}_{\delta_1 j_1}(x^{b_1}/t)$ so that

(d) On $R^{n^{b_1}}$

$$|(\nabla_{x^{b_1}} \Phi^{b_1}_{\delta_1 j_1})(x^{b_1}/t)| \leq C$$

for some constant $C > 0$ independent of $\tau_0, \sigma_0, \delta_0^0, \tau_0(\lambda)$, and $\theta_j^{b_1}$.

As in (I.2.24) we define a selfadjoint operator $L^{b_1,\delta_1,j_1}_{b_2,\lambda_2}(t)$ for $1 \leq |b_1| \leq N$, $b_2 < b_1$, $\lambda_2 \in \mathcal{T}^{\tau_0}_{\Lambda^{b_1}_{j_1}}$ by

(1.19) $$L^{b_1,\delta_1,j_1}_{b_2,\lambda_2}(t) = G^{b_1}_{b_2}(t)G^{b_1}_{b_2}(t)^*,$$

where

$$G^{b_1}_{b_2}(t) = P^{\delta_0^0}_{b_2}(t)\varphi^{\Lambda^{b_1}_{j_1}-\lambda_2}_{b_2 b_1}(q^{b_1}_{b_2})\varphi_{\delta_0^0}(|T_{b_1} - (\Lambda_{j_0} - \Lambda^{b_1}_{j_1})| < 2\delta_0^0)(g \cdot \tilde{\varphi}^{\Lambda^{b_1}_{j_1}-\lambda_2}_{\theta^{b_1}_{|b_2|}})(2T^{b_1}_{b_2})$$

$$\times \varphi_{\sigma_0}(||x|^2/t^2 - 2M| < \tau_2(\lambda_2))\varphi_{\sigma_1}(|\gamma^2 - 2M| < \tau_1(\lambda_2))\phi_{\sigma_1}(2\sigma_1 < \gamma < 2\tilde{\tilde{M}}^1_1).$$

Here $M = \Lambda_{j_0} - \lambda_2$ with $\lambda_2 \in \mathcal{T}^{\tau_0}_{\Lambda^{b_1}_{j_1}}$, $\tau_1(\lambda_2) = 2\tau_0(\lambda_2) + 4\delta_0^0$, $\tau_2(\lambda_2) = 2\tau_0(\lambda_2) + 2\sigma_0$, $0 < 2\sigma_1 < \min\{\sigma_0, \sqrt{\tau_0}\}$, $\tilde{M}^1_1 = \sup\{\Lambda_{j_0} - \lambda_2|\Lambda^{b_1}_{j_1} - \lambda_2 > 0, \lambda_2 \in \mathcal{T}_{b_1}\}$, $\tilde{\tilde{M}}^1_1 = \sup\{\sqrt{2(\Lambda_{j_0} - \lambda_2)}|\Lambda_{j_0} \geq \lambda_2, \lambda_2 \in \mathcal{T}_{b_1}\}$, and $g \in C^\infty_0(R^1)$ with $0 \leq g \leq 1$, $g(\tau) = 1$ ($|\tau| \leq 6\tilde{M}^1_1$), and $g(\tau) = 0$ ($|\tau| \geq 8\tilde{M}^1_1$).

Proposition I.2.3′ . *Choose $\tau_0 > 0$ and $\delta_0^0 > 0$ so small that $8\tau_0 << \min(\{\delta_0\} \cup \{M_{0j_1}\rho^{b_1}_N\})(> 0)$ and that (1.3) and $\delta_0^0 \max(\{\tilde{M}^1_1\} \cup \{\Lambda_{j_0} - \Lambda^{b_1}_{j_1}| j_1\}) << \tau_0^2$ hold. Then we have the following asymptotic relation as $n \to \infty$:*

(1.20)

$$(I - \psi^{\delta_1}_1(H^{b_1}))\psi_{j_0}(H)e^{-it_n H}\psi(H)f \sim \sum^{\text{finite}}_{j_1} \psi_{j_1}(H^{b_1})\psi_{j_0}(H)e^{-it_n H}\psi(H)f$$

$$\sim \Phi_{\delta_0 j_0}(x/t_n)^{-1} \sum_{2 \leq |b_1| \leq N} \sum_{\lambda_1 \in \mathcal{T}^{\tau_0}_{\Lambda_{j_0}}} \sum^{\text{finite}}_{j_1} \psi_{j_1}(H^{b_1})L^{\delta_0,j_0}_{b_1,\lambda_1}(t_n)\psi_{j_0}(H)e^{-it_n H}\psi(H)f$$

$$\sim \Phi_{\delta_0 j_0}(x/t_n)^{-1} \sum_{2 \leq |b_1| \leq N} \sum_{\lambda_1 \in \mathcal{T}^{\tau_0}_{\Lambda_{j_0}}} \sum^{\text{finite}}_{j_1} \Phi^{b_1}_{\delta_1 j_1}(x^{b_1}/t_n)^{-1}$$

$$\times \sum_{b_2 < b_1} \sum_{\lambda_2 \in \mathcal{T}^{\tau_0}_{\Lambda^{b_1}_{j_1}}} L^{b_1,\delta_1,j_1}_{b_2,\lambda_2}(t_n)\psi_{j_1}(H^{b_1}) \cdot L^{\delta_0,j_0}_{b_1,\lambda_1}(t_n)\psi_{j_0}(H)e^{-it_n H}\psi(H)f.$$

PROOF: Similar to that of Theorem I.2.3. We repeat however some points which should be taken care of when the argument of Sec.I.3 is adapted to the present case. The first point which should be noted is that we again introduce the decomposition (I.1.15) or (I.3.1) into (1.20):

$$(1.21) \qquad e^{-it_n H}\psi(H)f = \sum_{2\leq|d|\leq N} P_d^{M_{|d|}^n} \tilde{P}_{a_1 d}^{M_{a_1 d}^n} e^{-it_n H}\psi(H)f,$$

where $|a_1| = 1$. Then applying (I.3.9) and (I.3.6) to $L_{b_1,\lambda_1}^{\delta_0,j_0}(t_n)$, we remove the terms with $\sim (d \leq b_1)$ from (1.20) asymptotically.

To obtain the factor $\varphi_{\sigma_0}(||x|^2/t_n^2 - 2M| < \tau_2(\lambda_2))$ in the definition (1.19) of $L_{b_2,\lambda_2}^{b_1,\delta_1,j_1}(t_n)$, $(M = \Lambda_{j_0} - \lambda_2, \lambda_2 \in T_{b_1}^{\tau_0})$, we note by $d \leq b_1$ that $|x|^2/t_n^2 = |x_d|^2/t_n^2 + |x^d|^2/t_n^2$ is asymptotically equal to $2T_d$ on each state of (1.20). Since $T_d = H_d - H^d$ and $H_d \sim H_{b_1} \sim H$ by $d \leq b_1$, $2T_d \sim 2(H - H^d)$ asymptotically. Since on the state $\psi_{j_1}(H^{b_1})\psi_{j_0}(H)\tilde{P}_{a_1 d}^{M_{a_1 d}^n}e^{-it_n H}\psi(H)f$, $|H - \Lambda_{j_0}| < \delta_0^0$ and the eigenvalues of H^d corresponding to the projection $P_d^{M_{|d|}^n}$ for $d \leq b_1$ are included in $\bigcup_{\lambda \in T_{b_1}^{\tau_0}}[\lambda - \tau_0(\lambda), \lambda + \tau_0(\lambda)]$, we have from these that $||x|^2/t_n^2 - 2M| < 2\delta_0^0 + 2\tau_0(\lambda_2) < \tau_2(\lambda_2)$ asymptotically on each state of (1.20) for some $\lambda_2 \in T_{b_1}^{\tau_0}$.

The factor $\varphi_{\delta_0^0}(|T_{b_1} - (\Lambda_{j_0} - \Lambda_{j_1}^{b_1})| < 2\delta_0^0)$ in (1.19) is obtained if one notes that $T_{b_1} = H_{b_1} - H^{b_1}$ and $H_{b_1} \sim H$ asymptotically and that $|H - \Lambda_{j_0}| < \delta_0^0$ on $\psi_{j_0}(H)$ and $|H^{b_1} - \Lambda_{j_1}^{b_1}| < \delta_0^0$ on $\psi_{j_1}(H^{b_1})$.

In order to apply Lemma I.2.1, namely, to introduce $\Phi_{\delta_1 j_1}^{b_1}(x^{b_1}/t_n)$ in the asymptotic decomposition (1.20), it needs to see that $|x^{b_1}|^2/t_n^2 > 2M_{0j_1} - \tilde{\tau}_0$ for instance, since $\Phi_{\delta_1 j_1}^{b_1}$ merely decomposes the ring $\{x^{b_1} \in R^{n^{b_1}} | 2M_{0j_1} - \tilde{\tau}_0 < |x^{b_1}|^2/t_n^2 < 2M_{1j_1} + \tilde{\tau}_0\}$. Then, as we will see, one can insert $1 = \Phi_{\delta_1 j_1}^{b_1}(x^{b_1}/t_n)^{-1} \cdot \Phi_{\delta_1 j_1}^{b_1}(x^{b_1}/t_n)$ into the second line of (1.20) between $\sum_{j_1}^{\text{finite}}$ and $\psi_{j_1}(H^{b_1})$ and get the third line using Proposition I.1.1. To see $|x^{b_1}|^2/t_n^2 > 2M_{0j_1} - \tilde{\tau}_0$, we note $|x^{b_1}|^2/t_n^2 = |x|^2/t_n^2 - |x_{b_1}|^2/t_n^2 \sim |x|^2/t_n^2 - 2T_{b_1}$ asymptotically since $d \leq b_1$. But by the factors $\varphi_{\sigma_0}(||x|^2/t_n^2 - 2M| < \tau_2(\lambda_2))$ and $\varphi_{\delta_0^0}(|T_{b_1} - (\Lambda_{j_0} - \Lambda_{j_1}^{b_1})| < 2\delta_0^0)$ obtained above, we have

$$|(|x|^2/t_n^2 - 2T_{b_1}) - 2(\Lambda_{j_1}^{b_1} - \lambda_2)| < \tau_2(\lambda_2) + \sigma_0 + 6\delta_0^0 = 2\tau_0(\lambda_2) + 3\sigma_0 + 6\delta_0^0,$$

which implies $||x^{b_1}|^2/t_n^2 - 2(\Lambda_{j_1}^{b_1} - \lambda_2)| < \tilde{\tau}_0(\lambda_2) - 2\delta_0^0 = 2\tau_0(\lambda_2) + 4\sigma_0 \leq \tilde{\tau}_0 - 2\delta_0^0$ since $\delta_0^0 << \sigma_0$. It remains to show that $\lambda_2 \in T_{\Lambda_{j_1}^{b_1}}^{\tau_0}$ for $\lambda_2 \in T_{b_1}^{\tau_0}$, namely, $\lambda_2 < \Lambda_{j_1}^{b_1}$. If $\Lambda_{j_1}^{b_1} \leq \lambda_2$, then $\Lambda_{j_1}^{b_1} - \lambda_2 \leq -\delta_1 \leq -8\tau_0$ by the definitions of $\Lambda_{j_1}^{b_1}$ and $\psi_1^{\delta_1}$. Then $|x^{b_1}|^2/t_n^2 < 2(\Lambda_{j_1}^{b_1} - \lambda_2) + \tilde{\tau}_0(\lambda_2) \leq -8\tau_0 + \tilde{\tau}_0(\lambda_2) \leq -8\tau_0 + 2\tau_0 + 4\sigma_0 + 2\delta_0^0 < 0$ by $0 < \sigma_0 < \tau_0$ and $0 < \delta_0^0 << \tau_0$, a contradiction. Thus $\Lambda_{j_1}^{b_1} - \lambda_2 \geq \delta_1 \geq 8\tau_0$, and $|x^{b_1}|^2/t_n^2 \geq 2(\Lambda_{j_1}^{b_1} - \lambda_2) - \tilde{\tau}_0(\lambda_2) > 2M_{0j_1} - \tilde{\tau}_0$.

In particular we have proved that as $n \to \infty$

$$(1.22)$$
$$\Phi_{\delta_1 j_1}^{b_1}(x^{b_1}/t_n)\psi_{j_1}(H^{b_1})L_{b_1,j_1}^{\delta_0,j_0}(t_n)\psi_{j_0}(H)e^{-it_n H}\psi(H)f$$

$$\sim \sum_{d \le b_1} \sum_{\lambda_2 \in T_{\Lambda_{j_1}^{b_1}}^{\tau_0}} \sum_{b_2 < b_1} \varphi_{b_2}^{\Lambda_{j_1}^{b_1} - \lambda_2}(x_{b_2}^{b_1}/t_n)^2 h_{b_2}^{\Lambda_{j_1}^{b_1} - \lambda_2}(x_{b_2}^{b_1}/t_n)$$

$$\times \varphi_{\delta_0^0}(|T_{b_1} - (\Lambda_{j_0} - \Lambda_{j_1}^{b_1})| < 2\delta_0^0)^2 \varphi_{\delta_0^0}(||x^{b_1}|^2/t_n^2 - 2(\Lambda_{j_1}^{b_1} - \lambda_2)| < \tau_2(\lambda_2) + 2\sigma_0)$$

$$\times \varphi_{\sigma_0}(||x|^2/t_n^2 - 2M| < \tau_2(\lambda_2))^2 \varphi_{\sigma_1}(|\gamma^2 - 2M| < \tau_1(\lambda_2))^2$$

$$\times \phi_{\sigma_1}(2\sigma_1 < \gamma < 2\tilde{M}_1^1)^2$$

$$\times \psi_{j_1}(H^{b_1}) L_{b_1,j_1}^{\delta_0,j_0}(t_n) \psi_{j_0}(H) \tilde{P}_{a_1 d}^{M_{a_1 d}^n} e^{-it_n H} \psi(H) f.$$

By $|x^{b_1}|^2 = |x^{b_2}|^2 + |x_{b_2}^{b_1}|^2$ for $b_2 < b_1$, we have, on the support of $h_{b_2}^{\Lambda_{j_1}^{b_1} - \lambda_2}(x_{b_2}^{b_1}/t_n)$ $\varphi_{\delta_0^0}(||x^{b_1}|^2/t_n^2 - 2(\Lambda_{j_1}^{b_1} - \lambda_2)| < \tau_2(\lambda_2) + 2\sigma_0)$, that $|x^{b_2}|^2/t_n^2 < L \equiv 2(\Lambda_{j_1}^{b_1} - \lambda_2)\theta_{|b_2|}^{b_1} + 2\tau_2(\lambda_2) + 6\delta_0^0 + 6\sigma_0 + \tau_0$. Proposition I.1.1-(I.1.16), (I.2.12), and $L < 4(\Lambda_{j_1}^{b_1} - \lambda_2)\theta_{|b_2|}^{b_1}$ imply that

$$(1.23) \qquad \|\varphi_{b_2}^{\Lambda_{j_1}^{b_1} - \lambda_2}(x_{b_2}^{b_1}/t_n)^2 \phi(|x^{b_2}|^2/t_n^2 < L) L_{b_1,j_1}^{\delta_0,j_0}(t_n) \tilde{P}_{a_1 d}^{M_{a_1 d}^n} e^{-it_n H} \psi(H) f\| \to 0$$

as $n \to \infty$ for $\sim (d \le b_2)$ and $d \le b_1$. Thus we obtain

(1.24)

$$\Phi_{\delta_1 j_1}^{b_1}(x^{b_1}/t_n) \psi_{j_1}(H^{b_1}) L_{b_1,j_1}^{\delta_0,j_0}(t_n) \psi_{j_0}(H) e^{-it_n H} \psi(H) f$$

$$\sim \sum_{b_2 < b_1} \sum_{d \le b_2} \sum_{\lambda_2 \in T_{\Lambda_{j_1}^{b_1}}^{\tau_0}} \varphi_{b_2}^{\Lambda_{j_1}^{b_1} - \lambda_2}(x_{b_2}^{b_1}/t_n)^2 h_{b_2}^{\Lambda_{j_1}^{b_1} - \lambda_2}(x_{b_2}^{b_1}/t_n)$$

$$\times \varphi_{\delta_0^0}(|T_{b_1} - (\Lambda_{j_0} - \Lambda_{j_1}^{b_1})| < 2\delta_0^0)^2 \varphi_{\delta_0^0}(||x^{b_1}|^2/t_n^2 - 2(\Lambda_{j_1}^{b_1} - \lambda_2)| < \tau_2(\lambda_2) + 2\sigma_0)$$

$$\times \varphi_{\sigma_0}(||x|^2/t_n^2 - 2M| < \tau_2(\lambda_2))^2 \varphi_{\sigma_1}(|\gamma^2 - 2M| < \tau_1(\lambda_2))^2$$

$$\times \phi_{\sigma_1}(2\sigma_1 < \gamma < 2\tilde{M}_1^1)^2$$

$$\times \psi_{j_1}(H^{b_1}) L_{b_1,j_1}^{\delta_0,j_0}(t_n) \psi_{j_0}(H) \tilde{P}_{a_1 d}^{M_{a_1 d}^n} e^{-it_n H} \psi(H) f.$$

as $n \to \infty$. Further by a simple calculus of ψdo's, $d \le b_2$, and Proposition I.1.1-(I.1.18), we have that

(1.25)

$$\Phi_{\delta_1 j_1}^{b_1}(x^{b_1}/t_n) \psi_{j_1}(H^{b_1}) L_{b_1,j_1}^{\delta_0,j_0}(t_n) \psi_{j_0}(H) e^{-it_n H} \psi(H) f$$

$$\sim \sum_{b_2 < b_1} \sum_{\lambda_2 \in T_{\Lambda_{j_1}^{b_1}}^{\tau_0}} \sum_{d \le b_2} P_{b_2}^{\delta_0^0}(t_n) P_{b_2}^{\delta_0^0}(t_n)^* \varphi_{b_2 b_1}^{\Lambda_{j_1}^{b_1} - \lambda_2}(q_{b_2}^{b_1})^2 h_{b_2}^{\Lambda_{j_1}^{b_1} - \lambda_2}(x_{b_2}^{b_1}/t_n)$$

$$\times \varphi_{\delta_0^0}(|T_{b_1} - (\Lambda_{j_0} - \Lambda_{j_1}^{b_1})| < 2\delta_0^0)^2 \varphi_{\delta_0^0}(||x^{b_1}|^2/t_n^2 - 2(\Lambda_{j_1}^{b_1} - \lambda_2)| < \tau_2(\lambda_2) + 2\sigma_0)$$

$$\times \varphi_{\sigma_0}(||x|^2/t_n^2 - 2M| < \tau_2(\lambda_2))^2 \varphi_{\sigma_1}(|\gamma^2 - 2M| < \tau_1(\lambda_2))^2$$

$$\times \phi_{\sigma_1}(2\sigma_1 < \gamma < 2\tilde{M}_1^1)^2$$

$$\times \psi_{j_1}(H^{b_1})L^{\delta_0,j_0}_{b_1,j_1}(t_n)\psi_{j_0}(H)\tilde{P}^{M^n_{a_1 d}}_{a_1 d}e^{-it_n H}\psi(H)f.$$

as $n \to \infty$. (Note that every two factors on the left of $\tilde{P}^{M^n_{a_1 d}}_{a_1 d}$ commute with each other asymptotically by Proposition I.1.1 and ψdo calculus.) Since $|x^{b_1}_{b_2}|^2/t^2_n$ asymptotically equals $2T^{b_1}_{b_2} = 2(T_{b_2} - T_{b_1})$ and $2T^{b_1}_{b_2} \leq 2T_{b_2} \leq 2T_d \leq 2M + \tau_2(\lambda_2) + \sigma_0$ asymptotically on each state on the right hand side of (1.25), we can replace the factor $h^{\Lambda^{b_1}_{j_1}-\lambda_2}_{b_2}(x^{b_1}_{b_2}/t_n)$ asymptotically by $(g \cdot \tilde{\varphi}^{\Lambda^{b_1}_{j_1}-\lambda_2}_{\theta^{b_1}_{|b_2|}})(2T^{b_1}_{b_2})^2$ on each state in (1.25).

To conclude the proof of (1.20), we first remove asymptotically the factor $\varphi_{\delta^0_0}(||x^{b_1}|^2/t^2_n - 2(\Lambda^{b_1}_{j_1} - \lambda_2)| < \tau_2(\lambda) + 2\sigma_0)$ by using the factors $\varphi_{\sigma_0}(||x|^2/t^2_n - 2M| < \tau_2(\lambda_2))$ and $\varphi_{\delta^0_0}(|T_{b_1} - (\Lambda_{j_0} - \Lambda^{b_1}_{j_1})| < 2\delta^0_0)$. Then we rearrange the order of the factors on the left of $\tilde{P}^{M^n_{a_1 d}}_{a_1 d}$ as in (1.19) asymptotically, and doing a simple calculus of ψdo's, we note that with $\delta = \delta^0_0$

(1.26)

$$\left\|P^\delta_{b_2}(t)g(2T^{b_1}_{b_2})\varphi^{\Lambda^{b_1}_{j_1}-\lambda_2}_{b_2 b_1}(q^{b_1}_{b_2})(\{\prod_{k=1}^{k^{b_1}_{b_2}}\phi_\sigma(|z^{b_1}_k|^2/t^2 > 2(\Lambda^{b_1}_{j_1} - \lambda_2)\rho^{b_1}_{|b_2|} - \sigma - 9\sqrt{\delta\tilde{M}^1_1})\} - 1)\right\|$$
$$\leq C_\ell < t >^{-\ell},$$

$$\left\|P^\delta_{b_2}(t)g(2T^{b_1}_{b_2})\tilde{\varphi}^{\Lambda^{b_1}_{j_1}-\lambda_2}_{\theta^{b_1}_{|b_2|}}(2T^{b_1}_{b_2})\right.$$
$$\left.\times \phi_\sigma(|x^{b_1}_{b_2}|^2/t^2 < 2(\Lambda^{b_1}_{j_1} - \lambda_2)(1 - \theta^{b_1}_{|b_2|}) - \tilde{\tau}_0(\lambda) - 2\sigma - 9\sqrt{\delta\tilde{M}^1_1})\right\|$$
$$\leq C_\ell < t >^{-\ell},$$

$$\left\|P^\delta_{b_2}(t)\varphi_\delta(|T_{b_1} - (\Lambda_{j_0} - \Lambda^{b_1}_{j_1})| < 2\delta)\right.$$
$$\left.\times \phi_\sigma(|x_{b_1}|^2/t^2 < 2(\Lambda_{j_0} - \Lambda^{b_1}_{j_1}) - 8\delta - 9\sqrt{\delta|\Lambda_{j_0} - \Lambda^{b_1}_{j_1}|})\right\| \leq C_\ell < t >^{-\ell}$$

for any $\ell \geq 1$. Thus by $|x|^2 = |x_{b_1}|^2 + |x^{b_1}_{b_2}|^2 + |x^{b_2}|^2$ for $b_2 < b_1$, (I.2.12), (1.23), the factor $\varphi_{\sigma_0}(||x|^2/t^2_n - 2M| < \tau_2(\lambda_2))L^{\delta_0,j_0}_{b_1,j_1}(t_n)$, and our choice of $\sigma, \delta^0_0, \sigma_0, \tau_0$, we can return the sum over $d \leq b_2$ with $d < a_1$ to the one over all $d < a_1$ asymptotically. Then using (I.1.15) and $(I - P^{M^n_{|a_1|}}_{a_1})\psi(H) = \psi(H)$, we conclude the proof of (1.20). $\square$

Continuing this procedure and noting that $P_a = I$ for $|a| = N$, we obtain the following equality:

(1.27)
$$\psi(H)f$$
$$= s - \lim_{\delta_N \downarrow 0} \cdots s - \lim_{\delta_0 \downarrow 0} s - \lim_{n\to\infty} e^{it_n H}\times$$

$$\times \sum_{\ell=1}^{N} \sum_{k=0}^{\ell-1} \sum_{j_0} \Phi_{\delta_0 j_0}(x/t_n)^{-1} \sum_{2 \le |b_1| \le N} \sum_{\lambda_1 \in T_{\Lambda_{j_0}}^{\tau_0}} \sum_{j_1} \Phi_{\delta_1 j_1}^{b_1}(x^{b_1}/t_n)^{-1}$$

$$\times \sum_{b_2 < b_1} \sum_{\lambda_2 \in T_{\Lambda_{j_1}^{b_1}}^{\tau_0}} \sum_{j_2} \cdots \sum_{j_k} \Phi_{\delta_k j_k}^{b_k}(x^{b_k}/t_n)^{-1} \sum_{b_{k+1} < b_k} \sum_{\lambda_{k+1} \in T_{\Lambda_{j_k}^{b_k}}^{\tau_0}}$$

$$\times \psi_{k+1}^{\delta_{k+1}}(H^{b_{k+1}}) L_{b_{k+1},\lambda_{k+1}}^{b_k,\delta_k,j_k}(t_n) \psi_{j_k}(H^{b_k}) \cdots L_{b_1,\lambda_1}^{\delta_0,j_0}(t_n) \psi_{j_0}(H) e^{-it_n H} \psi(H) f,$$

where $|b_\ell| = N$ and $\psi_\ell^{\delta_\ell}(H^{b_\ell}) = I$. Using Proposition I.1.1 and the decomposition (1.21), it is easy to see that $\Phi_{\delta_k j_k}^{b_k}(x^{b_k}/t_n)^{-1}$ on the RHS of (1.27) asymptotically equals $\Phi_{\delta_k j_k}^{b_k}(x^{b_{k+1}}/t_n, q_{b_{k+1}}^{b_k})^{-1}$ as $n \to \infty$. Further by the properties (b) and (d) of $\Phi_{\delta_k j_k}^{b_k}$ in the above, we have

$$(1.28) \qquad |\Phi_{\delta_k j_k}^{b_k}(x^{b_{k+1}}/t_n, q_{b_{k+1}}^{b_k})^{-1} - \Phi_{\delta_k j_k}^{b_k}(0, q_{b_{k+1}}^{b_k})^{-1}|$$

$$\le c_0^{-2} t_n^{-1} |x^{b_{k+1}}| \left| \int_0^1 (\nabla_{x^{b_{k+1}}} \Phi_{\delta_k j_k}^{b_k})(\theta x^{b_{k+1}}/t_n, q_{b_{k+1}}^{b_k}) d\theta \right|$$

$$\le C |x^{b_{k+1}}|/t_n$$

for some constant $C > 0$ independent of $\tau_0, \sigma_0, \delta_0^0, \tau_0(\lambda)$, and $\theta_j^{b_k}$. But by the factors $P_{b_{k+1}}^{\delta_0^0}(t)$, $\varphi_{\delta_0^0}(|T_{b_k} - (\Lambda_{j_0} - \Lambda_{j_k}^{b_k})| < 2\delta_0^0)$, $\tilde{\varphi}_{\theta_{|b_{k+1}|}^{b_k}}^{\Lambda_{j_k}^{b_k} - \lambda_{k+1}}(2T_{b_{k+1}}^{b_k})$, and $\varphi_{\sigma_0}(||x|^2/t^2 - 2(\Lambda_{j_0} - \lambda_{k+1})| < \tau_2(\lambda_{k+1}))$ in the definition of $L_{b_{k+1},\lambda_{k+1}}^{b_k,\delta_k,j_k}(t)$, one has, using (1.26),

$$(1.29) \qquad |x^{b_{k+1}}|^2/t_n^2 \le 2(\Lambda_{j_k}^{b_k} - \lambda_{k+1})\theta_{|b_{k+1}|}^{b_k} + 10\tau_0$$

on each state on the RHS of (1.27). Since the above constant C in (1.28) is independent of $\theta_j^{b_k}$ and $0 < \tau_0 << M_{0j_k}\theta_j^{b_k}$ (see Lemma I.2.1), we can let $\theta_j^{b_k} \to 0$ as $\delta_0 \downarrow 0$ (recall that $8\tau_0 < \delta_0$). Hence as $n \to \infty$ and $\delta_0 \downarrow 0$, the function $\Phi_{\delta_k j_k}^{b_k}(x^{b_k}/t_n)^{-1}$ in (1.27) can be replaced asymptotically by some smooth function $\tilde{\Phi}_{\delta_k j_k}^{b_k}(q_{b_{k+1}}^{b_k})^{-1}$ of $q_{b_{k+1}}^{b_k}$ with $c_0 \le \tilde{\Phi}_{\delta_k j_k}^{b_k} \le C_0$.

We remark here that the smooth terms added to $\Phi_{\delta_k j_k}^{b_k}(x^{b_k}/t)$ in the above at the property (d) can be dealt with similarly. Namely that the supports of those terms are near the boundary of the sets $T_b^{b_k}(2M'\rho_{|b|}^{b_k}, 2M'\theta_{|b|}^{b_k})$ ($b = b_{k+1} < b_k, M' = \Lambda_{j_k}^{b_k} - \lambda_{k+1}$) yields that $|x^b|^2/t^2$ is near $2M'\theta_{|b|}^{b_k}$, which and that these added terms are the functions of $x_b^{b_k}/t$ near the boundary of the set $T_b^{b_k}(2M'\rho_{|b|}^{b_k}, 2M'\theta_{|b|}^{b_k})$ enable one to argue similarly to the above, and conclude that those terms can be regarded as smooth functions of $q_b^{b_k}$ asymptotically as $n \to \infty$ and $\delta_0 \downarrow 0$. Thus in taking the limit with respect to $t \to \infty$ and δ's $\downarrow 0$ in the following, we can argue in quite the same way as for the other original terms in the definition of $\Phi_{\delta_k j_k}^{b_k}(x^{b_k}/t)$. (Note that the added terms correspond to the factor $\varphi_{bb_k}^M(q_b^{b_k})\varphi_{\theta_{|b|}^{b_k}}^M(2T_b^{b_k})$ in $L_{b,\lambda_{k+1}}^{b_k,\delta_k,j_k}(t)$. Other factors in $L_{b,\lambda_{k+1}}^{b_k,\delta_k,j_k}(t)$ are obtained by the arguments similar to the ones after (1.21) independently of those added terms, once the terms with $\sim (d \leq b_k)$ have been removed from (1.21) by using the factor $L_{b_k,\lambda_k}^{b_{k-1},\delta_{k-1},j_{k-1}}(t) \cdots L_{b_1,\lambda_1}^{\delta_0,j_0}(t).)$

By properties of $L_{b_{k+1},\lambda_{k+1}}^{b_k,\delta_k,j_k}(t)\psi_{j_k}(H^{b_k}) \cdots L_{b_1,\lambda_1}^{\delta_0,j_0}(t)\psi_{j_0}(H)$ corresponding to (1.11)-(1.12) above, which can be proved similarly to the proof of Theorem I.4.3, we can easily see that the strong limit with respect to $t \to \infty$ of each term with $b_{k+1} = b$ on the RHS of (1.27) exists. (We remark that I_{b_1} is taken care of by $L_{b_1,\lambda_1}^{\delta_0,j_0}(t)$, $I_{b_2}^{b_1} = I_{b_2} - I_{b_1}$ is by $L_{b_2,\lambda_2}^{b_1,\delta_1,j_1}(t)$, and so on.) Further returning the sum over j_m in each term on the RHS of (1.27) to $I - \psi_m^{\delta_m}(H^{b_m})$ asymptotically as $n \to \infty$ and $\delta_0 \downarrow 0$ by (1.20) and an argument reverse to the above on $\tilde{\Phi}_{\delta_k j_k}^{b_k}(q_{b_{k+1}}^{b_k})^{-1}$, and arguing quite similarly to [I.6, Lemma 4.8], we can prove that the strong limit with respect to δ's of each term of (1.27) exists. (Use that $I - \psi_m^{\delta_m}(H^b)$ is monotonically increasing when $\delta_m \downarrow 0$ and that $(\psi_m^{\delta_2}(H^b) - \psi_m^{\delta_1}(H^b))(\psi_m^{\delta_0}(H^b) - \psi_m^{\delta_3}(H^b)) = 0$ if $0 < \delta_1 \leq \delta_2 << \delta_3 < \delta_0$, by the definition of ψ_m^{δ}.)

Summing up, we have proved the following

Theorem 1.1. *Let Assumptions 1 and 2 be satisfied. Let $f \in \mathcal{H}$, and $\psi \in C_0^\infty(R^1)$ with supp ψ bounded away from $\mathcal{T}$. Let the notation be as above, and set*

$$(1.30) \qquad \Psi_{\delta_0,\cdots,\delta_k}^{b_1,\cdots,b_k}(q_{b_{k+1}}) = \tilde{\Phi}_{\delta_k j_k}^{b_k}(q_{b_{k+1}}^{b_k})^{-1} \cdots \tilde{\Phi}_{\delta_0 j_0}(q_{b_1})^{-1},$$

$$\tilde{L}_{\delta_0,\cdots,\delta_k}^{b_1,\cdots,b_{k+1}}(t) = L_{b_{k+1},\lambda_{k+1}}^{b_k,\delta_k,j_k}(t)\psi_{j_k}(H^{b_k}) \cdots L_{b_1,\lambda_1}^{\delta_0,j_0}(t)\psi_{j_0}(H).$$

Then the strong limit

$$(1.31) \qquad \Omega_b^\psi f = s - \lim_{\delta_N \downarrow 0} \cdots s - \lim_{\delta_0 \downarrow 0} s - \lim_{t \to \infty} \sum e^{itH}\psi_{k+1}^{\delta_{k+1}}(H^{b_k+1})$$
$$\times \Psi_{\delta_0,\cdots,\delta_k}^{b_1,\cdots,b_k}(q_{b_{k+1}})\tilde{L}_{\delta_0,\cdots,\delta_k}^{b_1,\cdots,b_{k+1}}(t)e^{-itH}\psi(H)f$$

exists, where the summation is taken over all sums in (1.27) with $b_{k+1} = b$, and the following equality holds:

$$(1.32) \qquad \psi(H)f = \bigoplus_{2 \leq |b| \leq N} \Omega_b^\psi f.$$

PROOF: There remains to show the orthogonality in (1.32). Letting $b \neq b'$, we have to show $(\Omega_b^\psi f, \Omega_{b'}^\psi g) = 0$ for all $f, g \in \mathcal{H}$. We may assume $j = |b| \leq |b'| = j'$. Then either $b' \not\leq b$ or $b' < b$.

When $b' \not\leq b$, the result follows from the definition of $\tilde{L}_{\cdots}^{\cdots}(t)$ by

$$(\Omega_b^\psi f, \Omega_{b'}^\psi g)$$
$$= \sum \lim_{t \to \infty} (\psi_{k+1}^{\delta_{k+1}}(H^b)\Psi_{\cdots}^{\cdots}(q_{b_k})\tilde{L}_{\delta_0,\cdots,\delta_k}^{b_1,\cdots,b}(t)e^{-itH}\psi(H)f,$$
$$\psi_{k'+1}^{\delta_{k'+1}}(H^{b'})\Psi_{\cdots}^{\cdots}(q_{b_{k'}})\tilde{L}_{\delta_0,\cdots,\delta_{k'}}^{b_1,\cdots,b'}(t)e^{-itH}\psi(H)g).$$

In fact, $\tilde{L}_{\delta_0,\cdots,\delta_{k'}}^{b_1,\cdots,b'}(t)$ and $\tilde{L}_{\delta_0,\cdots,\delta_k}^{b_1,\cdots,b}(t)$ bound $x^{b'}$ and x^b, through $\psi^\delta(H^b)$'s and $\Psi_{\cdots}^{\cdots}(q_b)$'s with errors of $O(t^{-1})$, as $|x^{b'}|^2/t^2 \leq M\theta_{j'}^{a'}$ and $|x^b|^2/t^2 \leq M\theta_j^a$ respectively by (1.29), where $M > 0$ is some constant and $b' < a'$, $b < a$. Let $b \cup b'$ be the minimal cluster decomposition such that $b, b' \leq b \cup b'$. Then these estimates yield $|x^{b \cup b'}|^2/t^2 \leq M(\theta_{j'}^{a'} + \theta_j^a)$. Since $b' \not\leq b$, the coordinate $x^{b \cup b'}$ contains at least one z_ℓ that connects the centers of mass of 2 different clusters in b. Hence $|z_\ell|^2/t^2 \leq M(\theta_{j'}^{a'} + \theta_j^a)$. But $\tilde{L}_{\delta_0,\cdots,\delta_k}^{b_1,\cdots,b}(t)$ separates z_ℓ as $|z_\ell|^2/t^2 \geq M'\rho_j^d$ for some $M' > 0$ and $|d| \geq 1$. In Lemma I.2.1 and the arguments before the theorem, we may take ρ_j^d and $\theta_{j'}^{a'}$ such that $\min_d \rho_j^d \gg \max_{a'} \theta_{j'}^{a'}$ since $j \leq j'$. Thus we can let $|z_\ell|^2/t^2 \geq M'\rho_j^d > M(\theta_{j'}^{a'} + \theta_j^a) \geq |z_\ell|^2/t^2$, a contradiction.

We next consider the case $b' < b$. Then by the construction, $\Omega_{b',}^\psi g$ is included in the range of the limit (1.31) with $b_{k+1} = b$ and $\psi_{k+1}^{\delta_{k+1}}(H^{b_{k+1}})$ replaced by $I - \psi_{k+1}^{\delta_{k+1}}(H^b)$, where we have used the same argument as the one just before Theorem 1.1. Denoting this limit by $\Omega_b^{c\psi}g$, we thus have to show $(\Omega_b^\psi f, \Omega_b^{c\psi} g) = 0$. By the definitions of Ω_b^ψ and $\Omega_b^{c\psi}$, and the relation $(I - \psi^\delta(H^b))\psi^\delta(H^b) = \psi^\delta(H^b) - \psi^\delta(H^b)^2 (\geq 0)$, $(\Omega_b^\psi f, \Omega_b^{c\psi} g)$ is equal to $(\Omega_b^\psi f, \tilde{\Omega}_b^\psi g) - (\hat{\Omega}_b^\psi f, \tilde{\Omega}_b^\psi g)$, where $\hat{\Omega}_b^\psi f$ and $\tilde{\Omega}_b^\psi g$ are defined by (1.31) with $\psi^\delta(H^b)$ replaced by $\psi^\delta(H^b)^2$ and the identity, respectively. But since $\psi^\delta(H^b)^2 \leq \psi^\delta(H^b)$ and $\psi^{\delta'}(H^b) \leq \psi^\delta(H^b)^2$ for any $0 < \delta' \ll \delta$, the limits $\Omega_b^\psi f$ and $\hat{\Omega}_b^\psi f$ are equal. This completes the proof of the theorem. $\Box$

It is easy to see by using (1.30) and the definition of modified wave operators that $\Omega_b^\psi W_b^D f = \psi(H) W_b^D f$ for $f \in \mathcal{H}$. Hence $\mathcal{R}(\Omega_b^\psi) \supset \mathcal{R}(\psi(H)W_b^D)$. Thus (1.32) gives a necessary and sufficient condition for the asymptotic completeness to hold as stated in the introduction.

The asymptotic completeness part of Theorem 3 for the short-range case follows, if we have for any $2 \leq |b| \leq N$ and some $g_b \in \mathcal{H}$

$$(1.33) \qquad \Omega_b^\psi f = W_b^D g_b.$$

This and the asymptotic clustering for the long-range case will be shown in the next section.

2. Proof of Theorem 3

We begin with proofs of the existence of some strong limits. As in the introduction we define $U_b(t)$ as a unitary propagator generated by $H_b(t) = H_b + I_b^L(t, x)$:

$$(2.1) \qquad (\partial_t + iH_b(t))U_b(t) = 0, \qquad U_b(0) = I.$$

We first prove for $\Psi_{\cdots}^{\cdots}(q_b)$ and $\tilde{L}_{\cdots}^{\cdots}(t)$ defined by (1.30) with $b_{k+1} = b$ the following proposition, which and Theorem 1.1 prove the asymptotic clustering for the long-range case in Theorem 3 with

$$f_b^\pm = s - \lim_{\delta's \downarrow 0} s - \lim_{t \to \pm\infty} \sum_{b_{k+1}=b} U_{b_{k+1}}(t)^* \psi_{k+1}^{\delta_{k+1}}(H^{b_{k+1}}) \Psi_{\cdots}^{\cdots}(q_{b_{k+1}}) \tilde{L}_{\cdots}^{\cdots}(t) e^{-itH} \psi(H) f.$$

Proposition 2.1. *There exists the strong limit*

(2.2) $$\tilde{\Omega}_b^\psi f \equiv s - \lim_{\delta's \downarrow 0} s - \lim_{t \to \infty} U_b(t)^* \Psi_{\cdots}^{\cdots}(q_b) \tilde{L}_{\cdots}^{\cdots}(t) e^{-itH} \psi(H) f.$$

The limit also exists with $\Psi_{\cdots}^{\cdots}(q_b)$ *removed or replaced by* $\psi^\delta(H^b)\Psi_{\cdots}^{\cdots}(q_b)$.

PROOF: Since $\|[\tilde{L}_{\cdots}^{\cdots}(t), \Psi_{\cdots}^{\cdots}(q_b)]\| \leq Ct^{-1}$ by the constructions of $\tilde{L}_{\cdots}^{\cdots}(t)$ and $\Psi_{\cdots}^{\cdots}(q_b)$, we have to prove the existence of

(2.3) $$s - \lim_{t \to \infty} U_b(t)^* \tilde{L}_{\cdots}^{\cdots}(t) \Psi_{\cdots}^{\cdots}(q_b) e^{-itH} \psi(H) f.$$

We differentiate the state in (2.3) with respect to t:

(2.4)
$$\frac{d}{dt} U_b(t)^* \tilde{L}_{\cdots}^{\cdots}(t) \Psi_{\cdots}^{\cdots}(q_b) e^{-itH} \psi(H) f$$
$$= U_b(t)^* \{i(I_b^L(t, x_b, x^b) - I_b^L(x_b, x^b)) \tilde{L}_{\cdots}^{\cdots}(t) \Psi_{\cdots}^{\cdots}(q_b) + E(t)\} e^{-itH} \psi(H) f,$$

where $E(t) = i[H_b + I_b^L, \tilde{L}_{\cdots}^{\cdots}(t) \Psi_{\cdots}^{\cdots}(q_b)] - i\tilde{L}_{\cdots}^{\cdots}(t) \Psi_{\cdots}^{\cdots}(q_b) I_b^S + \partial_t \tilde{L}_{\cdots}^{\cdots}(t) \cdot \Psi_{\cdots}^{\cdots}(q_b)$. The first term on the RHS of (2.4) vanishes for sufficiently large t up to an error of $O(t^{-\ell})$ for any $\ell \geq 1$ due to the factor $\tilde{L}_{\cdots}^{\cdots}(t)$. (See (1.30), (1.19), (5), and use (I.3.9). Note that $b_{k+1} = b$ in (1.27) in the present case.) To treat the second term, we notice that

(2.5) $$E(t) = \{i[I_b^L, \tilde{L}_{\cdots}^{\cdots}(t) \Psi_{\cdots}^{\cdots}(q_b)] - i\tilde{L}_{\cdots}^{\cdots}(t) \Psi_{\cdots}^{\cdots}(q_b) I_b^S\} + D_t^b \tilde{L}_{\cdots}^{\cdots}(t) \cdot \Psi_{\cdots}^{\cdots}(q_b).$$

Some commutator estimates and the definition of $\tilde{L}_{\cdots}^{\cdots}(t)$ yield that the first term is of $O(t^{-1-\epsilon'})$ in operator norm.

To treat the second term in (2.5), arguing similarly to the proof of Theorem I.4.3, we use the properties of $\tilde{L}_{\cdots}^{\cdots}(t)$ corresponding to (1.11)-(1.12) which was used to prove the existence of (1.13). The positivity corresponding to that of (I.4.11) and (I.4.12) holds true also in the present case of $\tilde{L}_{\cdots}^{\cdots}(t)$. To treat the estimates corresponding to those for (I.4.13) and (I.4.14), arguing quite similarly to (I.4.18), we can show that the terms corresponding to (I.4.13) and (I.4.14) in our case can be written as

(2.6) $$\Phi(\gamma) \frac{1}{\sqrt{<x>}} B(t) \frac{1}{\sqrt{<x>}} \Phi(\gamma) + S(t),$$

where $\|B(t)(H + i)^{-2}\| \leq C$, $\|S(t)(H + i)^{-2}\| \leq C/t^2$, $t \geq 1$, and $\Phi(\gamma)$ is a real-valued C_0^∞ function of γ such that $0 \leq \Phi \leq 1$ and supp $\Phi \cap \Sigma(\Lambda_{j_0}) = \emptyset$. ($\Sigma(\Lambda_{j_0}) = \{\pm\sqrt{2(\Lambda_{j_0} - \lambda)} \mid \Lambda_{j_0} \geq \lambda, \lambda \in \mathcal{T}\}$. See Lemma I.4.1.) Thus by Lemma I.4.1, we have

(2.7) $$\int_{-\infty}^\infty \left\| \frac{1}{\sqrt{<x>}} \Phi(\gamma) e^{-itH} \psi(H) f \right\|^2 dt \leq C\|f\|^2.$$

But the following lemma yields

(2.8) $$\int_{-\infty}^\infty \left\| \frac{1}{\sqrt{<x>}} \Phi(\gamma) \tilde{\psi}(H_b) U_b(t) g \right\|^2 dt \leq C\|g\|^2,$$

where $\tilde{\psi}$ is a function similar to ψ with $\tilde{\psi}\psi = \psi$. Therefore we can argue similarly to the proof of Theorem I.4.3 after (I.4.21), and conclude the proof of the existence of the limit (2.3). We remark that the commutators $[H_b, \psi_{j_\ell}(H^{b_\ell})]$ ($b < b_\ell$) are treated similarly to the first term on the RHS of (2.5), and that the limits with respect to δ's in (2.2) are dealt with as in [I.6, Lemma 4.8].

Lemma 2.2. *(2.8) holds.*

PROOF: Let $\Psi(\gamma) = \int_{-\infty}^{\gamma} \Phi(\tau)^2 d\tau \leq C_1(< \infty)$. Then

$$
(2.9) \qquad 2C_1 \|g\|^2 \geq \|\sqrt{\Psi(\gamma)}\tilde{\psi}(H_b)U_b(T)g\|^2 - \|\sqrt{\Psi(\gamma)}\tilde{\psi}(H_b)U_b(-T)g\|^2
$$

$$
= \int_{-T}^{T} \frac{d}{dt} \|\sqrt{\Psi(\gamma)}\tilde{\psi}(H_b)U_b(t)g\|^2 dt
$$

$$
= \int_{-T}^{T} (g, U_b(t)^* i[H_b(t), \tilde{\psi}(H_b)\Psi(\gamma)\tilde{\psi}(H_b)]U_b(t)g)dt.
$$

By [I.13, (A.39), (A.40)],

$$
i[H_b(t), \tilde{\psi}(H_b)\Psi(\gamma)\tilde{\psi}(H_b)] = \tilde{\psi}(H_b)\Phi(\gamma)(i[H_b, \gamma] + i[I_b^L(t, x_b, x^b), \gamma])\Phi(\gamma)\tilde{\psi}(H_b)
$$

$$
+ i\{[I_b^L(t, x_b, x^b), \tilde{\psi}(H_b)]\Psi(\gamma)\tilde{\psi}(H_b) + \tilde{\psi}(H_b)\Psi(\gamma)[I_b^L(t, x_b, x^b), \tilde{\psi}(H_b)]\}
$$

$$
+ \tilde{\psi}(H_b)R\tilde{\psi}(H_b),
$$

where $\| < x > R < x > \| < \infty$ (cf. [I.3, I.13]). The second and third terms are of $O(< t >^{-1-\epsilon'})$ by (6). The first term is bounded from below by $\alpha\tilde{\psi}(H_b)\Phi(\gamma)\frac{1}{<x>}\Phi(\gamma)\tilde{\psi}(H_b)$ with $\alpha > 0$ and some error E similar to R (cf. [I.13, Sec.7]).

Replacing $\Psi(\gamma)$ in (2.9) by $< A >^{-2\rho}$ with $\rho > 0$ and $A = \frac{1}{2}(x \cdot D_x + D_x \cdot x)$, and arguing similarly to the above with using Mourre estimate (see [2]) and taking supp ψ and supp $\tilde{\psi}$ small enough, one can get

$$
(2.10) \qquad \int_{-\infty}^{\infty} \| < x >^{-1/2-\rho} \tilde{\psi}(H_b)U_b(t)g\|^2 dt \leq C\|g\|^2
$$

for some constant $C > 0$ independent of $g \in \mathcal{H}$.

Summarizing the above arguments with applying (2.10) to the term $R + E$, we now arrive at

$$
\int_{-T}^{T} \left\| \frac{1}{\sqrt{< x >}}\Phi(\gamma)\tilde{\psi}(H_b)U_b(t)g \right\|^2 dt \leq C\|g\|^2
$$

for some constant $C > 0$ independent of $g \in \mathcal{H}$ and $T > 1$. This completes the proof of the lemma.

As in the introduction $U_b^D(t)$ is a propagator generated by $H_b^D(t)$:

$$
U_b^D(t) = \exp\{-i[S_b(D_b, t) + tH^b]\}.
$$

Proposition 2.3. *There exists the limit for $1 \leq M < \infty$*

$$
(2.11) \qquad s - \lim_{t \to \infty} U_b^D(t)^* P_b^M U_b(t).
$$

PROOF: Since $[U_b^D(t), P_b^M] = 0$, we have using (3)

(2.12)
$$\frac{d}{dt} U_b^D(t)^* P_b^M U_b(t)$$
$$= U_b^D(t)^* P_b^M (\partial_t S_b(D_b, t) + H^b - H_b - I_b^L(t, x)) U_b(t)$$
$$= U_b^D(t)^* P_b^M (I_b^L(t, \nabla_{\xi_b} S_b(D_b, t), 0) - I_b^L(t, x_b, x^b)) U_b(t).$$

The difference on the RHS is equal to, up to an error of $O(t^{-1-\epsilon'})$

(2.13)
$$\int_0^1 \nabla_{x_b} I_b^L(t, x_b + \theta(\nabla_{\xi_b} S_b(D_b, t) - x_b), 0) d\theta \cdot (\nabla_{\xi_b} S_b(D_b, t) - x_b)$$
$$- x^b \cdot \int_0^1 \nabla_{x^b} I_b^L(t, x_b, \theta x^b) d\theta$$
$$= O(t^{-1-\epsilon'}) \cdot (\nabla_{\xi_b} S_b(D_b, t) - x_b) + O(|x^b|) O(t^{-1-\epsilon'}).$$

We prepare a lemma.

Lemma 2.4. *There exists a constant $C > 0$ independent of $t > 2$ such that*

(2.14)
$$\|P_b^M (\nabla_{\xi_b} S_b(D_b, t) - x_b) U_b(t) < x_b >^{-1} \| \le C.$$

PROOF: This is a stronger version of [14, Theorem 5.1]. We have only to show

(2.15)
$$\| < x_b >^{-1} I(t) < x_b >^{-1} \| \le C,$$

where $I(t) = U_b(t)^* P_b^M (\nabla_{\xi_b} S_b(D_b, t) - x_b)^2 U_b(t)$. We compute noting that $[P_b^M, \nabla_{\xi_b} S_b(D_b, t) - x_b] = 0$ and $(P_b^M)^2 = P_b^M$

(2.16)
$$\frac{d}{dt} I(t)$$
$$= 2U_b(t)^* \text{Re}\{P_b^M \{i[H_b(t), \nabla_{\xi_b} S_b(D_b, t) - x_b] + \partial_t \nabla_{\xi_b} S_b(D_b, t)\}$$
$$\times P_b^M (\nabla_{\xi_b} S_b(D_b, t) - x_b)\} U_b(t)$$
$$+ 2U_b(t)^* \text{Re}\{i[I_b^L(t, x), P_b^M](\nabla_{\xi_b} S_b(D_b, t) - x_b)^2 P_b^M\} U_b(t)$$
$$= 2U_b(t)^* \text{Re}\{P_b^M \{i[I_b^L(t, x_b, x^b), \nabla_{\xi_b} S_b(D_b, t)] - i[H_b, x_b]$$
$$+ \partial_t \nabla_{\xi_b} S_b(D_b, t)\} P_b^M (\nabla_{\xi_b} S_b(D_b, t) - x_b)\} U_b(t)$$
$$+ O(t^{-1-\epsilon'}) I(t) + I(t) O(t^{-1-\epsilon'})$$
$$= 2U_b(t)^* \text{Re}\{P_b^M \{-\nabla_{x_b} I_b^L(t, x_b, x^b) \cdot \nabla_{\xi_b}^2 S_b(D_b, t) - \nabla_{\xi_b} T_b(D_b)$$
$$+ \partial_t \nabla_{\xi_b} S_b(D_b, t)\} P_b^M (\nabla_{\xi_b} S_b(D_b, t) - x_b)\} U_b(t)$$
$$+ \text{Re}\{O(t^{-1-\epsilon'}) I(t)\} + \text{Re}\{O(t^{-1-\epsilon'}) \sqrt{I(t)}\},$$

where Re $T = (T + T^*)/2$ for an operator T in $\mathcal{H}$, and $O(t^{-1-\epsilon'})$ is an operator $R(t)$ with $\|R(t)\| \leq Ct^{-1-\epsilon'}$ for $t > 2$. (The error $\mathrm{Re}\{O(t^{-1-\epsilon'})\sqrt{I(t)}\}$ in the last step arises from ψdo calculus.) To treat the first term on the RHS of (2.16), we calculate the difference

$$
\begin{aligned}
(2.17) \qquad & \nabla_{x_b} I_b^L(t, x_b, x^b) - \nabla_{x_b} I_b^L(t, \nabla_{\xi_b} S_b(D_b, t), x^b) \\
&= \int_0^1 \nabla_{x_b}^2 I_b^L(\cdots) d\theta \cdot (x_b - \nabla_{\xi_b} S_b(D_b, t)) \\
&= O(t^{-2-\epsilon'}) \cdot (x_b - \nabla_{\xi_b} S_b(D_b, t)).
\end{aligned}
$$

Thus by (4) the contribution from this difference is $\mathrm{Re}\{O(t^{-1-\epsilon'})\left(I(t) + \sqrt{I(t)}\right)\}$ when we include the factor $\nabla_{\xi_b}^2 S_b(D_b, t) P_b^M (\nabla_{\xi_b} S_b(D_b, t) - x_b) U_b(t)$.

The remaining terms are P_b^M times

$$
\begin{aligned}
(2.18) \qquad & -\nabla_{x_b} I_b^L(t, \nabla_{\xi_b} S_b(D_b, t), x^b) \cdot \nabla_{\xi_b}^2 S_b(D_b, t) - \nabla_{\xi_b} T_b(D_b) + \partial_t \nabla_{\xi_b} S_b(D_b, t) \\
&= \nabla_{\xi_b}(\partial_t S_b(D_b, t) - T_b(D_b) - I_b^L(t, \nabla_{\xi_b} S_b(D_b, t), x^b)) \\
&= \nabla_{\xi_b}(I_b^L(t, \nabla_{\xi_b} S_b(D_b, t), 0) - I_b^L(t, \nabla_{\xi_b} S_b(D_b, t), x^b)) \\
&= -x^b \cdot \int_0^1 \nabla_{x^b} \nabla_{x_b} I_b^L(t, \nabla_{\xi_b} S_b(D_b, t), \theta x^b) d\theta \cdot \nabla_{\xi_b}^2 S_b(D_b, t) \\
&= O(|x^b|) O(t^{-1-\epsilon'})
\end{aligned}
$$

by (3)-(4), which yields a term $\mathrm{Re}\{O(t^{-1-\epsilon'})\sqrt{I(t)}\}$ by Assumption 2 if the factor P_b^M $(\nabla_{\xi_b} S_b(D_b, t) - x_b) U_b(t)$ is included. Thus we have using the inequality $2ab \leq a^2 + b^2$ $(a, b \in R^1)$

$$
(2.19) \qquad \frac{d}{dt} \| < x_b >^{-1} I(t) < x_b >^{-1} \|^2 = O(t^{-1-\epsilon'}) + O(t^{-1-\epsilon'}) \| < x_b >^{-1} I(t) < x_b >^{-1} \|^2.
$$

Integrating this differential inequality with using $\| < x_b >^{-1} I(0) < x_b >^{-1} \| < \infty$, we get (2.15). This finishes the proof of the lemma.

Combining this lemma with (2.12)-(2.13) and using Assumption 2, we have the existence of the limit (2.11). This completes the proof of the proposition.

The following obvious proposition will be used to prove the completeness for the short-range case.

Proposition 2.5. Let $V_\alpha^L(x_\alpha) \equiv 0$ for all α in Assumption 1. Then for $U_b(t)$ and $\psi^\delta(H^b) = \psi_{k+1}^\delta(H^b)$ above, we have

$$
(2.20) \qquad [\psi^\delta(H^b), U_b(t)^*] = 0.
$$

(2.20) also holds with $\psi^\delta(H^b)$ replaced by P_b^M with $1 \leq M < \infty$.

Remark 2.6. According to Dereziński [4], Proposition 2.5 holds in the form

$$\lim_{M\to\infty}\lim_{\delta\downarrow0}\lim_{t\to\infty}\|(\psi^\delta(H^b)-P_b^M)U_b(t)\tilde{\Omega}_b^\psi f\|=0$$

for the long-range case if $\epsilon > \sqrt{3}-1$. This is seen if one notes that his proof of [4, Proposition 6.8] implies

$$\lim_{R\to\infty}\lim_{t\to\infty}\|E_{|x^b|}([R,\infty))U_b(t)\tilde{\Omega}_b^\psi f\|=0.$$

This proves the asymptotic completeness for such long-range cases in the same way as below.

To prove the short-range part of Theorem 3, we first see

Proposition 2.7. Let $V_\alpha^L(x_\alpha)\equiv0$ for all α in Assumption 1. Then we have

$$(2.21)\qquad \lim_{M\to\infty}\lim_{\delta\downarrow0}\limsup_{t\to\infty}\|(\psi^\delta(H^b)-P_b^M)\Psi_{\cdots}^{\cdots}(q_b)\tilde{L}_{\cdots}^{\cdots}(t)e^{-itH}\psi(H)f\|=0.$$

PROOF: We rewrite the lim sup in (2.21) as, using $U_b(t)$

$$(2.22)\qquad \limsup_{t\to\infty}\|U_b(t)^*(\psi^\delta(H^b)-P_b^M)\Psi_{\cdots}^{\cdots}(q_b)\tilde{L}_{\cdots}^{\cdots}(t)e^{-itH}\psi(H)f\|.$$

By Propositions 2.1 and 2.5, (2.22) equals

$$(2.23)\qquad \|(\psi^\delta(H^b)-P_b^M)\; s\text{-}\lim_{t\to\infty}U_b(t)^*\Psi_{\cdots}^{\cdots}(q_b)\tilde{L}_{\cdots}^{\cdots}(t)e^{-itH}\psi(H)f\|.$$

This tends to zero as $\delta\downarrow0$ and $M\to\infty$ by $s\text{-}\lim_{\delta\downarrow0}\psi^\delta(H^b)=P_b=s\text{-}\lim_{M\to\infty}P_b^M$, which completes the proof of the proposition.

Proposition 2.8. Let $V_\alpha^L(x_\alpha)\equiv0$ for all α in Assumption 1. Then there exists the limit

$$(2.24)\quad g_b\equiv s-\lim_{\delta_N\downarrow0}\cdots s-\lim_{\delta_0\downarrow0}s-\lim_{t\to\infty}\sum U_b^D(t)^*\psi^\delta(H^b)\Psi_{\cdots}^{\cdots}(q_b)\tilde{L}_{\cdots}^{\cdots}(t)e^{-itH}\psi(H)f,$$

and we have $g_b=P_bg_b$, where $\psi^\delta(H^b)=\psi_{k+1}^{\delta_{k+1}}(H^b)$ and the sum is taken over $b_{k+1}=b$ in the situation of Theorem 1.1.

PROOF: By Proposition 2.7, we have only to prove the existence of (2.24) with $\psi^\delta(H^b)$ replaced by P_b^M. But this limit exists by Propositions 2.1 and 2.3. The relation $g_b=P_bg_b$ is then obvious by $[P_b,U_b^D(t)]=0$. This finishes the proof of the proposition.

By Proposition 2.8, $\Omega_b^\psi f$ in (1.31) is equal to

$$(2.25)\qquad \Omega_b^\psi f=s-\lim_{t\to\infty}e^{itH}U_b^D(t)P_b\cdot g_b,$$

where g_b is given by (2.24). The strong limit in (2.25) exists by the next Sec.3, and equals $W_b^Dg_b$, which concludes the proof of (1.33) for the short-range case.

Theorem 3 has been proved.

3. A proof of the existence of wave operators

In this section, we give a proof of the existence of the modified channel wave operators

$$(3.1) \qquad W_b^D f = s - \lim_{t \to \infty} e^{itH} U_b^D(t) P_b f$$

defined by (7) under Assumptions 1 and 2.

Since $P_b = s - \lim_{M \to \infty} P_b^M$, we may assume that P_b is 1-dimensional: $P_b f = (f, \varphi)\varphi$, where $\varphi \in \mathcal{H}^b$ satisfies $|x^b|^2 \varphi(x^b) \in \mathcal{H}^b$ by Assumption 2. Further it suffices to consider the state f with $E_{|q_k|^2}((\delta^2, \infty))f = f$ for any $k = 1, \cdots, k_b$ and an arbitrarily fixed $\delta > 0$, where $E_{|q_k|^2}(\Delta)$ is the spectral measure for $|q_k|^2$. (For the definition of q_k, see (I.1.8).)

We differentiate the state on the RHS of (3.1) with respect to t using (3):

$$(3.2) \qquad \frac{d}{dt} e^{itH} U_b^D(t) P_b f$$

$$= e^{itH} i(H - \partial_t S_b(D_b, t) - H^b) U_b^D(t) P_b f$$

$$= e^{itH} i(I_b^S(x_b, x^b) + I_b^L(x_b, x^b) - I_b^L(t, \nabla_{\xi_b} S_b(D_b, t), 0)) U_b^D(t) P_b f.$$

Since $S_b(\xi_b, t)$ satisfies (3) and (4), we have $U_b^D(t) = e^{-i[S_b(D_b, t) + tH^b]}$, hence we can apply the method of stationary phase (see e.g. [8]) to the state $U_b^D(t) P_b f = U_b^D(t) P_b \prod_{k=1}^{k_b} E_{|q_k|^2}((\delta^2, \infty))f$ and see that

$$(3.3) \qquad \|\chi_{\{|z_k| < \delta t/2\}} U_b^D(t) P_b f\| \le C_\ell < t >^{-\ell}$$

for any $\ell \ge 1$ and $k = 1, \cdots, k_b$. Here χ_S denotes the characteristic function of a set S. Further by Assumption 2 we have for $1 > \rho > 1/2$

$$(3.4) \qquad \|\chi_{\{|x^b| > t^\rho\}} U_b^D(t) P_b f\| = \|\chi_{\{|x^b| > t^\rho\}} P_b U_b^D(t) f\| \le C < t >^{-2\rho} \in L^1(R_t^1).$$

Since, for any pair α with $\sim (\alpha \le b)$, $|x_\alpha| \ge c_1 |z_k| - c_2 |x^b|$ for some constants $c_1, c_2 > 0$ and some $1 \le k \le k_b$, we thus have for some constants $C, c > 0$ and for sufficiently large t that

$$(3.5) \qquad \|\chi_{\{|x_\alpha| < c\delta t\}} U_b^D(t) P_b f\|$$

$$\le \|\chi_{\{|x_\alpha| < c\delta t\}} \chi_{\{|x^b| \le t^\rho\}} U_b^D(t) P_b f\| + C < t >^{-2\rho}$$

$$\le \|\chi_{\{|z_k| < \delta t/2\}} U_b^D(t) P_b f\| + C < t >^{-2\rho}$$

$$\le C_\ell < t >^{-\ell} + C < t >^{-2\rho} \in L^1(R_t^1).$$

Therefore we have only to consider $\prod_{\sim(\alpha \leq b)} \chi_{\{|x_\alpha|>c\delta t\}} U_b^D(t) P_b f \equiv \Psi U_b^D(t) P_b f$. Then $I_b^S(x_b, x^b)$ on the RHS of (3.2) decays integrably with respect to t on $\Psi U_b^D(t) P_b f$, and $I_b^L(x_b, x^b)\Psi = I_b^L(t, x_b, x^b)\Psi$ for large $|t| > 1$. So we calculate the difference

(3.6)
$$I_b^L(t, x_b, x^b) - I_b^L(t, \nabla_{\xi_b} S_b(D_b, t), 0)$$
$$= x^b \cdot \int_0^1 \nabla_{x^b} I_b^L(t, x_b, \theta x^b) d\theta$$
$$- \int_0^1 \nabla_{x_b} I_b^L(t, x_b + (1-\theta)(\nabla_{\xi_b} S_b(D_b, t) - x_b), 0) d\theta \cdot (\nabla_{\xi_b} S_b(D_b, t) - x_b),$$

which holds up to an error of $O(<t>^{-1-\epsilon'}) \in L^1(R_t^1)$ by (4), (6) and ψdo calculus. The first term times $U_b^D(t) P_b f = P_b U_b^D(t) f$ is of order $O(<t>^{-1-\epsilon'})$ by Assumption 2 and (6), hence is integrable. The second term times $U_b^D(t) P_b f$ is bounded by

(3.7)
$$O(<t>^{-1-\epsilon'}) \|(\nabla_{\xi_b} S_b(D_b, t) - x_b) U_b^D(t) P_b f\|.$$

Computing similarly to (2.16) of Lemma 2.4, we have

(3.8)
$$\frac{d}{dt} U_b^D(t)^* (\nabla_{\xi_b} S_b(D_b, t) - x_b) U_b^D(t) P_b f$$
$$= U_b^D(t)^* \{i[\partial_t S_b(D_b, t) + H^b, \nabla_{\xi_b} S_b(D_b, t) - x_b]$$
$$+ \partial_t \nabla_{\xi_b} S_b(D_b, t)\} U_b^D(t) P_b f$$
$$= U_b^D(t)^* \{-i[T_b(D_b) + I_b^L(t, \nabla_{\xi_b} S_b(D_b, t), 0), x_b]$$
$$+ \nabla_{\xi_b}(T_b(D_b) + I_b^L(t, \nabla_{\xi_b} S_b(D_b, t), 0))\} U_b^D(t) P_b f$$
$$= 0.$$

Thus (3.6) times $U_b^D(t) P_b f$ is integrable with respect to t. This completes the proof of the existence of $W_b^D f$.

The orthogonality of $W_b^D f$'s for different b's is proved similarly to (I.5.5) of Sec.I.5 by using Assumption 2.

4. Remarks on the asymptotic completeness

In this section we prove that certain parts of $\Omega_b^\psi f$ in (1.31) belong to $\mathcal{R}(W_b^D)$. Further we show that the asymptotic completeness holds when $V_\alpha \geq 0$ for all α and $\epsilon > 1/2$ in Assumption 1.

We recall that the function $\psi^\delta(H^b) \equiv \psi_{k+1}^\delta(H^b)$ in (1.31) satisfies $\psi^\delta(\lambda) = 1$ for λ, $|\lambda - \lambda_2| \leq \delta = \delta_{k+1}$ with $\lambda_2 \in \mathcal{T}_b^{\tau_0}$ and $\psi^\delta(\lambda) = 0$ away from those neighborhoods, where $\mathcal{T}_b^{\tau_0} = \{\lambda \in \mathcal{T}^F | \exists \tilde{\lambda} \in \mathcal{T}_b \text{ s.t. } |\lambda - \tilde{\lambda}| \leq \tau_0(\lambda)\}$. Since $0 < \tau_0(\lambda) \leq \tau_0 < \delta_0$, supp ψ^δ shrinks at the points in $\mathcal{T}_b$ when $\delta_0 \downarrow 0$ and $\delta = \delta_{k+1} \downarrow 0$. We decompose $\psi^\delta = \sum_{\ell=1}^K \psi_\ell^\delta$ such that supp ψ_ℓ^δ lies in a neighborhood of $\lambda_\ell \in \mathcal{T}^F = \{\lambda_\ell\}_{\ell=1}^K$, which is defined in (1.1). Note that K depends on $\tau_0 < \delta_0$, and ψ_ℓ^δ vanishes identically if $\lambda_\ell \notin T_b^{\tau_0}$. We denote the numbers to which supp ψ_ℓ^δ converges as $\delta_0, \delta \downarrow 0$ by $\{E_j\}$. Then $\{E_j\} = T_b = \{0\} \cup \bigcup_{d \leq b} \sigma_p(H^d)$. For convenience, we write ψ_ℓ^δ as ψ_j^δ when supp ψ_ℓ^δ converges to E_j, and write $\psi^\delta = \sum_j \psi_j^\delta$. Note that the latter sum is actually a finite one for any fixed δ, $\delta_0 > 0$. We define $\Omega_{bj}^\psi f$ by (1.31) with $\psi_{k+1}^{\delta_{k+1}}(H^{b_{k+1}})$ replaced by $\psi_j^{\delta_{k+1}}(H^{b_{k+1}})$. Then we have

$$(4.1) \qquad \Omega_b^\psi f = \sum_j \Omega_{bj}^\psi f.$$

This is generally an infinite sum, since we take the limits with respect to δ's in (1.31). We divide T_b as

$$(4.2) \qquad T_b = \tilde{T}_b \cup (\sigma_p(H^b) - \tilde{T}_b),$$

where $\tilde{T}_b = \{0\} \cup \bigcup_{d < b} \sigma_p(H^d)$ is the set of threshold of H^b. We consider the case $E_j \in \sigma_p(H^b) - \tilde{T}_b$ in the followings.

We decompose

$$(4.3) \qquad \Omega_{bj}^\psi f = e^{it_n H} P_b^{M_{|b|}^n} e^{-it_n H} \Omega_{bj}^\psi f + e^{it_n H}(I - P_b^{M_{|b|}^n}) e^{-it_n H} \Omega_{bj}^\psi f, \qquad f \in \mathcal{H},$$

where $t_n \to \infty$ $(n \to \infty)$ and $M_{bd}^n = (M_{|b|}^n, \cdots, M_{|d|}^n)$ $(d < b)$ are the sequences specified in Proposition I.1.1, which are determined by $\Omega_{bj}^\psi f$.

Proposition 4.1. Let Assumptions 1 and 2 be satisfied. Let $2 \leq |b| \leq N$ and $E_j \in \sigma_p(H^b) - \tilde{T}_b$. Then the following limit exists and we have

$$(4.4) \qquad s - \lim_{n \to \infty} e^{it_n H} P_b^{M_{|b|}^n} e^{-it_n H} \Omega_{bj}^\psi f \in \mathcal{R}(W_b^D).$$

PROOF: When $E_j \in \sigma_p(H^b) - \tilde{T}_b$, the eigenspace of H^b corresponding to eigenvalue E_j is of finite dimension. Thus on $\psi_j^\delta(H^b)$ which can be taken out from $g = \Omega_{bj}^\psi f$ to the left of $e^{-it_n H} g$, $P_b^{M_{|b|}^n}$ can be replaced by P_b if $M_{|b|}^n$ is large and δ is small enough. Thus by the definition of $g = \Omega_{bj}^\psi f$, the limit in (4.4) equals

$$(4.5) \qquad s - \lim_{n \to \infty} e^{it_n H} U_b^D(t_n) P_b \cdot U_b^D(t_n)^* P_b^M U_b(t_n) \cdot U_b(t_n)^* P_b^M U_b(t_n) \cdot \tilde{\Omega}_{bj}^\psi f$$

for some $1 \leq M < \infty$, if this exists. Here $\tilde{\Omega}_{bj}^\psi f$ is defined by (1.31) with e^{itH} replaced by $U_b(t)^*$ as in (2.2).

It is easy by estimating the commutator $[I_b(t, x), P_b^M]$ to see that

$$(4.6) \qquad s - \lim_{n \to \infty} U_b(t_n)^* P_b^M U_b(t_n)$$

exists. By Proposition 2.3, the limit

$$(4.7) \qquad s - \lim_{n \to \infty} U_b^D(t_n)^* P_b^M U_b(t_n)$$

exists, and by Sec.3, the limit

$$(4.8) \qquad W_b^D = s - \lim_{n \to \infty} e^{it_n H} U_b^D(t_n) P_b$$

exists. Thus (4.5) exists and equals $W_b^D g_b$ for some $g_b \in \mathcal{H}$. $\square$

Remark 4.2. For $|b| = N$, (4.4) holds with $P_b^{M_{|b|}^n} = I$. Thus $\mathcal{R}(\Omega_b^\psi) = \mathcal{R}(\psi(H)W_b^D)$ for $|b| = N$.

We next consider $(I - P_b^{M_{|b|}^n})e^{-it_n H}\Omega_{bj}^\psi f$ in (4.3). We again consider the case $E_j \in \sigma_p(H^b) - \tilde{T}_b$. By (I.1.15)

$$(4.9) \qquad I - P_b^{M_{|b|}^n} = \sum_{d < b} \tilde{P}_{bd}^{M_{bd}^n} = \sum_{d < b} P_d^{M_{|d|}^n} \hat{P}_{b,|d|-1}^{\hat{M}_{bd}^n},$$

where

$$(4.10)$$

$$\hat{P}_{b,k}^{\hat{M}_{bd}^n} = \left(I - \sum_{|d|=k, d<b} P_d^{M_k^n} \right) \cdots \left(I - \sum_{|d|=|b|+1, d<b} P_d^{M_{|b|+1}^n} \right) \left(I - P_b^{M_{|b|}^n} \right),$$

$$k = |b|, \cdots, N-1.$$

By Proposition I.1.1-(1.17), on each state $\tilde{P}_{bd}^{M_{bd}^n} e^{-it_n H}\Omega_{bj}^\psi f$, $H^b = H^d + I_d^b + T_d^b$ ($I_d^b = I_d - I_b$, $T_d^b = T_d - T_b$) is asymptotically equal to $H^d + T_d^b$, which equals $E + T_d^b$ for some $E \in \tilde{T}_b$ on that state by the factor $P_d^{M_{|d|}^n}$ in $\tilde{P}_{bd}^{M_{bd}^n}$. On the other hand, by the factor $\psi_j^\delta(H^b)$ in $\Omega_{bj}^\psi f$, H^b is close to E_j on $e^{-it_n H}\Omega_{bj}^\psi f$ as $t_n \to \infty$. Thus $T_d^b \sim E_j - E \neq 0$ by our assumption: $E_j \in \sigma_p(H^b) - \tilde{T}_b$. Therefore in each $\tilde{P}_{bd}^{M_{bd}^n} e^{-it_n H}\Omega_{bj}^\psi f$, only the components with $E_j - E > 0$ remain. Since $E_j \in \sigma_p(H^b) - \tilde{T}_b$, $\delta \equiv \inf\{E_j - E | E_j > E, E \in \tilde{T}_b\} > 0$. Thus $H_0^b \geq T_d^b \geq \delta/2 > 0$ on the state $(I - P_b^{M_{|b|}^n})e^{-it_n H}\Omega_{bj}^\psi f$ asymptotically as $n \to \infty$. We can then repeat the arguments in Sec.1 on this state to get

$$(4.11) \qquad (I - P_b^{M_{|b|}^n})e^{-it_n H}\Omega_{bj}^\psi f \sim \sum_{b' < b} e^{-it_n H}\Omega_{b'}^{I-P_b}\Omega_{bj}^\psi f.$$

Here for $g = \Omega_{bj}^\psi f$

$$(4.12)$$

$$\Omega_{b'}^{I-P_b} g$$

$$= s - \lim_{\delta's \downarrow 0} s - \lim_{t \to \infty} \sum e^{itH} \psi_{k+1}^{\delta_{k+1}}(H^{b_{k+1}}) \Psi_{\delta_0, \cdots, \delta_k}^{b_1, \cdots, b_k}(q_{b_{k+1}}) \tilde{L}_{\delta_0, \cdots, \delta_k}^{b_1, \cdots, b_{k+1}}(t)$$

$$\times (I - P_b)e^{-itH} g$$

as in (1.31), where $b_1 = b$ and $b_{k+1} = b'$, and we have used the assumption $E_j \in \sigma_p(H^b) - \tilde{T}_b$ to replace $P_b^{M_{|b|}^n}$ and t_n by P_b and t. The constants appearing during the construction of $\Psi^{...}_{...}(q_{b'})$ and $\tilde{L}^{...}_{...}(t)$ are chosen suitably for the present situation. For the reason that makes the decomposition possible, see the explanation after (1.10). The limit (4.12) exists at least on $g = \Omega_{bj}^{\psi}f$. The proof is done similarly to that of (1.31) by approximating g by the terms on the RHS of (1.31).

The state $\Omega_{b'}^{I-P_b}g$ in (4.11)-(4.12) can be further decomposed as in (4.1):

$$(4.13) \qquad \Omega_{b'}^{I-P_b}g = \sum_{j'} \Omega_{b'j'}^{I-P_b}g,$$

where j' designates $E_{j'} \in T_{b'}$ and $\psi_{j'}^{\delta}$, so that $\psi_{k+1}^{\delta}(H^{b'}) = \sum_{j'} \psi_{j'}^{\delta}(H^{b'})$ as before. Since the limit (4.12) exists without taking a subsequence t_n when $E_j \in \sigma_p(H^b) - \tilde{T}_b$, we can further decompose $\Omega_{b'j'}^{I-P_b}g$ as in (4.3) and thereafter.

Continuing this procedure, we obtain a decomposition for $E_j \in \sigma_p(H^b) - \tilde{T}_b$

$$(4.14)$$

$$\Omega_{bj}^{\psi}f = \sum_{\ell=0}^{N-|b|} \sum_{b^\ell < \cdots < b^1 < b} \sum_{E_{j^\ell} \in \sigma_p(H^{b^\ell}) - \tilde{T}_{b^\ell}} \cdots \sum_{E_{j^1} \in \sigma_p(H^{b^1}) - \tilde{T}_{b^1}} W_{b^\ell}^D g_{b^\ell \cdots b^1 b}^{j^\ell \cdots j^1 j}$$

$$+ \text{ remainder terms.}$$

The first terms correspond to the state on which all subsystem eigenvalues are not in its threshold set. These terms except for $W_b^D g_b^j$ are orthogonal to $\Omega_{bj}^{\psi}f$. The rest are the remainder terms on which some subsystem eigenvalue is included in its threshold set.

Therefore when we consider the asymptotic completeness, only the states $\Omega_{bj}^{\psi}f$ with $E_j \in \tilde{T}_b$ remain to be determined. On these states we have the following preliminary result. We assume $V_\alpha^S = 0$ for all α.

Proposition 4.3. Let Assumptions 1 and 2 with $\epsilon > 1/2$ be satisfied. Let $V_\alpha^L(x_\alpha) \geq 0$ for all pairs α. Let $2 \leq |b| \leq N$, and $E_j \in \tilde{T}_b$. Then $\mathcal{R}(\Omega_{bj}^{\psi}) \subset \mathcal{R}(W_b^D)$. The conclusion also holds for $\mathcal{R}(\Omega_{b'j'}^{I-P_b}\Omega_{bj}^{\psi})$, and so forth.

PROOF: By the arguments in Propositions 2.7 and 2.8, we have only to show

$$(4.15) \qquad \lim_{M \to \infty} \lim_{\delta \downarrow 0} \lim_{t \to \infty} \|(I - P_b^M)\psi_j^{\delta}(H^b)U_b(t)\tilde{\Omega}_{bj}^{\psi}f\| = 0,$$

where $U_b(t)$ is defined as in (2.1), and $\tilde{\Omega}_{bj}^{\psi}f$ is defined by (1.31) with e^{itH} replaced by $U_b(t)^*$ as in (2.2). To see (4.15) it suffices, as in Remark 2.6, to prove

$$(4.16) \qquad \lim_{R \to \infty} \lim_{t \to \infty} \|\chi_{\{|x^b|>R\}}(I - P_b^M)\psi_j^{\delta}(H^b)U_b(t)\tilde{\Omega}_{bj}^{\psi}f\| = 0.$$

In fact if we have (4.16), we can approximate the norm in (4.15) by

$$(4.17) \qquad \|\chi_{\{|x^b|<R\}}(I - P_b^M)\psi_j^\delta(H^b)U_b(t)\tilde{\Omega}_{bj}^\psi f\|$$

for any large $t \gg 1$ and some large $R > 1$ with an arbitrarily given error. But since $\chi_{\{|x^b|<R\}}\psi_j^\delta(H^b)$ is compact on $\mathcal{H}^b$, we have

$$(4.18)$$
$$\lim_{M\to\infty}\lim_{\delta\downarrow 0}\|\chi_{\{|x^b|<R\}}\psi_j^\delta(H^b)(I - P_b^M)\| = \lim_{M\to\infty}\lim_{\delta\downarrow 0}\|\chi_{\{|x^b|<R\}}\psi_j^\delta(H^b)(P_b - P_b^M)\| = 0.$$

We prove (4.16). We first note that the limit

$$(4.19) \qquad \Omega f \equiv s - \lim_{\delta\downarrow 0} s - \lim_{t\to\infty} U_b(t)^*(I - P_b^M)\psi_j^\delta(H^b)U_b(t)\tilde{\Omega}_{bj}^\psi f$$

exists by Proposition 2.1. Since supp ψ_j^δ shrinks at $E = E_j \in \tilde{T}_b$, we have

$$(4.20) \qquad \|(H^b - E)U_b(t)\Omega f\| \to 0 \qquad (t \to \infty).$$

Note that $E \leq 0$ by [6].

We need a lemma, whose proof will be given later.

Lemma 4.4. *Let* $V_\alpha^L(x_\alpha) \geq 0$ *for all* α. *Then*

$$(4.21) \qquad \|(H^b - E)U_b(t)\Omega f\| \leq Ct^{-2\epsilon'},$$

where $C > 0$ *is independent of* $t > 1$, *and* ϵ' *is some constant such that* $1/2 < \epsilon' < \epsilon < 1$ *with* ϵ *being the decay order in Assumption 1.*

Define $\tilde{U}_b(t)$ as a unitary propagator generated by

$$(4.22) \qquad T_b + I_b^L(t,x) + E.$$

Then since $\epsilon' > 1/2$ and

$$(4.23) \qquad \left\|\frac{d}{dt}\tilde{U}_b(t)^*U_b(t)\Omega f\right\| = \|\tilde{U}_b(t)i(E - H^b)U_b(t)\Omega f\| \leq Ct^{-2\epsilon'}$$

by Lemma 4.4, the limit

$$(4.24) \qquad s - \lim_{t\to\infty} \tilde{U}_b(t)^*U_b(t)\Omega f$$

exists. By the definition (4.19) of Ωf, this is equal to

$$(4.25) \qquad h \equiv s - \lim_{t\to\infty} \tilde{U}_b(t)^*(I - P_b^M)\psi_j^\delta(H^b)U_b(t)\tilde{\Omega}_{bj}^\psi f.$$

Since $\|\chi_{\{|x^b|>R\}}h\| \to 0$ $(R \to \infty)$ and $\tilde{U}_b(t)$ commutes with x^b, we obtain (4.16). Proposition 4.3 has been proved except Lemma 4.4.

Proof of Lemma 4.4. We first note the inequality

(4.26)
$$\||D^b|U_b(t)g\|^2 \le c(H_0^b U_b(t)g, U_b(t)g)$$
$$\le c((H^b - E)U_b(t)g, U_b(t)g) + c((E - V_b)U_b(t)g, U_b(t)g) \le c\|(H^b - E)U_b(t)g\|\|g\|,$$

where we have used $E \le 0$ and $V_b \ge 0$.

To prove (4.21), we set with $g = \Omega f$

(4.27)
$$f(t) = \|(H^b - E)U_b(t)g\|^2.$$

Then

(4.28)
$$f'(t) = \frac{df}{dt}(t) = \frac{d}{dt}((H^b - E)^2 U_b(t)g, U_b(t)g)$$
$$= (i[I_b^L(t, x), (H^b - E)^2]U_b(t)g, U_b(t)g).$$

The commutator is equal to

$$2\text{Re}\{(H^b - E)(\sum_{\sim(\alpha \le b)} c_\alpha \nabla_{x^b} V_\alpha^L(t, x_\alpha) \cdot D^b + O(t^{-2-\epsilon'}))\}.$$

Thus

(4.29)
$$|f'(t)| \le Ct^{-1-\epsilon'}\|(H^b - E)U_b(t)g\|\||D^b|U_b(t)g\| + Ct^{-2-\epsilon'}\|(H^b - E)U_b(t)g\|.$$

By (4.26), this is bounded by

$$\le Ct^{-1-\epsilon'}f(t)^{3/4} + Ct^{-2-\epsilon'}f(t)^{1/2}.$$

Since (4.20) yields

(4.30)
$$f(t) = -\int_t^\infty f'(\tau)d\tau,$$

and $f(t)$ is uniformly bounded, we first get

(4.31)
$$|f(t)| \le Ct^{-\epsilon'}.$$

We insert this into (4.29), and get $|f(t)| \le Ct^{-\epsilon'-3\epsilon'/4} + Ct^{-1-\epsilon'-\epsilon'/2}$. Repeating this procedure, we get

(4.32)
$$|f(t)| \le Ct^{-4\epsilon'}$$

for some $1/2 < \epsilon' < \epsilon$. This proves the lemma. $\square$

For the 3-body case, $2 \le |b| \le 3$. The case $|b| = 3$ has been treated in Remark 4.2, and there remains the case $|b| = 2$. In this case we have the following

Proposition 4.5. *Let Assumptions 1 and 2 be satisfied. Let $N = 3$ and $|b| = 2$. Let $E = E_j \in \mathcal{T}_b - \{0\}$. Then $\mathcal{R}(\Omega_{bj}^\psi) \subset \mathcal{R}(W_b^D)$.*

PROOF: This is clear, because $T_b - \{0\} = \sigma_p(H^b) - \{0\}$ in the present case. Namely from this follows that $E = E_j$ is an isolated eigenvalue of 2-body subsystem H^b. Hence $\psi_j^\delta(H^b)$ is equal to the eigenprojection corresponding to the eigenvalue E when $\delta > 0$ is sufficiently small, and we can apply Proposition 4.1. $\square$

Appendix A
A solution of Hamilton-Jacobi equation

In this appendix we prove the existence of a solution $S_b(\xi_b, t)$ of Hamilton-Jacobi equation

(A.1) $$\partial_t S_b(\xi_b, t) = T_b(\xi_b) + I_b^L(t, \nabla_{\xi_b} S_b(\xi_b, t), 0), \qquad t \geq 0$$

such that $S_b(\xi_b, 0) = 0$ and

(A.2) $$|\partial_{\xi_b}^\alpha \{S_b(\xi_b, t) - t T_b(\xi_b)\}| \leq C_{\alpha\epsilon'} < t >^{1-\epsilon'}$$

for $t \geq 0$ and all α and ϵ' with $1 > \epsilon > \epsilon' > 0$.

We begin with the solution $\tilde{S}_b(x^b; \xi_b, t)$ of

(A.3) $$\partial_t \tilde{S}_b(x^b; \xi_b, t) = T_b(\xi_b) + I_b^L(t, \nabla_{\xi_b} \tilde{S}_b(x^b; \xi_b, t), x^b),$$
$$\tilde{S}_b(x^b; \xi_b, 0) = 0.$$

To solve this equation we set

$$H(t, x^b; x_b, \xi_b) = T_b(\xi_b) + I_b^L(t, x_b, x^b)$$

and consider the Hamilton equation for $t \geq s \geq 0$

$$\frac{dq_b}{ds}(s, t) = \nabla_{\xi_b} T_b(p_b(s, t)), \qquad \frac{dp_b}{ds}(s, t) = -\nabla_{x_b} I_b^L(s, q_b(s, t), x^b)$$

with initial condition

$$q_b(s, s) = x_b, \qquad p_b(s, s) = \xi_b.$$

Since $I_b^L(t, x_b, x^b)$ satisfies

(A.4) $$|\partial_x^\alpha I_b^L(t, x_b, x^b)| \leq C_\alpha < t >^{-|\alpha|-\epsilon'} \qquad (\epsilon > \epsilon' > 0)$$

for all α by the definition of $I_b^L(t, x_b, x^b)$, similarly to [10, Sec.2], we obtain several estimates for these $q_b(s, t, x^b; x_b, \xi_b)$ and $p_b(s, t, x^b; x_b, \xi_b)$. In particular we can solve the equation

$$q_b(s, t, x^b; y_b, \xi_b) = x_b$$

with respect to $y_b = y_b(s, t, x^b; x_b, \xi_b)$ for $t \geq s \geq 0$, and get the estimates

(A.5) $$|\partial_{\xi_b}^\alpha \partial_{x^b}^\beta \{y_b(s, t, x^b; 0, \xi_b) - (t-s)\nabla_{\xi_b} T_b(\xi_b)\}| \leq C_{\alpha\beta} < t - s >$$

for all α, β. (Note that we can take $C_\alpha > 0$ in (A.4) small enough by cutting off a part of each $V_\alpha^L(x_\alpha)$ near $x_\alpha = 0$ and transferring that part into $V_\alpha^S(x_\alpha)$ in the decomposition $V_\alpha(x_\alpha) = V_\alpha^L(x_\alpha) + V_\alpha^S(x_\alpha)$. Then we can apply [10, Proposition 2.2] with $T = 0$ to the present case.) Now as in [10, Sec.3], the solution $\tilde{S}_b(x^b; \xi_b, t)$ of (A.3) is constructed as follows:
Set

$$L(t, x^b; x_b, \xi_b) = \xi_b \cdot \nabla_{\xi_b} H(t, x^b; x_b, \xi_b) - H(t, x^b; x_b, \xi_b),$$

and

$$u(0, t, x^b; y_b, \eta_b)$$
$$= y_b \cdot \eta_b + \int_t^0 L(\tau, x^b; q_b(\tau, t, x^b; y_b, \eta_b), p_b(\tau, t, x^b; y_b, \eta_b))d\tau.$$

Then the solution $\tilde{S}_b(x^b; \xi_b, t)$ of (A.3) is given by

(A.6) $$\tilde{S}_b(x^b; \xi_b, t) = u(0, t, x^b; y_b(0, t, x^b; 0, \xi_b), \xi_b),$$

and satisfies

(A.7) $$\nabla_{\xi_b} \tilde{S}_b(x^b; \xi_b, t) = y_b(0, t, x^b; 0, \xi_b).$$

From (A.3), (A.4), (A.5) and (A.7) follows the estimate

(A.8) $$|\partial_{\xi_b}^\alpha \partial_{x^b}^\beta \{\tilde{S}_b(x^b; \xi_b, t) - tT_b(\xi_b)\}| \leq C_{\alpha\beta} < t >^{1-\epsilon'}$$

for all α, β.

Now the desired solution $S_b(\xi_b, t)$ is obtained by setting $S_b(\xi_b, t) = \tilde{S}_b(0; \xi_b, t)$.

Appendix B
Polynomial decay of eigenvectors

In this appendix, we prove that general polynomial decay of eigenvectors holds for 3-body Hamiltonians with homogeneous pair potentials $V_\alpha(x_\alpha) = c_\alpha |x_\alpha|^{-\epsilon}$, $c_\alpha \in R^1$, $2 > \epsilon > 0$. On the decay property of eigenvectors, there is a work of Froese-Herbst [6], who proved the exponential decay of non-threshold eigenvectors. Also Agmon's method [1] implies some exponential decay of eigenvectors in some cones if the eigenvalues are less than zero. We prove general polynomial decay of all eigenvectors for such homogeneous potentials at the expense of strong decay like exponential ones. We remark that our result for the 2- and 3-body cases serves for the scattering problem up to 4-body cases.

Our Hamiltonian H is

(B.1) $$H = H_0 + V, \qquad V = \sum_{\alpha=(i,j)} V_\alpha, \qquad V_\alpha = c_\alpha |x_\alpha|^{-\epsilon}, \qquad 2 > \epsilon > 0$$

in $L^2(R^n)$, $n = (N-1)\nu$, $\nu \geq 1$, $N = 3$. Here H_0 is the free Hamiltonian $-\sum_{i=1}^{N} \frac{\Delta_i}{2m_i}$ with the center of mass removed, and V_α defined above satisfies with $h_0 = -\Delta$ in $L^2(R_{x_\alpha}^\nu)$

$$(B.2) \qquad V_\alpha(1 + h_0)^{-1} \text{ is compact,}$$

$$(B.3) \qquad \nabla V_\alpha(1 + h_0)^{-2} \text{ is compact.}$$

Since the values m_i of masses are unessential, we may and shall assume $H_0 = -\Delta$ in $L^2(R^n)$ in the followings.

We denote by $E(\Delta) = E_H(\Delta)$ for Borel set Δ of R^1 the spectral measure for H. Let A and A_b for a cluster decomposition b be the generators of dilations:

$$(B.4) \qquad A = \frac{1}{2i}(x \cdot \nabla_x + \nabla_x \cdot x),$$

$$A_b = \frac{1}{2i}(x_b \cdot \nabla_{x_b} + \nabla_{x_b} \cdot x_b),$$

and let

$$(B.5)$$
$$i[H, A] = i(HA - AH) = 2H_0 - x \cdot \nabla_x V = 2H_0 + \epsilon V = (2 - \epsilon)H_0 + \epsilon H,$$
$$i[T_b + I_b, A] = 2T_b - x \cdot \nabla_x I_b = 2T_b + \epsilon I_b = (2 - \epsilon)T_b + \epsilon(T_b + I_b)$$

be the forms on $\mathcal{D}(H_0) \times \mathcal{D}(H_0)$, where we have used the homogeneity of pair potentials in (B.1).

We quote the Mourre estimate from [2, p.85, (4.43)]: Let

$$\tilde{d}(\lambda) = \text{dist}\,(\lambda, (\{\text{thresholds of } H\} \cup \sigma_p(H)) \cap (-\infty, \lambda]) \,(\geq 0)$$

for $\lambda \in R^1$ and $\tilde{d}^\delta(\lambda) = \tilde{d}(\lambda + \delta) - \delta$ for $\lambda \in R^1$ and $\delta > 0$. Then for every $\lambda \in R^1$ and $\delta > 0$, there exists an open interval Δ around λ such that

$$(B.6) \qquad E(\Delta)i[H, A]E(\Delta) \geq 2(\tilde{d}^\delta(\lambda) - \delta)E(\Delta) \geq -4\delta E(\Delta).$$

Theorem B.1. Let $H = -\Delta + V$ in $L^2(R^n)$ and $N = 3$, where $V = \sum V_\alpha$ is a sum of real-valued homogeneous functions defined by (B.1). Assume $c_\alpha + c_\beta \neq 0$ for any pairs $\alpha \neq \beta$ in (B.1). Let $\langle x \rangle = (1 + |x|^2)^{1/2}$. Suppose $H\psi = E\psi$ with $\psi \in L^2(R^n)$ and $E \in R^1$. Then for any $\lambda > 0$, $\langle x \rangle^\lambda \psi \in L^2(R^n)$.

PROOF: We first show the following proposition, which holds for general N-body Hamiltonians with homogeneous pair potentials.

Proposition B.2. Let $H\psi = E\psi$ for some $\psi \in L^2(R^n)$, $\psi \neq 0$. Then $E < 0$.

Proof. By (B.5) and the virial theorem, we have

(B.7)
$$0 = (\psi, i[H, A]\psi) = (2 - \epsilon)(\psi, H_0\psi) + \epsilon(\psi, H\psi)$$
$$= (2 - \epsilon)(\psi, H_0\psi) + \epsilon E\|\psi\|^2.$$

Therefore $E = -\frac{2-\epsilon}{\epsilon}(\psi, H_0\psi)/\|\psi\|^2 < 0$, since $\psi \neq 0$. $\square$

We return to the 3-body case. We assume the polynomial decay of 2-body eigenvectors, which holds by Proposition B.2 and the result of [6].

To prove the theorem, it now suffices to derive $E \geq 0$ from the assumption:

$$\exists \lambda > 0, \qquad \langle x \rangle^\lambda \psi \notin L^2(R^n).$$

The conclusion $E \geq 0$ follows, if we have

(B.8)
$$\exists \lambda > 0, \quad \forall \alpha = (i, j), \qquad \langle x_\alpha \rangle^\lambda \psi \notin L^2(R^n).$$

In fact the following lemma holds under (B.8), which is an extension of [6, (2.11)].

Lemma B.3. *Let the condition (B.8) hold. Define*

(B.9)
$$\rho_\delta = \prod_\alpha \rho_{\delta\alpha}, \quad \rho_{\delta\alpha} = \left(\frac{\langle x_\alpha \rangle}{1 + \delta\langle x_\alpha \rangle} \right)^\lambda, \quad \delta > 0,$$

(B.10)
$$\psi_\delta = \rho_\delta \psi, \quad \Psi_\delta = \psi_\delta/\|\psi_\delta\|.$$

Then for any $R > 0$ and α

(B.10)
$$\lim_{\delta \downarrow 0} \int_{|x_\alpha| < R} |\Psi_\delta(x)|^2 dx = 0.$$

Assuming this lemma and using (B.2) and $\psi \in \mathcal{D}(H)$, we have

(B.11)
$$\lim_{\delta \downarrow 0} \|V\Psi_\delta\| = 0.$$

Further since $(H - E)\Psi_\delta = i[H_0, \rho_\delta]\rho_\delta^{-1}\Psi_\delta$ by $H\psi = E\psi$, and

(B.12)
$$i[H_0, \rho_\delta]\rho_\delta^{-1} = 2\lambda \sum_\alpha \text{Re}\left(\frac{1}{1 + \delta\langle x_\alpha \rangle} \frac{1}{\langle x_\alpha \rangle^2} A_\alpha \right) = \sum_\alpha O(\langle x_\alpha \rangle^{-1} D_\alpha),$$

we have from (B.11) and the lemma that

(B.13)
$$\lim_{\delta \downarrow 0}(\Psi_\delta, (H_0 - E)\Psi_\delta) = \lim_{\delta \downarrow 0}(\Psi_\delta, (H - E)\Psi_\delta) = 0.$$

Thus $E = E\|\Psi_\delta\|^2 = \lim_{\delta \downarrow 0}(\Psi_\delta, H_0\Psi_\delta) \geq 0$.

Therefore we have only to show (B.8) under the assumption

(B.14)
$$\exists \lambda > 0, \qquad \langle x \rangle^\lambda \psi \notin L^2(R^n).$$

Lemma B.3 will be proved at the end of this appendix.

To obtain (B.8) from (B.14), assuming the contrary

$$(\text{B.15}) \qquad |\exists b| = 2, \quad \forall \lambda > 0, \qquad \langle x^b \rangle^\lambda \psi \in L^2(R^n),$$

we derive a contradiction in the following.

(B.14) and (B.15) yield

$$(\text{B.16}) \qquad \langle x_b \rangle^\lambda \psi \notin L^2(R^n), \qquad \psi \neq 0.$$

We define as in Lemma B.3

$$(\text{B.17}) \qquad \rho_\delta = \left(\frac{\langle x_b \rangle}{1 + \delta \langle x_b \rangle} \right)^\lambda, \qquad \delta > 0,$$

$$(\text{B.18}) \qquad \psi_\delta = \rho_\delta \psi, \quad \Psi_\delta = \psi_\delta / \|\psi_\delta\|, \quad \|\Psi_\delta\| = 1.$$

Further letting P_b be the eigenprojection of the subsystem H^b with eigenvalue E, we define

$$(\text{B.19}) \qquad \varphi_\delta = P_b \psi_\delta = P_b \rho_\delta \psi = \rho_\delta P_b \psi, \qquad \Phi_\delta = \varphi_\delta / \|\varphi_\delta\|.$$

Then as in Lemma B.3 with using the condition (B.15), we have

$$(\text{B.20}) \qquad \lim_{\delta \downarrow 0} \|(H_b - E)\Psi_\delta\| = \lim_{\delta \downarrow 0} \|(H - E)\Psi_\delta\| = 0.$$

Therefore for any interval $\Delta = (E - a, E + a), a > 0,$

$$(\text{B.21}) \qquad \lim_{\delta \downarrow 0} \|(I - E_{H_b}(\Delta))\Psi_\delta\| = \lim_{\delta \downarrow 0} \|(I - E_H(\Delta))\Psi_\delta\| = 0.$$

(B.5) and Mourre estimate (B.6) thus yield

$$(\text{B.22}) \qquad \lim_{\delta \downarrow 0}(\Psi_\delta, i[H_b, A]\Psi_\delta) = \lim_{\delta \downarrow 0}(\Psi_\delta, i[H, A]\Psi_\delta) \geq 0.$$

On the other hand, by $H\psi = E\psi$ and a calculation, we have

(B.23)

$$Ai(H - E)\Psi_\delta = Ai[T_b, \rho_\delta]\rho_\delta^{-1}\Psi_\delta = 2\lambda \text{Re}\left(A\frac{1}{1 + \delta\langle x_b \rangle}\frac{1}{\langle x_b \rangle^2}A_b \right)\Psi_\delta$$
$$= O(\langle x_b \rangle^{-1}D_b)\Psi_\delta + Q\Psi_\delta,$$

where we have used (B.15), and $Q = 2\lambda \text{Re}\left(A_b \frac{1}{1+\delta\langle x_b\rangle} \frac{1}{\langle x_b\rangle^2}A_b \right)$ is a positive operator up to an error $R = O(\langle x_b \rangle^{-1}D_b)$. Thus

$$(\text{B.24}) \qquad (\Psi_\delta, i[H, A]\Psi_\delta) = -2(Q\Psi_\delta, \Psi_\delta) + (O(\langle x_b \rangle^{-1}D_b)\Psi_\delta, \Psi_\delta).$$

The second term tends to 0 as $\delta \downarrow 0$ as in (B.11). This and (B.22) give

(B.25)
$$\lim_{\delta \downarrow 0} (\Psi_\delta, i[H_b, A]\Psi_\delta) = 0.$$

Since $i[H_b, A] = 2T_b + i[H^b, A^b]$, $A^b = (x^b \cdot D^b + D^b \cdot x^b)/2$, applying Mourre estimate and (B.21) with $H^b = H_b - T_b$ to $i[H^b, A^b]$, we have

(B.26)
$$0 = \lim_{\delta \downarrow 0} (\Psi_\delta, i[H_b, A]\Psi_\delta) \geq \limsup_{\delta \downarrow 0} (\Psi_\delta, 2T_b\Psi_\delta) \geq 0.$$

Thus

(B.27)
$$\lim_{\delta \downarrow 0} \|T_b \Psi_\delta\| = 0.$$

This and (B.20) give

(B.28)
$$\lim_{\delta \downarrow 0} \|(H^b - E)\Psi_\delta\| = 0,$$

namely for any $\Delta = (E - a, E + a)$, $a > 0$

(B.29)
$$\lim_{\delta \downarrow 0} \|(I - E_{H^b}(\Delta))\Psi_\delta\| = 0.$$

Since $\langle x^b \rangle^\lambda \Psi_\delta \in L^2(R^n)$ uniformly in $\delta > 0$ by (B.15) and $\lim_{a \downarrow 0} \|(E_{H^b}(\Delta) - P_b)\langle x^b \rangle^{-\lambda}\| = 0$, we have from (B.29) that

(B.30)
$$\lim_{\delta \downarrow 0} \|\Psi_\delta - P_b \Psi_\delta\| = 0.$$

This and $\|\Psi_\delta\| = 1 \neq 0$ imply $P_b\Psi_\delta \neq 0$, hence $P_b\psi \neq 0$. Therefore Φ_δ in (B.19) is well-defined, and $\|\Phi_\delta\| = 1$. By (B.15), $\lim_{M \to \infty} \|P_b^M \Phi_\delta - \Phi_\delta\| = \lim_{M \to \infty} \|(P_b^M - P_b)\Phi_\delta\| = 0$ uniformly in $\delta > 0$, where P_b^M is defined in the introduction. Hence there is an integer $M \geq 1$ such that $\Phi_\delta^M = P_b^M \Phi_\delta$ satisfies

(B.31)
$$1 \geq \sup_{\delta > 0} \|\Phi_\delta^M\| \geq \inf_{\delta > 0} \|\Phi_\delta^M\| \geq \frac{1}{2}.$$

These Φ_δ^M and P_b^M will be denoted Φ_δ and P_b in the following unless otherwise specified.

As in (B.23) we have using the polynomial decay of subsystem eigenvectors

(B.32)
$$i(T_b + I_b)\Phi_\delta = i[T_b, \rho_\delta]\rho_\delta^{-1}\Phi_\delta + i[I_b, P_b]\tilde{\Phi}_\delta$$
$$= O(\langle x_b \rangle^{-1} D_b)\Phi_\delta + O(\langle x_b \rangle^{-1-\epsilon})\tilde{\Phi}_\delta,$$

where $\tilde{\Phi}_\delta = \rho_\delta \psi / \|\varphi_\delta\|$, hence $\Phi_\delta = P_b \tilde{\Phi}_\delta$. Setting

(B.33)
$$\tau = \tau_\delta = \frac{\langle x_b \rangle}{1 + \delta\langle x_b \rangle},$$

and using (B.32), the virial theorem $P_b i[H^b, A^b]P_b = 0$, and $\tau\Phi_\delta = P_b\tau\Phi_\delta$, we compute for $\sigma \in R^1$,

(B.34)
$$
\begin{aligned}
(\Phi_\delta, \tau^\sigma i[H - E, A]\tau^\sigma \Phi_\delta) &= (\Phi_\delta, \tau^\sigma i[T_b + I_b, A]\tau^\sigma \Phi_\delta) \\
&= -2\mathrm{Re}(\tau^\sigma Ai[T_b, \tau^\sigma]\Phi_\delta, \Phi_\delta) - 2\mathrm{Re}(\tau^\sigma A\tau^\sigma i(T_b + I_b)\Phi_\delta, \Phi_\delta) \\
&= -4\sigma\mathrm{Re}\left(\tau^\sigma A\frac{1}{1 + \delta\langle x_b\rangle}\frac{1}{\langle x_b\rangle^2}A_b\tau^\sigma\Phi_\delta, \Phi_\delta\right) \\
&\quad -4\lambda\mathrm{Re}\left(\tau^\sigma A\tau^\sigma\frac{1}{1 + \delta\langle x_b\rangle}\frac{1}{\langle x_b\rangle^2}A_b\Phi_\delta, \Phi_\delta\right) - 2\mathrm{Re}(\tau^\sigma A\tau^\sigma O(\langle x_b\rangle^{-1-\epsilon})\tilde\Phi_\delta, P_b\tilde\Phi_\delta).
\end{aligned}
$$

The last term equals

$$
\begin{aligned}
&-2\mathrm{Re}(\tilde\Phi_\delta, O(\langle x_b\rangle^{-1-\epsilon})\tau^\sigma A_b\tau^\sigma\Phi_\delta) + (O(\tau^{2\sigma}\langle x_b\rangle^{-1-\epsilon})\tilde\Phi_\delta, \tilde\Phi_\delta) \\
&= (O(\tau^{2\sigma-\epsilon}D_b)\Phi_\delta, \tilde\Phi_\delta) + (O(\tau^{2\sigma-1-\epsilon})\tilde\Phi_\delta, \tilde\Phi_\delta).
\end{aligned}
$$

Thus similarly to (B.24), (B.34) is equal to
(B.35)
$$
-\text{positive term} + (O(\tau^{2\sigma-1}D_b)\Phi_\delta, \Phi_\delta) + (O(\tau^{2\sigma-\epsilon}D_b)\Phi_\delta, \tilde\Phi_\delta) + (O(\tau^{2\sigma-1-\epsilon})\tilde\Phi_\delta, \tilde\Phi_\delta).
$$

On the other hand, from the relation which follows from (B.32)

(B.36)
$$
\begin{aligned}
\tau^\sigma(H - E)\tau^\sigma\Phi_\delta &= \tau^\sigma(T_b + I_b)\tau^\sigma\Phi_\delta \\
&= \tau^\sigma[T_b, \tau^\sigma]\Phi_\delta + \tau^{2\sigma}(T_b + I_b)\Phi_\delta \\
&= O(\tau^{2\sigma-1}D_b)\Phi_\delta + O(\tau^{2\sigma-1-\epsilon})\tilde\Phi_\delta,
\end{aligned}
$$

we have for any $\Delta = (E - a, E + a)$, $a > 0$

(B.37)
$$
\begin{aligned}
\|\tau^\sigma(I - E_H(\Delta))\tau^\sigma\Phi_\delta\| &= \|O_\Delta(\tau^{2\sigma-1}D_b)\Phi_\delta\| + \|O_\Delta(\tau^{2\sigma-1-\epsilon})\tilde\Phi_\delta\| \\
&\le C_\Delta\|\tau^{2\sigma-1}D_b\Phi_\delta\| + C_\Delta\|\tau^{2\sigma-1-\epsilon}\tilde\Phi_\delta\|.
\end{aligned}
$$

Since $\tau^\sigma\Phi_\delta \in L^2(R^n)$, this with (B.5) and Mourre estimate yield

(B.38)
$$
\text{(B.34)} \ge -\alpha_\Delta + (O_\Delta(\tau^{2\sigma-1}D_b)\Phi_\delta, \Phi_\delta) + (O_\Delta(\tau^{2\sigma-1-\epsilon})\tilde\Phi_\delta, \Phi_\delta)
$$

for some $\alpha_\Delta > 0$ converging to 0 as $|\Delta| = 2a \downarrow 0$. (B.38) and (B.34)-(B.35) give with $\epsilon_0 = \min\{\epsilon, 1\}$

(B.39)
$$
|(\Phi_\delta, \tau^\sigma i[H - E, A]\tau^\sigma\Phi_\delta)| \le \alpha_\Delta + \|O_\Delta(\tau^{2\sigma-\epsilon_0}D_b)\Phi_\delta\| + \|O_\Delta(\tau^{2\sigma-1-\epsilon})\tilde\Phi_\delta\|,
$$

where we have used $\sup_{\delta>0}\|\tilde\Phi_\delta\| < \infty$, which follows from $\|P_b\tilde\Phi_\delta\| = 1$ and $\|\tilde\Phi_\delta - P_b\tilde\Phi_\delta\| = \|(I - P_b)\tilde\Phi_\delta\| \to 0$ as $\delta \downarrow 0$ by (B.30). (Here P_b is not P_b^M but the whole projection.)

(B.5) also gives

(B.40)
$$(\Phi_\delta, \tau^\sigma i[H - E, A]\tau^\sigma \Phi_\delta) = (2 - \epsilon)(\Phi_\delta, \tau^\sigma T_b \tau^\sigma \Phi_\delta) + \epsilon(\Phi_\delta, \tau^\sigma (T_b + I_b)\tau^\sigma \Phi_\delta)$$
$$= (2 - \epsilon)(\Phi_\delta, \tau^\sigma T_b \tau^\sigma \Phi_\delta) + \epsilon \mathrm{Re}(\Phi_\delta, \tau^\sigma [T_b, \tau^\sigma]\Phi_\delta) + \epsilon(\Phi_\delta, \tau^{2\sigma}(T_b + I_b)\Phi_\delta)$$
$$= (2 - \epsilon)(\Phi_\delta, \tau^\sigma T_b \tau^\sigma \Phi_\delta) + (O(\tau^{2\sigma-1} D_b)\Phi_\delta, \Phi_\delta) + (O(\tau^{2\sigma-1-\epsilon})\Phi_\delta, \tilde\Phi_\delta).$$

In the last step, we have used (B.32).

We get from (B.39) and (B.40) that

(B.41)
$$|(\Phi_\delta, \tau^\sigma T_b \tau^\sigma \Phi_\delta)| \le \alpha_\Delta + \|O_\Delta(\tau^{2\sigma-\epsilon_0} D_b)\Phi_\delta\| + \|O_\Delta(\tau^{2\sigma-1-\epsilon})\tilde\Phi_\delta\|,$$

or

(B.42)
$$\|\tau^\sigma D_b \Phi_\delta\| \le c\sqrt{\alpha_\Delta} + C_\Delta \|\tau^{2\sigma-\epsilon_0} D_b \Phi_\delta\|^{1/2} + C_\Delta \|\tau^{2\sigma-1-\epsilon_0} \tilde\Phi_\delta\|^{1/2}.$$

Here $\sigma \in R^1$ is arbitrary. Hence we can iterate these estimates ℓ times to get

(B.43)
$$\|\tau^\sigma D_b \Phi_\delta\| + (\Phi_\delta, \tau^\sigma T_b \tau^\sigma \Phi_\delta)^{1/2}$$
$$\le C_{\Delta\ell}\sqrt{\alpha_\Delta} + C_{\Delta\ell}\|\tau^{2^{\ell+1}\sigma-(2^{\ell+1}-1)\epsilon_0}\Phi_\delta\|^{\frac{1}{2^{\ell+1}}} + \sum_{k=0}^{\ell} C_{\Delta k}\|\tau^{2^{k+1}\sigma-(2^{k+1}-1)\epsilon_0-1}\tilde\Phi_\delta\|^{\frac{1}{2^{k+1}}}.$$

To make the powers of τ on the RHS of (B.43) negative, it suffices, by letting ℓ large enough, to choose σ as

(B.44)
$$\begin{cases} \sigma < 1 & \text{when} & \epsilon \ge 1, \\ \sigma < \epsilon & \text{when} & 0 < \epsilon < 1. \end{cases}$$

In this case, the RHS of (B.43) tends to $C_{\Delta\ell}\sqrt{\alpha_\Delta}$ as $\delta \downarrow 0$. Noting that $\alpha_\Delta \to 0$ as $|\Delta| = 2a \downarrow 0$ and the LHS of (B.43) does not depend on Δ, we thus have

(B.45)
$$\lim_{\delta \downarrow 0}\{\|\tau^\sigma D_b \Phi_\delta\| + (\Phi_\delta, \tau^\sigma T_b \tau^\sigma \Phi_\delta)\} = 0,$$

if (B.44) holds.

Returning to the relation (B.32), we now get

(B.46) $\quad |(\Phi_\delta, \tau^\epsilon I_b \Phi_\delta)| \le |(\Phi_\delta, \tau^{\epsilon/2} T_b \tau^{\epsilon/2} \Phi_\delta)| + |(O(\tau^{\epsilon-1} D_b)\Phi_\delta, \Phi_\delta)| + |(O(\tau^{-1})\tilde\Phi_\delta, \tilde\Phi_\delta)|.$

By (B.45) and (B.44), the RHS tends to 0 as $\delta \downarrow 0$, if $0 < \epsilon < 2$.

On the other hand, the LHS of (B.46) is bounded from below by some positive constant when $\delta \downarrow 0$: To see this, we have only to consider the function $\tau^\epsilon I_b |\Phi_\delta|^2$ in the region $\{|x^b| < R, |x_b| > LR\}$ for $L \gg 1$. In this region $I_b \approx \frac{c}{|x_b|^\epsilon}$, $c = c_\alpha + c_\beta \neq 0$, if $b = \gamma \neq \alpha, \beta$, $\alpha \neq \beta$. Hence $\tau^\epsilon I_b \approx c(1 + \delta\langle x_b \rangle)^{-\epsilon}$. Now the ratio

$$\lim_{\delta \downarrow 0} \frac{\int (1 + \delta\langle x_b \rangle)^{-\epsilon} \rho_\delta^2 |P_b^M \psi|^2 \, dx}{\int \rho_\delta^2 |P_b \psi|^2 \, dx}$$

equals

$$\lim_{L \to \infty} \frac{\int_{|x_b| < L} \rho_0^2 |P_b^M \psi|^2 \, dx}{\int_{|x_b| < L} \rho_0^2 |P_b \psi|^2 \, dx} = \lim_{\delta \downarrow 0} \|\Phi_\delta^M\|^2,$$

which is bounded from below by 1/4 by (B.31). Thus we see that $\lim_{\delta \downarrow 0} |(\Phi_\delta, \tau^\epsilon I_b \Phi_\delta)| \geq |c|/4 > 0$. This contradicts with (B.46).

The proof of the theorem is complete except Lemma B.3.

Proof of Lemma B.3. We have to show for $b = \alpha$ and $R > 0$

$$(\text{B.47}) \qquad \lim_{\delta \downarrow 0} \frac{\iint_{|x_\alpha| < R} \rho_\delta^2 |\psi|^2 \, dx_\alpha dx_b}{\iint \rho_\delta^2 |\psi|^2 \, dx_\alpha dx_b}$$

$$= \lim_{L \to \infty} \frac{\int_{|x_b| < L} \int_{|x_\alpha| < R} \rho_0^2 |\psi|^2 \, dx_\alpha dx_b}{\int_{|x_b| < L} \int_{|x_\alpha| < L} \rho_0^2 |\psi|^2 \, dx_\alpha dx_b}$$

$$\leq \lim_{R' \to \infty} \lim_{L \to \infty} \frac{\int_{|x_b| < L} \int_{|x_\alpha| < R} \rho_0^2 |\psi|^2 \, dx_\alpha dx_b}{\int_{|x_b| < L} \int_{|x_\alpha| < R'} \rho_0^2 |\psi|^2 \, dx_\alpha dx_b} = 0.$$

If the numerator remains bounded as $L \to \infty$, then one may assume that it goes to ∞ by replacing the numerator by the sum of the numerator and denominator, and subtracting 1 from the whole ratio. In this case the RHS of (B.47) equals

$$(\text{B.48}) \qquad \lim_{R' \to \infty} \lim_{L \to \infty} \frac{L^{\nu-1} \int_{S^{\nu-1}} \int_{|x_\alpha| < R} \rho_0^2 |\psi|^2 \, dx_\alpha d\omega_b \big|_{|x_b| = L}}{L^{\nu-1} \int_{S^{\nu-1}} \int_{|x_\alpha| < R'} \rho_0^2 |\psi|^2 \, dx_\alpha d\omega_b \big|_{|x_b| = L}},$$

where $d\omega_b$ is the surface element on the sphere $S^{\nu-1}$, and we have differentiated the numerator and denominator on the RHS of (B.47) w.r.t. L. In the region $|x_\alpha| < R$ or $|x_\alpha| < R'$, $\rho_0^2 \approx L^{4\lambda} \langle x_\alpha \rangle^{2\lambda}$ when $L \to \infty$. Thus (B.48) equals

$$(\text{B.49}) \qquad \lim_{R' \to \infty} \lim_{L \to \infty} \frac{\int_{S^{\nu-1}} \int_{|x_\alpha| < R} \langle x_\alpha \rangle^{2\lambda} |\psi|^2 \, dx_\alpha d\omega_b \big|_{|x_b| = L}}{\int_{S^{\nu-1}} \int_{|x_\alpha| < R'} \langle x_\alpha \rangle^{2\lambda} |\psi|^2 \, dx_\alpha d\omega_b \big|_{|x_b| = L}}.$$

Integrating the numerator and denominator w.r.t. L in reverse to the above argument, we get

$$(\text{B.50}) \qquad (\text{B.49}) = \lim_{R' \to \infty} \lim_{L \to \infty} \frac{\int_{|x_b| < L} \int_{|x_\alpha| < R} \langle x_\alpha \rangle^{2\lambda} |\psi|^2 \, dx_\alpha dx_b}{\int_{|x_b| < L} \int_{|x_\alpha| < R'} \langle x_\alpha \rangle^{2\lambda} |\psi|^2 \, dx_\alpha dx_b}.$$

The numerator goes to a finite limit since $\psi \in L^2(R^n)$ and $\langle x_\alpha \rangle^{2\lambda} \leq \langle R \rangle^{2\lambda}$ in the region $|x_\alpha| < R$. The denominator goes to ∞ as $L \to \infty$ and $R' \to \infty$ by (B.8). The proof is complete. $\square$

Acknowledgements. This work was initiated when the author was staying at the University of Alabama at Birmingham in 1990. He expresses his sincere gratitude to Professor Y. Saitō for unceasing encouragement and kind hospitality during his stay there. He also expresses his thanks to the financial support from the Mathematics Component of NSF EPSCoR in Alabama, and to the referee for helpful suggestions which improved the description of Sec.1.

References

1. S. Agmon, "Lectures on Exponential Decay of Solutions of Second-order Elliptic Equations," Mathematical Notes 29, Princeton University Press and University of Tokyo Press, 1982.
2. H. L. Cycon, R.G. Froese, W. Kirsch, and B. Simon, "Schrödinger Operators," Springer-Verlag, Berlin-Heidelberg-New York-London-Paris-Tokyo, 1987.
3. J. Dereziński, *Algebraic approach to the N-body long-range systems*, Rev. in Math. Phys. **3** (1991), 1-62.
4. J. Dereziński, *Asymptotic completeness of long-range N-body quantum systems*, preprint, 1991.
5. V. Enss, *Quantum scattering theory for two- and three-body systems with potentials of short and long range*, in "Schrödinger Operators" ed. S. Graffi, Lect. Note in Math. 1159, Springer-Verlag, Berlin-Heidelberg-New York-London-Paris-Tokyo, 1985, pp.39-176.
6. R. Froese and I. Herbst, *Exponential bounds and absence of positive eigenvalues for N-body Schrödinger operators*, Commun. Math. Phys. **87** (1982), 429-447.
7. C. Gérard and J. Dereziński, *A remark on asymptotic clustering for N-particle quantum systems*, preprint, 1991.
8. L. Hörmander, *The existence of wave operators in scattering theory*, Math. Z. **146** (1976), 69-91.
9. H. Kitada, *Asymptotic completeness of N-body wave operators I. Short-range quantum systems*, Rev. in Math. Phys. **3** (1991), 101-124.
10. H. Kitada, *Scattering theory for Schrödinger equations with time-dependent potentials of long-range type*, J. Fac. Sci. Univ. Tokyo, Sec. IA, **29** (1982), 353-369.
11. H. Kitada and K. Yajima, *A scattering theory for time-dependent long-range potentials*, Duke Math. J. **49** (1982), 341-376.
12. H. Kitada, *Asymptotic completeness for N-body Schrödinger operators*, preprint (unpublished), 1984.
13. S.P. Merkuriev, *On the three-body Coulomb scattering problem*, Ann. Phys. **130** (1980), 395-426.
14. I. M. Sigal, *On long-range scattering*, Duke Math. J. **60** (1990), 473-496.
15. I.M. Sigal and A. Soffer, *Long-range many-body scattering. Asymptotic clustering for Coulomb-type potentials*, preprint, 1987.
16. I.M. Sigal and A. Soffer, *Asymptotic completeness of 3- and 4-body long-range systems*, preprint, 1988.
17. X.P. Wang, *On the three-body long range scattering problems*, preprint, 1991.

18. D. Yafaev, *Radiation condition and scattering theory for three-particle Hamiltonians*, preprint, 1991.

Semigroups of locally Lipschitzian operators and applications

Yoshikazu Kobayashi
Department of Applied Mathematics
Faculty of Engineering, Niigata Univ.
Niigata 950-21, Japan

Shinnosuke Oharu
Department of Mathematics
Faculty of Science, Hiroshima Univ.
Higashi-Hiroshima 724, Japan

1. Introduction

In the recent work [10] of the authors, a class of locally Lipschitzian operators has been introduced and the generation of such operator semigroups, as well as their characterization in terms of the corresponding generators, has been discussed. The class is considered in a general Banach space $(X, |\cdot|)$ and the local Lipschitz continuity is restricted by means of lower semicontinuous functionals on X. More precisely, let φ be a nonnegative lower semicontinuous functional on X, which is not necessary convex, and D a subset of X. We say that an operator semigroup $\{S(t)\}_{t\geq 0}$ on D is locally quasi-contractive with respect to the functional φ, or shortly that $\{S(t)\}_{t\geq 0}$ is in the class $\mathcal{S}(D, \varphi)$, if it satisfies the following condition:

(LQC) $D = D(\varphi) = \{v \in X : \varphi(v) < \infty\}$, and for $\alpha \geq 0$ and $\tau \geq 0$ there exists $\omega = \omega(\alpha, \tau) \in \boldsymbol{R}$ such that

$$|S(t)v - S(t)w| \leq e^{\omega t}|v - w| \qquad \text{for } v, w \in D_\alpha \text{ and } t \in [0, \tau],$$

where $D_\alpha = \{v \in D : \varphi(v) \leq \alpha\}$ for $\alpha \geq 0$.

This framework has been considered to treat systems of nonlinear partial differential equations and the lower semicontinuous functionals are supposed to be constructed according to the nature of the nonlinear systems. The objective of the present paper is twofold. First we outline the main points of the semigroup theory mentioned above in a more general setting than our original argument given in [10]. In this setting, we may even treat existence problems of local solutions to nonlinear evolution equations. Secondly, it is important to show that this abstract theory is applicable to a variety of nonlinear systems of partial differential equations. We here choose three typical types of nonlinear evolution problems and exhibit how the generation theorem may be applied to those problems.

The theory of semigroups of locally Lipschitzian operators introduced above contains four features. First, an appropriate growth condition is imposed on a semigroup $\{S(t)\}_{t\geq 0}$ in the class $\mathcal{S}(D, \varphi)$ by means of nonnegative functions $\varphi(S(\cdot)v)$, $v \in D$. Namely, given a semigroup $\{S(t)\}_{t\geq 0}$ in $\mathcal{S}(D, \varphi)$ we think of the following condition which we call an exponential growth condition:

(EG) For $v \in D$ and $t \in [0, \infty)$,

$$\varphi(S(t)v) \leq \varphi(v)e^{at} + b \int_0^t e^{a(t-s)} \, ds \qquad \text{for } v \in D \text{ and } t \geq 0,$$

where a and b are nonnegative constants.

Semigroups of this type arise as families of solution operators to the Cauchy problem for an evolution equation in X of the form

(CP) $$\qquad\qquad\qquad (d/dt)u(t) \in Au(t), \quad t > 0; \quad u(0) = u_0,$$

where A is a possibly multi-valued operator in X.

Secondly, the notion of quasi-dissipativity of operator A appearing in (CP) is localized in the following way: An operator A in X is said to be *locally quasi-dissipative* with respect to the functional φ or belong to the *class* $\mathcal{G}(D, \varphi)$, if it satisfies the following condition:

(LQD) $D(A) \subset D$, and for each $\alpha \geq 0$ there exists $\omega = \omega(\alpha) \in \mathbf{R}$ such that

$$[v_1 - v_2, w_1 - w_2]_- \leq \omega|v_1 - v_2|$$

for $v_1, v_2 \in D(A) \cap D_\alpha$, $w_1 \in Av_1$ and $w_2 \in Av_2$, where

$$[v, w]_- = \lim_{\lambda \uparrow 0}(|v + \lambda w| - |v|)/\lambda \qquad \text{for } v, w \in X.$$

Thirdly, we construct semigroups in the class $\mathcal{S}(D, \varphi)$ under the local quasi-dissipativity condition (LQD) and the *range condition* for A below:

(R) For $\varepsilon > 0$ and $u_0 \in D$ there exist $\delta \in (0, \varepsilon]$, $u_\delta \in D(A)$ and $w_\delta \in X$ which satisfy $|w_\delta| < \varepsilon$ and the two relations below:

$$\delta^{-1}(u_\delta - u_0) \in Au_\delta + w_\delta,$$
$$\delta^{-1}(\varphi(u_\delta) - \varphi(u_0)) \leq g(\varphi(u_\delta)) + \varepsilon,$$

where $g(r) = ar + b$ for $r \geq 0$.

Finally, the differentiability of semigroups in the class $\mathcal{S}(D, \varphi)$ can be investigated under the assumption that X is reflexive, the norm $|\cdot|$ is uniformly Gâteaux differentiable and that φ is convex.

As seen from the subsequent discussions, our generation theory extends the generation theorems given in [2, 3, 7, 9, 21] and the thought of the localization with respect to a lower semicontinuous functional is affected by the Lyapunov method. If in particular $a = b = 0$ in (EG), then φ becomes a Lyapunov function for A and asymptotic behaviors of $\{S(t)\}_{t \geq 0}$ can also be discussed. See [16]. The existence of such a functional φ is a priori assumed in our argument. However, given a concrete partial differential equations, appropriate functionals have to be chosen so that the generation theory may be applied to the associated problems. Although it is not necessarily straightforward to find adequate Lyapunov function for given partial differential equations, we here make an attempt to discuss typical well-posed problems of various types in the framework of the theory of semigroups of locally Lipschitzian operators.

In Section 2 we outline the above-mentioned generation theory and present the main results in a little more general form than those stated in [10]. Section 3 is concerned with the application to the FitzHugh-Nagumo equation which is a typical weakly coupled reaction-diffusion systems. We here employ a functional φ suggested by the geometrical shape of positively invariant regions. Section 4 discusses the Klein-Gordon equation from the point of view of our generation theory. Naturally, we consider a functional φ which represents an energy function for the semilinear wave equation. Finally, Section 5 is devoted to the semigroup approach to the Navier-Stokes equations in two space dimension. We eventually find a Lyapunov function for the basic equation for the flow of viscous incompressible fluid and show that the generation theory is also applicable to the Navier-Stokes problems.

2. The Abstract Theory

Let X be a real Banach space with norm $|\cdot|$. An *operator* A in X means a possibly multi-valued operator with domain $D(A)$ and range $R(A)$ in X. The identity operator on X is denoted by I. For $u, v \in X$, we define $[u, v]_\lambda = \lambda^{-1}(|u + \lambda v| - |u|)$ for $\lambda \in \boldsymbol{R} - \{0\}$, $[u, v]_+ = \inf_{\lambda>0}[u, v]_\lambda = \lim_{\lambda\downarrow0}[u, v]_\lambda$ and $[u, v]_- = -[u, -v]_+$. See [12, 5, 11] for the basic properties of the functionals $[\cdot, \cdot]_\pm$. If in particular X is a Hilbert space with inner product $(\cdot, \cdot)$, then $(u, v) = |u|[u, v]_\pm$.

Let D be a subset of X and let $\varphi : X \to [0, \infty]$ be a lower semi-continuous functional on X such that $D = D(\varphi) = \{u \in X : \varphi(u) < \infty\}$. For each $\alpha \geq 0$ the level set of φ is defined by

$$D_\alpha = D_\alpha(\varphi) = \{u \in D : \varphi(u) \leq \alpha\}. \tag{2.1}$$

We then introduce a class of nonlinear operators in X that are quasi-dissipative in a local sense with respect to the functional φ.

DEFINITION 1. An operator A in X is said to *be locally quasi-dissipative* with respect to the functional φ, if it satisfies condition (LQD) for the functional φ. In what follows, we simply say that A belongs to the *class* $\mathcal{G}(D, \varphi)$.

Let $\psi : X \to [0, \infty]$ be lower semi-continuous and satisfy $D = D(\psi)$. It is obvious that $A \in \mathcal{G}(D, \varphi)$ is equivalent to $A \in \mathcal{G}(D, \psi)$, provided that any ψ-bounded set of X is φ-bounded set, and *vice versa*. In this case, we say that the functional ψ is *equivalent* to φ.

A one-parameter family $\{S(t)\}_{t\geq0}$ of possibly nonlinear operators from D into itself is called a (*nonlinear*) *semigroup* on D if it has the two properties below:

(S1) For $s, t \geq 0$ and $v \in D$, $S(0)v = v$ and $S(s + t)v = S(s)S(t)v$.

(S2) For $v \in D$, $u(\cdot) = S(\cdot)v$ is continuous on $[0, \infty)$.

As will be seen later, semigroups generated by operators in the class $\mathcal{G}(D, \varphi)$ satisfy the local Lipschitz condition (LQC). This leads us to the following:

DEFINITION 2. A semigroup $\{S(t)\}_{t \geq 0}$ on D is said to be *locally quasi-contractive* with respect to the functional φ, if condition (LQC) is satisfied for the functional φ. In what follows, we say that $\{S(t)\}_{t \geq 0}$ belongs to the class $\mathcal{S}(D, \varphi)$.

Let A be an operator in the class $\mathcal{G}(D, \varphi)$ and consider the differential inclusion of the form

(DI) $$\qquad\qquad (d/dt)u(t) \in Au(t), \quad t > 0.$$

We here discuss a semigroup in the class $\mathcal{S}(D, \varphi)$ which provides solutions in a generalized sense to the differential inclusion (DI).

Prior to introducing a notion of mild solution to (DI), we recall the notion of strong solution of (DI). Let τ be an arbitrary but fixed positive number.

DEFINITION 3. A function $u : [0, \tau] \to X$ is said to be a *strong solution* of (DI) on $[0, \tau]$, if it is Lipschitz continuous over $[0, \tau]$, differentiable a.e. in $(0, \tau)$ and the strong derivative $u'(t)$ exists, $u(t) \in D(A)$, and belongs to the set $Au(t)$ for a.a. $t \in (0, \tau)$.

In the case that X is a general Banach space, the differential inclusion (DI) does not necessarily admit strong solutions even though the initial values lie in $D(A)$. We here adopt a notion of solution which refers directly to the approximation method used to establish the existence of solutions, so-called *method of discretization in time*.

DEFINITION 4. Let $\varepsilon > 0$. A piecewise constant function $v : [0, \tau] \to X$ is said to be an ε-approximate solution of (DI) on $[0, \tau]$, if there exists a partition $\{0 = t_0 < t_1 < \cdots < t_N\}$ of the interval $[0, t_N]$ and a finite sequence $\{(u_i, w_i)\}_{i=1}^{N}$ with the three properties below:

(ε.1) $$\qquad\qquad v(t) = \begin{cases} u_0 & \text{for } t = 0 \\ u_i & \text{for } t \in (t_{i-1}, t_i] \cap [0, \tau] \end{cases}$$

and

$$(t_i - t_{i-1})^{-1}(u_i - u_{i-1}) \in Au_i + w_i, \quad i = 1, \cdots, N,$$

(ε.2) $$\qquad t_i - t_{i-1} \leq \varepsilon, \quad i = 1, 2, \cdots \quad \text{and} \quad \tau \leq t_N < \tau + \varepsilon,$$

(ε.3) $$\qquad\qquad \sum_{i=1}^{N}(t_i - t_{i-1})|w_i| \leq \varepsilon t_N.$$

DEFINITION 5. A continuous function $u : [0, \tau] \to X$ is said to be a *mild solution* of (DI) on $[0, \tau]$, provided that for each $\varepsilon > 0$ there is an ε-approximate solution v^ε of (DI) on $[0, \tau]$ such that $|u(t) - v^\varepsilon(t)| \leq \varepsilon$ for $t \in [0, \tau]$. If there is a constant $\alpha \in [0, \infty)$ such that $v^\varepsilon(t) \in D_\alpha$ for $\varepsilon > 0$ and $t \in [0, \tau]$, then we say that the mild solution u is φ-*bounded* or *confined to* D_α on the interval $[0, \tau]$.

Notice that if u is a mild solution on $[0, \tau]$ confined to D_α then $u(t) \in D_\alpha$ for $t \in [0, \tau]$ since D_α is closed in X. A mild solutions confined to some D_α is therefore a uniform limit of approximate solutions confined to D_α. A strong solution $u(t)$ confined to some D_α is a mild solution confined to D_α.

PROPOSITION 1. *If $u : [0, \tau] \to X$ is a strong solution of* (DI) *on $[0, \tau]$, then it is a mild solution of* (DI) *on $[0, \tau]$. If in addition $u(t) \in D_\alpha$ for $t \in [0, \tau]$ and some $\alpha > 0$, then the mild solution u is confined to D_α.*

The proof is not obvious. See [1] and [11] for the detailed proof.

We have the following type of uniqueness theorem for φ-bounded mild solutions.

UNIQUENESS THEOREM. *Let $\alpha \geq 0$ and let $u : [0, \tau] \to X$ and $v : [0, \tau] \to X$ be mild solutions of* (DI) *on $[0, \tau]$ confined to D_α and D_β, respectively. Then there is $\omega = \omega(\alpha, \beta) \in [0, \infty)$ such that*

$$|v(t) - u(t)| \leq e^{\omega t}|v(0) - u(0)| \qquad for \qquad t \in [0, \tau].$$

If in particular $v(0) = u(0)$, then $v(t) \equiv u(t)$ on $[0, \tau]$.

It is convenient to employ a notion of *locally* φ-bounded mild solutions defined on semi-open intervals. To this end, we write σ for an arbitrary but fixed extended real number in $(0, \infty]$.

DEFINITION 6. Let $u : [0, \sigma) \to X$ be continuous over $[0, \sigma)$. We say that u is a *locally φ-bounded* mild solution of (DI) on $[0, \sigma)$, if to each $\tau \in [0, \sigma)$ there corresponds $\alpha \in [0, \infty)$ such that the restriction of u to $[0, \tau]$ gives a mild solution of (DI) on $[0, \tau]$ confined to D_α. A locally φ-bounded mild solution on $[0, \infty)$ is called locally φ-bounded *global* mild solution.

The next result is an immediate consequence of the uniqueness theorem.

COROLLARY 2. *Let $u : [0, \sigma) \to X$ and $v : [0, \sigma) \to X$ be mild solutions of* (DI). *Then for each $\tau \in [0, \sigma)$ there is $\omega \in [0, \infty)$ such that*

$$|u(t) - v(t)| \leq e^{\omega t}|u(0) - v(0)| \qquad for \ t \in [0, \tau].$$

If in particular $v(0) = u(0)$, then $v(t) \equiv u(t)$ on $[0, \sigma)$.

Suppose that for each $v \in D$ there is a *locally φ-bounded* global mild solution $u(\cdot; v)$ of (DI) satisfying $u(0; v) = v$. Then one can define for each $t \in [0, \infty)$ an operator $S(t) : D \to D$ by

$$S(t)v = u(t; v) \qquad \text{for} \quad v \in D. \tag{2.2}$$

To assert that the family $\{S(t)\}_{t \geq 0}$ forms a semigroup belonging to the class $\mathcal{S}(D, \varphi)$, we need φ-boundedness condition (C) below:

(C) For each $\alpha \in [0, \infty)$ and each $\tau \in [0, \sigma)$ there is $\beta \in [0, \infty)$ such that for $v \in D_\alpha$ the restriction of the associated global mild solution $u(\cdot; v)$ to $[0, \tau]$ is confined to D_β.

PROPOSITION 3. *Let $\{S(t)\}_{t \geq 0}$ be a family of self maps of D defined by (2.2). Then $\{S(t)\}_{t \geq 0}$ forms a semigroup on D. Assume further that condition (C) holds. Then the semigroup $\{S(t)\}_{t \geq 0}$ belongs to the class $\mathcal{S}(D, \varphi)$.*

We now introduce a *growth condition* with respect to the functional φ to define a natural class of semigroups on D satisfying condition (C) and then present a generation theorem for semigroups in the class $\mathcal{S}(D, \varphi)$ satisfying such a growth condition.

Let g be a continuous function defined on the interval $[0, \infty)$ such that $g(r) \geq 0$ for $r \in [0, \infty)$, which we call a *comparison function*. We write $m(\cdot; \alpha)$ for the non-extendable maximal solution of the initial value problem

$$r'(t) = g(r(t)), \quad r \geq 0; \quad r(0) = \alpha,$$

where α is a nonnegative number. The interval of existence of the non-extendable maximal solution $m(\cdot; \alpha)$ is denoted by $[0, \sigma_\infty(\alpha))$, where $\sigma_\infty(\alpha) \in (0, \infty]$.

Let A be an operator in X belonging to the class $\mathcal{G}(D, \varphi)$. We consider the following condition (R) which we call the *range condition* for the operator A.

(R) For $\varepsilon > 0$ and $u_0 \in D$ there exist $\delta \in (0, \varepsilon]$, $u_\delta \in D(A)$ and $w_\delta \in X$ satisfying $|w_\delta| < \varepsilon$ and the two relations below:

$$\delta^{-1}(u_\delta - u_0) \in Au_\delta + w_\delta,$$
$$\delta^{-1}(\varphi(u_\delta) - \varphi(u_0)) \leq g(\varphi(u_\delta)) + \varepsilon.$$

Then the following existence theorem for locally φ-bounded mild solutions is obtained:

EXISTENCE THEOREM. *Let $A \in \mathcal{G}(D, \varphi)$. Suppose that $D \subset \overline{D(A)}$, and that the range condition (R) holds. Let $u_0 \in D$, $\alpha_0 = \varphi(u_0)$ and $\sigma_0 = \sigma_\infty(\alpha_0)$. Then, there exists a locally φ-bounded mild solution $u(t)$ of the differential equation (DI) on $[0, \sigma_0)$ satisfying $u(0) = u_0$ and*

$$\varphi(u(t)) \leq m(t, \alpha_0) \quad \text{for} \quad t \in [0, \sigma_0).$$

In view of the above result, we consider the following growth condition with respect to φ on a semigroup $\{S(t)\}_{t \geq 0}$ in the class $\mathcal{S}(D, \varphi)$.

(G) For $\alpha \in [0, \infty)$, $\sigma_\infty(\alpha) = \infty$, $u_0 \in D$ and $t \in [0, \infty)$,

$$\varphi(S(t)u_0) \leq m(t; \varphi(u_0))$$

If in particular $g(r) = ar + b$ for some nonnegative constants a and b, (G) is called the *exponential growth condition* (EG). In this case, the non-extendable maximal solution $m(\cdot; \alpha)$ can be explicitly represented as

$$m(t; \alpha) = \alpha e^{at} + b \int_0^t e^{a(t-s)}\, ds$$

for $t \in [0, \infty)$ and $\alpha \in [0, \infty)$.

Our generation theorem is then stated as follows:

GENERATION THEOREM. *Let $A \in \mathcal{G}(D, \varphi)$. Suppose that $D \subset \overline{D(A)}$, the range condition (R) holds, and that $\sigma_\infty(\alpha) = \infty$ for any $\alpha \in [0, \infty)$. Then there exists a semigroup $\{S(t)\}_{t \geq 0}$ in the class $\mathcal{S}(D, \varphi)$ such that the growth condition (G) holds and for each $u_0 \in D$ the function $u(\cdot) = S(\cdot)u_0$ gives a unique locally φ-bounded global mild solution of (CP).*

In order to establish the basic results mentioned above, we first prepare Lemma 4 below and apply the Convergence Theorem which follows readily from the convergence theorems due to Kobayashi [9], Crandall and Evans [4] and Kobayasi, Kobayashi and Oharu [11]. In particular, Lemma 4 together with its proof contains fundamental estimates in our generation theory.

For each $\varepsilon > 0$ we write $m_\varepsilon(t; \alpha)$ for the maximal solution of the initial-value problem

$$r'(t) = g_\varepsilon(r(t)), \qquad t > 0; \qquad r(0) = \alpha,$$

where g_ε is defined by

$$g_\varepsilon(r) = g(r) + \varepsilon, \qquad r \in [0, \infty).$$

The maximal interval of existence of the non-extendable solution $m_\varepsilon(t; \alpha)$ is denoted by $[0, \sigma_\infty^\varepsilon(\alpha))$. If in particular $g(r) = ar + b$, it is seen that $m_\varepsilon(t; \alpha)$ is represented as

$$m_\varepsilon(t; \alpha) = \alpha e^{at} + (b + \varepsilon) \int_0^t e^{a(t-s)}\, ds.$$

LEMMA 4. *Let $A \in \mathcal{G}(D, \varphi)$. Suppose that $D \subset \overline{D(A)}$ and the range condition (R) holds. Let $u_0 \in D$. Then for each $\varepsilon > 0$ there exists a sequence $\{(h_n, u_n, v_n)\}_{n=1}^\infty$ in $(0, \varepsilon] \times D(A) \times X$ with the following properties:*

$$\sigma_\infty^\varepsilon(\varphi(x_0)) \leq \sum_{n=1}^\infty h_n,$$

$$v_n \in Au_n, \qquad n = 1, 2, \cdots,$$

$$|u_n - u_{n-1} - h_n v_n| \leq \varepsilon h_n, \qquad n = 1, 2, \cdots,$$

$$\varphi(u_n) \leq m_\varepsilon(h_n; \varphi(u_{n-1})), \qquad n = 1, 2, \cdots.$$

CONVERGENCE THEOREM. *Let A be an operator in the class $\mathcal{G}(D, \varphi)$ satisfying $D \subset \overline{D(A)}$. Let $\tau > 0, \alpha > 0$ and let $u_0 \in D_\alpha$. Suppose that there exists a positive number ε_0, and that for each $\varepsilon \in (0, \varepsilon_0)$ there is an ε-approximate solution $u^\varepsilon : [0, \tau] \to X$ such that $u^\varepsilon(t) \in D_\alpha$ for $t \in [0, \tau]$. If $\lim_{\varepsilon \downarrow 0} u^\varepsilon(0) = u_0$, then there exists a unique mild solution u of (CP) on $[0, \tau]$ confined to D_α and*

$$\lim_{\varepsilon \downarrow 0}\left(\sup\{|u^\varepsilon(t) - u(t)| : t \in [0, \tau]\}\right) = 0.$$

3. Reaction Diffusion Systems

In this section we make an attempt to describe the application of the generation theorem to reaction diffusion systems. For this purpose we take the so-called FitzHugh-Nagumo equations. The initial value problem for the FitzHugh-Nagumo equation on the real line $\boldsymbol{R}$ may be formulated as follows:

$$u_t = \nu u_{xx} + \sigma v - \gamma u, \quad v_t = v_{xx} + f(v) - u, \qquad (x, t) \in \boldsymbol{R} \times (0, \infty), \qquad (3.1)$$

$$u(x, 0) = u_0(x), \quad v(x, 0) = v_0(x), \qquad x \in \boldsymbol{R}, \qquad (3.2)$$

where $\nu, \sigma, \gamma > 0$ and $f(v)$ has the qualitative form of cubic polynomial $f(v) = -v(v - a)(v - b)$, $0 < a < b$.

For the weakly coupled reaction-diffusion system (3.1), positively invariant regions can be found and *a priori* estimates with respect to maximum norms for the solutions are obtained. Therefore it is natural to convert the problem (3.1)—(3.2) to (CP) in the Banach space $X = BUC(\boldsymbol{R})^2$ of $\boldsymbol{R}^2$-valued bounded uniformly continuous functions on $\boldsymbol{R}$. The norm of the space X is defined by

$$\|[u, v]\| = \sup_{x \in \boldsymbol{R}} |u(x)| \vee \sup_{x \in \boldsymbol{R}} |v(x)|.$$

Furthermore, it is expected that appropriate combinations of the maximum norms of u and v will be employed as functionals φ appearing in the growth condition (G).

We define a linear operator L on the domain $D(L) = \{[u, v]; [u_{xx}, v_{xx}] \in X\}$ by

$$L[u, v] = [\nu u_{xx}, v_{xx}]$$

and a nonlinear operator F on $D(F) = X$ by

$$F[u, v] = [\sigma v - \gamma u, f(v) - u].$$

Set $D(A) = D(L)$ and $A = L + F$. Then the problem (3.1)-(3.2) is transformed into (CP) in X and the following theorem is obtained.

THEOREM 5. *The operator A in the space X is densely defined in X and satisfies conditions (LQD) and (R) for the comparison function $g(r) \equiv 0$ and the functional φ defined by*

$$\varphi(u, v) = \left(\gamma|u|_{BUC(\boldsymbol{R})}\right) \vee \left(\sigma|v|_{BUC(\boldsymbol{R})}\right) \vee R_0, \qquad (3.3)$$

where R_0 is a positive number which may depend upon a, b, σ and γ. Therefore the problem (3.1)-(3.2) has a unique norm-bounded global mild solution of (CP) *in X for all $[u_0, v_0] \in X$.*

PROOF. Since $D(L)$ is dense in X, so is $D(A)$. It is well known that L is m-dissipative in X. Furthermore, it is easily seen that F is Lipschitz continuous on any $|\cdot|$-bounded set of X. Therefore, A is locally quasi-dissipative with respect to the norm $|\cdot|$ of X. Observing that the norm $|\cdot|$ is equivalent to the functional φ defined by (3.3), we see that A belongs to $\mathcal{G}(X, \varphi)$.

To show the range condition (R), take any $[u_0, v_0] \in X$ and set

$$[u_\delta, v_\delta] = (I - \delta L)^{-1}([u_0, v_0] + \delta F[u_0, v_0]) \tag{3.4}$$

for $\delta > 0$. Equation (3.4) can be written as

$$([u_\delta, v_\delta] - [u_0, v_0]) - \delta A[u_\delta, v_\delta] = \delta(F[u_0, v_0] - F[u_\delta, v_\delta]).$$

Since $[u_\delta, v_\delta] \to 0$ as $\delta \downarrow 0$ and F is Lipschitz continuous on $|\cdot|$-bounded sets, we have

$$\lim_{\delta \downarrow 0} \left(\delta^{-1}([u_\delta, v_\delta] - [u_0, v_0]) - A[u_\delta, v_\delta] \right) = 0.$$

We then claim that for a suitably chosen number $R_0 > 0$

$$\limsup_{\delta \downarrow 0} \delta^{-1}(\varphi([u_\delta, v_\delta]) - \varphi([u_0, v_0]) \le 0.$$

If this would hold, then we would see that A satisfies the range condition (R) with $g(r) \equiv 0$ for the continuous functional φ. Set

$$[\bar{u}_\delta, \bar{v}_\delta] = [u_0, v_0] + \delta F[u_0, v_0]$$

for $\delta > 0$. The resolvent $(I - \delta L)^{-1}$ is nonexpansive on $BUC(\boldsymbol{R})$ and so $\varphi([u_\delta, v_\delta]) \le \varphi([\bar{u}_\delta, \bar{v}_\delta])$. Thus, it is sufficient to show that

$$\limsup_{\delta \downarrow 0} \delta^{-1}(\varphi([\bar{u}_\delta, \bar{v}_\delta]) - \varphi([u_0, v_0])) \le 0 \tag{3.5}$$

for some number $R_0 > 0$. To this end, we first observe that

$$\bar{u}_\delta(x) = u_0(x) + \delta(\sigma \bar{v}_\delta(x) - \gamma \bar{u}_\delta(x)) + \delta(\sigma(v_0(x) - \bar{v}_\delta(x)) - \gamma(u_0(x) - \bar{u}_\delta(x))). \tag{3.6}$$

Multiplying both sides of (3.6) by $\text{sign}(\bar{u}_\delta(x))$, we have

$$
\begin{aligned}
|\bar{u}_\delta(x)| &= u_0(x)\,\text{sign}(\bar{u}_\delta(x)) + \delta(\sigma \bar{v}_\delta(x) - \gamma \bar{u}_\delta(x))\,\text{sign}(\bar{u}_\delta(x)) \\
&\quad + \delta(\sigma(v_0(x) - \bar{v}_\delta(x)) - \gamma(u_0(x) - \bar{u}_\delta(x)))\,\text{sign}(\bar{u}_\delta(x)) \\
&\le |u_0(x)| + \delta(\sigma|\bar{v}_\delta(x)| - \gamma|\bar{u}_\delta(x)|) \\
&\quad + \delta(\sigma|v_0(x) - \bar{v}_\delta(x)| + \gamma|u_0(x) - \bar{u}_\delta(x)|).
\end{aligned} \tag{3.7}
$$

A similar argument implies

$$|\bar{v}_\delta(x)| \leq |v_0(x)| + \delta(|\bar{u}_\delta(x)| - |\bar{v}_\delta(x)|(\bar{v}_\delta(x) - a)(\bar{v}_\delta(x) - b))$$
$$+ \delta(|u_0(x) - \bar{u}_\delta(x)| + |f(v_0(x)) - f(\bar{v}_\delta(x))|). \tag{3.8}$$

Let $\varepsilon > 0$ and choose $\delta_0 > 0$ so that

$$\gamma\left(\sigma|v_0(x) - \bar{v}_\delta(x)| + \gamma|u_0(x) - \bar{u}_\delta(x)|\right) \leq \varepsilon$$
$$\sigma\left(|u_0(x) - \bar{u}_\delta(x)| + |f(v_0(x)) - f(\bar{v}_\delta(x))|\right) \leq \varepsilon$$

for $x \in \mathbf{R}$ and $\delta \in (0, \delta_0)$. Then it follows from (3.7) and (3.8) that for $x \in \mathbf{R}$ and $\delta \in (0, \delta_0)$

$$\gamma|\bar{u}_\delta(x)| \leq \gamma|u_0(x)| + \delta\gamma(\sigma|\bar{v}_\delta(x)| - \gamma|\bar{u}_\delta(x)|) + \delta\varepsilon \tag{3.9}$$

and

$$\sigma|\bar{v}_\delta(x)| \leq \sigma|v_0(x)| + \delta\sigma\left(|\bar{u}_\delta(x)| - |\bar{v}_\delta(x)|(\bar{v}_\delta(x) - a)(\bar{v}_\delta(x) - b)\right) + \delta\varepsilon. \tag{3.10}$$

Choose a number R_0 so large that

$$\sigma/\gamma - (r - a)(r - b) \leq 0 \quad \text{for} \quad r \in \mathbf{R} \quad \text{with} \quad \sigma|r| \geq R_0. \tag{3.11}$$

In view of these estimates (3.9) and (3.10) we obtain

$$\varphi([\bar{u}_\delta, \bar{v}_\delta]) \leq \varphi([u_0, v_0]) + \delta\varepsilon, \tag{3.12}$$

for $\delta \in (0, \delta_0)$. In fact, we will show (3.12) in the following three cases:

(a) $(\gamma|\bar{u}_\delta(x)|) \vee (\sigma|\bar{v}_\delta(x)|) \leq R_0$,

(b) $(\gamma|\bar{u}_\delta(x)|) \vee (\sigma|\bar{v}_\delta(x)|) \geq R_0$ and $\gamma|\bar{u}_\delta(x)| \geq \sigma|\bar{v}_\delta(x)|$,

(c) $(\gamma|\bar{u}_\delta(x)|) \vee (\sigma|\bar{v}_\delta(x)|) \geq R_0$ and $\gamma|\bar{u}_\delta(x)| \leq \sigma|\bar{v}_\delta(x)|$.

In the first case (a),

$$(\gamma|\bar{u}_\delta(x)|) \vee (\sigma|\bar{v}_\delta(x)|) \vee R_0 = R_0 \leq \varphi([u_0, v_0]) \leq \varphi([u_0, v_0]) + \delta\varepsilon.$$

In the second case (b), we have $\sigma|\bar{v}_\delta(x)| - \gamma|\bar{u}_\delta(x)| \leq 0$, and so (3.9) implies

$$(\gamma|\bar{u}_\delta(x)|) \vee (\sigma|\bar{v}_\delta(x)|) \vee R_0 = \gamma|\bar{u}_\delta(x)| \leq \gamma|\bar{u}_0(x)| + \delta\varepsilon \leq \varphi([u_0, v_0]) + \delta\varepsilon.$$

In the last case (c), $\sigma|\bar{v}_\delta(x)| - \gamma|\bar{u}_\delta(x)| \geq 0$ and $\sigma|\bar{v}_\delta(x)| = (\sigma|\bar{v}_\delta(x)|) \vee (\gamma|\bar{u}_\delta(x)|) \geq R_0$. Therefore, it follows from (3.11) that

$$|\bar{u}_\delta(x)| - |\bar{v}_\delta(x)|(\bar{v}_\delta(x) - a)(\bar{v}_\delta(x) - b) \leq (\sigma/\gamma)|\bar{v}_\delta(x)| - |\bar{v}_\delta(x)|(\bar{v}_\delta(x) - a)(\bar{v}_\delta(x) - b)$$
$$= |\bar{v}_\delta(x)|((\sigma/\gamma) - (\bar{v}_\delta(x) - a)(\bar{v}_\delta(x) - b)) \leq 0$$

Thus, (3.10) implies that

$$(\gamma|\bar{u}_\delta(x)|) \vee (\sigma|\bar{v}_\delta(x)|) \vee R_0 = \sigma|\bar{v}_\delta(x)| \leq \sigma|\bar{v}_0(x)| + \delta\varepsilon \leq \varphi([u_0, v_0]) + \delta\varepsilon.$$

In any case we obtain (3.12) and *a fortiori* (3.5). This completes the proof of Theorem 3.
□

REMARK. We have used a functional φ defined by (3.3). This approach is based on the theory of positively invariant regions for systems of nonlinear diffusion equations. See Smoller's book [20].

4. Nonlinear Wave Equations

We here treat the initial-boundary value problem for the Klein–Gordon equation

$$u_t = v, \quad v_t = \Delta u - |u|^2 u, \qquad (x,t) \in \Omega \times (0,\infty), \tag{4.1}$$

$$u(x,t) = 0, \qquad (x,t) \in \partial\Omega \times (0,\infty), \tag{4.2}$$

$$u(x,0) = u_0(x), \quad v(x,0) = v_0(x), \qquad x \in \Omega, \tag{4.3}$$

where Ω is a domain in $\mathbf{R}^3$ with smooth compact boundary $\partial\Omega$. For the hyperbolic system (4.1)-(4.3) energy functions can be found and *a priori* estimates for the solutions are obtained in terms of the energy functions. Therefore a natural approach is what is called energy method. The problem (4.1)-(4.3) can be interpreted as (CP) in the Hilbert space $X = H_0^1(\Omega) \times L^2(\Omega)$ with inner product

$$|([u,v],[\hat{u},\hat{v}])| = \int_\Omega (u\hat{u} + \nabla u \cdot \nabla \hat{u} + v\hat{v})\, dx$$

and a natural functional φ may be defined by

$$\varphi(u,v) = \int_\Omega (\tfrac{1}{2}|u|^2 + \tfrac{1}{2}|\nabla u|^2 + \tfrac{1}{4}|u|^4 + \tfrac{1}{2}|v|^2)\, dx, \quad [u,v] \in X. \tag{4.4}$$

We then define the wave operator L in the space X by

$$D(L) = (H_0^1(\Omega) \cap H^2(\Omega)) \times H_0^1(\Omega) \qquad \text{and} \qquad L[u,v] = [v, \Delta u]$$

and a nonlinear operator F in X by

$$D(F) = X \qquad \text{and} \qquad F[u,v] = [0, -|u|^2 u].$$

Set $D(A) = D(L)$ and $A = L + F$. Then the problem (4.1)-(4.3) is transformed into (CP) in X and A satisfies (LQD) and (R) for the functional φ, as stated in the following theorem.

THEOREM 6. *The operator A in the Banach space X is densely defined in X and satisfies conditions* (LQD) *and* (R) *with the comparison function $g(r) \equiv r$ for the functional φ. The problem (4.1)-(4.3) has a unique φ-bounded global mild solution for all $[u_0, v_0] \in X$.*

PROOF. To investigate the nonlinear operator F, we take a pair u, $\hat{u} \in L^2(\Omega)$. Using Hölder's inequality, we have

$$\int_\Omega ||u|^2 u - |\hat{u}|^2 \hat{u}|^2\, dx = \int_\Omega (p \int_0^1 |u + \theta(\hat{u} - u)|^2\, d\theta)^2 |u - \hat{u}|^2\, dx$$

$$\leq \left(\int_\Omega \left((p \int_0^1 |u + \theta(\hat{u} - u)|^2\, d\theta)^2 \right)^3 dx \right)^{2/3} \left(\int_\Omega |u - \hat{u}|^6\, dx \right)^{1/3}.$$

Since $H_0^1(\Omega)$ is continuously embedded in $L^6(\Omega)$, we have

$$\left(\int_\Omega |u - \hat{u}|^6 \, dx \right)^{1/3} \leq C \|u - \hat{u}\|_{H_0^1(\Omega)}^2$$

for some constant C. Similarly, we have

$$\left(\int_\Omega \left(\left(p \int_0^1 |u + \theta(\hat{u} - u)|^2 \, d\theta \right)^2 \right)^3 dx \right)^{2/3}$$

$$\leq p^2 \left(\int_\Omega (|u| + |\hat{u}|)^6 \, dx \right)^{2/3}$$

$$\leq C(\|u\|_{H_0^1(\Omega)} + \|\hat{u}\|_{H_0^1(\Omega)})^4.$$

Combining the above estimates, we obtain

$$\||u|^2 u - |\hat{u}|^2 \hat{u}\|_{L^2(\Omega)} \leq C(\|u\|_{H_0^1(\Omega)} + \|\hat{u}\|_{H_0^1(\Omega)})^2 \|u - \hat{u}\|_{H_0^1(\Omega)}. \tag{4.5}$$

This implies that $D(F) = X$ and F is Lipschitz continuous on $|\cdot|$-bounded sets. From this it follows that F is Lipschitz continuous on subsets which are bounded with respect to the functional φ defined by (4.4).

Note that a similar argument implies that

$$\||u|^4 - |\hat{u}|^4\|_{L^1(\Omega)} \leq C(\|u\|_{H_0^1(\Omega)} + \|\hat{u}\|_{H_0^1(\Omega)})^3 \|u - \hat{u}\|_{L^2(\Omega)}.$$

This means that $D(\varphi) = X$ and φ is continuous on X.

As well known, $L - I$ is densely defined and m-dissipative in X. Hence, $A = L + F$ is densely defined and locally quasi-dissipative with respect to the functional φ.

It now remains to show that A satisfies the range condition (R). To this end, let $[u_0, v_0] \in X$ and set

$$[u_\delta, v_\delta] = (I - \delta L)^{-1}([u_0, v_0] + \delta F[u_0, v_0]) \tag{4.6}$$

for $\delta \in (0, 1)$. Since $L - I$ is m-dissipative, $(I - \delta L)^{-1}$ is well-defined for $\delta \in (0, 1)$. Then (4.6) can be written as

$$([u_\delta, v_\delta] - [u_0, v_0]) - \delta A[u_\delta, v_\delta] = \delta(F[u_0, v_0] - F[u_\delta, v_\delta]).$$

As easily seen, $[u_\delta, v_\delta] \to [u_0, v_0]$ as $\delta \downarrow 0$. Since F is Lipschitz continuous on $|\cdot|$-bounded subsets of X, it follows that

$$\lim_{\delta \downarrow 0} \left(\delta^{-1}([u_\delta, v_\delta] - [u_0, v_0]) - A[u_\delta, v_\delta] \right) = 0.$$

By the definition of $[u_\delta, v_\delta]$ we have

$$u_\delta - \delta v_\delta = u_0 \tag{4.7}$$

and

$$v_\delta - \delta \Delta u_\delta = v_0 - \delta |u_0|^2 u_0. \qquad (4.8)$$

Multiplying both sides of (4.8) by v_δ, we have

$$v_\delta^2 - \delta \Delta u_\delta v_\delta = v_0 v_\delta - \delta |u_0|^2 u_0 v_\delta \le \frac{1}{2}(v_0^2 + v_\delta^2) - \delta |u_0|^2 u_0 v_\delta.$$

Hence,

$$\frac{1}{2}(v_\delta^2 - v_0^2) - \delta \Delta u_\delta v_\delta \le -\delta |u_0|^2 u_0 v_\delta.$$

Since $\delta v_\delta = u_\delta - u_0$ by (4.7), we get

$$\frac{1}{2}(v_\delta^2 - v_0^2) - \Delta u_\delta (u_\delta - u_0) \le -\delta |u_0|^2 u_0 v_\delta. \qquad (4.9)$$

Integrating both sides of (4.9) over Ω, we have

$$\int_\Omega \left(\frac{1}{2}(v_\delta^2 - v_0^2) + \nabla u_\delta \cdot \nabla(u_\delta - u_0) \right) dx \le \int_\Omega -\delta |u_0|^2 u_0 v_\delta \, dx.$$

Since $2^{-1}(|\nabla u_\delta|^2 - |\nabla u_0|^2) \le \nabla u_\delta \cdot \nabla(u_\delta - u_0)$ and $4^{-1}(|u_\delta|^4 - |u_0|^4) \le |u_\delta|^2 u_\delta(u_\delta - u_0)$, the application of (4.7) implies

$$\int_\Omega \left(\frac{1}{2}(v_\delta^2 - v_0^2) + \frac{1}{2}(|\nabla u_\delta|^2 - |\nabla u_0|^2) + \frac{1}{4}(|u_\delta|^4 - |u_0|^4) \right) dx$$

$$\le \int_\Omega (|u_\delta|^2 u_\delta - |u_0|^2 u_0)(u_\delta - u_0) \, dx$$

$$\le \||u_\delta|^2 u_\delta - |u_0|^2 u_0\|_{L^2(\Omega)} \|u_\delta - u_0\|_{L^2(\Omega)}$$

$$\le C(\|u_\delta\|_{H_0^1(\Omega)} + \|u_0\|_{H_0^1(\Omega)})^2 \|u_\delta - u_0\|_{H_0^1(\Omega)} \|u_\delta - u_0\|_{L^2(\Omega)}. \qquad (4.10)$$

On the other hand, (4.7) implies

$$u_\delta^2 - \delta v_\delta u_\delta = u_0 u_\delta \le \frac{1}{2}(u_\delta^2 + u_0^2)$$

and hence

$$\frac{1}{2}(u_\delta^2 - u_0^2) \le \delta v_\delta u_\delta \le \frac{\delta}{2}(u_\delta^2 + v_\delta^2).$$

Therefore

$$\int_\Omega \frac{1}{2}(u_\delta^2 - u_0^2) \, dx \le \int_\Omega \frac{\delta}{2}(u_\delta^2 + v_\delta^2) \, dx \le \delta \varphi([u_\delta, v_\delta]). \qquad (4.11)$$

Combining (4.10) and (4.11), we obtain

$$\varphi([u_\delta, v_\delta]) - \varphi([u_0, v_0]) \le \delta \varphi([u_\delta, v_\delta])$$

$$+ C(\|u_\delta\|_{H_0^1(\Omega)} + \|u_0\|_{H_0^1(\Omega)})^2 \|u_\delta - u_0\|_{H_0^1(\Omega)} \|u_\delta - u_0\|_{L^2(\Omega)}. \qquad (4.12)$$

Since u_δ converges to u_0 in $H_0^1(\Omega)$ and $\delta^{-1}(u_\delta - u_0) = v_\delta$ is bounded in $L^2(\Omega)$ as $\delta \downarrow 0$, the inequality (4.12) implies

$$\limsup_{\delta \downarrow 0} \left(\delta^{-1}(\varphi([u_\delta, v_\delta]) - \varphi([u_0, v_0])) - \varphi([u_\delta, v_\delta]) \right) \le 0.$$

Thus, it is concluded that the operator A satisfies range condition (R) with $g(r) = r$. The proof is now complete. $\qquad \square$

5. Fluid Mechanics

The dynamical equations for the flow of viscous incompressible fluid is the Navier–Stokes equation. We here describe the nonlinear semigroup approach to the initial-boundary value problem for the Navier–Stokes equation

$$u_t + (u \cdot \nabla)u - \nu \Delta u + \nabla p = 0, \quad \text{div } u = 0, \qquad (x, t) \in \Omega \times (0, \infty), \qquad (5.1)$$

$$u(x, t) = 0, \qquad (x, t) \in \partial\Omega \times (0, \infty), \qquad (5.2)$$

$$u(x, 0) = u_0(x), \qquad x \in \Omega, \qquad (5.3)$$

in the case that Ω is a bounded domain in $\mathbf{R}^2$ with smooth boundary $\partial\Omega$ and $\nu > 0$ is kinetic viscosity. The unknown function $u = u(x, t)$ on $\Omega \times (0, \infty)$ is an $\mathbf{R}^2$-valued function $[u_1, u_2]$ which represents the velocity components and p stands for the pressure which is eventually determined by u.

For the problem (5.1)-(5.3) numerous estimates for the weak solutions have been obtained and various properties of the nonlinear terms appearing in (5.1) have been investigated. Referring to the well-known results, we define two spaces X and V by taking the closures of $\{u \in C_0^\infty(\Omega)^2; \text{div } u = 0\}$ in $L^2(\Omega)^2$ and in $H_0^1(\Omega)^2$, respectively. The inner products of the spaces X and V are defined by

$$(u, v) = \sum_{i=1}^2 \int_\Omega u_i v_i \, dx \quad \text{and} \quad ((u, v)) = \sum_{i=1}^2 \int_\Omega (\partial_{x_i} u_i)(\partial_{x_i} v_i) \, dx,$$

respectively, and $|\cdot|$ and $\|\cdot\|$ denote the associated norms of X and of V, respectively. Then the Stokes operator L is defined in X as a self-adjoint positive operator, namely, $w = Lv$ if and only if $v \in V$ and $w \in X$ and $(w, f) = ((v, f))$ for any $f \in V$.

The main point in this approach is to introduce an adequate nonlinear operator A representing the semilinear differential operator $\nu \Delta u - (u \cdot \nabla)u$ and convert the problem (5.1)-(5.3) to (CP) in the Hilbert space X. The operator A is defined as follows: We write $w = Av$ if and only if $v \in V$ and $w \in X$ and

$$(w, f) + \nu((v, f)) + b(v, v, f) = 0 \qquad \text{for any } f \in V,$$

where $b(\cdot, \cdot, \cdot)$ is a trilinear form on $V \times V \times V$ defined by

$$b(u, v, w) = \sum_{i,j=1}^2 \int_\Omega u_i (\partial_{x_i} v_j) w_j \, dx.$$

The problem (5.1)–(5.3) is then transformed into (CP) in the Hilbert space X. In this formulation of (CP), $D = V$ and the functional φ appearing in (G) is given by

$$\varphi(u) = \begin{cases} |u|^4 + c_0 \nu^4 \log(1 + \|u\|^2), & u \in V, \\ \infty, & u \in X \setminus V, \end{cases} \qquad (5.4)$$

where c_0 is a suitably chosen positive number which depends upon Ω. Note that φ is *not* convex. Since there is a complete orthonormal system of X consisting of the eigenfunctions of the Stokes operator L, the Faedo–Galerkin method can be applied to obtain the following theorem.

THEOREM 7. *The domain $D(A)$ is dense in X and the operator A in X satisfies conditions (LQD) and (R) for the comparison function $g(r) \equiv 0$ and the functional φ defined by (5.4). The problem (5.1)-(5.3) has a unique $\| \cdot \|$-bounded global mild solution for all $u_0 \in V$.*

PROOF. First we note that the trilinear form b is well-defined and continuous on $X \times X \times X$, and that

$$b(u, v, w) = -b(u, w, v), \quad b(u, v, v) = 0 \tag{5.5}$$

for $u, v, w \in V$. Furthermore, we obtain the estimates

$$|b(u, v, w)| \leq \sqrt{2} |u|^{1/2} \|u\|^{1/2} \|v\| \, |w|^{1/2} \|w\|^{1/2} \tag{5.6}$$

for $u, v, w \in V$. (See Lemma 3.4, Chap. III, [22]). We here divide the proof into three steps.

Step 1. We first show that A is locally quasi-dissipative with respect to the functional φ defined by (5.4). Obviously, $D(A) \subset V$. Let $u_1, u_2 \in D(A)$. By the definition of the operator A

$$(Au_1 - Au_2, v) + \nu((u_1 - u_2, v)) = b(u_2, u_2, v) - b(u_1, u_1, v) \tag{5.7}$$

for $v \in V$. We have

$$b(u_1, u_1, u_1 - u_2) = b(u_1, u_2, u_1 - u_2)$$

and

$$b(u_2, u_2, u_1 - u_2) - b(u_1, u_2, u_1 - u_2) = b(u_2 - u_1, u_2, u_1 - u_2).$$

Therefore, setting $v = u_1 - u_2$ in (5.7), we get

$$(Au_1 - Au_2, u_1 - u_2) + \nu((u_1 - u_2, u_1 - u_2)) = b(u_2 - u_1, u_2, u_1 - u_2). \tag{5.8}$$

Applying (5.6) gives

$$|b(u_2 - u_1, u_2, u_1 - u_2)| \leq \sqrt{2} \|u_1 - u_2\| \, \|u_2\| \, |u_1 - u_2| \leq \nu \|u_1 - u_2\|^2 + \frac{1}{2\nu} \|u_2\|^2 |u_1 - u_2|^2.$$

Combining this with (5.8), we obtain

$$(Au_1 - Au_2, u_1 - u_2) \leq \frac{1}{2\nu} \|u_2\|^2 |u_1 - u_2|^2.$$

Thus, we conclude that A is locally quasi-dissipative with respect to the stronger norm $\| \cdot \|$.

Step 2. We next discuss the existence of stationary problems and give some a priori estimates for the solutions. Let $u_0 \in V$ and $\delta > 0$. We show that there exists $u_\delta \in D(A)$ such that $\delta^{-1}(u_\delta - u_0) = Au_\delta$ and

$$\frac{1}{2\delta}(\|u_\delta\|^2 - \|u_0\|^2) + \frac{\nu}{2}|Lu_\delta|^2 \leq \frac{C}{\nu^3}|u_\delta|^2\|u_\delta\|^4 \tag{5.9}$$

for some positive constant C which may depend on Ω. To this end, we follow the argument of Lions–Prodi [14] and apply the Faedo–Galerkin method by means of the a complete orthonormal system of X consisting of the eigenfunctions of L. It is well–known that L is self–adjoint positive operator such that L^{-1} is compact. (See §2.6 ,Chap. I in [22]). Hence, there is a complete system $\{w_j\}_{j=1}^\infty$ consisting of the eigenfunctions of L such that

$$w_j \in V, \quad Lw_j = \lambda_j w_j, \quad \lambda_j > 0, \quad j = 1, 2, \cdots,$$

$$\lambda_j \to \infty \quad \text{as} \quad j \to \infty$$

and

$$(w_j, w_k) = \delta_{jk} \quad \text{and} \quad ((w_j, w_k)) = \lambda_j \delta_{jk}.$$

Let V_m be the subspace of V spanned by $w_1, w_2, \cdots, w_m$ and define $P_m : V_m \to V_m$ by

$$((P_m(u), v)) = \delta^{-1}(u - u_0, v) + \nu((u, v)) + b(u, u, v)$$

for $u, v \in V_m$. We wish to find a fixed point in V_m of the operator P_m. In fact if $u_m = \sum_{k=1}^M \xi_{km} w_k$, $\xi_{km} \in \mathbb{R}$, is a fixed point of P_m, then u_m satisfies

$$\delta^{-1}(u_m - u_0, w_k) + \nu((u_m, w_k)) + b(u_m, u_m, w_k) = 0 \tag{5.10}$$

for $k = 1, 2, \cdots, m$. First the operator P_m is continuous. Furthermore, we have

$$\begin{aligned}((P_m u, u)) &= \delta^{-1}(u - u_0, u) + \nu((u, u)) + b(u, u, u) \\ &= \delta^{-1}(|u|^2 - (u_0, u)) + \nu\|u\|^2 \\ &\geq -\delta^{-1}\|u_0\|_{V'}\|u\| + \nu\|u\|^2 \\ &\geq \delta^{-1}\|u\|(\nu\delta\|u\| - \|u_0\|_{V'}).\end{aligned}$$

Therefore, taking $k > (\nu\delta)^{-1}\|u_0\|_{V'}$, we see that

$$((P_m u, u)) > 0 \quad \text{for } u \in V_m \quad \text{such that} \quad \|u\| = k.$$

By virtue of Lemma 1.4 ,§1, Chap. II in [22], we conclude that there exist $u_m \in V_m$ such that $P_m u_m = u_m$.

Let $u_m = \sum_{k=1}^m \xi_{km} w_k$ satisfy (5.10) for $k = 1, 2, \cdots, m$. Multiplying (5.10) by ξ_{km} and summing up the corresponding equations with respect to $k = 1, 2, \cdots, m$, we get

$$\delta^{-1}(u_m - u_0, u_m) + \nu((u_m, u_m)) + b(u_m, u_m, u_m) = 0.$$

Now the application of (5.5) yields,

$$\delta^{-1}(u_m - u_0, u_m) + \nu\|u_m\|^2 = 0. \tag{5.11}$$

Therefore we have the estimates

$$|u_m|^2 + \delta\nu\|u_m\|^2 \le (u_0, u_m) \le \|u_0\|_{V'}\|u_m\|, \qquad m = 1, 2, \cdots.$$

From this we see that the sequence $\{u_m\}$ is bounded in V. Let $\{u_{m'}\}$ be a subsequence of $\{u_m\}$ such that $u_{m'}$ converges weakly to u in V. Since the injection of V into X is compact, we can assume that $u_{m'}$ is converges strongly to u in X. Let $v \in \mathcal{V} = \{u \in C_0^\infty(\Omega)^2; \nabla \cdot u = 0\}$. Then, it is easily seen that

$$\lim_{m' \to \infty} b(u_{m'}, u_{m'}, v) = -\lim_{m' \to \infty} \sum_{i,j=1}^{2} \int_\Omega u_{m',i} u_{m',j} \partial_{x_i} v_j \, dx$$

$$= -\sum_{i,j=1}^{2} \int_\Omega u_i u_j \partial_{x_i} v_j \, dx = b(u, u, v). \tag{5.12}$$

Since $\|u_m\|$ is bounded, we see that (5.12) holds for any $v \in V$. Therefore, letting $m = m' \to \infty$ in (5.10), we get

$$\delta^{-1}(u - u_0, w_k) + \nu((u, w_k)) + b(u, u, w_k) = 0 \tag{5.13}$$

for $k = 1, 2, \cdots$. Since $w_k/\sqrt{\lambda_k}$ forms complete orthonormal system in V, (5.13) implies that

$$\delta^{-1}(u - u_0, v) + \nu((u, v)) + b(u, u, v) = 0 \tag{5.14}$$

for $v \in V$. Hence $v \in D(A)$ and $\delta^{-1}(v - u_0) = Av$. Putting $v = u$ in (5.14), we have

$$\delta^{-1}(u - u_0, u) + \nu\|u\|^2 = 0. \tag{5.15}$$

On the other hand, letting $m = m' \to \infty$ in (5.11) yields

$$\delta^{-1}(u - u_0, u) + \nu \lim_{m' \to \infty} \|u_{m'}\|^2 = 0.$$

Therefore, $\lim_{m' \to \infty} \|u_{m'}\| = \|u\|$ and hence $u_{m'}$ converges strongly to u in V. Multiplying (5.10) by $\lambda_k \xi_{ik}$ and adding up the corresponding equations with respect to $k = 1, 2, \cdot, m$, we have

$$\delta^{-1}(u_m - u_0, Lu_m) + \nu((u_m, Lu_m)) + b(u_m, u_m, Lu_m) = 0$$

or

$$\delta^{-1}((u_m - u_0, u_m)) + \nu|Lu_m|^2 + b(u_m, u_m, Lu_m) = 0. \tag{5.16}$$

By Lemma 3.8 in Chap. III of [22], we have

$$|b(u_m, u_m, Lu_m)| \le C|u_m|^{1/2}\|u_m\| \cdot |Lu_m|^{3/2} \le \frac{C}{\nu^3}|u_m|^2\|u_m\|^4 + \frac{\nu}{2}|Lu_m|^2.$$

Therefore, (5.16) implies

$$\frac{1}{2m}(\|u_m\|^2 - \|u_0\|^2) + \frac{\nu}{2}|Lu_m|^2 \le \frac{C}{\nu^3}|u_m|^2\|u_m\|^4. \tag{5.17}$$

Letting $m = m' \to \infty$ in (5.17), we see that the limit u satisfies (5.9) with $u = u_\delta$.

Step 3. Finally we show that A satisfies the range condition (R). Let $u_\delta \in D(A)$ satisfy $\delta^{-1}(u_\delta - u_0) = Au_\delta$ and (5.9). The existence of such an element u_δ has just been shown in Step 2. First, we show that u_δ converges strongly in V to u_0 as $\delta \downarrow 0$. By assumption we have

$$(u_\delta - u_0, v) + \delta\nu((u_\delta, v)) + \delta(u_\delta, u_\delta, v) = 0 \tag{5.18}$$

for $v \in V$. Putting $v = u_\delta$ in (5.18) gives

$$(u_\delta - u_0, u_\delta) + \delta\nu\|u_\delta\|^2 = 0 \tag{5.19}$$

and

$$|u_\delta|^2 + 2\delta\nu\|u_\delta\|^2 \leq |u_0|^2. \tag{5.20}$$

Therefore, there exists a subsequence $\{u_{\delta(n)}\}$ such that $u_{\delta(n)}$ converges weakly in H to some element $w_0 \in H$. Furthermore, (5.20) implies that $\sqrt{\delta}\|u_\delta\|$ is bounded, and so that $\delta\|u_\delta\| \to 0$ as $\delta \downarrow 0$. Hence, letting $\delta = \delta(n) \to 0$ in (5.18), we get

$$(w_0 - u_0, v) = 0 \qquad \text{for} \quad v \in V.$$

Hence, $w_0 = u_0$ and (5.20) implies

$$|u_0| \leq \liminf_{n \to \infty} |u_{\delta(n)}| \leq \limsup_{n \to \infty} |u_{\delta(n)}| \leq |u_0|.$$

This means that $|u_{\delta(n)}| \to |u_0|$ as $n \to 0$. From this and the weak convergence of $\{u_{\delta(n)}\}$ it follows that u_δ converges strongly to u_0 in H. Now (5.9) implies

$$\|u_\delta\|^2 - \|u_0\|^2 + \nu\delta|Lu_\delta|^2 \leq \frac{2C}{\nu^3}\delta|u_\delta|^2\|u_\delta\|^4$$

or

$$\|u_\delta\|^2\left(1 - \frac{2C}{\nu^3}\delta|u_\delta|^2\|u_\delta\|^2\right) + \nu\delta|Lu_\delta|^2 \leq \|u_0\|^2. \tag{5.21}$$

Since $\delta\|u_\delta\| \to 0$ as $\delta \downarrow 0$, (5.21) implies that

$$\limsup_{\delta\downarrow 0} \|u_\delta\|^2 \leq \|u_0\|^2.$$

But u_δ converges weakly to u_0 in V, it turns out that u_δ converges strongly to u_0 in V as $\delta \downarrow 0$.

Multiplying both sides of (5.20) by $|u_\delta|^2$, we have

$$|u_\delta|^4 + 2\delta\nu\|u_\delta\|^2\,|u_\delta|^2 \leq |u_0|^2|u_\delta|^2 \leq \frac{1}{2}(|u_0|^4 + |u_\delta|^4)$$

or

$$\delta^{-1}(|u_\delta|^4 - |u_0|^4) + 4\nu\|u_\delta\|^2\,|u_\delta|^2 \leq 0. \tag{5.22}$$

Since $\xi \geq 1 + \log \xi$ for $\xi > 0$, we apply (5.9) to get

$$\delta^{-1}(\log(1 + \|u_\delta\|^2) - \log(1 + \|u_0\|^2))\frac{2\nu^4}{C}$$

$$\leq \delta^{-1}(\|u_\delta\|^2 - \|u_0\|^2)\frac{2\nu^4}{C(1 + \|u_0\|^2)}$$

$$\leq \frac{4\nu|u_\delta|^2\|u_\delta\|^4}{(1 + \|u_0\|^2)}. \tag{5.23}$$

Combining (5.22) and (5.23), we obtain

$$\delta^{-1}(\varphi(u_\delta) - \varphi(u_0)) \leq \frac{4\nu|u_\delta|^2\|u_\delta\|^2(\|u_\delta\|^2 - \|u_0\|^2 - 1)}{1 + \|u_0\|^2}.$$

where we have chosen $2/C$ as the number c_0 in the definition (5.4) of the functional φ. Since $\|u_\delta\| \to \|u_0\|$ as $\delta \downarrow 0$, the above estimates imply

$$\limsup_{\delta \downarrow 0} \left(\delta^{-1}(\varphi(u_\delta) - \varphi(u_0))\right) \leq -\frac{4\nu|u_0|^2\|u_0\|^2}{1 + \|u_0\|^2} \leq 0.$$

Finally, it is easily seen that the functional φ is lower semi-continuous on V. The proof is now complete.

$\square$

REMARK. A nonlinear semigroup approach to Navier Stokes equations was also discussed by Z. Yoshida and Y. Giga [23]. In their paper, the nonlinear term is truncated to get a family of globally Lipschitz operators and the nonlinear semigroup theory is applied to each of the truncated Navier Stokes operators to construct the solutions of the problem (4.1)-(4.3).

Contrary to their results, our semigroup theory can be directly applied to the original Navier Stokes operator A and provides us with unique global *generalized* solutions to the problem (4.1)-(4.3) in the abstract framework.

References

[1] Ph. Bénilan, Equations d'évolution dans un espace de Banach quelconque et applications, Thèsis, Orsay, 1972.

[2] J. Chambers and S. Oharu, Semi-groups of local Lipschitzians in a Banach space, Pacific J. Math., **39**(1971), 89–112.

[3] M. Crandall and T. Liggett, Generation of semi-groups of nonlinear transformations on general Banach spaces, Amer. J. Math., **93**(1971), 265–298.

[4] M. Crandall and L. Evans, On the relation of operator $\partial/\partial s + \partial/\partial \tau$ to evolution governed by accretive operators, Israel J. math., **21**(1975), 261–278.

[5] M. Crandall, Nonlinear semigroups and evolution governed by accretive operators, Proc. Symp. in Pure Math., **43**, Part 2, Amer. Math. Soc., Providence, R. I., 1986.

[6] T. Kato, Accretive operators and nonlinear evolution equations in Banach spaces, in Nonlinear Functional Analysis, Proc. Symp. in Pure Math., **18**, Amer. Math. Soc., Providence, R. I., 1970, 138–161.

[7] N. Kenmochi and S. Oharu, Difference approximation of nonlinear evolution equations and semigroups of nonlinear operators, Publ. R. I. M. S., Kyoto Univ., **10**(1974), 147–207.

[8] Y. Kōmura, Nonlinear semigroups in Hilbert spaces, J. Math. Soc. Japan, **19**(1967), 503–520.

[9] Y. Kobayashi, Difference approximation of Cauchy of problems for quasi-dissipative operators and generation of nonlinear semigroups, J. Math. Soc. Japan, **27**(1975), 640–665.

[10] Y. Kobayashi and S. Oharu, Semigroups of locally Lipschitzian operators in Banach spaces, Hiroshima Math. J., **20**(1990), 573–611.

[11] K. Kobayasi, Y. Kobayashi and S. Oharu, Nonlinear evolution operators in Banach spaces, Osaka J. Math., **21**(1984), 281–310.

[12] V. Lakshmikantham and S. Leela, Nonlinear Differential Equations in Abstract Spaces, International Series in Nonlinear Mathematics, **2**, Pergamon Press, 1981.

[13] J. L. Lions, Quelques Méthodes de Résolution des Problème aux Limites Nonlinéairs, Paris Dunod, Gauthiers-Villars, 1969.

[14] J. L. Lions and G. Prodi, Un théorème d'existence et d'unicité dans les équations de Navier–Stokes en dimension 2, C. R. Acad. Sci., Paris, **248**(1959), 3519-3521.

[15] S. Oharu and T. Takahashi, Locally Lipschitz continuous perturbations of linear dissipative operators and nonlinear semigroups, Proc. Amer. Math. Soc., **100**(1987), 187–194.

[16] A. Pazy, Semigroups of nonlinear contractions and their asymptotic behavior, in Nonlinear Analysis and Mechanics, edited by R. J. Knops, Heriot-Watt Sympos., Vol. 3, Pitman Research Notes in math., **30**, 1979, 36–134.

[17] M. Pierre, Génération et perturbation de semi-groupes de contractions nonlinéaires, Thèse de Docteur de 3 é cycle, Université de Paris VI, 1976.

[18] M. Pierre, Un théoreme général de generation de semigroupes Nonlinéaires, Israel J. Math., **23**(1976), 189–199.

[19] M. Schechter, Interpolation of nonlinear partial differential operators and generation of differential evolutions, J. Diff. Eq., 46(1982), 78–102.

[20] J. Smoller, Shock Waves and Reaction–Diffusion Equations, Springer–Verlag, 1983.

[21] T. Takahashi, Convergence of difference approximations of nonlinear evolution equations and generation of semigroups, J. Math. Soc. Japan, **28**(1976), 96–113.

[22] R. Temam, Navier–Stokes Equations, Theory and Numerical Analysis, North–Holand, 1979.

[23] Z. Yoshida and Y. Giga, A nonlinear semigroup approach to the Navier Stokes system, Comm. in Partial Differential Equations, **9**(1984), 215-230.

[24] K. Yosida, Functional Analysis, 6th edition, Springer–Verlag, 1980.

[25] J. Walker, Dynamical Systems and Evolutions, Theory and Applications, Prenum Press, 1980.

Operational Calculus and Semi-groups of Operators

Hikosaburo Komatsu

Department of Mathematics, Faculty of Science, University of Tokyo
Hongo, Tokyo, 113 Japan

1. Introduction.

It has always been controversial whether or not Operational Calculus is a mathematics. As far as we know every textbook on Operational Calculus starts with a discussion on this issue. When the Mathematical Society of Japan revised its Encyclopedic Dictionary of Mathematics about ten years ago, the chief editor Professor K. Itô asked opinions of foreign scholars about the then second edition. In his reply a French mathematician wrote "Operational Calculus has no value of being mentioned ; it is a bad succedaneum of distributions and is very far from being useful."

Incidentally Professor Yosida was the author of that item. He liked Operational Calculus very much on the contrary. As soon as J. Mukusiński published the book [24] on a new foundation of Operational Calculus, he arranged its translation into Japanese. A section of his famous textbook Functional Analysis [34] is devoted to Mikusiński's theory. He also wrote four papers [35, 37, 38, 39] and two versions of a book, one in Japanese [36] and the other in English [40].

His fondness for Operational Calculus comes probably from his belief that a good mathematics must be not only beautiful but also useful. We imagine that it also comes from his experience. In his doctoral thesis [30] published in 1936, he proved that a locally compact group embedded in a Banach algebra is a Lie group. Actually Banach algebras were introduced by him and Nagumo [25] at that time. They defined inverses $(1-A)^{-1}$, exponentials $\exp A$ and logarithms $\log(1+A)$ by power series expansions and proved among others that any uniformly continuous semi-group is $\exp tA$ for an element A in the Banach algebra. Nagumo [25] went further to obtain the Jordan decomposition of a compact operator at a spectrum different from zero by complex integration of resolvents. They did not notice, however, the Gelfand representation [6] which appeared five years later. Since then Professor Yosida applied the Gelfand representation to obtain the simultaneous spectral decomposition of a commutative family of normal operators [31] and other results.

Although he employed again power series expansions in his proof of the Hille-Yosida theorem [32], he was interested in the complex method of Operational Calculus after he knew Hille's work [7], and wrote several papers including one [33] on fractional powers of operators. Our series of papers [8, 9, 10, 11, 12, 13] succeeded it.

In this talk we give a unified theory of the complex method of operational calculus, semi-groups of operators and operator valued sine functions. The method of Laplace transforms is believed to apply only to functions of exponential growth. However, we can overcome this difficulty by extending the definition of Laplace transforms to hyperfunctions. Thus we have a new foundation of operational calculus. Considering the case where hyperfunctions have values in Banach spaces, we obtain characterizations of closed linear operators in a Banach space which generate semi-groups and sine functions in distributions, in ultradistributions of class (M_p) and $\{M_p\}$ under suitable conditions on the sequence M_p, and in hyperfunctions, respectively. Thus we extend Chazarain's result [3] in the case of distributions and ultradistributions of Gevrey classes $(p^s), 1 < s < \infty$, and Ciorănescu-Zsidó's [4] in the case of ultradistributions of class (M_p).

2. Reduction to a convolution equation.

A typical problem Operational Calculus deals with is the Cauchy problem for the ordinary differential equation:

$$(1) \qquad \begin{cases} P(d/dx)u(x) = f(x), & 0 \leq x < \infty, \\ u^{(j)}(0) = g_j, & 0 \leq j < m, \end{cases}$$

where

$$(2) \qquad P(d/dx) = a_m(d/dx)^m + a_{m-1}(d/dx)^{m-1} + \cdots + a_0,$$

with $a_j \in \mathbf{C}$ and $a_m \neq 0$, and the data $f(x) \in C([0,\infty))$ and $g_j \in \mathbf{C}$.

Let $\theta(x)$ be the *Heaviside function*

$$(3) \qquad \theta(x) = \begin{cases} 0, & x < 0, \\ 1, & x \geq 0. \end{cases}$$

In view of Green's formula

$$(4) \qquad \frac{d^j}{dx^j}(\theta u) = \theta(x) u^{(j)}(x) + u^{(j-1)}(0)\delta(x) + \cdots + u(0)\delta^{(j-1)}(x)$$

valid for $u \in C^j((-\varepsilon, \infty))$, $\varepsilon > 0$, we may reduce Problem (1) to the convolution equation

$$(5) \qquad p * (\theta u) = (\theta f) + (a_m g_{m-1} + \cdots + a_1 g_0)\delta(x) + \cdots + a_m g_0 \delta^{(m-1)}(x)$$

in the space $\mathcal{D}'_{[0,\infty)}$ of distributions with support in $[0,\infty)$, where

$$(6) \qquad p(x) = P(d/dx)\delta(x).$$

Since a semi-group of operators $T(t)$ is the solution of

$$(7) \qquad \begin{cases} dT(t)/dt = AT(t), & 0 < t < \infty, \\ T(0) = I, \end{cases}$$

with a closed linear operator A in a Banach space E, a similar reduction is possible. We need, however, more care about domains of A and $T(t)$ and others.

The Heaviside calculus is a symbolic method to solve (5). Its justification by Bromwich [1], Carson [2], P. Lévy [21], Doetsch [5] et al. may be interpreted as follows. Taking the Laplace transform

$$(8) \qquad \hat{u}(\lambda) = \int_0^\infty e^{-\lambda x} u(x) dx,$$

of both sides of (5), we have

$$(9) \qquad P(\lambda)\hat{u}(\lambda) = \hat{f}(\lambda) + (a_m g_{m-1} + \cdots + a, g_0) + \cdots + a_m g_0 \lambda^{m-1}.$$

Since the original function $u(x)$ is retrieved from its Laplace transform $\hat{u}(\lambda)$ by the *Bromwich inversion formula*

$$(10) \qquad u(x) = \frac{1}{2\pi i} \int_{\Lambda - i\infty}^{\Lambda + i\infty} e^{\lambda x} \hat{u}(\lambda) d\lambda,$$

the solution $u(x)$ of (5) is obtained by applying (10) to

$$(11) \qquad \hat{u}(\lambda) = \frac{1}{P(\lambda)} \{ \hat{f}(\lambda) + (a_m g_{m-1} + \cdots + a_1 g_0) + \cdots + a_m g_0 \lambda^{m-1} \}.$$

The trouble is that we have to assume the exponential type condition

$$(12) \qquad |u(x)| \leqq C e^{Hx}, \ 0 \leqq x < \infty,$$

on the solution $u(x)$ and its derivatives $u^{(j)}(x)$ in order that the Laplace transform (8) converge and that the inversion formula (10) hold for $\Lambda > H$. To avoid this difficulty Mikusiński [24] gave a new foundation of the operational calculus based on the Titchmarsh theorem saying that the convolution algebra $C([0,\infty))$ has no divisors of zero.

3. Laplace hyperfunctions and their Laplace transforms.

A few years ago we discovered another way to evade this difficulty [17, 18, 19]. Let

$$(13) \qquad \mathbf{O} = \mathbf{C} \cup S^1 \infty$$

be the radial compactification of the complex plane, and $\mathcal{O}^{\exp}$ the sheaf over $\mathbf{O}$ of holomorphic functions of exponential type, that is, if V is an open set in $\mathbf{C}$, then $\mathcal{O}^{\exp}(V)$ is the space $\mathcal{O}(V)$ of all holomorphic functions on V ; if an open set $V \subset \mathbf{O}$ contains points at infinitely, then $\mathcal{O}^{\exp}(V)$ is the space of all $F(z) \in \mathcal{O}(V \cap \mathbf{C})$ such that on each sector $\Sigma \Subset V$ we have

$$(14) \qquad |F(z)| \leqq C e^{H|z|}, \quad z \in \Sigma \cap \mathbf{C}$$

with constants H and C.

Definition 1. Let $-\infty < a \leqq \infty$. The space $\mathcal{B}^{\exp}_{[a,\infty]}$ of *Laplace hyperfunctions* with support in $[a, \infty]$ is defined to be the quotient space

$$(15) \qquad \mathcal{B}^{\exp}_{[a,\infty]} := \mathcal{O}^{\exp}(\mathbf{O}\backslash[a,\infty])/\mathcal{O}^{\exp}(\mathbf{O}).$$

If $F(z)$ is in $\mathcal{O}^{\exp}(\mathbf{O}\backslash[a,\infty])$, its class $f(x) \in \mathcal{B}^{\exp}_{[a,\infty]}$ is denoted by

$$(16) \qquad f(x) = F(x+i0) - F(x-i0).$$

Compare this with the definition of ordinary *hyperfunctions* with support in $[a, \infty)$:

$$(17) \qquad \mathcal{B}_{[a,\infty)} := \mathcal{O}(\mathbf{C}\backslash[a,\infty))/\mathcal{O}(\mathbf{C}).$$

The inclusion mappings $\mathcal{O}^{\exp}(\mathbf{O}\backslash[a,\infty]) \to \mathcal{O}(\mathbf{C}\backslash[a,\infty))$ and $\mathcal{O}^{\exp}(\mathbf{O}) \to \mathcal{O}(\mathbf{C})$ induce the natural mapping

$$(18) \qquad \rho : \mathcal{B}^{\exp}_{[a,\infty]} \to \mathcal{B}_{[a,\infty)}.$$

It may be a surprise to have the following.

Theorem 1. *The mapping ρ is surjective.*

See [17, 18, 19] for proofs. The elementary proof in [19] employs only the Mittag-Leffler arguments.

We call ρ the restriction mapping. Since the kernel of ρ coincides with $\mathcal{B}^{\exp}_{\{\infty\}}$, we have the representation

$$(19) \qquad \mathcal{B}_{[a,\infty)} = \mathcal{B}^{\exp}_{[a,\infty]}/\mathcal{B}^{\exp}_{\{\infty\}}.$$

More generally let for $-\infty < a < b \leqq \infty$ $\mathcal{B}_{[a,b)}$ be the space of all hyperfunctions f on $(-\infty, b)$ with support in $[a, b)$. Then every $f \in \mathcal{B}_{[a,b)}$ may first be extended to a hyperfunction in $\mathcal{B}_{[a,\infty)}$ by the flabbiness of the hyperfunctions, and then to a Laplace hyperfunction in $\mathcal{B}^{\exp}_{[a,\infty]}$. Hence we have the representation:

$$(20) \qquad \mathcal{B}_{[a,b)} = \mathcal{B}^{\exp}_{[a,\infty]}/\mathcal{B}^{\exp}_{[b,\infty]}.$$

Definition 2. The *Laplace transform* $\hat{f}(\lambda)$ of a Laplace hyperfunction $f(x) \in \mathcal{B}^{\exp}_{[a,\infty]}$ is defined in terms of its defining function $F(z)$ by the integral

$$(21) \qquad \hat{f}(\lambda) := \int_C e^{-\lambda z} F(z) dz$$

where the integration path C is composed of a ray from $e^{i\alpha}\infty, -\pi/2 < \alpha < 0$, to a point $c < a$ and a ray from c to $e^{i\beta}\infty, 0 < \beta < \pi/2$.

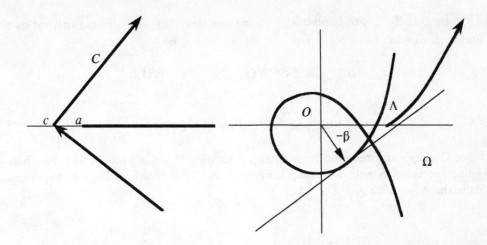

Theorem 2. The Laplace transform $\hat{f}(\lambda)$ of an $f \in \mathcal{B}^{\exp}_{[a,\infty)}$ does not depend on the defining function $F(z) \in \mathcal{O}^{\exp}(\mathbf{O}\backslash[a,\infty])$. The Laplace transformation $\mathcal{L}$ is an isomorphism from $\mathcal{B}^{\exp}_{[a,\infty]}$ onto the space $\mathcal{L}\mathcal{B}^{\exp}_{[a,\infty]}$ of all holomorphic functions $\hat{f}(\lambda)$ of exponential type on a neighborhood Ω of the semi-circle $\{e^{i\theta}\infty; |\theta| < \pi/2\}$ at infinity with the estimates

$$(22) \qquad \varlimsup_{\rho \to \infty} \frac{\log |\hat{f}(\rho e^{i\theta})|}{\rho} \leqq -a\cos\theta, \quad |\theta| < \pi/2.$$

If $\hat{f}(\lambda)$ is in $\mathcal{L}\mathcal{B}^{\exp}_{[a,\infty]}$, a defining function $F(z)$ of the inverse image $\mathcal{L}^{-1}\hat{f}$ is obtained as the integral

$$(23) \qquad F(z) = \frac{1}{2\pi i} \int_\Lambda^\infty e^{\lambda z} \hat{f}(\lambda) d\lambda,$$

where Λ is a fixed point in $\Omega \cap \mathbf{C}$ and the integral is along a convex curve in $\Omega \cap \mathbf{C}$.

The proof [17] is essentially due to the classical results of Pólya [29] and Macintyre [23].

In view of the representation (20) we can define the Laplace transform of $f(x) \in \mathcal{B}_{[a,b)}$ as a class in $\mathcal{L}\mathcal{B}^{\exp}_{[a,\infty]}/\mathcal{L}\mathcal{B}^{\exp}_{[b,\infty]}$ and obtain the isomorphism

$$(24) \qquad \mathcal{L} : \mathcal{B}_{[a,b)} \cong \mathcal{L}\mathcal{B}^{\exp}_{[a,\infty]}/\mathcal{L}\mathcal{B}^{\exp}_{[b,\infty]}.$$

Let E be a complex Banach space and denote by ${}^E\mathcal{O}$ (resp. ${}^E\mathcal{O}^{\exp}$) the space of holomorphic functions (resp. of exponential type) with values in E. Then the space

${}^{E}\mathcal{B}^{\exp}_{[a,\infty]}$ (resp. ${}^{E}\mathcal{B}_{[a,\infty)}$) of Laplace hyperfunctions (resp. hyperfunctions)with values in E and with support in $[a,\infty]$ (resp. $[a,\infty)$) is defined by

$$(15)' \qquad {}^{E}\mathcal{B}^{\exp}_{[a,\infty]} := {}^{E}\mathcal{O}^{\exp}(\mathbf{O}\backslash[a,\infty])/{}^{E}\mathcal{O}^{\exp}(\mathbf{O}),$$

$$(17)' \qquad {}^{E}\mathcal{B}_{[a,\infty)} := {}^{E}\mathcal{O}(\mathbf{C}\backslash[a,\infty))/{}^{E}\mathcal{O}(\mathbf{C})).$$

Since Cauchy's theorem and Runge's approximation theorem hold for E-valued holomorphic functions as well, we have Theorems 1 and 2 for E-valued Laplace hyperfunctions without any changes of proofs.

4. Convolutions of hyperfunctions.

Definition 3. The *convolution* $f * g$ of Laplace hyperfunctions $f \in \mathcal{B}^{\exp}_{[a,\infty]}$ and $g \in \mathcal{B}^{\exp}_{[b,\infty]}$ is the Laplace hyperfunction in $\mathcal{B}^{\exp}_{[a+b,\infty]}$ defined by

$$(25) \qquad f * g = \mathcal{L}^{-1}(\hat{f}(\lambda)\,\hat{g}(\lambda)).$$

Thus the convolution

$$(26) \qquad * : \mathcal{B}^{\exp}_{[a,\infty]} \times \mathcal{B}^{\exp}_{[b,\infty]} \to \mathcal{B}^{\exp}_{[a+b,\infty]}$$

is a bilinear mapping, whose Laplace transform is the pointwise multiplication

$$(27) \qquad \mathcal{L}\mathcal{B}^{\exp}_{[a,\infty]} \times \mathcal{L}\mathcal{B}^{\exp}_{[b,\infty]} \to \mathcal{L}\mathcal{B}^{\exp}_{[a+b,\infty]}.$$

Suppose that D, E and L are complex Banach spaces such that a canonical bilinear mapping

$$(28) \qquad L \times D \to E$$

is defined with the property

$$(29) \qquad \|l \cdot d\|_{E} \leqq \|l\|_{L}\,\|d\|_{D}.$$

In particular, let $L = L(D, E)$ be the space of continuous linear mappings $l : D \to E$. Then the convolution

$$(30) \qquad {}^{L}\mathcal{B}^{\exp}_{[a,\infty]} \times {}^{D}\mathcal{B}^{\exp}_{[b,\infty]} \to {}^{E}\mathcal{B}^{\exp}_{[a+b,\infty]}$$

is defined by the pointwise multiplication

$$(31) \qquad \mathcal{L}^{L}\mathcal{B}^{\exp}_{[a,\infty]} \times \mathcal{L}^{D}\mathcal{B}^{\exp}_{[b,\infty]} \to \mathcal{L}^{E}\mathcal{B}^{\exp}_{[a+b,\infty]}.$$

Let $a, b \in \mathbf{R}$ and $0 < c \leqq \infty$. Then it follows from (20) that the convolution induces the bilinear mapping

$$(32) \qquad * : {}^{L}\mathcal{B}_{[a,a+c)} \times {}^{D}\mathcal{B}_{[b,b+c)} \to {}^{E}\mathcal{B}_{[a+b,a+b+c)}$$

which coincides with the convolution of ordinary hyperfunctions. If $p \in {}^{L}\mathcal{B}_{[a,a+c)}$ and $u \in {}^{D}\mathcal{B}_{[b,b+c)}$, then their convolution $p * u \in {}^{E}\mathcal{B}_{[a+b,a+b+c)}$ is the restriction to $(-\infty, a+b+c)$ of the convolution $\tilde{p} * \tilde{u} \in {}^{E}\mathcal{B}^{\exp}_{[a+b,\infty]}$ of arbitrary extensions $\tilde{p} \in {}^{L}\mathcal{B}^{\exp}_{[a,\infty]}$ and $\tilde{u} \in {}^{D}\mathcal{B}^{\exp}_{[b,\infty]}$ of p and u respectively.

Thus the convolution equation

$$(33) \qquad p * u = f$$

in $u \in {}^{D}\mathcal{B}_{[b,b+c)}$ is equivalent to the congruence equation

$$(34) \qquad \hat{p}(\lambda) \cdot \hat{u}(\lambda) \equiv \hat{f}(\lambda) \bmod \mathcal{L}^{E}\mathcal{B}^{\exp}_{[a+b+c,\infty]},$$

where $\hat{p}(\lambda), \hat{u}(\lambda)$ and $\hat{f}(\lambda)$ are Laplace transforms of arbitrary extensions $\tilde{p}(x), \tilde{u}(x)$ and $\tilde{f}(x)$ of p, u and f respectively.

We remark here that the Titchmarsh-Lions theorem for ultradistribution

$$(35) \qquad \inf \operatorname{supp} p * u = \inf \operatorname{supp} p + \inf \operatorname{supp} u$$

does not hold in general for scalar valued hyperfunctions [19]. In the vector valued case this does not hold either for distributions or ultradistributions.

5. Regular solvability.

To solve convolution equations, we start with the scalar case. Let $p(x) \in \mathcal{B}_{[0,c)}$, $0 < c \leqq \infty$, and consider the equation

$$(36) \qquad p * u = f$$

in $u(x) \in \mathcal{B}_{[0,c)}$.

Theorem 3. *Equation (36) has always a solution $u(x) \in \mathcal{B}_{[0,c)}$ for any $f \in \mathcal{B}_{[0,c)}$ if and only if the Laplace transform $\hat{p}(\lambda)$ of some (and any) extension $\tilde{p}(x) \in \mathcal{B}^{\exp}_{[0,\infty]}$ of $p(x)$ satisfies*

$$(37) \qquad \hat{p}(\lambda)^{-1} \in \mathcal{L}\mathcal{B}^{\exp}_{[0,\infty]}.$$

Proof. Suppose that there is a $q(x) \in \mathcal{B}_{[0,c)}$ such that $p * q = \delta$ in $\mathcal{B}_{[0,c)}$. Then we have

$$(38) \qquad \tilde{p} * \tilde{q} = \delta - r$$

with an $r \in \mathcal{B}^{\text{exp}}_{[c,\infty]}$ for any extensions $\tilde{p}$ and $\tilde{q}$ in $\mathcal{B}^{\text{exp}}_{[0,\infty]}$ of p and q respectively. Then

$$(1 - \hat{r}(\lambda))^{-1} = \sum_{n=0}^{\infty} \hat{r}(\lambda)^n$$

converges in $\mathcal{LB}^{\text{exp}}_{[0,\infty]}$. Hence it follows that

$$\hat{p}(\lambda)^{-1} = \hat{q}(\lambda)(1 - \hat{r}(\lambda))^{-1} \in \mathcal{LB}^{\text{exp}}_{[0,\infty]}.$$

Conversely suppose that (37) holds for an extension $\tilde{p}(x) \in \mathcal{B}^{\text{exp}}_{[0,\infty]}$. Then the inverse Laplace transform $q \in \mathcal{B}^{\text{exp}}_{[0,\infty]}$ of $\hat{p}(\lambda)^{-1}$ satisfies

$$\tilde{p} * q = q * \tilde{p} = \delta.$$

Hence $u = q * \tilde{f}|_{(-\infty,c)} \in \mathcal{B}_{[0,c)}$ is a unique solution of (36).

Then we say that equation (36) is *regularly solvable*. The same proof shows that if $p_0 \in \mathcal{B}_{[0,c)}$ is regularly invertible and if $p_1 \in \mathcal{B}_{[0,c)}$ satisfies inf supp $p_1 > 0 = $ inf supp p_0, then $p = p_0 + p_1$ is regularly invertible. Hence the regular solvability of equation (36) is decided only by the behavior of $p(x)$ near the origin.

If equation (36) is regularly solvable, then $\hat{p}(\lambda)$ cannot have zeros on a neighborhood of the semi-circle $\{e^{i\theta}\infty; |\theta| < \pi/2\}$ at infinity. If p has support only at the origin, or if p is an ultradistribution near the origin, this is also sufficient (see [19], Theorem 4).

6. Hyperfunction semi-groups and hyperfunction sines.

In the vector valued case, where $L = L(D, E)$, we say that equation (36) is *regularly solvable* if for some (and any) extension $\tilde{p}(x) \in {}^L\mathcal{B}^{\text{exp}}_{[0,\infty]}$ of $p(x)$ there is a $q \in {}^H\mathcal{B}^{\text{exp}}_{[0,\infty]}$ such that

$$(39) \qquad \begin{cases} \tilde{p} * q = 1_E \otimes \delta(x), \\ q * \tilde{p} = 1_D \otimes \delta(x), \end{cases}$$

where $H = L(E, D)$. Then

$$(40) \qquad u = q * \tilde{f}|_{(-\infty,c)}$$

is a unique solution of (36).

By the same proof we have the following.

Theorem 3$'$. *Equation* (36) *is regularly solvable if and only if some (and any) extension* $\tilde{p}(x) \in {}^L\mathcal{B}^{\text{exp}}_{[0,\infty]}$ *of* $p(x) \in {}^L\mathcal{B}_{[0,c)}$ *has the Laplace transform* $\hat{p}(\lambda)$ *such that*

$$(37)' \qquad \hat{p}(\lambda)^{-1} \in \mathcal{L}\,{}^H\mathcal{B}^{\text{exp}}_{[0,\infty]}.$$

We apply this to the equation

$$(41) \qquad \left(\frac{d}{dx} - A\right) u(x) = f(x),$$

where A is a closed linear operator with domain $D(A) = D$ in a Banach space E. If we endow D with the graph norm of A, then A and the identity mapping 1_L of D into E are in L and we have equation (36) with

$$(42) \qquad p(x) = 1_L \otimes \delta'(x) - A \otimes \delta(x).$$

Since we have

$$(43) \qquad \hat{p}(\lambda) = \lambda 1_L - A,$$

equation (41) is regularly solvable if and only if the resolvent $(\lambda - A)^{-1}$ exists on a neighborhood Ω of the semi-circle $\{e^{i\theta}\infty; |\theta| < \pi/2\}$ and satisfies estimates

$$(44) \qquad \varlimsup_{\rho \to \infty} \frac{\log \|(\rho e^{i\theta} - A)^{-1}\|_H}{\rho} \leqq 0, \quad |\theta| < \pi/2.$$

Since we have

$$\|(\lambda - A)^{-1}\|_{L(E)} \leqq \|(\lambda - A)^{-1}\|_H \leqq \|(\lambda - A)^{-1}\|_{L(E)} + \|A(\lambda - A)^{-1}\|_{L(E)}$$
$$\leqq 1 + (1 + |\lambda|)\|(\lambda - A)^{-1}\|_{L(E)},$$

the norm $\| \quad \|_H$ in (44) may be replaced by $\| \quad \|_{L(E)}$.

Thus we have the following characterization of the generators of hyperfunction semi-groups $T(x) = \exp(Ax)$ by Ōuchi[28].

Theorem 4. *Let A be a closed linear operator in a complex Banach space E with domain $D(A) = D$. There is an operator valued hyperfunction $T(x) \in {}^H\mathcal{B}_{[0,\infty)}$ (or ${}^H\mathcal{B}_{[0,c)}$ for a $c > 0$) such that*

$$(45) \qquad \begin{cases} T'(x) - AT(x) = 1_E \otimes \delta(x) \\ T'(x) - T(x)A = 1_D \otimes \delta(x) \end{cases}$$

if and only if the resolvent $(\lambda - A)^{-1}$ exists on a neighborhood Ω of the semi-circle $\{e^{i\theta}\infty, |\theta| < \pi/2\}$ at infinity and satisfies estimates

$$(46) \qquad \varlimsup_{\rho \to \infty} \frac{\log \|(\rho e^{i\theta} - A)^{-1}\|}{\rho} \leqq 0, \quad |\theta| < \pi/2.$$

Then for any $g \in E$ $u(x) = T(x)g$ is a unique solution of

$$(47) \qquad u'(x) - Au(x) = 0, \quad 0 < x < \infty,$$

satisfying the initial condition $u(0) = g$ in the sense that

$$(48) \qquad u'(x) - Au(x) = g\,\delta(x) \text{ on } \mathbf{R}.$$

Similarly we have the following characterization of the generators of hyperfunction sines $S(x) = \sin(\sqrt{-A}x)/\sqrt{-A}$.

Theorem 5. *Let A be a closed linear operator in a complex Banach space E with domain $D(A) = D$. There is an operator valued hyperfunction $S(x) \in {}^H\mathcal{B}_{[0,\infty)}$ (or ${}^H\mathcal{B}_{[0,c)}$ for a $c > 0$) such that*

(49)
$$\begin{cases} S''(x) - AS(x) = 1_E \otimes \delta(x) \\ S''(x) - S(x)A = 1_D \otimes \delta(x) \end{cases}$$

if and only if the resolvent $(\mu - A)^{-1}$ exists on a neighborhood of $\{e^{i\theta}\infty; |\theta| < \pi\}$ and satisfies

(50)
$$\varlimsup_{\rho \to \infty} \frac{\|(\rho^2 e^{i\theta} - A)^{-1}\|}{\rho} \leqq 0, \quad |\theta| < \pi.$$

Then for any g_0 and $g_1 \in E$ $u(x) = S(x)g_1 + S'(x)g_0$ is a unique solution of

(51)
$$u''(x) - Au(x) = 0, \quad 0 < x < \infty,$$

satisfying the initial conditions $u(0) = g_0$ and $u'(0) = g_1$ in the sense that

(52)
$$u''(x) - Au(x) = g_1\,\delta(x) + g_0\,\delta'(x) \quad \text{on } \mathbf{R}.$$

To our regret people are not so much interested in hyperfunction solutions. We will therefore consider conditions under which the solutions are distributions or ultradistributions.

7. Ultradistributions.

Let M_p, $p = 0, 1, 2, \cdots$, be a sequence of positive numbers satisfying the following conditions with constants A and B :

(M.0) $\qquad\qquad\qquad M_0 = M_1 = 1;$

(M.1) $\qquad\qquad\qquad M_p^2 \leqq M_{p-1}M_{p+1};$

(M.2) $\qquad\qquad\qquad M_p \leqq AB^p M_q M_{p-q}, \; 0 \leqq q \leqq p;$

(M.3)
$$\sum_{q=p+1}^{\infty} \frac{M_{q-1}}{M_q} \leqq Ap\frac{M_p}{M_{p+1}}.$$

The Gevrey sequences $M_p = p!^s$, $1 < s < \infty$, satisfy all the conditions.

Let U be an open set in $\mathbf{R}$ (or $\mathbf{R}^n$). An infinitely differentiable function $\varphi \in \mathcal{D}(U)$ with compact support is called an *ultradifferentiable function* of class (M_p) (resp. $\{M_p\}$) if for any $h > 0$ (resp. some h) there is a constant C such that

(53)
$$\sup |\partial^\alpha \varphi(x)| \leqq Ch^{|\alpha|} M_{|\alpha|}, \quad |\alpha| = 0, 1, \cdots.$$

We denote by $\mathcal{D}^{(M_p)}(U)$ (resp. $\mathcal{D}^{\{M_p\}}(U)$) the space of such functions.

From now on $*$ stands for one of ϕ, (M_p) and $\{M_p\}$. If K is a compact set in $\mathbf{R}^n$, we write

$$
(54) \qquad \mathcal{D}_K^* = \{\varphi \in \mathcal{D}^*(\mathbf{R}^n); \operatorname{supp} \varphi \subset K\}.
$$

This space has the natural locally convex topology determined by the following family of semi-norms (see [16] Proposition 3.5)

$$
(55) \qquad p_N(\varphi) := \sup_{|\alpha| \leqq N} |\partial^\alpha \varphi(x)|, \ N = 0, 1, \cdots \ \text{ if } * = \phi;
$$

$$
(56) \qquad p_H(\varphi) := \sup_\alpha \frac{|\partial^\alpha \varphi(x)|}{H_{|\alpha|} M_{|\alpha|}},
$$

where

$$
(57) \qquad H_p = \begin{cases} h^p, & h > 0, \text{ if } * = (M_p); \\ h_1 \cdots h_p, & 0 < h_p \nearrow \infty, \text{ if } * = \{M_p\}. \end{cases}
$$

The space $\mathcal{D}^*(U)$ is represented as the inductive limit

$$
(58) \qquad \mathcal{D}^*(U) = \varinjlim_{K \Subset U} \mathcal{D}_K^*
$$

and is endowed with the strict inductive limit locally convex topology.

Its dual i.e. the space of all continuous linear functionals is defined to be the space of *ultradistributions* of class $*$ on U:

$$
(59) \qquad \mathcal{D}^{*\prime}(U) := (\mathcal{D}^*(U))'.
$$

If E is a Banach space, the space ${}^E\mathcal{D}^{*\prime}(U)$ of ultradistributions of class $*$ with values in E is defined to be the space of continuous linear mappings on $\mathcal{D}^*(U)$ into E:

$$
(60) \qquad {}^E\mathcal{D}^{*\prime}(U) = L_\beta(\mathcal{D}^*(U), E) \cong \mathcal{D}^{*\prime}(U) \hat{\otimes} E
$$

(see [16] for vector valued ultradistributions).

In order to characterize the ultradistributions of class $*$ we make use of *weight functions* $\omega(\lambda), \lambda \in \mathbf{C}$, of class $*$ and *growth functions* $\gamma(t), t > 0$, of class $*$ defined by

$$
(61) \qquad \omega(\lambda) = \begin{cases} N \log(|\lambda| + 1), & \text{if } * = \phi, \\ \sup_p \log \dfrac{|\lambda|^p}{H_p M_p}, & \text{if } * = (M_p) \text{ or } \{M_p\}, \end{cases}
$$

and

$$
(62) \qquad \gamma(t) = \begin{cases} -N \log t, & \text{if } * = \phi, \\ \sup_p \log \dfrac{p!}{t^p H_p M_p}, & \text{if } * = (M_p) \text{ or } \{M_p\}, \end{cases}
$$

where N is a constant and H_p is a sequence defined by (57) for some $h > 0$ or some sequence $0 < h_p \nearrow \infty$. If $* = (p!^s), 1 < s < \infty$, then we have

$$
(63) \qquad \omega(\lambda) \sim (|\lambda|/h)^{1/s},
$$

$$
(64) \qquad \gamma(t) \sim (ht)^{-1/(s-1)}.
$$

The following is the Paley-Wiener theorem for ultradistributions.

Theorem 6. If $f(x) \in {}^E\mathcal{D}_{[a,b]}^{*\prime}$ is an ultradistribution of class $*$ with support in the compact interval $[a, b]$ and with values in the complex Banach space E, then its Laplace transform

$$(65) \qquad \hat{f}(\lambda) = \int_{\mathbf{R}} e^{-\lambda x} f(x) dx$$

is an entire function satisfying the estimates

$$(66) \qquad \|\hat{f}(\lambda)\|_E \leqq C \exp\{\omega(\lambda) + H(\lambda)\}$$

for a weight function $\omega(\lambda)$ of class $*$ and a constant C, where

$$(67) \qquad H(\lambda) = \max\{-a\operatorname{Re}\lambda, -b\operatorname{Re}\lambda\}.$$

Conversely if $\hat{f}(\lambda) \in {}^E\mathcal{O}(\mathbf{C})$ satisfies estimates (66), then it is the Laplace transform of a unique $f(x) \in {}^E\mathcal{D}_{[a,b]}^{*\prime}$.

Proof. Let $K = [a - \varepsilon, b + \varepsilon]$ be a compact neighborhood of $[a, b]$. Then it is easy to see that $\|\hat{f}(\lambda)\|_E$ is bounded by

$$C \sup_{\substack{x \in K \\ p}} |d^p e^{-\lambda x}/dx^p|/H_p M_p$$

$$= C \sup_p (|\lambda|^p / H_p M_p) \cdot \sup_{x \in K} |e^{-\lambda x}|.$$

To eliminate $\varepsilon > 0$, we employ the following lemma of the Phragmén-Lindelöf type.

Lemma 1. Let $\varphi(\zeta)$ be a subharmonic function on the upper half plane $\operatorname{Im} \zeta \geqq 0$ satisfying estimates

$$(68) \qquad \varphi(\zeta) \leqq \varepsilon|\zeta| + C_\varepsilon, \quad \operatorname{Im}\zeta \geqq 0,$$

for any $\varepsilon > 0$ with a constant C_ε and

$$(69) \qquad \int_{-\infty}^{\infty} \frac{\varphi(\xi)}{\xi^2 + 1} d\xi < \infty.$$

Then we have

$$(70) \qquad \varphi(\xi + i\eta) \leqq \frac{1}{\pi} \int_{-\infty}^{\infty} \frac{\eta\varphi(\tau)}{(\xi - \tau)^2 + \eta^2} d\tau.$$

Applying this lemma to

$$\varphi(\zeta) = \log\|e^{-ia\zeta}\hat{f}(-i\zeta)\| \quad \text{and} \quad \log\|e^{ib\zeta}\hat{f}(i\zeta)\|,$$

we have (66) with $\omega(\lambda)$ replaced by

$$\tilde{\omega}(\eta - i\xi) = \frac{1}{\pi} \int_{-\infty}^{\infty} \frac{|\eta|}{(\xi - \tau)^2 + \eta^2} \omega(\tau) d\tau.$$

Condition (M.3) for M_p is equivalent to saying that the left hand side is bounded by another weight function $\omega_1(\eta - i\xi) + \text{const}$ ([15] Proposition 4.4).

The proof of the converse is standard ([16] Theorem 1.1).

Incidentally condition (M.2) is equivalent to the assertion that for any weight functions $\omega_1(\lambda)$ and $\omega_2(\lambda)$ there are a third one $\omega_3(\lambda)$ and a constant C such that

$$(71) \qquad \omega_1(\lambda) + \omega_2(\lambda) \leqq \omega_3(\lambda) + C$$

or that the ultradistributions of class $*$ with compact support are closed under convolution ([15] Proposition 3.6).

Lemma 1 is derived from the Nevanlinna formula [26, 27]

$$\varphi(\xi + i\eta) \leqq \frac{1}{\pi} \int_{-R}^{R} \left\{ \frac{\eta}{(\xi - \tau)^2 + \eta^2} - \frac{R^2 \eta}{(R^2 - \tau\xi)^2 + \tau^2 \eta^2} \right\} \varphi(\tau) dt$$
$$+ \frac{2R\eta}{\pi} \int_0^\pi \frac{(R^2 - (\xi^2 + \eta^2)) \sin\theta}{|R^2 e^{2i\theta} - 2Re^{i\theta}\xi + \xi^2 + \eta^2|^2} \varphi(Re^{i\theta}) d\theta$$

as $R \longrightarrow \infty$. By (68) the limes superior of the second term is less than or equal to 0. The first term tends to the right hand side of (70) by the following lemma which is an easy consequence of the Labesgue convergence theorem and the Fatou lemma.

Lemma 2. Let $f_n(x)$ be a sequence of measurable functions tending to $\lim f_n(x)$ almost everywhere. If $0 \leqq f_n(x) \leqq \lim f_n(x)$, a.e., then

$$\lim_{n \to \infty} \int f_n(x) d\mu(x) = \int \lim_{n \to \infty} f_n(x) d\mu(x).$$

8. Characterization of ultradistributions among hyperfunctions.

Let U be an open set in $\mathbf{R}$, and E a complex Banach space. The space $^E\mathcal{B}(U)$ of *hyperfunctions* on U with values in E is defined by

$$(72) \qquad {}^E\mathcal{B}(U) = {}^E\mathcal{O}(V \setminus U)/{}^E\mathcal{O}(V),$$

where V is an open set in $\mathbf{C}$ containing U as a closed set. It follows from the Mittag-Leffler theorem that $^E\mathcal{B}(U)$ does not depend on V.

If U_1 is an open subset of U, a natural restriction mapping $\rho_{U_1}^U : {}^E\mathcal{B}(U) \to {}^E\mathcal{B}(U_1)$ is defined. The hyperfunctions $^E\mathcal{B}(U)$, $U \subset \mathbf{R}$, form a flabby sheaf under the restriction mappings. Let K be a compact set in $\mathbf{R}$. Then the space $^E\mathcal{B}_K$ of all E-valued hyperfunctions with support in K is represented as

$$(73) \qquad {}^E\mathcal{B}_K = {}^E\mathcal{O}(V \setminus K)/{}^E\mathcal{O}(V) \cong L(\mathcal{A}(K), E),$$

where V is an open neighborhood of K in $\mathbf{C}$, and $\mathcal{A}(K)$ the space of all germs of real analytic functions defined on a neighborhood of K with the inductive limit topology under the representation

$$(74) \qquad \mathcal{A}(K) = \varinjlim_{W \supseteq K} \mathcal{O}(W).$$

If $F(z) \in {}^E\mathcal{O}(V \setminus K)$ represents $f(x) \in {}^E\mathcal{B}_K$ and $\varphi(x) \in \mathcal{A}(K)$ has an analytic continuation to W, then the isomorphism (73) is defined by the integral

$$(75) \qquad \int f(x)\varphi(x)dx := -\int_C F(z)\varphi(z)dz,$$

where C is a simple closed curve in $(V \setminus K) \cap W$ encircling K once. This is compatible with the intuitive interpretation of hyperfunctions as boundary values of defining holomorphic functions:

$$(76) \qquad f(x) = F(x + i0) - F(x - i0).$$

If $f(x) \in {}^E\mathcal{B}_{cp}(\mathbf{R})$ is an E-valued hyperfunction with compact support, then a defining holomorphic function is obtained by the integral

$$(77) \qquad F(z) = \frac{-1}{2\pi i} \int_\mathbf{R} \frac{1}{t - z} f(t)dt,$$

which is called the *standard defining function* of $f(x)$. It is also called the *Cauchy transform* of $f(x)$. Then

$$(78) \qquad F(x + iy) - F(x - iy) = \frac{1}{\pi} \int_\mathbf{R} \frac{y}{(x - t)^2 + y^2} f(t)dt$$

is the *Poisson integral* of $f(x)$.

If $f(x) \in {}^E\mathcal{D}_K^{*\prime}$ is an E-valued ultradistribution with support in K, then $\int f(x)\varphi(x)dx$, $\varphi \in \mathcal{A}(K)$, is defined as a coupling on a neighborhood of K in $\mathbf{R}$. Hence we have a natural embedding ${}^E\mathcal{D}_K^{*\prime} \subset {}^E\mathcal{B}_K$. Since this correspondence preserves the support, it follows that the sheaf ${}^E\mathcal{D}^{*\prime}$ of E-valued ultradistributions is naturally embedded in the sheaf ${}^E\mathcal{B}$ of E-valued hyperfunctions. The following theorem characterizes the ultradistributions of class $*$ in terms of defining holomorphic functions.

Theorem 7. *Let $U \subset V$ be open sets in $\mathbf{R}$ and $\mathbf{C}$ respectively such that U is closed in V. The following conditions are equivalent for an E-valued holomorphic function in ${}^E\mathcal{O}(V \setminus U)$:*
(a) *The hyperfunction $F(x + i0) - F(x - i0)$ belongs to $\mathcal{D}^{*\prime}(U)$;*
(b) *$F(x + iy) - F(x - iy)$ converges in $\mathcal{D}^{*\prime}(U)$ as $y \searrow 0$;*
(c) *For any compact set K in U there are a growth function $\gamma(t)$ of class $*$ and constants C and $\delta > 0$ such that*

$$(79) \qquad \sup_{x \in K} \|F(x + iy)\| \leqq C \exp \gamma(|y|), \quad |y| < \delta.$$

Then $F(x + iy) - F(x - iy)$ converges in $\mathcal{D}^{*\prime}(U)$ to the ultradistribution $F(x + i0) - F(x - i0)$ as $y \searrow 0$.

Proof. (a) $\Rightarrow$ (b). Since the topology of $\mathcal{D}^{*\prime}(U)$ is of local character, it suffices to prove the convergence in ${}^E\mathcal{D}^{*\prime}(U_1)$ for an arbitrary relatively compact open subset U_1 of U. Take a $g(x) \in {}^E\mathcal{D}^{*\prime}_{cp}(U)$ which coincides with $F(x+i0) - F(x-i0)$ on a neighborhood of the closure $[U_1]$ of U_1, and let $G(z)$ be its standard defining function. Then $H(z) := F(z) - G(z)$ is holomorphic on a neighborhood of U_1. It is easy to see that the Poisson integral $G(x+iy) - G(x-iy)$ converges as $y \searrow 0$ to $g(x)$ in ${}^E\mathcal{D}^{*\prime}(\mathbf{R})$ and hence to $F(x+i0) - F(x-i0)$ in ${}^E\mathcal{D}^{*\prime}(U_1)$. On the other hand, $H(x+iy) - H(x-iy)$ converges to zero in ${}^E\mathcal{A}(U_1)$ and hence in ${}^E\mathcal{D}^{*\prime}(U_1)$.

(a)$\Rightarrow$(c). We may consider as above only an $f(x) \in {}^E\mathcal{D}^{*\prime}_{[a,b]}$ with compact support and its standard defining function $F(z)$. Then $F(z)$ is represented as the integral

$$(80) \qquad F(z) = \frac{1}{2\pi i} \int_0^\infty e^{\lambda z} \hat{f}(\lambda) d\lambda$$

of the Laplace transform $\hat{f}(\lambda)$, where the path of integral is an arbitrary ray from the origin. If we take the upper imaginary axis, then the integral converges on the upper half plane $\operatorname{Im} z > 0$. In view of estimate (66) we have

$$(81) \qquad \|F(x+iy)\| \leqq \frac{C}{2\pi} \int_0^\infty e^{-\tau y + \omega(\tau)} d\tau, \quad y > 0,$$

for a weight function $\omega(\lambda)$ of class $*$.

Suppose that $\omega(\lambda)$ is defined by the second formula of (61). Then we have

$$(82) \qquad \sup_{0 \leqq \tau < \infty} (\omega(\tau) - \tau y) = \sup_\tau \sup_p \left(\log \frac{\tau^p}{H_p M_p} - \tau y \right)$$

$$= \sup_p \log \frac{p^p e^{-p}}{y^p H_p M_p}.$$

In view of Stirling's formula this is bounded by the growth function $\gamma(y)$ defined by (62) plus a constant.

Since we have

$$\int_0^\infty e^{-\tau y + \omega(\tau)} d\tau \leqq e^{\sup(\omega(\tau) - \tau y/2)} \int_0^\infty e^{-\tau y/2} d\tau$$

$$\leqq \sup_p \log \frac{2^{p+1} p^p e^{-p}}{y^{p+1} H_p M_p}$$

and $M_{p+1} \leqq AB^{p+1} M_p$ by (M.3), the left hand side of (81) is bounded by a constant times $\exp \gamma_1(y)$ for a growth function $\gamma_1(y)$ of class $*$.

If $\omega(\lambda)$ is given by the first formula of (61), then we have

$$\int_0^\infty e^{-\tau y + \omega(\tau)} d\tau = \Gamma(N+1) t^{-N-1}.$$

(c)$\Rightarrow$(b). As in the scalar case [15] we can construct an ultradifferential operator $P(d/dz)$ of class $*$ and an E-valued holomorphic function $G(z)$ which is continuous up to the boundary U_1 both from above and below such that $F(z) = P(d/dz)G(z)$. The function $G(z)$ is obtained by applying to $F(z)$ the Green function of $P(d/dz)$ which has a decay order compensating the growth order $\exp \gamma(|y|)$ of $F(z)$. Then $G(x+iy)-G(x-iy)$ converges in $^E C(U_1)$. Hence $F(x + iy) - F(x - iy)$ converges in $^E \mathcal{D}^{*\prime}(U_1)$.

(b)$\Rightarrow$(a). Let U_1 be as above an arbitrary relatively compact open subset of U. Choose a $g(x) \in \mathcal{D}_{cp}^{*\prime}(U)$ which coincides on U_1 with the limit $F(x + iy) - F(x - iy)$ in $\mathcal{D}^{*\prime}(U)$, and let $G(z)$ be its standard defining function. The following generalization of the Painlevé theorem shows that $F(z) - G(z)$ is analytically continued to U_1. Hence $F(x + i0) - F(x - i0)$ is equal to the ultradistribution $g(x)$ on U_1.

Lemma 3. *Let $U \subset \mathbf{R}$ and $V \subset \mathbf{C}$ be as in the definition of hyperfunctions. We assume that V is symmetric in the imaginary variable y. An E-valued holomorphic function $F(z) \in {}^E\mathcal{O}(V \setminus U)$ is analytically continued to V if and only if $F(x + iy) - F(x - iy)$ converges to zero in $\mathcal{D}^{*\prime}(U)$ as $y \searrow 0$.*

Proof. It follows from the reflection principle in $\mathcal{D}^{*\prime}$ [20] that the E-valued harmonic function

$$H(x,y) = F(x + iy) - F(x - iy)$$

has an analytic continuation to V. Hence $\partial H = F'$ has a holomorphic continuation to V. Thus it follows that its holomorphic primitive $F(z)$ on the upper half part of V (resp. on the lower half part of V) has an analytic continuation to V. Two continuations coincide on U by the assumption.

The theorem shows that the boundary value representation (76) holds topologically for any defining holomorphic function $F(z)$ if $f(x)$ is an ultradistribution.

Condition (c) is required on $F(z)$ on the upper and lower parts of $V \setminus U$ respectively. Hence it follows that if the conditions of the theorem hold, then respective topological limits $F(x \pm i0)$ exist in $\mathcal{D}^{*\prime}(U)$. In this sense the Hilbert transformation is continuous in $\mathcal{D}^{*\prime}(U)$ although it is decided only up to a difference of a real analytic function on U.

9. A condition for a Laplace hyperfunction to be an ultradistribution.
Suppose that

$$f(x) = F(x + i0) - F(x - i0)$$

is a Laplace hyperfunction in $^E\mathcal{B}_{[a,\infty]}^{\exp}$. Then its restriction to $\mathbf{R}$ is an ultradistribution in $\mathcal{D}_{[a,\infty)}^{*\prime}$ if and only if the defining function $F(z) \in {}^E\mathcal{O}^{\exp}(\mathbf{O} \setminus [a, \infty])$ satisfies the conditions of the preseding theorem. We will give an almost necessary, sufficient condition in terms of its Laplace transform $\hat{f}(\lambda)$.

A domain Ω in $\mathbf{C}$ is said to be *of class* $*$ if it includes the closed region

(83) $$\Omega_0 := \{\lambda \in \mathbf{C}; \operatorname{Re}\lambda \geqq \omega_0(\lambda) + C_0\}$$

for a weight function $\omega_0(\lambda)$ of class $*$ and a costant C_0.

It follows from (M.3) that $\omega_0(\lambda) = o(\lambda)$. Hence the union of Ω and the semi-circle $\{e^{i\theta}\infty; |\theta| < \pi/2\}$ at infinity is a neighborhood of the semi-circle in $\mathbf{O}$. Let $z = x + iy$

with $y > 0$ and consider the integral (23) along the upper half Γ of the boundary $\partial\Omega_0$. We have on Γ

$$\begin{aligned}
(84) \qquad \mathrm{Re}(\lambda z) &= x\,\mathrm{Re}\lambda - y\,\mathrm{Im}\lambda \\
&= x\,\omega_0(\lambda) - y\,\mathrm{Im}\lambda + xC_0
\end{aligned}$$

Its supremum on Γ may be estimated as (82). It is natural to assume a Paley-Wiener type condition on $\hat{f}(\lambda)$.

Theorem 8. *Suppose that $f(x)$ is an E-valued Laplace hyperfunction in ${}^E\mathcal{B}^{\exp}_{[a,\infty]}$. If its Laplace transform $\hat{f}(\lambda)$ is defined on a domain Ω of class $*$, and if it is estimated for any $\varepsilon > 0$ as*

$$(85) \qquad \|\hat{f}(\lambda)\| \leq C_\varepsilon \exp\{(-a+\varepsilon)\mathrm{Re}\lambda + \omega_\varepsilon(\lambda)\}, \quad \lambda \in \Omega_0$$

with a weight function $\omega_\varepsilon(\lambda)$ of class $*$ and a constant C_ε, then the restriction $f|_{\mathbf{R}}$ is an ultradistribution in ${}^E\mathcal{D}^{*\prime}_{[a,\infty)}$.

Proof. In view of (84) the integrand of (23) is bounded by

$$\exp\{(x - a + \varepsilon)\omega_0(\lambda) + \omega_\varepsilon(\lambda) - y\,\mathrm{Im}\lambda + x\,C_0 + \log C_\varepsilon\}.$$

If $d < \infty$ is fixed, we can find by (71) a weight function $\omega(\lambda)$ and a constant C such that

$$(x - a + \varepsilon)\omega_0(\lambda) + \omega_\varepsilon(\lambda) \leq \omega(\lambda) + C, \quad x \leq d.$$

Since $\mathrm{Im}\lambda$ is equivalent to $|\lambda|$ on Γ, we have the desired estimate

$$\sup_{x \leq d}\|F(x + iy)\| \leq C \exp \gamma(y), \quad 0 < y < \delta$$

for a growth function $\gamma(t)$ of class $*$ in the same way as the proof of Theorem 7.

10. Ultradistribution solutions of vector valued convolution equations.

Let D and E be complex Banach spaces. We set

$$(86) \qquad L := L(D, E), \quad H := L(E, D)$$

and consider the convolution equation

$$(87) \qquad p * u = f$$

in $u \in {}^D\mathcal{D}^{*\prime}_{[b,b+c]}$, where $p \in {}^L\mathcal{D}^{*\prime}_{[a,a+c)}$ and $f \in {}^E\mathcal{D}^{*\prime}_{[a+b,a+b+c)}$.

For the sake of simplicity we consider the case where $p \in {}^L\mathcal{D}^{*\prime}_{[0,k]}$ with compact support. Equation (87) is called *regularly solvable* in $\mathcal{D}^{*\prime}$ if there exists a $q \in {}^H\mathcal{D}^{*\prime}_{[0,\infty)}$ such that

$$(88) \qquad \begin{cases} p * q = 1_E \otimes \delta(x), \\ q * p = 1_D \otimes \delta(x) \end{cases}$$

Theorem 9. *The following conditions are equivalent for* $p \in {}^L\mathcal{D}^{*\prime}_{[0,k]}$, $k < \infty$:

(a) *The convolution equation* (87) *is regularly solvable in* $\mathcal{D}^{*\prime}$;

(b) *There is a* $q \in {}^H\mathcal{D}^{*\prime}_{[0,c)}$ *for a* $c > 0$ *such that* (88) *hold in* ${}^{L(E)}\mathcal{D}^{*\prime}_{[0,c)}$ *and* ${}^{L(D)}\mathcal{D}^{*\prime}_{[0,c)}$ *respectively*;

(c) *The inverse* $\hat{p}(\lambda)^{-1}$ *of the Laplace transform* $\hat{p}(\lambda)$ *of* $p(x)$ *exists on a domain* Ω *of class* $*$ *and is estimated as*

$$\|\hat{p}(\lambda)^{-1}\|_H \leq C \exp \omega(\lambda), \quad \lambda \in \Omega, \tag{89}$$

with a weight function $\omega(\lambda)$ *of class* $*$ *and a constant* C;

(d) *There is a domain* Ω *of class* $*$ *on which* $\hat{p}(\lambda)^{-1} \in {}^H\mathcal{O}(\Omega)$, *and for each* $\varepsilon > 0$ *there are a weight function* $\omega_\varepsilon(\lambda)$ *of class* $*$ *and a constant* C_ε *such that*

$$\|\hat{p}(\lambda)^{-1}\|_H \leq C_\varepsilon \exp\{\varepsilon \operatorname{Re}\lambda + \omega_\varepsilon(\lambda)\}, \quad \lambda \in \Omega. \tag{90}$$

Proof. (a)$\Rightarrow$(b) is trivial.

(b)$\Rightarrow$(c). Let $0 < a < b < c$ and choose an ultradifferentiable function $\chi(x) \in \mathcal{E}^*(\mathbf{R})$ such that

$$\chi(x) = \begin{cases} 1, & -\infty < x \leq a, \\ 0, & b \leq x. \end{cases}$$

Then the product $\chi q \in \mathcal{D}^{*\prime}_{[0,b]}$ satisfies

$$\begin{cases} p * (\chi q) = 1_E \otimes \delta - r, \\ (\chi q) * p = 1_D \otimes \delta - l, \end{cases}$$

with an $r \in {}^{L(E)}\mathcal{D}^{*\prime}_{[a,b+k]}$ and an $l \in {}^{L(D)}\mathcal{D}^{*\prime}_{[a,b+k]}$. Their Laplace transforms are

$$\begin{cases} \hat{p}(\lambda)(\chi q)^\wedge(\lambda) = 1_E - \hat{r}(\lambda), \\ (\chi q)^\wedge(\lambda)\hat{p}(\lambda) = 1_D - \hat{l}(\lambda). \end{cases}$$

The Paley-Wiener Theorem 6 asserts that

$$\|\hat{r}(\lambda)\|_{L(E)} \leq C_1 \exp\{\omega_1(\lambda) - a\operatorname{Re}\lambda\}, \quad \lambda \in \mathbf{C},$$

for a weight function $\omega_1(\lambda)$ of class $*$ and a constant C_1. In view of condition (M.2)

$$\Omega := \{\lambda \in \mathbf{C}; \operatorname{Re}\lambda \geq a^{-1}\{\omega_1(\lambda) + \log(2C_1)\}$$

is a domain of class $*$, and we have

$$\|\hat{r}(\lambda)\|_{L(E)} \leq 1/2, \quad \lambda \in \Omega.$$

Hence the Neumann series $(1 - \hat{r}(\lambda))^{-1}$ converges on Ω and is bounded by 2 in norm of $L(E)$. Thus we have

$$\hat{p}(\lambda)\{(\chi q)^\wedge(\lambda)(1 - \hat{r}(\lambda))^{-1}\} = 1_E.$$

Similarly $(1 - \hat{l}(\lambda))^{-1}$ exists on a domain of class $*$ and is bounded there. Hence

$$\hat{p}(\lambda)^{-1} = (\chi q)^\wedge(1 - \hat{r})^{-1} = (1 - \hat{l})^{-1}(\chi q)^\wedge$$

satisfies condition (c).

(c)$\Rightarrow$(d) is trivial.

(d)$\Rightarrow$(a) is an immediate consequence of Theorem 8.

11. Ultradistribution semi-groups and ultradistribution sines.

Applying the preceding theorem to the equations

(91)
$$\left(\frac{d}{dx} - A\right) u(x) = f(x)$$

and

(92)
$$\left(\frac{d^2}{dx^2} - A\right) u(x) = f(x),$$

we obtain the following theorems characterizing the generators of ultradistribution semi-groups and those of ultradistribution sines respectively.

Theorem 10. *Let A be a closed linear operator in a complex Banach space E with domain $D(A) = D$. There is an operator valued ultradistribution $T(x) \in {}^H D^{*\prime}_{[0,\infty)}$ such that*

(93)
$$\begin{cases} T'(x) - AT(x) = 1_E \otimes \delta(x) \\ T'(x) - T(x)A = 1_D \otimes \delta(x) \end{cases}$$

if and only if the resolvent $(\lambda - A)^{-1}$ exists on a domain Ω of class $*$ and satisfies the estimates

(94)
$$\|(\lambda - A)^{-1}\|_{L(E)} \leqq C \exp \omega(\lambda), \quad \lambda \in \Omega,$$

with a weight function $\omega(\lambda)$ of class $*$ and a constant C.
Then for any $g \in E$ $u(x) = T(x)g$ is a unique solution of

(95)
$$u'(x) - Au(x) = 0, \ 0 < x < \infty,$$

satisfying the initial condition $u(0) = g$ in the sense that

(96)
$$u'(x) - Au(x) = g\,\delta(x) \text{ on } \mathbf{R}.$$

Estimate (94) may be replaced by the weaker estimates

(97)
$$\|(\lambda - A)^{-1}\|_{L(E)} \leqq C_\varepsilon \exp\{\varepsilon \operatorname{Re}\lambda + \omega_\varepsilon(\lambda)\}$$

for any $\varepsilon > 0$. The theorem is due to Chazarain [3] and Ciorănescu-Zsido [4] in the case where $* = (M_p)$.

Theorem 11. *Let A be a closed linear operator in a complex Banach space E with domain $D(A) = D$. There is an operator valued ultradistribution $S(x) \in {}^H \mathcal{D}^{*\prime}_{[0,\infty)}$ such that*

(98)
$$\begin{cases} S''(x) - AS(x) = 1_E \otimes \delta(x), \\ S''(x) - S(x)A = 1_D \otimes \delta(x), \end{cases}$$

if and only if the resolvent $(\lambda^2 - A)^{-1}$ *exists when* λ *is in a domain* Ω *of class* $*$ *and satisfies the estimates*

$$(99) \qquad \|(\lambda^2 - A)^{-1}\|_{L(E)} \leqq C \, \exp \omega(\lambda), \ \lambda \in \Omega,$$

with a weight function $\omega(\lambda)$ *of class* $*$ *and constant* C.

Then for any g_0 *and* $g_1 \in E$ $u(x) = S(x)g_1 + S'(x)g_0$ *is a unique solution of*

$$(100) \qquad u''(x) - Au(x) = 0, \ 0 < x < \infty,$$

satisfying the initial conditions $u(0) = g_0$ *and* $u'(0) = g_1$ *in the sense that*

$$(101) \qquad u''(x) - Au(x) = g_1\delta(x) + g_0\delta'(x) \text{ on } \mathbf{R}.$$

We may again replace the right hand side of (99) by that of (97).

References

1. T. J. I. A. Bromwich, *Normal coordinates in dynamical systems*, Proc. London Math. Soc., **15** (1916), 401–448.
2. J. R. Carson, *The Heaviside operational calculus*, Bull. Amer. Math. Soc., **32** (1926), 43–68.
3. J. Chazarain, *Problèmes de Cauchy abstracts et applications à quelques problèmes mixtes*, J. Functional Anal., **7** (1971), 386–446.
4. I. Cioranescu and L. Zsidó, *ω-ultradistributions and their applications to the operator theory*, "Spectral Theory", Banach Center Publications, **8** (1982), 77–220.
5. G. Doetsch, *Theorie und Anwendung der Laplace-Transformation*, Springer, Berlin, 1937.
6. I. M. Gelfand, *Normierte Ringe*, Mat. Sb., **9** (1941), 3–24.
7. E. Hille, *Functional Analysis and Semi-Groups*, Amer. Math. Soc., Providence, 1948.
8. H. Komatsu, *Fractional powers of operators*, Pacific J. Math. **19** (1966), 285–346.
9. H. Komatsu, *Fractional powers of operators, II, Interpolation spaces*, Pacific J. Math., **21** (1967), 89–111.
10. H. Komatsu, *Fractional powers of operators, III, Negative Powers*, J. Math. Soc. Japan, **21** (1969), 205–220.
11. H. Komatsu, *Fractional powers of operators, IV, Potential operators*, J. Math. Soc. Japan, **21** (1969), 221–228.
12. H. Komatsu, *Fractional powers of operators, V, Dual operators*, J. Fac. Sci. Univ. Tokyo, Sect. I, **17** (1970), 373–396.
13. H. Komatsu, *Fractional powers of operators, VI, Interpolation of non-negative operators and imbedding theorems*, J. Fac. Sci. Univ. Tokyo, Sect. IA, Math., **19** (1972), 1–63.
14. H. Komatsu, *Ultradistributions, I, Structure theorems and a characterization*, J. Fac. Sci. Univ. Tokyo, Sect. IA, Math., **20** (1973), 25–105.

15. H. Komatsu, *Ultradistributions, II, The kernel theorem and ultradistributions with support in a submanifold*, J. Fac. Sci. Univ. Tokyo, Sect. IA, Math., **24** (1977), 607–628.

16. H. Komatsu, *Ultradistributions, III, Vector valued ultradistributions and the theory of kernels*, J. Fac. Sci. Univ. Tokyo, Sect. IA, Math., **29** (1982), 653–718.

17. H. Komatsu, *Laplace transforms of hyperfunctions – A new foundation of the Heaviside calculus*, J. Fac. Sci., Univ. Tokyo, Sect. IA, Math., **34** (1987), 805–820.

18. H. Komatsu, *Laplace transforms of hyperfunctions : another foundation of the Heaviside operational calculus*, "Generalized Functions, Convergence Structures and their Applications", Plenum Press, New York-London, 1988, pp. 57–70.

19. H. Komatsu, *Operational calculus, hyperfunctions and ultradistributions*, "Algebraic Analysis", Vol. 1, Academic Press, 1988, pp.357–372.

20. H. Komatsu, *Microlocal analysis in Gevrey classes and in complex domains*, Lecture Notes in Math., **1495** (1991), 161–236.

21. P. Lévy, *Le calcul symbolique d'Heaviside*, Bull. Sci. Math., **(2) 50** (1926), 174–192.

22. J. L. Lions, *Les semi-groupes distributions*, Portugal. Math., **19** (1960), 141–164.

23. A. J. Macintyre, *Laplace's transformation and integral functions*, Proc. London Math. Soc., (2) **45** (1938), 1–20.

24. J. Mikusiński, *Rachunek Operatorów*, Warszawa, 1953; English Edition, Pergamon, London, 1957.

25. M. Nagumo, *Einige analytische untersuchungen in linearen metrischen Ringen*, Japan J. Math., **13** (1936), 61–80.

26. F. and R. Nevanlinna, *Über die Eigenschaften analytischer Funktionen in der Umgebung einer singulären Stelle oder Linie*, Acta Soc. Sci. Fennicae, **5** no.5 (1922).

27. R. Nevanlinna, *Über die Eigenschaften meromorphier Funktionen in einer Winkelraum*, Acta Soc. Sci., Fennicae, **5** no. 12 (1925).

28. S. Ōuchi, *hyperfunction solutions of the abstract Cauchy problem*, Proc. Japan Acad., **47** (1971), 541–544.

29. G. Pólya, *Untersuchungen über Lücken und Singularitäten von Potenzreichen*, Math. Z., **29** (1929), 231-273.

30. K. Yosida, *On the group embedded in the metrical complete ring*, Japan J. Math., **13** (1936), 7–26.

31. K. Yosida, *Normed rings and spectral theorems*, Proc. Imp. Acad. Tokyo, **19** (1943), 356–359.

32. K. Yosida, *On the differentiability and the representation of one-parameter semigroup of linear operators*, J. Math. Soc. Japan, **1** (1948), 15–21.

33. K. Yosida, *Fractional powers of infinitesimal generators and the analyticity of the semi-groups generated by them*, Proc. Japan Acad., **36** (1960), 86–89.

34. K. Yosida, *Functional Analysis*, Springer, 1965.

35. K. Yosida and S. Okamoto, *A note on Mikusiński's operational calculus*, Proc. Japan Acad., **56** (1980), 1–3.

36. K. Yosida, *Enzanshi-ho*, Univ. of Tokyo Press, 1982 (in Japanese).

37. K. Yosida, *The algebraic derivative and Laplace's differential equation*, Proc. Japan Acad. Ser. A, **59** (1983), 1–4.

38. K. Yosida, *A simple complement to Mikusiński's operational calculus*, Studia Math., **11** (1983), 95–98.
39. K. Yosida and S. Matsuura, *A note on Mikusiński's proof of the Titchmarsh convolution theorem*, "Conference in Modern Analysis and Probability", Contemp. Math., **26**, Amer. Math. Soc., 1984, pp. 423-425.
40. K. Yosida, *Operational Calculus A Theory of Hyperfunctions*, Springer, 1984.

Wave equations in nonreflexive spaces

Yukio Kōmura and Kiyoko Furuya
Ochanomizu University

1 Introduction

The theory of evolution equations based on the Hille-Yosida theorem has been developed in general (not necessarily reflexive) Banach spaces and very successful for parabolic equations, while nonparabolic partial differential equations is generally discussed in L^2 spaces. As far as nonparabolic equations are concerned, there seem to be some gap between the abstract theory and the concrete theory.

In what kind of function spaces are nonparabolic equations wellposed?
(Here wellposedness means that the family of solutions $\{u(t)\}$ defines a C_0-semigroup $\{T_t\}$: $\|T_t\| \le Me^{\omega t}$, $T_t u(0) = u(t)$). We shall study this problem in the case of the wave or Schrödinger equations.

2 Simple examples

2.1 The Schrödinger equation in R^N

The Schrödinger equation $\partial_t u = i\Delta u$ is transformed to $\partial_t \hat{u} = -i|\xi|^2 \hat{u}$ by Fourier transformation, whose solution is expressed as

$$\hat{u}(t) = e^{-i|\xi|^2 t}\hat{u}(0).$$

We consider a Banach space Y of complex measures satisfying the following condition:

(A) $\qquad f \in Y, h \in L^\infty \Rightarrow hf \in Y, \|h \cdot f\|_Y \le \|h\|_\infty \cdot \|f\|_Y.$

Obviously $\|h \cdot f\|_Y = \|f\|_Y$, if $|h(\xi)| \equiv 1$.

Theorem 1 *The solutions to the Schrödinger equation define an isometric semigroup $\{T_t\}$ in the (inverse) Fourier transform $Z(= \hat{Y})$ of Y: $\|T_t\|_Z = 1$.*

As for the wave equation we have similar results. Thus it is wellposed in Z.

2.2 The wave equation in a bounded domain

Let Ω be a bounded domain with smooth boundary $\partial\Omega$. $\{e_n\}$ and $\{\lambda_n\}$ denote the eigenfunctions and eigenvalues of Δ with Dirichlet condition : $\Delta e_n = \lambda_n e_n$ in Ω, $e_n \mid_{\partial\Omega} = 0$.

We denote by $\Lambda^p = \Lambda^p(\Omega)$ the space $\{\sum \xi_n e_n \mid (\xi_n) \in l^p\}$ isomorphic to l^p. The formal solution

$$\begin{cases} u(t,x) = \sum[\xi_n \cos(\sqrt{-\lambda_n}t) + \eta_n \sin(\sqrt{-\lambda_n}t)]e_n \\ u_t(t,x) = -\sum[\xi_n\sqrt{-\lambda_n}\sin(\sqrt{-\lambda_n}t) - \eta_n\sqrt{-\lambda_n}\cos(\sqrt{-\lambda_n}t)]e_n \end{cases}$$

defines an isometric semigroup $\{T_t\}$ in $\Lambda^p \times \Lambda^{p(1)}$, where $\Lambda^{p(1)} = \{\sum \xi_n e_n \mid (\sqrt{-\lambda_n}\xi) \in l^p\}$. Note that Λ^p is a non-reflexive Banach space with the Radon-Nikodym property.

The same results hold for wave equations with variable coefficients :

$$\partial_t^2 u = \sum \partial_j(a_{jk}(x)\partial_k u).$$

The space Λ^p depends on the coefficients $a_{jk}(x)$.

3　Tensor products of Banach spaces

We shall give a very short introduction to the theory of tensor products of Banach spaces. Let $F_x = \{f(x)\}$ be a vector space of some functions with the variable x. A space of functions of two variables is constructed as follows:

$$H_{xy} = \left\{h(x,y) = \sum a_{ij}f_i(x)g_j(y) \mid f_i \in F_x, g_j \in G_y\right\}. \tag{1}$$

In an abstract form we use the notation of tensor products:

$$H = F \otimes G = \left\{\sum a_{ij}f_i \otimes g_j \mid f_i \in F, g_j \in G\right\}. \tag{2}$$

F' and G' denote the dual of F and G respectively. The duality of $F \otimes G$ and $F' \otimes G'$ is defined by

$$\langle h, h'\rangle = \sum a_{ij}b_{kl}\langle f_i, f_k'\rangle\langle g_j, g_l'\rangle \tag{3}$$

for $h = \sum a_{ij}f_i \otimes g_j \in F \otimes G$, $h' = \sum b_{kl}f_k' \otimes g_l' \in F' \otimes G'$.

The tensor product $F \otimes G$ of two normed spaces F and G has infinitely many compatible norms. (The norm $\|\cdot\|_\alpha$ is said to be *compatible* if $\|f \otimes g\|_\alpha = \|f\|_F \cdot \|g\|_G$.) The strongest one is the π-norm and the weakest one the ε-norm:

$$\|h\|_\pi = \inf\left\{\sum |b_{ij}| \cdot \|u_i\|_F \cdot \|v_j\|_G \mid \sum b_{ij}u_i \otimes v_j = h\right\} \tag{4}$$

for $h = \sum a_{ij}x_i \otimes y_j \in F \otimes G$.

The ε-norm of $F \otimes G$ is defined as the dual of $F'\otimes_\pi G'$. $F\hat{\otimes}_\alpha G$ means the completion of $F \otimes G$ with respect to a compatible norm $\|\cdot\|_\alpha$. The π-norm and ε-norm are not equivalent unless one of F and G is of finite dimensions:

$$F\hat{\otimes}_\pi G \neq F\hat{\otimes}_\varepsilon G.$$

For instance, we have

$$L^1(R^2) = L^1(R^1)\hat{\otimes}_\pi L^1(R^1) \underset{\neq}{\subseteq} L^1(R^1)\hat{\otimes}_\varepsilon L^1(R^1)$$

$$C_0(R^2) = C_0(R^1)\hat{\otimes}_\varepsilon C_0(R^1) \underset{\neq}{\supseteq} C_0(R^1)\hat{\otimes}_\pi C_0(R^1)$$

where $C_0(R^N) = \{f \in C(R^N) \mid \lim_{x\to\infty} f(x) = 0\}$.

4 Wellposed function spaces

Let Y_j satisfy condition (A), and let $W_\pi = \hat{Y}_1 \hat{\otimes}_\pi \cdots \hat{\otimes}_\pi \hat{Y}_N$, and $W_\varepsilon = \hat{Y}_1 \hat{\otimes}_\varepsilon \cdots \hat{\otimes}_\varepsilon \hat{Y}_N$.

Theorem 2 *The Schrödinger operator $i\triangle$ generates an isometric semigroup in W_π and W_ε.*

Proof. For $\varepsilon > 0$ and $u \in \hat{Y}_1 \otimes_\pi \cdots \otimes_\pi \hat{Y}_N$, there exist by the definition of π-norm $a_{ij} \in \mathbf{C}$ and $f_i \in Y_i$ such that

$$u = \sum a_{i_1 \cdots i_N} f_{i_1} \otimes \cdots \otimes f_{i_N},$$

$$\sum |a_{i_1 \cdots i_N}| \cdot \|f_{i_1}\|_{Y_1} \cdots \|f_{i_N}\|_{Y_N} < (1+\varepsilon)\|u\|.$$

Since $\mathcal{F}(T_t(a_{i_1 \cdots i_N} f_{i_1} \otimes \cdots \otimes f_{i_N})) = a_{i_1 \cdots i_N} e^{-i\xi_1^2 t} f_{i_1} \cdots e^{-i\xi_N^2 t} f_{i_N}$, we have

$$T_t(\hat{Y}_1 \otimes \cdots \hat{Y}_N) \subset \hat{Y}_1 \otimes \cdots \otimes \hat{Y}_N,$$

$$\|T_t u\|_\pi \le \sum |a_{i_1 \cdots i_N}| \cdot \|e^{-i\xi_1^2 t} f_{i_1}\|_{Y_1} \cdots \|e^{-i\xi_N^2 t} f_{i_N}\|_{Y_N} \le (1+\varepsilon)\|u\|_\pi.$$

Since ε is arbitrary, we have $\|T_t\|_\pi = 1$ and hence $T_t(W_\pi) \subset W_\pi$.

Put $Y_i^* = Y_i' \cap M^1$. Y_i^* also satisfies condition (A). Note that ${}^t T_t = T_{-t}, T_{-t}(Y_1^* \otimes \cdots \otimes Y_N^*) \subset Y^* \otimes \cdots \otimes Y_N^*$ and that $\|T_{-t} u'\|_\pi = \|u'\|_\pi$ for $u' \in Y_1^* \otimes \cdots \otimes Y_N^*$. Hence we have

$$\|T_t u\|_\varepsilon = \sup_{\|u'\|_\pi \le 1} |\langle u, {}^t T_t u' \rangle| = \|u\|_\varepsilon.$$

The proof is complete.

As far as we know now, the space $U = \hat{M}^1 \hat{\otimes}_\varepsilon \cdots \hat{\otimes}_\varepsilon \hat{M}^1$ is nearest the space C_0 among wellposed function spaces. If an evolution equation is wellposed in two Banach spaces V and W, it is also wellposed in an interpolation space of V and W. We can say more about the wave equation, but omit it.

5 Applications

A) *Wave equations with oblique conditions.* For simplicity we discuss wave equations with two space variables:

$$\partial_t^2 u = \triangle u \quad \text{in } \{(x,y) \mid x > 0\},$$

$$u_x + \alpha u_y = 0, \ u(0,x,y) = f(x,y), \ u_t(0,x,y) = g(x,y).$$

This equation is wellposed in $\hat{U}^{(1)} \times \hat{U}$, where $U = M_x^1 \hat{\otimes}_\pi M_y^1$ and $\hat{U}^{(1)} = \{\hat{f}(\xi,\eta) \in U \mid (|\xi| + |\eta|)\hat{f}(\xi,\eta) \in U\}$. Note that this is not wellposed in $H_0^1 \times L^2$ (solvable in some Sobolev spaces of high order derivatives).

B) *Semilinear equations.* Our theory is related to semilinear equations. For instance, the semilinear Schrödinger equation

$$\partial_t u = i\triangle u + f(u),$$

where $f(u) = u^p$ or an entire function of u, has local solutions in $\hat{M}^1$. Local solutions are easily obtained in some Sobolev spaces of high order derivatives. But we cannot apply this method if the perturbation is of the form $f(x) = a(x)u^p$, $a \in \hat{M}^1$ and $\notin C^1$.

C) *Quasilinear equations.* If we could apply our theory to quasilinear wave equations, we might obtain local solutions without using higher order derivatives. For this purpose, we need to generalize them to linear equations of the form:

$$\partial_t^2 u = \sum \partial_j(a_{jk}(t,x)\partial_k u).$$

But it seems somewhat complicated: The space Λ^p in **2.2** depends on t for $p \neq 2$.

D) *Feynman path integral.* Since the heat or two dimensional Dirac equations are wellposed in C_0, related path integrals induce measures, as is known. Let U be the space in **4**. Using the U-wellposedness of the Schrödinger or Dirac equations, the path integrals related to these equations are expressed by some measurelike elements. That is, the space U is somewhat near C_0.

References

[1] Yukio Kōmura. *Lectures on Evolution Equations, in preparation.*

Remarks on Systems with Incompletely given Initial Data and Incompletely given Part of the Boundary

J.L. LIONS

Collège de France
3, rue d'Ulm, 75005 Paris, France

DEDICATED TO THE MEMORY OF PROFESSOR K. YOSIDA

1. Introduction.

We consider a distributed system, i. e. a system whose state, denoted by y, is given by the solution of a Partial Differential Equation, assumed here to be **scalar** (only to simplify slightly the exposition) and of the **dissipative type** (parabolic type).

Usually, the equation is defined in **a domain** $\Omega \subset \mathbf{R}^n$ (where $n = 2, 3$ in the applications) and one adds to the equation the data corresponding to **initial conditions** and to **boundary conditions**.

We consider here situations where **the domain** Ω **itself is not completely known** and where the initial conditions are also not completely known.

We have first to define what is meant by the domain not being completely known.

We start with Ω given, with smooth boundary Γ .

Let Γ_0 be a (smooth) subset of Γ. We assume that Γ_0 is submitted in fact **to fluctuations "around"** Γ_0 . More precisely let $\nu(x)$ denote the unitary normal to Γ_0 directed toward the exterior of Ω. We are given m functions

$$(1.1) \qquad \alpha_1(x,t), ..., \alpha_m(x,t) \text{ defined on } \Sigma_0 = \Gamma_0 \times (0,T),$$

with the assumption that they are smooth enough and that

$$(1.2) \qquad \alpha_j(x,t) = 0 \quad \text{if } x \in \partial\Gamma_0 = \text{boundary of } \Gamma_0 \text{ in } \Gamma, \quad 1 \leq j \leq m.$$

For every $\sigma = \sigma_1, \cdots, \sigma_m \in \mathbf{R}^m$, with $|\sigma|$ small enough, we define

$$(1.3) \qquad \Gamma_{0,\sigma} = \{x + \sum_{i=1}^{m} \sigma_i \, \alpha_i(x,t) \, \nu(x) \mid x \in \Gamma_0\}.$$

Actually $\Gamma_{0,\sigma}$ depends on t, a fact that we shall recall by using the notation $\Gamma_{0,\sigma}(t)$. We observe that if we set

$$\Gamma_1 = \Gamma \setminus \Gamma_0$$

then the surface $\Gamma_1 \cup \Gamma_{0,\sigma}(t)$ is, for every fixed $t \in (0,T)$, **continuous** in $x \in \mathbf{R}^n$. If $|\sigma|$ is small enough there is no difficulty (multiply connected sets included) in defining

$$(1.4) \qquad \Omega_\sigma(t) = \text{"interior" of } \Gamma_1 \cup \Gamma_{0,\sigma}(t).$$

To fix ideas we shall assume that

$$(1.5) \qquad\qquad \alpha_j(x,0) = 0 \quad \forall j$$

so that

$$(1.6) \qquad\qquad \Omega_\sigma(0) = \Omega \quad \forall \sigma.$$

In what follows we assume the α_j's given but the **scalar parameters σ are not known.** It is only assumed that they are "small". $\square$

The state equation is now the following :

$$(1.7) \qquad\qquad \frac{\partial y}{\partial t} + Ay + f(y) = 0 \quad \text{in } Q_\sigma = \bigcup_{0<t<T} \Omega_\sigma(t),$$

where

A = second order elliptic operator with smooth enough coefficients (defined in a neighborhood of Ω),

f = not necessarily linear function, assumed to be C^1 and to satisfy other conditions indicated below (cf. (1.16)).

We now add **initial conditions and boundary conditions.** $\square$

Initial conditions.

We set $y(0) = "x \to y(x,0)"$. The initial condition is

$$(1.8) \qquad\qquad y(0) = y^0 + \tau \hat{y}^0,$$

where

$$(1.9) \qquad\qquad \begin{cases} y^0 \text{ is given in } L^2(\Omega), \\ \|y^0\|_{L^2(\Omega)} \le 1, \quad |\tau| \text{ is small.} \end{cases}$$

In other words the information that we have on $y(0)$ is that

$$(1.10) \qquad\qquad \|y(0) - y^0\|_{L^2(\Omega)} \le |\tau|. \qquad \square$$

Boundary conditions.

Let S be a given smooth subset of $\Gamma_1 = \Gamma \setminus \Gamma_0$ (cf. Fig. 1). We assume that

$$(1.11) \qquad\qquad y = g + \lambda \hat{g} \quad \text{on } S \times (0,T)$$

and

(1.12) $$y = 0 \quad \text{on } (\partial\Omega_\sigma(t) \setminus S) \times (0, T).$$

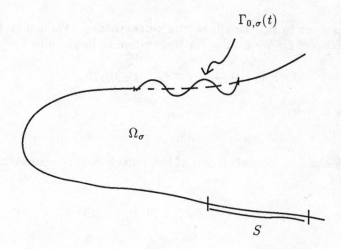

Figure 1
(The dotted line represents Γ_0)

In (1.11) we assume that

(1.13)
g is given smooth enough on $S \times (0, T)$
$\hat{g}$ is "bounded" in the corresponding topology
λ is "small" .

The term $\lambda\hat{g}$ corresponds to a pollution term we want to identify, or, at least, we want to obtain information on.

Conditions (1.12) can be rewritten as :

(1.14) $$y = 0 \quad \text{on } (\Gamma_1 \setminus S) \times (0, T)$$

and

(1.15) $$y(x + \sum_{i=1}^{m} \sigma_i \, \alpha_i(x, t) \, \nu(x), t) = 0 \quad \text{for } x \in \Gamma_0, \quad t \in (0, T). \qquad \square$$

It is assumed (this is essentially a hypothesis on f) that

(1.16)
if $\tau\hat{y}^0, \lambda\hat{g}$ and σ are given, problem (1.7) (1.8) (1.11) (1.12)
admits a unique solution, which is denoted by $y(x, t; \lambda, \sigma, \tau)$.

We want to find (or at least to have informations on) $\lambda \hat{g}$ **without trying to recover informations on** $\tau \hat{y}^0$ nor on σ.

Of course we need some further informations on y.　　□

Observatory $\mathcal{O}$.

Let $\mathcal{O}$ be an open set in Ω. It will be **the observatory**. We denote by $\chi_{\mathcal{O}}$ the characteristic function of $\mathcal{O}$. We assume the observation to be given by

$$(1.17) \qquad m_0(x,t) = y\,\chi_{\mathcal{O}}, \quad x \in \mathcal{O}, \quad t \in (0,T).$$

A few remarks are in order.

Remark 1.1.

The observatory could be located on part of Γ_1, outside S. The observation could in that case consist of

$$(1.18) \qquad m_0(x,t) = \frac{\partial y}{\partial \nu} \quad \text{on } \mathcal{O} \times (0,T). \qquad □$$

Remark 1.2.

One can extend what we are going to say to the cases where m_0 contains an added noise. For instance, in case (1.17), if $m_0 \in L^2(\mathcal{O} \times (0,T))$ we could have

$$(1.19) \qquad \begin{aligned} & y\,\chi_{\mathcal{O}} \in m_0 + K, \\ & K = \text{closed subspace of } L^2(\mathcal{O} \times (0,T)). \end{aligned}$$

We can then apply in the present situation the methods introduced in J.L. LIONS [1]. □

In what follows we introduce **sentinels** adapted to the present situation. Sentinels have been introduced by the author in [2] [3], with a more systematic presentation in [1].

We introduce Sentinels in Section 2, we show in Section 3 how they are connected with a controllability problem. In section 4 we construct sentinels and we end up in Section 5 by some remarks on **stealthiness**. The plan is as follows :

2.　　　Sentinels.
3.　　　Adjoint State. Controllability problem.
4.　　　Construction of sentinels.
5.　　　Stealthiness.
　　　　Bibliography.

2. Sentinels.

Let h_0 be **given** in $L^2(\mathcal{O} \times (0,T))$, and let w be a function **to be defined in** $L^2(\mathcal{O} \times (0,T))$. We introduce the functional

$$(2.1) \qquad S(\lambda, \sigma, \tau) = \iint\limits_{\mathcal{O} \times (0,T)} (h_0 + w)\, y(x, t; \lambda, \sigma, \tau)\, dx\, dt.$$

We want the following conditions to take place :

$$(2.2) \qquad \frac{\partial S}{\partial \sigma_i}(0,0,0) = 0, \qquad 1 \le i \le m,$$

$$(2.3) \qquad \frac{\partial S}{\partial \tau}(0,0,0) = 0,$$

$$(2.4) \qquad \|w\|_{L^2(\mathcal{O} \times (0,T))} = \text{minimum}.$$

Remark 2.1.

Conditions (2.2) (2.3) express **insensitivity** of S with respect to (small) variations of σ and τ. Condition (2.4) expresses the fact that we want the functional (2.1) to be "as close as possible" to the case $w = 0$, where one can think of (2.1) as an average (but where in general (2.2) (2.3) do not hold true). $\square$

Remark 2.2.

For the choice $w = -h_0$, (2.2) and (2.3) are true, but of course (2.1) is without any interest ! Therefore problem (2.2) (2.3) (2.4) **admits a unique solution**. We want **to compute** this solution and we want next to see when $w \ne -h_0$. $\square$

A functional (actually **the** functional) (2.1) (with conditions (2.2) (2.3) (2.4)) and when $w \ne -h_0$ is called a **sentinel**. We will address, after constructing the sentinel, the question of the **information on** $\lambda \hat{g}$ **given by the sentinel**, thus justifying the terminology introduced. $\square$

Remark 2.3.

Of course (2.2) (2.3) assume the derivatives to exist ! $\square$

Let us compute now $\dfrac{\partial y}{\partial \tau}(\lambda = 0, \sigma = 0, \tau = 0)$, in short $\dfrac{\partial y}{\partial \tau}$.

We shall set

$$(2.5) \qquad y\,(\lambda = 0, \sigma = 0, \tau = 0) = y_0.$$

Then (if $f'(\xi) = \dfrac{d}{d\xi} f(\xi)$)

(2.6)
$$\begin{cases} \dfrac{\partial}{\partial t}\left(\dfrac{\partial y}{\partial \tau}\right) + A\left(\dfrac{\partial y}{\partial \tau}\right) + f'(y_0)\dfrac{\partial y}{\partial \tau} = 0 & \text{in } \Omega \times (0,T) \\[2mm] \dfrac{\partial y}{\partial \tau}(0) = \hat{y}^0, \\[2mm] \dfrac{\partial y}{\partial \tau} = 0 & \text{on } \partial\Omega \times (0,T). \quad \square \end{cases}$$

We now compute
$$\frac{\partial y}{\partial \sigma_i}\,(\lambda = 0, \sigma = 0, \tau = 0) = \frac{\partial y}{\partial \sigma_i}.$$

Some informations are straightforward :

(2.7)
$$\begin{cases} \dfrac{\partial}{\partial t}\left(\dfrac{\partial y}{\partial \sigma_i}\right) + A\left(\dfrac{\partial y}{\partial \sigma_i}\right) + f'(y_0)\dfrac{\partial y}{\partial \sigma_i} = 0 & \text{in } \Omega \times (0,T) \\[2mm] \dfrac{\partial y}{\partial \sigma_i}(0) = 0 & \text{in } \Omega, \\[2mm] \dfrac{\partial y}{\partial \sigma_i} = 0 & \text{on } \Gamma_1 \times (0,T). \end{cases}$$

We rewrite (1.15) in the form

(2.8)
$$y\left(x + \sum_{i=1}^{m} \sigma_i\, \alpha_i\, \nu(x), t; \lambda, \sigma, \tau\right) = 0, \quad x \in \Gamma_0,$$

so that, expanding (2.8)

(2.9)
$$y(x,t;0,0,0) + \sum_{i=1}^{m} \sigma_i\, \alpha_i\, \frac{\partial}{\partial \nu} y(x,t;0,0,0) +$$
$$+ \sum_{i=1}^{m} \sigma_i \frac{\partial y}{\partial \sigma_i}(x,t) + \cdots = 0.$$

But $y(x,t;0,0,0) = y_0(x,t)$ and $y_0 = 0$ on $\Gamma_0 \times (0,T)$ so that (2.9) gives

(2.10)
$$\frac{\partial y}{\partial \sigma_i}(x,t) = -\alpha_i \frac{\partial y_0}{\partial \nu} \quad \text{on } \Sigma_0 = \Gamma_0 \times (0,T).$$

Therefore $\dfrac{\partial y}{\partial \sigma_i}$ is defined by (2.7) (2.10). $\square$

With the preceding notations, we have :

(2.11)
$$\frac{\partial \mathcal{S}}{\partial \sigma_i}(0,0,0) = \iint_{\mathcal{O} \times (0,T)} (h_0 + w) \frac{\partial y}{\partial \sigma_i}\, dx\, dt,$$

$$(2.12) \qquad \frac{\partial S}{\partial \tau}(0,0,0) = \iint_{\mathcal{O} \times (0,T)} (h_0 + w) \frac{\partial y}{\partial \tau} \, dx \, dt.$$

We now transform these expressions using the adjoint state.

3. Adjoint State. Controllability problem.

We define the adjoint state q by

$$(3.1) \qquad \begin{cases} -\dfrac{\partial q}{\partial t} + A^* q + f'(y_0) q = (h_0 + w) \chi_{\mathcal{O}} & \text{in } \Omega \times (0,T), \\ q(T) = 0, \quad q = 0 \quad \text{on } \Sigma = \Gamma \times (0,T), \end{cases}$$

where $A^* = $ adjoint of A.

Of course q depends on w **which is not yet chosen.**

If now we multiply (3.1) by $\dfrac{\partial y}{\partial \sigma_i}$ and $\dfrac{\partial y}{\partial \tau}$ and if we integrate by parts using (2.7) , (2.10) and (2.6), we obtain

$$(3.2) \qquad \frac{\partial S}{\partial \sigma_i}(0,0,0) = \int_{\Sigma_0} \alpha_i \frac{\partial y_0}{\partial \nu} \frac{\partial q}{\partial \nu_*} \, d\Sigma$$

(where $\dfrac{\partial}{\partial \nu_*}$ denotes the conormal derivative on Γ_0 with respect to A^*) , and

$$(3.3) \qquad \frac{\partial S}{\partial \tau}(0,0,0) = \int_{\Omega} q(x,0) \, \hat{y}^0 \, dx.$$

Therefore the problem is now the following : **find w such that the solution $q(x,t) = q(x,t;w)$ of (3.1) satisfies the conditions**

$$(3.4) \qquad \int_{\Sigma_0} \alpha_i \frac{\partial y_0}{\partial \nu} \frac{\partial q}{\partial \nu_*} \, d\Sigma = 0, \quad 1 \le i \le m,$$

$$(3.5) \qquad q(x,0) = 0 \quad \text{in } \Omega,$$

and such that

$$\|w\|_{L^2(\mathcal{O} \times (0,T))} = \min . \qquad \square$$

If we think of w as a **control** and $q = q(w)$ as a **state**, (3.4) (3.5) (3.6) is a problem of the **controllability type.** Indeed, we want to choose the control w such that the system is driven to the final state $q(0) = 0$ (the system runs backward in time) with minimum use of w (in the L^2 norm) and with the added conditions (3.4) (state constraints).

We now give the solution of this question.

4. Construction of sentinels.

We decompose q in

$$q = q_0 + z,$$

where q_0 is given by

(4.2)
$$\begin{cases} -\dfrac{\partial q_0}{\partial t} + A^* q_0 + f'(y_0)\, q_0 = h_0\, \chi_\mathcal{O}, \\ q_0(T) = 0, \quad q_0 = 0 \quad \text{on } \Sigma \end{cases}$$

and where z is given by

(4.3)
$$\begin{cases} \dfrac{-\partial z}{\partial t} + A^* z + f'(y_0)\, z = w\, \chi_\mathcal{O}, \\ z(T) = 0, \quad z = 0 \quad \text{on } \Sigma. \end{cases}$$

We want now to find w such that

(4.4)
$$\int_{\Sigma_0} \alpha_i \frac{\partial y_0}{\partial \nu} \frac{\partial z}{\partial \nu_*}\, d\Sigma = -\int_{\Sigma_0} \alpha_i \frac{\partial y_0}{\partial \nu} \frac{\partial q_0}{\partial \nu_*}\, d\Sigma, \quad 1 \le i \le m,$$

(4.5)
$$z(0) = -q_0(0),$$

and with (3.6).

The solution is given by the following construction. $\qquad \square$

Let ρ^0 be a function (to be found, in a suitable Hilbert space — see below) and let $\beta = \{\beta_i \mid i \le i \le m\}$ be a vector to be found in $\mathbf{R}^m$. We define $\rho = \rho(x, t; \rho^0, \beta)$ as the solution of

(4.6)
$$\frac{\partial \rho}{\partial t} + A\rho + f'(y_0)\, \rho = 0,$$

(4.7)
$$\rho(0) = \rho^0,$$

(4.8)
$$\begin{cases} \rho = \displaystyle\sum_{i=1}^{m} \beta_i\, \alpha_i \frac{\partial y_0}{\partial \nu} \quad \text{on } \Sigma_0, \\ \rho = 0 \quad \text{on } \Sigma_1 = \Gamma_1 \times (0, T). \end{cases}$$

We then define z by

(4.9)
$$\begin{cases} -\dfrac{\partial z}{\partial t} + A^* z + f'(y_0) z = \rho\, \chi_\mathcal{O}, \\ z(T) = 0, \quad z = 0 \quad \text{on } \Sigma. \end{cases}$$

Therefore we can define a linear operator Λ by

$$(4.10) \qquad \Lambda\{\rho^0,\beta\} = \left\{ z(0); \int_{\Sigma_0} \alpha_i \frac{\partial y_0}{\partial \nu} \frac{\partial z}{\partial \nu_*} \, d\Sigma, \ 1 \leq i \leq m \right\}.$$

If we define

$$(4.11) \qquad Mh_0 = \left\{ q_0(0), -\int_{\Sigma_0} \alpha_i \frac{\partial y_0}{\partial \nu} \frac{\partial q_0}{\partial \nu_*} \, d\Sigma, \ 1 \leq i \leq m \right\},$$

then (4.4) (4.5) is equivalent to

$$(4.12) \qquad \Lambda\{\rho^0,\beta\} = -Mh_0.$$

It can be proven (by methods similar to those used in J.L. LIONS [1]) that if (4.12) admits a (unique) solution, then **the function w we are interested in, is given by**

$$(4.13) \qquad w = \rho \chi_{\mathcal{O}}.$$

We prove below that this is indeed the case, so that (4.12) gives

$$(4.14) \qquad \{\rho^0,\beta\} = -\Lambda^{-1} M h_0.$$

One can verify easily that

$$(4.15) \qquad M^*\{\rho^0,\beta\} = \rho \chi_{\mathcal{O}}$$

so that (4.13) is given by

$$(4.16) \qquad w = -M^* \Lambda^{-1} Mh_0.$$

The sentinel defined by h_0 is given by

$$(4.17) \qquad \mathcal{S}(\lambda,\sigma,\tau) = \iint_{\mathcal{O}\times(0,T)} (h_0 - M^*\Lambda^{-1}Mh_0)\, y(\lambda,\sigma,\tau)\, dx\, dt.$$

It remains to prove **the inversibility of Λ in suitable function spaces.** □

Let us multiply (4.9) by ρ and let us integrate by parts. We obtain

$$(4.18) \qquad \langle \Lambda\{\rho^0,\beta\}, \{\rho^0,\beta\}\rangle = \iint_{\mathcal{O}\times(0,T)} \rho^2 \, dx\, dt.$$

This leads to the introduction of **new function spaces.** We define

$$(4.19) \qquad \|\{\rho^0,\beta\}\|_F = \left(\iint_{\mathcal{O}\times(0,T)} \rho^2 \, dx\, dt \right)^{1/2}.$$

Let us see if this prehilbertian semi-norm is a norm. If $\|\{\rho^0, \beta\}\|_F = 0$, then $\rho = 0$ on $\mathcal{O} \times (0, T)$.

But if $f'(y_0)$ and the coefficients of A are smooth enough, it follows, using S. MIZO-HATA [1] uniqueness theorem, that

$$\rho \equiv 0 \quad \text{in } \Omega \times (0, T).$$

Therefore

$$\rho^0 = 0$$

and

$$(4.20) \qquad \sum_{i=1}^{m} \beta_i \alpha_i \frac{\partial y_0}{\partial \nu} = 0 \quad \text{on } \Sigma_0.$$

We make the hypothesis:

(4.21) The functions $\alpha_i \dfrac{\partial y_0}{\partial \nu}$, $1 \le i \le m$, are linearly independent in $L^2(\Sigma_0)$.

Then (4.20) implies $\beta_i = 0$ $\forall i$ and (4.19) is a norm.

We introduce next

$(4.22) \qquad F = \textbf{completion of } L^2(\Omega) \times \mathbf{R}^m \textbf{ for the norm } (4.19).$

If F' denotes the dual space of F, we have

$(4.23) \qquad\qquad\qquad \Lambda$ is an isomorphism of F onto F'.

(We have also $\Lambda^* = \Lambda$). In order to obtain (4.14) it remains to show that

$(4.24) \qquad\qquad\qquad\qquad M\, h_0 \in F'.$

But

$$(4.25) \qquad \langle M\, h_0, \{\rho^0, \beta\}\rangle = \iint_{\mathcal{O} \times (0,T)} h_0\, \rho\, dx\, dt,$$

hence (4.24) follows, with in fact

$$(4.26) \qquad \|M\, h_0\|_{F'} \le \|h_0\|_{L^2(\mathcal{O} \times (0,T))}.$$

Remark 4.1.

We still have a solution of (4.12) if (4.21) does not hold true. In that case F is a **quotient** space, (4.24) remains true. $\square$

Remark 4.2.

Since w is given by (4.13), ρ as a solution of the equation (4.6) of parabolic type is smooth so that if h_0 is **not** smooth, $h_0 + \rho \not\equiv 0$. $\square$

We examine now the question of the **informations given by the sentinel defined by h_0**. This is the goal of the following section.

5. Stealthiness.

We use (4.17). We have, to the first order :

$$(5.1) \qquad \mathcal{S}(\lambda, \sigma, \tau) \simeq \mathcal{S}(0,0,0) + \lambda \frac{\partial \mathcal{S}}{\partial \lambda}(0,0,0),$$

i. e. using the observation (1.17)

$$(5.2) \qquad \lambda \frac{\partial \mathcal{S}}{\partial \lambda}(0,0,0) \simeq \iint_{\mathcal{O} \times (0,T)} (h_0 - M^* \Lambda^{-1} M h_0)(m_0 - y_0) \, dx \, dt.$$

We now observe that $\dfrac{\partial y}{\partial \lambda} \, (\lambda = 0, \sigma = 0, \tau = 0) = \dfrac{\partial y}{\partial \lambda}$ is defined by

$$(5.3) \qquad \begin{cases} \dfrac{\partial}{\partial t}(\dfrac{\partial y}{\partial \lambda}) + A \dfrac{\partial y}{\partial \lambda} + f'(y_0) \dfrac{\partial y}{\partial \lambda} = 0 \\[2mm] \dfrac{\partial y}{\partial \lambda}(0) = 0, \\[2mm] \dfrac{\partial y}{\partial \lambda} = \hat{g} \quad \text{on } S \times (0,T), \; 0 \text{ elsewhere on } \Sigma. \end{cases}$$

If we define

$$(5.4) \qquad q(h_0) = \text{solution of (3.1) for } w = -M^* \Lambda^{-1} M h_0,$$

then we obtain that

$$\lambda \frac{\partial \mathcal{S}}{\partial \lambda}(0,0,0) = -\lambda \int_{S \times (0,T)} \frac{\partial}{\partial \nu_*} q(h_0) \, \hat{g} \, d\Sigma.$$

Therefore (5.2) gives an estimate of

$$(5.6) \qquad \lambda \int_{S \times (0,T)} \frac{\partial q}{\partial \nu_*}(h_0) \, \hat{g} \, d\Sigma \simeq - \iint_{\mathcal{O} \times (0,T)} (h_0 - M^* \Lambda^{-1} M h_0)(m_0 - y_0) \, dx \, dt.$$

A pollution $\lambda \, \hat{g}$ will be **stealthy for the sentinel defined by** h_0 if

$$(5.7) \qquad \int_{S \times (0,T)} \frac{\partial q(h_0)}{\partial \nu_*} \hat{g} \, d\Sigma = 0.$$

Therefore one can say that

$$(5.8) \qquad \begin{array}{l} \textbf{one sentinel on } \mathcal{O} \textbf{ observes "something", but not "everything".} \\ \textbf{Some pollutions are stealthy for it.} \quad \square \end{array}$$

Remark (5.8) leads to the following question. We can consider a sequence

$$(5.9) \qquad h_{01}, h_{02}, \dots \quad \text{complete in } L^2(O \times (0,T))$$

and we can construct the corresponding sentinels (all on the **same** fixed observatory). One can prove (by using the methods introduced in J.L. LIONS [1]) that

(5.10) **there are no pollutions $\hat{g}$ which are stealthy for all sentinels defined (on $\mathcal{O}$) by the sequence $h_{01}, \ldots, h_{0m}, \ldots$ as in (5.9).** □

Remark 5.1.

In all what has been said, $\mathcal{O}$ and T **can be chosen arbitrarily small.** □

Remark 5.2.

Numerical applications are presented in J.P. KERNEVEZ [1] and J.P. KERNEVEZ and J.L. LIONS [1]. No computations have yet been made for the present situation where there are missing informations on the boundary of the domain. □

Remark 5.3.

The space F introduced in (4.22) can be "very large". In general it will contain elements which are **not** distributions on Ω . □

Remark 5.4.

Every uniqueness theorem allows to define families of new Hilbert (or Banach) spaces. This remark has been used here in (4.22). This technique was already introduced for classical type controllability problems in J.L. LIONS [4] [5]. □

Bibliography.

J.P. KERNEVEZ and Associates
[1]   Reports on Sentinels, U.T.C. (Université de Technologie de Compiègne), 1991.

J.P. KERNEVEZ and J.L. LIONS
[1]   To appear.

J.L. LIONS
[1]   *Sentinelles pour les systèmes distribués données incomplètes,* R.M.A. Masson, Paris, 1992.
[2]   Sur les sentinelles des systèmes distribués, C. R. Acad. Sci. **307** (1988), 819-823, 865-870.
[3]   Furtivité et sentinelles pour les systèmes distribués données incomplètes, C. R. Acad. Sci., 1990.
[4]   Exact controllability, stabilization and perturbations for distributed systems. SIAM Rev., **30**, 1, (1988), 1-68.
[5]   *Contrôlabilité exacte, stabilisation et perturbations des systèmes distribués,* R.M.A., Masson, Paris, Vol. 1 and 2, 1988.

S. MIZOHATA
[1]   Unicité du prolongement des solutions pour quelques opérateurs différentiels paraboliques. Mem. Coll. Sc. Univ., Kyoto, **31** (3), (1958), 219-239.

ON NON-CONVEX CURVES OF CONSTANT ANGLE

By

Shigetake MATSUURA
RIMS, Kyoto University

0. Introduction.

In the title, "non-convex" means, as usual, "not necessarily convex." And, to give a reasonably understandable exposition, it is preferable to start with the case of convex curves, because one has a good intuitive interpretation in that case.

The theory of curves of constant angle is a kind of generalization of that of curves of constant breadth. It is also an extensive generalization of the incomplete work of Blaschke [19] which generalized the classical example of ellipses, answering the problem posed by C. Neumann.

As we see in the definitions in §3, curves of constant angle depend on a circle C (call it the director circle) and an angle α. Intuitively they are curves (convex case) in the director circle C, spanning a constant angle α $(0 < \alpha < \pi)$ at every point on C. The limiting case $R \to \infty, \alpha \to 0$ corresponds to a curve of constant breadth.

In the case of constant angle R remains finite. For simplicity, we put $R = 1$.

The old paper of Blaschke [19] treats only the convex case with $\alpha = \dfrac{\pi}{2}$, his description is entirely intuitive and without proofs. Furthermore, his intuitive sketch has an error. He gives, however, a good example. The present work is not the continuation of Blaschke's. It has a different origin in the higher dimentional geometry. After having finished our work, we found Blaschke's paper, and its content is a part of ours.

To make complete the natural duality among curves of constant angle in general, we are obliged to include non-convex case. [see §8 (for general case)].

In view of the lack of space, we can scarcely give proofs of the results. We try, however, to make this exposition as readable as possible. We are obliged to omit many interesting and important results, especially solutions to various variational problems. But this brief sketch may be considered an introduction to the theory of curves of constant angle. Our subjects are a special kind of plane curves. However, from the beginning of definition of admissible curves, we need to make much use of functional analysis. Thus, our theory of curves of constant angle might be recognized as an example of *Functional-Analytic Geometry.*

1. Notations.

To simplify the notations in this article, we introduce the following: let α be a given angle $0 < \alpha < \pi$; put $\hat{\alpha} = \pi - \alpha$, and also use

$$c_1 = c_1(\alpha) = \sin\alpha, \qquad c_2 = c_2(\alpha) = \cos\alpha,$$
$$\tilde{c}_1 = \tilde{c}_1(\alpha) = \sin\frac{\alpha}{2}, \qquad \tilde{c}_2 = \tilde{c}_2(\alpha) = \cos\frac{\alpha}{2}.$$

2. Characteristic function χ_α and modified characteristic function $\tilde{\chi}_\alpha$.

This section collects fundamental formulas which are used in later sections. Readers may begin with Section 3.

Let $\Omega_\alpha = \min(\tilde{c}_1, \tilde{c}_2)$. The open intervals I_α, J_α are defined by

$$I_\alpha = (-\Omega_\alpha, \Omega_\alpha)$$
$$J_\alpha = \begin{cases} (\ 0\,, c_1) & \text{if} \quad 0 < \alpha \le \pi/2 \\ (-c_2,\ 1) & \text{if} \quad \pi/2 \le \alpha < \pi. \end{cases}$$

The *characteristic function* χ_α and the *modified characteristic function* $\tilde{\chi}_\alpha$ are defined by the formulas

$$\chi_\alpha(t) = c_1\sqrt{1 - t^2} - c_2 t, \qquad t \in J_\alpha$$
$$\tilde{\chi}_\alpha(s) = \tilde{c}_1\sqrt{1 - s^2} - \tilde{c}_2 s, \qquad s \in I_\alpha \quad \text{or} \quad s \in J_\alpha.$$

Proposition 2.1. *χ_α maps J_α onto J_α and is strictly monotone decreasing. χ_α has the only one fixed point $\tilde{c}_1$. And its inverse mapping χ_α^{-1} coincides with χ_α.*

Schematically

$$\chi_\alpha : J_\alpha \rightleftarrows J_\alpha.$$

Proposition 2.2. *$\tilde{\chi}_\alpha$ maps J_α onto I_α and is strictly monotone decreasing. $\tilde{\chi}_\alpha$ maps $\tilde{c}_1$ to 0. And its inverse $\tilde{\chi}_\alpha^{-1}$ has the same expression as $\tilde{\chi}_\alpha$.*

Schematically

$$\tilde{\chi}_\alpha : J_\alpha \rightleftarrows I_\alpha.$$

In both schemes, $\rightleftarrows$ means that the mappings are one-to-one, onto and that their inverses have the same expressions. $\tilde{\chi}_\alpha$ has the *linearization effect* on χ_α, in fact, we have the following

Proposition 2.3. *If $w \in I_\alpha$, $p \in J_\alpha$, $w = \tilde{\chi}_\alpha(p)$, then we have*

$$\tilde{\chi}_\alpha(\chi_\alpha(p)) = -w.$$

Fig. 2.1.   Graph of χ_α, $\eta = \chi_\alpha(\xi)$ Fig. 2.2.   Graph of $\tilde{\chi}_\alpha$, $p = \tilde{\chi}_\alpha(w)$

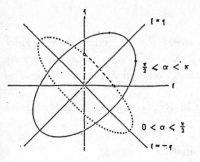

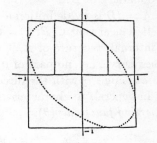

$$\xi^2 + \eta^2 + 2c_2\xi\eta = c_1^2 \qquad\qquad p^2 + \omega^2 + 2\tilde{c}_2 pw = \tilde{c}_1^2$$

3. Curves of constant angle α (convex case).

Let C be the circle of radius R with center at the orgin $O = (0,0)$ of the plane $\mathbf{R}^2$, and call it the *director circle*. [This terminology comes from the classical example of ellipses, $\alpha = \pi/2$.] Hereafter we assume $R = 1$, without loss of generality. Let A be a figure contained in C. A figure simply means here a subset of $\mathbf{R}^2$. For a point $P \in C$ and we put

$$C(P; A) = \{\text{ray; starting at } P, \text{ passing a point of } A\}$$

(ray = closed half line). $C(P; A)$ is called the *sight-cone* at P for A. We assume that $C(P; A)$ is a closed convex cone with angle α at the vertex P.

Fig. 3.1

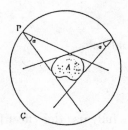

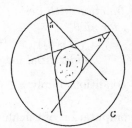

Suppose that the angle α is independent of P.

Then, there exists a closed convex set D with non-empty interior such that

$$\partial D \subseteq A \subseteq D$$

since it is not difficult to see that figures of constant angle is strictly convex. And the fact that $p(\theta) \in C^1$ is also easy to show.

(In fact, $D = \bigcap_{P \in C} C(P; A)$ and the origin $O \in \overset{\circ}{D}$: the interior of D). ∂D designates the boundary of D.

$\Lambda = \partial D$ is by definition a closed convex curve.

It is clear that $C(P; \Lambda) = C(P; A) = C(P; D)$ for every $P \in C$. Thus, if we neglect the internal structure of A, it is enough to study D or $\Lambda = \partial D$. Λ is in fact a *strictly convex* curve, i.e. no part of it is a straight line segment. We call Λ a *convex curve of constant angle* α with the director circle C.

In general, to characterize a closed convex curve in $\mathbf{R}^2$, it is enough to obtain its *supporting function* $p(\theta)$

$$p(\theta) = \sup_{(x,y) \in \Lambda} (x \cos \theta + y \sin \theta).$$

Since Λ is strictly convex, p is a C^1-function with period 2π. Λ has the following *parametric representation*

$$(3.1) \qquad \begin{cases} x = x(\theta) = p(\theta) \cos \theta - \dot{p}(\theta) \sin \theta \\ y = y(\theta) = p(\theta) \sin \theta + \dot{p}(\theta) \cos \theta \end{cases} \qquad (0 \le \theta \le 2\pi)$$

where $\dot{p}(\theta)$ means the derivative $dp(\theta)/d\theta$. This is a continuous closed curve, i.e. $(x(\theta), y(\theta))$ depends continuously on θ but not C^1 in θ. In fact the second derivative $\ddot{p}(\theta)$ in the sense of distribution of L. Schwartz $\mathcal{D}'$ satisfies the inequality

$$p + \ddot{p} \ge 0$$

in the present convex case. This means that the left hand side is a Radon measure ≥ 0.

Theorem 3.1. *For a continuous function $p(\theta)$ of θ to be the supporting function of a convex curve of constant angle α, it is necessary and sufficient that it satisfies the following four conditions:*

$(1°)$ $p(\theta)$ *is a function with period* 2π
$(2°)$ $p(\theta) \in J_\alpha$ *for all* θ
$(3°)$ $p(\theta + \pi - \alpha) = \chi_\alpha(p(\theta))$ *[functional equation]*
$(4°)$ $p + \ddot{p} \ge 0$ *[differential inequality].*

Remark 1. These conditions imply $p \in C^1(\mathbf{R})$.

Furthermore, computation of calculus of variations fully use this effect [18], [23].

Remark 2. For every α, $0 < \alpha < \pi$, there exists a convex curve of constant angle α. In fact, if we employ the function $p(\theta) \equiv \tilde{c}_1$, we get a circle of radius $\tilde{c}_1$ concentric with C the director circle. We call this the *trivial* curve of constant angle α.

The totality of functions p satisfying the four conditions of the theorem above is denoted by $\wp_\alpha$ or more precisely $\wp_\alpha^{\text{convex}}$. The case that $\wp_\alpha$ is a singleton $\wp_\alpha = \{\tilde{c}_1\}$ is not interesting. The following theorem is the answer to the question: "Is $\wp_\alpha = \{\tilde{c}_1\}$? "

Theorem 3.2. *(I) If $\alpha/\pi \notin \mathbf{Q}$, then the answer is "yes."*
(II) Suppose $\alpha/\pi \in \mathbf{Q}$, and let $\alpha/\pi = \dfrac{m}{n}$ be its irreducible fraction representation.
  (i) *If $mn =$ odd, then the answer is "yes."*
  (ii) *If $mn =$ even, then the answer is "no."*

Outline of the proof. It follows from Theorem 3.1 ($3°$) and Proposition 2.1 that $2\pi - 2\alpha$ is a period of $p(\theta)$. Hence 2π and 2α are periods. Thus the case $\alpha/\pi \notin \mathbf{Q}$ is trivial.

The case $\alpha/\pi = m/n$ with irreducible fraction: m, n being coprime, there exist integers k, l such that $1 = kn + lm$. Thus $2\pi/n = 2\pi k + 2\alpha l$ is a period.

When both m, n are odd, $n - m$ is even. Thus $\pi - \alpha = \pi \dfrac{n-m}{n}$ is a period. Now, a simple geometrical observation gives that $p(\theta) \equiv \tilde{c}_1$ by means of monotonicity of χ_α (see [1, Theorems 30, 31]).

Remark. In the case (II),(ii) $\wp_\alpha = \wp_\alpha^{\text{convex}}$ is an infinite dimensional set. This means that, if $\wp_\alpha^{\text{convex}}$ is suitably topologized, it contains a non-empty open set of an infinite dimensional normed space. To prove it linearization effect Proposition 2.3 is essential.

4. The space $B_{2,\sigma}$ and its open convex cone $\wp$.

Let $C(\mathbf{T})$ be the space of all real-valued continuous functions on the torus $\mathbf{T} = \mathbf{R}/2\pi\mathbf{Z}$ equipped with the uniform norm. Then, its dual space $\mathcal{M}(\mathbf{T}) = C(\mathbf{T})'$ is the space of all Radon measures on $\mathbf{T}$. Both $C(\mathbf{T})$ and $\mathcal{M}(\mathbf{T})$ are Banach spaces. Now, consider the space of functions

$$B_2 = \left\{ u; u \in C(\mathbf{T}), \dot{u} \in C(\mathbf{T}), \ddot{u} \in \mathcal{M}(\mathbf{T}) \right\}.$$

This becomes naturally a Banach space. But, for later applications, the norm topology of $\mathcal{M}(\mathbf{T})$ is too strong. Thus we replace it by the $*$-weak topology $\sigma(\mathcal{M}(\mathbf{T}), C(\mathbf{T}))$, i.e. the simple convergence topology. Then we denote the space by $\mathcal{M}_\sigma(\mathbf{T})$. Correspondingly, we weaken the topology of B_2. The new topology is the pull-back by the one-to-one mapping

$$B_2 \ni u \to (u, \dot{u}, \ddot{u}) \in C(\mathbf{T}) \times C(\mathbf{T}) \times \mathcal{M}_\sigma(\mathbf{T})$$

when the target space is equipped with the product topology. Thus topologized B_2 is denoted by $B_{2,\sigma}$ which is a locally convex topological vector space but no more a Banach space.

Now consider

$$\wp = \left\{ p; p \in B_{2,\sigma}, \min_{\theta \in \mathbf{T}} p(\theta) > 0 \right\}.$$

This is clearly an open convex cone in $B_{2,\sigma}$. The space $\wp$ plays an important role when we introduce non-convex curves of constant angle.

For every $p \in \wp$, we consider a closed continuous plane curve Λ with the following parametric representation

(4.1)
$$\begin{cases} x = x(\theta) = p(\theta) \cos \theta - \dot{p}(\theta) \sin \theta \\ y = y(\theta) = p(\theta) \sin \theta + \dot{p}(\theta) \cos \theta \end{cases}$$

p is called the *generator* of Λ.

Remark. When Λ is a strictly convex curve, p is its supporting function. But for general non-convex curves, the meaning of $p(\theta)$ will be discussed later.

We put

$$(4.2) \qquad T(\theta) = \begin{pmatrix} \cos\theta & -\sin\theta \\ \sin\theta & \cos\theta \end{pmatrix}.$$

Then (4.1) takes the form

$$(4.3) \qquad \begin{pmatrix} x(\theta) \\ y(\theta) \end{pmatrix} = T(\theta) \begin{pmatrix} p(\theta) \\ \dot{p}(\theta) \end{pmatrix}.$$

Since $T(\theta)$ is an orthogonal matrix and $T(\theta)^{-1} = T(-\theta)$, we get

$$(4.4) \qquad \begin{pmatrix} p(\theta) \\ \dot{p}(\theta) \end{pmatrix} = T(-\theta) \begin{pmatrix} x(\theta) \\ y(\theta) \end{pmatrix}$$

and

$$(4.5) \qquad x(\theta)^2 + y(\theta)^2 = p(\theta)^2 + \dot{p}(\theta)^2.$$

(4.3), (4.4) show that the generator p determines the curve Λ and conversely the curve Λ determines the generator p. Thus, we sometimes use the designations p_Λ and Λ_p.

When Λ is a strictly convex curve, its length $L[p]$ and the area $A[p]$ inscribed by Λ ($p = p_\Lambda$) are calculated by the formulas

$$(4.6) \qquad L[p] = \int (p + \ddot{p})[d\theta] = \int_0^{2\pi} p(\theta)d\theta$$

$$(4.7) \qquad A[p] = \frac{1}{2} \int p(p + \ddot{p})[d\theta] = \frac{1}{2} \int_0^{2\pi} (p^2 - \dot{p}^2)d\theta.$$

$$[d\theta] : \text{ integral by a general Radon measure}$$

$$d\theta : \text{ Riemann- Lebesgue measure}$$

Even for non-convex case, we employ these formulas as representing the *oriented length* and the *oriented area* respectively.

To derive these formulas, we use the Leibniz formulas

$$(4.8) \qquad (fg)\dot{} = \dot{f}g + f\dot{g}.$$

This holds in the sense of $\mathcal{D}'$, if $f \in C^1$ and g is a function of bounded variation, i.e. $\dot{g}$ is a Radon measure.

Both functional $L[p]$ and $A[p]$ are continuous even on $B_{2,\sigma}$. If $\alpha \neq \dfrac{n-1}{n}\pi$ $(n \geq 2)$, then $\wp_\alpha^{\text{convex}}$ is compact in $B_{2,\sigma}$. Therefore they attain their maxima and minima. To obtain the concrete solutions of these or other types of functionals, we are led to difficult *variational problems*. Some of them have been solved but many remain unsolved.

5. Duality and dual curves.

We start with the case of convex curves with the generators $\in \wp_\alpha^{convex}$. Given a function $p \in \wp_\alpha^{convex}$, let Λ be the corresponding convex curve of constant angle α, and P be a point on the director circle C. Then, Λ spans a sector region of angle α at P (see the figure below).

Fig. 5.1

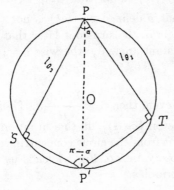

The two edge lines (rays) of the sector intersect with C at points S and T respectively. If we consider the point P' antipodal to P, i.e. the other end of the diameter of C passing P, then it is clear that the angle $SP'T = \hat{\alpha} = \pi - \alpha$. Now, let P move around the circle C, then P' also moves around C. In this situation, can or cannot the new sector of angle $\pi - \alpha$ with vertex P' with edge lines passing through S and T (respectively) be the sector corresponding to a curve $\hat{\Lambda}$ with a generator $\hat{p} \in \wp_{\hat{\alpha}}^{convex}$?

Remark. If $\alpha/\pi \notin \mathbf{Q}$, then $\hat{\alpha}/\pi \notin \mathbf{Q}$. Therefore $\wp_\alpha^{convex}, \wp_{\hat{\alpha}}^{convex}$ are the singletons $\{\tilde{c}_1(\alpha)\}$, $\{\tilde{c}_1(\hat{\alpha})\}$ [Theorem 3.2.]. In this case, we have trivially $\widehat{\tilde{c}_1(\alpha)} = \tilde{c}_1(\hat{\alpha})$.

Proposition 5.1. If $\alpha/\pi = \dfrac{m}{n}$ (irreducible fraction), $0 < \alpha \le \pi/2$, n : even (m : odd), then we have the duality mapping

$$\hat{} : \wp_\alpha^{convex} \ni p \to \hat{p} \in \wp_{\hat{\alpha}}^{convex}$$

with the following properties:
(i) $\hat{}$ is given by the formula

(5.1)
$$\hat{p}(\theta) = \sqrt{1 - p\left(\theta - \frac{\pi}{2}\right)^2} \, .$$

(ii) $\hat{}$ is an injective mapping such that $\hat{\hat{}}$ is the identity on $\wp_\alpha^{convex}$, i.e. $\hat{\hat{p}} = p$.
(iii) If $\alpha = \dfrac{\pi}{2}$, every element of $\wp_\alpha^{convex}$ is self-dual, i.e. $\hat{p} = p$. Therefore, $\hat{}$ is onto

mapping.

(iv)   *If* $0 < \alpha < \dfrac{\pi}{2}$ $\hat{\ }: \wp_\alpha^{convex} \to \wp_{\hat\alpha}^{convex}$ *is never onto.*

Remark. The last property (iv) shows that, if $\dfrac{\pi}{2} < \alpha < \pi$, there exist always an element $p \in \wp_\alpha^{convex}$ such that $\hat p$ defined by (5.1) is not the supporting function of a convex curve of constant angle $\hat\alpha$. It turns out later that this $\hat p$ can be the generator of a non-convex curve of constant angle $\hat\alpha$, otherwise $\hat p = (\hat p_1, \hat p_2)$ the generator of twin type curves (see below).

If $\alpha = \dfrac{m}{n}\pi$, n : odd, m : even, then $\hat\alpha = \dfrac{n-m}{n}\pi$. Thus both $n - m$ and n are odd. Therefore $\wp_{\hat\alpha}^{convex}$ is the singleton $\{\tilde c_1(\hat\alpha)\}$. In this situation, it is impossible, for a curve Λ of constant angle α different from the circle concentric with C, to have its dual curve $\hat\Lambda$ of single type. Natural observation of Fig 5.1 leads us to consider a pair functions $\hat p = (\hat p_1, \hat p_2)$ defined by the formulas

$$(5.2) \qquad \begin{cases} \hat p_1(\theta) = \sqrt{1 - p\big(\theta - \dfrac{\pi}{2}\big)^2} \\[2mm] \hat p_2(\theta) = \sqrt{1 - p\big(\theta + \dfrac{\pi}{2}\big)^2} \end{cases}$$

where p is the supporting function of Λ.

The pair of curves $\hat\Lambda = (\hat\Lambda_1, \hat\Lambda_2)$, $\hat\Lambda_j$ being the curve defined by the generator $\hat p_j$, would be the dual of Λ. This situation will be justified after we introduce the *admissible non-convex curves* of constant angle α of *twin type*. Then, it turns out that

$$(5.3) \qquad \hat p = (\hat p_1, \hat p_2) \in \wp_{\hat\alpha}^{twin}.$$

By the formula (5.2), $\hat p_2(\theta) = \hat p_1(\theta + \pi)$, thus $\hat\Lambda_2$ is the symmetric image of $\hat\Lambda_1$ with respect to the center of symmetry the origin $O = (0,0)$.

Conversely, given $p = (p_1, p_2) \in \wp_\alpha^{twin}$, thus satisfying

$$(5.4) \qquad p_2(\theta) = p_1(\theta + \pi),$$

its dual $\hat p = (\hat p_1, \hat p_2)$ is defined by

$$(5.5) \qquad \begin{cases} \hat p_1(\theta) = \sqrt{1 - p_2\big(\theta - \dfrac{\pi}{2}\big)^2} \\[2mm] \hat p_2(\theta) = \sqrt{1 - p_1\big(\theta + \dfrac{\pi}{2}\big)^2}. \end{cases}$$

By the formula (5.4) $\hat p_1(\theta) = \hat p_2(\theta)$. Thus we get a single function, therefore we identify the pair $\hat p = (\hat p_1, \hat p_2)$ with this single function, denote it again by the same letter $\hat p\,(= \hat p_1 = \hat p_2)$. Then, even $p \in \wp_\alpha^{convex}$, $\alpha = \dfrac{m}{n}\pi$ (m : odd $\quad n$: even), we get $\hat{\hat p} = p$. The similar argument shows that, starting with $p = (p_1, p_2) \in \wp_\alpha^{twin}$, $\alpha = \dfrac{m}{n}\pi(mn : \text{odd})$, we get $\hat{\hat p} = p$ by (5.4). (5.5) indicates the possibility of curves of constant angle α of double type and the space of generators $\wp_\alpha^{\text{double}}$ and the theory of them was developed by my student Miss. J. SHEN [23].

6. Admissible curves and their generators.

We are obliged to treat non-convex curves and to seek for something which plays the role of supporting lines to convex curves. They are the *generalized tangent lines* to admissible curves defined below.

Definition 6.1. *An admissible curve Λ and its generalized tangent lines $\ell(\theta), \theta \in \mathbf{T}$, are defined by the following three conditions:*

(1) Λ is a closed continuous curve parametrized by $\theta \in \mathbf{T}$

$$\begin{cases} x = x(\theta) \\ y = y(\theta) \end{cases}$$

(2) Λ is rectifiable, i.e. for any partion Δ of $[0, 2\pi]$

$$\Delta : 0 = \theta_0 < \theta_1 < \cdots < \theta_N = 2\pi$$

consider the length of the closed polygon connecting the points $Q_j = (x(\theta_j), y(\theta_j))$ in the natural order.

$$L_\Delta(\Lambda) = \sum_{j=1}^{N} \sqrt{\left(x((\theta_j) - x((\theta_{j-1}))\right)^2 + \left(y(\theta_j) - y((\theta_{j-1}))\right)^2}.$$

Then, these quantities are bounded as Δ runs over all partions. The length of Λ is defined by

(6.1) $$\sup_\Delta L_\Delta(\Lambda).$$

(3) For every fixed θ, consider the straight line ℓ_θ the equation of which takes the canonical form

(6.2) $$\ell_\theta : x \cos\theta + y \sin\theta = p(\theta).$$

Let $e_\theta = (\cos\theta, \sin\theta), \tilde{e}_\theta = (-\sin\theta, \cos\theta)$, e_θ represents the direction perpendicular to the line ℓ_θ, $\tilde{e}_\theta$ represents a direction parallel to ℓ_θ and determines the orientation (or direction) of ℓ_θ. $p(\theta)$ can be considered the distance between the origin O and the oriented line ℓ_θ.

We assume always the inequality

(6.3) $$p(\theta) > 0$$

holds for all θ. This is a non-trivial restriction on $p(\theta)$.

Further we assume that for every fixed θ_0 the line ℓ_{θ_0} is a generalized tangent line to the curve Λ at the point $(x(\theta_0), y(\theta_0))$ of Λ. Notice that, since Λ is not C^1–regular curve in general, the ordinary tangent lines may not exist. Thus we define the generalized *tangent line in the following way: the tangent vector at the point $((x(\theta_0), y(\theta_0))$, if it exists, should be $(\dot{x}(\theta_0), \dot{y}(\theta_0))$. Therefore, we require the condition*

$$(6.4) \qquad \dot{x}(\theta)\cos\theta + \dot{y}(\theta)\sin\theta = 0.$$

The derivatives $\dot{x}(\theta), \dot{y}(\theta)$ are taken in the sense of $\mathcal{D}'(\mathbf{T})$. Thus, (6.4) should be an equation in an open neighbourhood of the fix value θ_0. Since θ_0 is arbitrary, it is the same to require that the equation (6.4) holds globally.

Remark. *Since Λ is rectifiable, the functions $x(\theta), y(\theta)$ are of bounded variation. Therefore, $\dot{x}(\theta), \dot{y}(\theta)$ are Radon measures.*

Since the point $(x(\theta), y(\theta))$ is on the line ℓ_θ, we have

$$(6.5) \qquad x(\theta)\cos\theta + y(\theta)\sin\theta = p(\theta).$$

This equality shows that $p(\theta)$ is continuous in θ. Differentiating both sides of this equality, we get

$$\dot{p}(\theta) = -x(\theta)\sin\theta + y(\theta)\cos\theta + \dot{x}(\theta)\cos\theta + \dot{y}(\theta)\sin\theta.$$

Using (6.4) we get

$$(6.6) \qquad \dot{p}(\theta) = -x(\theta)\sin\theta + y(\theta)\cos\theta.$$

This shows that $\dot{p}(\theta)$ is continuous, i.e. $p \in C^1(\mathbf{T})$. Differentiating again both sides of (6.6), we get

$$\ddot{p}(\theta) = -x(\theta)\cos\theta - y(\theta)\sin\theta - \dot{x}(\theta)\sin\theta + \dot{y}(\theta)\cos\theta.$$

Thus we get

$$(6.7) \qquad p(\theta) + \ddot{p}(\theta) = -\dot{x}(\theta)\sin\theta + \dot{y}(\theta)\cos\theta.$$

This shows $p + \ddot{p}$ is a Radon measure on $\mathbf{T}$. From (6.5) and (6.6) we get

$$(6.8) \qquad \begin{pmatrix} p \\ \dot{p} \end{pmatrix} = T(-\theta) \begin{pmatrix} x \\ y \end{pmatrix}$$

or

$$(6.9) \qquad \begin{pmatrix} x(\theta) \\ y(\theta) \end{pmatrix} = T(\theta) \begin{pmatrix} p(\theta) \\ \dot{p}(\theta) \end{pmatrix},$$

where $T(\theta)$ is given by (4.2). This is just the parametric representation of Λ, and $p(\theta)$ is its generator (6.4). (6.7) gives that

$$\begin{pmatrix} 0 \\ p + \ddot{p} \end{pmatrix} = T(-\theta) \begin{pmatrix} \dot{x} \\ \dot{y} \end{pmatrix}$$

or

$$(6.10) \qquad \begin{pmatrix} \dot{x}(\theta) \\ \dot{y}(\theta) \end{pmatrix} = T(\theta) \begin{pmatrix} 0 \\ p + \ddot{p} \end{pmatrix}.$$

Therefore $\dot{x}, \dot{y}$ are Radon measures if $p + \ddot{p}$ is such and Λ is rectifiable. Thus we get

Theorem 6.1. *An admissible curve is a curve with a positive C^1 generator $p(\theta)$, such that $p + \ddot{p}$ is a Radon measure, i.e. $p \in \wp$.*

Remark. If the admissible curve Λ has a true tangent line at some point on it, that tangent line should coincide with some ℓ_θ, i.e. a generalized tangent line.

Proposition 6.2. Λ *is an admissible curve contained in C if and only if its generator $p \in \wp$ satisfies*

$$p(\theta)^2 + \dot{p}(\theta)^2 < 1.$$

7. Incoming and outgoing generalized tangent lines of admissible curves (to and from) a point P on the director circle C.

Let Λ be an admissible curve in the director circle C. Given θ, let ℓ_θ be the generalized tangent line directed (or oriented) by the vector $\tilde{e}_\theta = (-\sin\theta, \cos\theta)$.

Fig 7.1

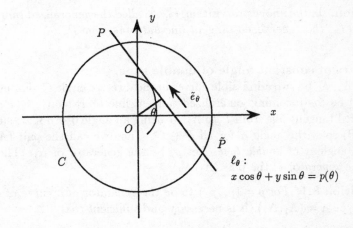

$$\ell_\theta :$$
$$x\cos\theta + y\sin\theta = p(\theta)$$

Let the intersecting points of ℓ_θ and C be P and $\tilde{P}$. P is the target point and $\tilde{P}$ be the source point of the directed segment $\overline{P\tilde{P}}$. For simplicity, this directed segment is also designated by ℓ_θ. We write the coordinates $P = (\cos\varphi, \sin\varphi)$, $\tilde{P} = (\cos\tilde{\varphi}, \sin\tilde{\varphi})$, P and $\tilde{P}$ being on ℓ_θ, the equation of ℓ_θ

$$x\cos\theta + y\sin\theta = p(\theta)$$

leads to the equations

(7.1) $$\begin{cases} \cos(\varphi - \theta) = p(\theta) \\ \cos(\tilde{\varphi} - \theta) = p(\theta). \end{cases}$$

Since $0 < \varphi - \theta < \pi$, $\pi < \tilde{\varphi} - \theta < 2\pi$, we have

(7.2) $$\begin{cases} \varphi = \theta + \mathrm{ACS}(p(\theta)) \\ \tilde{\varphi} = \theta + \pi - \mathrm{ACS}(p(\theta)) \end{cases}$$

where ACS is the *principal value* of the inverse cosine function $\cos^{-1}$. Thus we get

(7.3)
$$\begin{cases} \dot{\varphi} = 1 - \dfrac{\dot{p}}{\sqrt{1 - p^2}} & > 0 \\[3mm] \dot{\check{\varphi}} = 1 + \dfrac{\dot{p}}{\sqrt{1 - p^2}} & > 0 \end{cases}$$

because $p^2 + \dot{p}^2 < 1$. This result can be stated as

Proposition 7.1. *As the parameter θ goes around in the couter-clock-wise the unit circle $\mathbf{T}$, the target point P and the source point $\check{P}$ of ℓ_θ both go round C in the couter-clock-wise with speeds greater than a positive constant.*

In other words:

Proposition 7.2. *Given a point $P \in C$, there exist uniquely a pair of values (θ_1, θ_2) such that P is the target point of ℓ_{θ_1} and the source point of ℓ_{θ_2}.*

Definition. *In the above proposition, ℓ_{θ_1} is called the generalized tangent line incoming to P and ℓ_{θ_2} the generalized tangent line outgoing from P.*

8. Curves of constant angle of double type.

Let Λ_1, Λ_2 be two admissible curves in the director circle C. For every point $P \in C$, let $\ell_1(P)$ be the incoming generalized tangent line to P to Λ_1, $\ell_2(P)$ be the outgoing generalized tangent line from P to Λ_2. α being an angle $0 < \alpha < \pi$, suppose that $\ell_1(P)$ and $\ell_2(P)$ span the angle α for all $P \in C$. Then we call the pair (Λ_1, Λ_2) a curve of constant angle α of *double type*. Let p_j be the generator of Λ_j. Then $p = (p_1, p_2)$ is called the generator of the pair (Λ_1, Λ_2).

Proposition 8.1. *For $p = (p_1, p_2)$ to be the generator of a curve of constant angle of double type $\Lambda = (\Lambda_1, \Lambda_2)$, it is necessary and sufficient that*

(i) $p_j(\theta) \in J_\alpha, j = 1, 2 \quad (\theta \in \mathbf{T})$
(ii) $p_j(\theta)^2 + \dot{p}_j(\theta)^2 < 1, j = 1, 2 \quad (0 \in \mathbf{T})$
(iii) $p_2(\theta + \pi - \alpha) = \chi_\alpha(p_1(\theta)), \theta \in \mathbf{T}$
(iv) $p_j + \ddot{p}_j \in \mathcal{M}(\mathbf{T}), j = 1, 2 \quad$ (since $p_j \in \wp; \quad j = 1, 2$).

Definition 8.1. *We write $p = (p_1, p_2) \in \wp_\alpha^{double}$ if the conditions of Prop.8.1 are satisfied.*

Definition 8.2. *If $p_1(\theta) \equiv p_2(\theta)$, i.e. $\Lambda_1 = \Lambda_2$, then the curve $\Lambda(= \Lambda_1 = \Lambda_2)$ is called a curve of constant angle α of single type and we write p, instead of (p, p) : Notation $\wp_\alpha^{single}$.*

Definition 8.3. *If $p_2(\theta) = p_1(\theta + \pi)$, we call $\Lambda = (\Lambda_1, \Lambda_2)$ a curve of constant angle α of twin type : Notation $\wp_\alpha^{twin}$.*

Definition 8.4. *If $p = (p_1, p_2) \in \wp_\alpha^{double}$ and $\overset{\vee}{p} = (p_2, p_1) \in \wp_\alpha^{double}$, we say that $p = (p_1, p_2)$ or $\Lambda = (\Lambda_1, \Lambda_2)$ is symmetric double type and write $p \in \wp_\alpha^{double,sym}$.*

Remark. $\wp_\alpha^{twin} \subseteq \wp_\alpha^{double,sym}$.

9. Linearization by $\tilde{\chi}_\alpha$, space $\mathcal{W}_\alpha^{double}$ and the duality.

Since $\wp_\alpha^{single} \subseteq \wp_\alpha^{double}, \wp_\alpha^{twin} \subseteq \wp_\alpha^{double}$, we explain the situation for the most general case, i.e. the case of double type.

We define the space $\mathcal{W}_\alpha^{double}$ by

$$(9.1) \qquad \mathcal{W}_\alpha^{double} = \{w = (w_1, w_2); w_j = \tilde{\chi}_\alpha(p_j), j = 1, 2, p = (p_1, p_2) \in \wp_\alpha^{double}\}$$

when $p = (p_1, p_2) \in \wp_\alpha^{double}$, its dual $\hat{p} = (\hat{p}_1, \hat{p}_2) \in \wp_{\hat{\alpha}}^{double}$ is defined as in the case of twin type

$$(9.2) \qquad \begin{cases} \hat{p}_1(\theta) = \sqrt{1 - p_2(\theta - \frac{\pi}{2})^2} \\ \hat{p}_2(\theta) = \sqrt{1 - p_1(\theta + \frac{\pi}{2})^2} \ . \end{cases}$$

And define $\hat{w} = (\hat{w}_1, \hat{w}_2)$ by

$$(9.3) \qquad \hat{w}_j = \tilde{\chi}_{\hat{\alpha}}(\hat{p}_j), \ j = 1, 2.$$

Easy calculations show

Theorem 9.1. $\hat{w}_j$ are calculated as follows:

$$(9.4) \qquad \begin{cases} \hat{w}_1(\theta) = -w_2(\theta - \frac{\pi}{2}) \\ \hat{w}_2(\theta) = -w_1(\theta + \frac{\pi}{2}) \ . \end{cases}$$

And it is clear that $\hat{\hat{w}} = w$.

Remark 1. $\hat{\alpha} = \pi - \alpha$, $J_{\hat{\alpha}}$ is different from J_α in general. But $I_{\hat{\alpha}} = I_\alpha$ always. This simplifies the situation.

Remark 2. It is easy to see that the union of curves of constant angle of single type and of twin type is also *stable* under the duality map $\hat{\ }$.

10. Geometry of admissible curves.

Let Λ be an admissible curve. Decompose the open set $\mathbf{R}^2 \setminus \Lambda$ into its connected components. The unbounded component U_∞ is the outside of Λ. The component containing the origin O, denoted by U_0, is the *core* of Λ.

Theorem 10.1. U_0 *is a convex set and it is the interior of the compact convex set* Δ *which is defined by*

$$\Delta = \bigcap_{\theta \in \mathbf{T}} \Delta(\theta)$$

where $\Delta(\theta)$ *is the closed half plane defined by*

$$\Delta(\theta) = \{(x, y); x \cos\theta + y \sin\theta \leq p(\theta)\},$$

$p(\theta)$ being the generator of Λ.

Remark. $\Lambda^0 = \partial\Delta$ is a closed convex curve, its supporting function is denoted by $p^0(\theta)$.

When p is real analytic, a corner point of Λ^0 is a self-intersection point of Λ. Such points are finite in number.

To each corner point of Λ^0, a swallow tail (possibly of complicated shape) is attached.

Proposition 10.2. *(i) The oriented length of swallow tails are always non-negative. (ii) The oriented area of swallow tails are always non-positive.*

Remark. For general generator $p \in \wp$, we may approximate by C^∞ generator by regularization. The latter again approximated by analytic generator by using its Fourier expansion, and exploiting a suitable initial finite sum. Thus we can reduce the problem to anlytic case.

11. Non oriented length $|L|[p]$ and non-oriented area $|A|[p]$.

The length of an admissible curve is defined by (6.1). But more manageable new formula is given by

Theorem 11.1. *For a rectifiable closed continuous curve Λ*

$$\begin{cases} x = x(\theta) \\ y = y(\theta) \end{cases}$$

the (non-oriented) length of Λ can be given by the formula

$$(11.1) \qquad |L|[\Lambda] = \sup_{\substack{\varphi^2+\psi^2\leq 1 \\ \varphi,\psi \in C(\mathbf{T})}} (\dot{x}[\varphi] + \dot{y}[\psi]).$$

Remark. This is a *new formula of arc-length* and it can easily be generalized for a curve in $\mathbf{R}^n$. Here, $C(\mathbf{T})$ can be replaced by $C^\infty(\mathbf{T})$ since the latter is dense in the former.

Using this new formula, we get

Proposition 11.2. *For a admissible curve Λ with generator p, the length $|L|[p]$ is given by the formula*

$$(11.2) \qquad |L|[p] = \|p + \ddot{p}\|_{\mathcal{M}(\mathbf{T})}.$$

Remark. This functional $|L|[p]$ is not continuous on $\wp$, but lower semicontinuous. For any $\alpha \in \mathbf{Q}, 0 < \alpha < \pi, |L|[p]$ is unbounded on $\wp_\alpha^{\text{single,twin or double}}$.

Using the results in the preceding section, we define the non-oriented area $|A|[p]$ by

$$(11.3) \qquad |A|[p] = 2A[p^0] - A[p]$$

where p^0 is the supporting function of Λ^0 (or of the convex core $\overset{\circ}{\Delta}$).

Remark. This functional $|A|[p]$ is continuous in $\wp$.

By constructing explicitly an extremalizing sequence $p_k \in \wp_\alpha^{\text{single}}$ $k = 1, 2, \cdots$, we have

Theorem 11.3. $\alpha/\pi = \dfrac{m}{n}$ *(irreducible fraction)*, $mn = $ even, $0 < \alpha \leq \pi/2$. *Then we have*

(11.4)
$$\sup_{p \in \wp_\alpha^{\text{single}}} |A|[p] = \pi.$$

Remark. Every curve with generator $p \in \wp_\alpha^{\text{single}}$ is contained in C. But π is the area of C. Thus, this result may be somewhat amazing. In the case $\pi/2 < \alpha < \pi$, the problem remains unsolved.

12. Local regularity of solutions to simultaneous relations.

To simplify the situations, we first limit ourselves to the case of single type. Let $u(\theta)$ be an unknown function. In order for u to be the supporting function of a convex curve of constant angle α with respect to the director circle C (its radius being 1), it is necessary and sufficient to satisfy the following four conditions (Theorem 3.1).

(1°) $u(\theta)$ is a continuous function with period 2π
(2°) $u(\theta) \in J_\alpha$ for all $\theta \in \mathbf{R}$
(3°) $u(\theta + \pi - \alpha) = \chi_\alpha(u(\theta))$
(4°) $u + \ddot{u} \geq 0$.

Then, what can we say about the local regularity of u?

Theorem 12.1. *If a distribution $u \in \mathcal{D}'(\mathbf{R})$ satisfies the simultaneous relations* (1°), (2°), (3°), (4°), *then $u \in C^1(\mathbf{R})$ and $\ddot{u} \in L^\infty(\mathbf{R})$. This result is the best possible one as the local regularity of u for general angle α.*

Remark 1. The same results hold for the twin type curves if they are convex.
Remark 2. Thus, there exist strictly convex curves whose radius of curvature is everywhere essentially discontinuous.

For non-convex curves of constant angle, the local regularity is weaker.
The characterizing conditions are the following five ones.

(1*) $u(\theta)$ is a C^1-function with period 2π
(2*) $u(\theta) \in J_\alpha$ for all $\theta \in \mathbf{R}$
(3*) $u(\theta)^2 + \dot{u}(\theta)^2 < 1$
(4*) $u(\theta + \pi - \alpha) = \chi_\alpha(u(\theta))$
(5*) $u + \ddot{u} \in \mathcal{M}(\mathbf{R})$

Then, what we can say?

Theorem 12.2. *If an element $u \in \mathcal{D}'(\mathbf{R})$ satisfies the simultaneous relations* (1*), (2*), (3*), (4*), (5*), *then $u \in C^1(\mathbf{R})$ and $\ddot{u}$ is an atom-free Radon measure. This result is the best possible one as the local regularity of u for general angle α.*

Remark 1. "atom-free" means that every one point set $\{\theta\}$ is of measure 0 with respect to $\ddot{u}$.
Remark 2. The same results hold for the twin type curves if they are non-convex.
Remark 3. Thus, there exist non-convex curves whose oriented radius of curvature is an unbounded L^1 function, a continuous singular measure or their sum.

13. Examples.

Because our subjects are special kind of plane curves, it would not be good for us to state only theoretical facts without giving examples. Thus we give in this section two types of examples. .

(1) We give here a few examples of convex and non-convex curves of constant angle α for $\alpha = \dfrac{\pi}{4}$ and for $\alpha = \dfrac{\pi}{6}$. The convex ones are embeded in the family of non-convex ones.

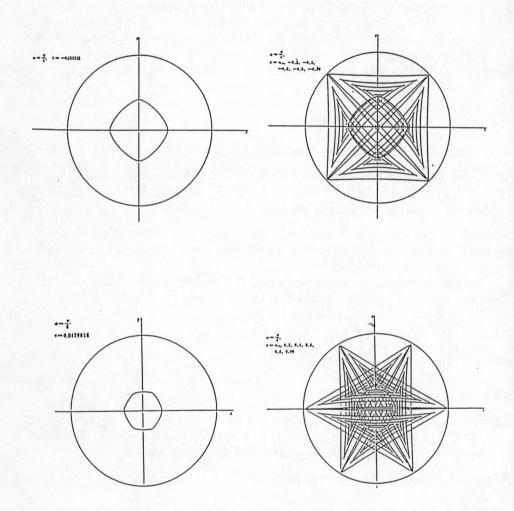

(2) Here we give two examples of a curve of constant angle $\alpha = \dfrac{2}{3}\pi$ of single type and its dual curves of constant angle $\alpha = \dfrac{1}{3}\pi$ of twin type of convex and nonconvex type.

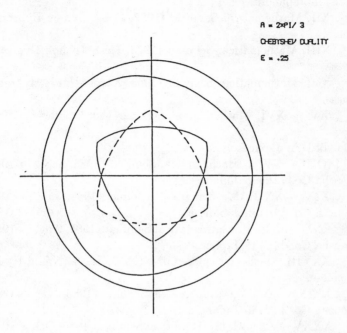

A = 2*PI/ 3
CHEBYSHEV DUALITY
E = .25

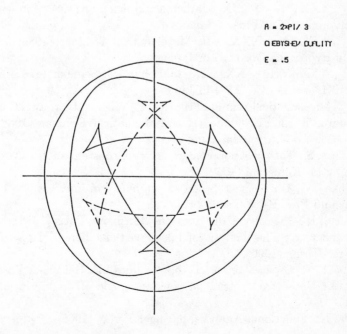

A = 2*PI/ 3
CHEBYSHEV DUALITY
E = .5

References

[1] Matsuura, S., I~ VII, Mathematical Seminar (1981, August ~ 1982, Feb.) (Nihon Hyōronsha Co. Ltd) in Japanese.

[2] _____, VIII~ XI, Mathematical Seminar (1982, July ~ 1982, Oct.) (Nihon Hyōronsha Co. Ltd) in Japanese.

[3] _____, XII, Mathematical Seminar (1982, Dec.) (Nihon Hyōronsha Co. Ltd) in Japanese.

[4] _____, XIII, Mathematical Seminar (1983, Feb.) (Nihon Hyōronsha Co. Ltd) in Japanese.

[5] _____, XVI, Mathematical Seminar (1983, Apr.) (Nihon Hyōronsha Co. Ltd) in Japanese.

[6] _____, XV ~ XVI, Mathematical Seminar (1983, June ~ 1983, July) (Nihon Hyōron
sha Co. Ltd) in Japanese.

[7] _____, XVII ~ XVIII, Mathematical Seminar (1983, Sept. ~ 1983, Oct.) (Nihon Hyōronsha Co. Ltd) in Japanese.

[8] _____, XIX ~ XXIII, Mathematical Seminar (1984, Jan. ~ 1984, May)(Nihon Hyōronsha Co. Ltd) in Japanese.

[9] _____, XXIV ~ XXVII, Mathematical Seminar (1984, Aug. ~ 1984, Nov.) (Nihon Hyōronsha Co. Ltd) in Japanese.

[10] _____, XXVIII, Mathematical Seminar (1985, Jan.) (Nihon Hyōronsha Co. Ltd) in Japanese.

[11] _____, XXIX ~ XXXIV, Mathematical Seminar (1985, Mar. ~1985, Aug.) (Nihon Hyōronsha Co. Ltd) in Japanese.

[12] _____, XXXV ~ XXXVI, Mathematical Seminar (1985, Oct. ~1985, Nov.) (Nihon Hyōronsha Co. Ltd) in Japanese.

[13] _____, XXXVII ~ XXXX, Mathematical Seminar (1986, Jan. ~1986, Apr.) (Nihon Hyōronsha Co. Ltd) in Japanese.

[14] _____, XXXXI ~ XXXXVII, Mathematical Seminar (1986, June ~1986, Dec) (Nihon Hyōronsha Co. Ltd) in Japanese.

[15] _____, XXXXVIII ~ XXXXIX, Mathematical Seminar (1987, Feb.~1987, Mar.) (Nihon Hyōronsha Co. Ltd) in Japanese.

[16] _____, Mathematical Seminar (1987, Apr.) (Nihon Hyōronsha Co. Ltd) in Japanese

[17] Matsuura, S. and Kasahara, K., Mathematical Seminar (1988, Mar.) (Nihon Hyōronsha Co. Ltd) in Japanese.

[18] Matsuura, S., Theory of non-convex curves of constant angle (22 expositions), Notes taken by M. Kametani (1990).

[19] Blaschke, W., Über eine Ellipsen eigenschaft und über gewisse Eilinien, Ark. der Maths und Phy. III, Reihe 26 (1917).

[20] Schwartz, L., Théorie des distributions, Hermann, Paris (1966).

[21] Hörmander, L., The analysis of linear partial differential operators I (2nd ed.), Springer-Verlag (1990).

[22] Bourbaki, N., Espaces vectoriels topologiques I,II, Hermann, Paris (1966).

[23] Shen, J., Non-convex curves of constant angle of double type (Master Thesis) (1991).

[24] Yosida, K., Functional Analysis, Springer-Verlag (1968).

Asymptotic behavior of weak solutions of the convection equation

MORIMOTO, Hiroko

School of Science and Technology

Meiji University

Let Ω be a bounded domain in $\mathbf{R}^n (2 \leq n \leq 4)$ with the boundary $\partial\Omega$ satisfying

$$\partial\Omega = \Gamma_1 \cup \Gamma_2, \Gamma_1 \cap \Gamma_2 = \phi.$$

We consider the following initial boundary value problem:

$$\begin{cases} \dfrac{\partial u}{\partial t} + (u \cdot \nabla)u = -\dfrac{1}{\rho}\nabla p + \nu\Delta u + \beta g\theta, \\ \mathrm{div}\, u = 0, \qquad\qquad\qquad x \in \Omega,\ t > 0, \\ \dfrac{\partial\theta}{\partial t} + (u \cdot \nabla)\theta = \chi\Delta\theta, \end{cases} \tag{1}$$

$$\begin{cases} u(x,t) = 0, \quad \theta(x,t) = \xi(x,t), & x \in \Gamma_1,\ t > 0, \\ u(x,t) = 0, \quad \dfrac{\partial}{\partial n}\theta(x,t) = \eta(x,t), & x \in \Gamma_2,\ t > 0, \end{cases} \tag{2}$$

$$\begin{cases} u(x,0) = a_0(x), \\ \theta(x,0) = \tau_0(x), \end{cases} \qquad x \in \Omega, \tag{3}$$

where $u = (u_1, u_2, \ldots, u_n)$ is the fluid velocity, p is the pressure, θ is the temperature, $u \cdot \nabla = \sum_{j=1}^{n} u_j \dfrac{\partial}{\partial x_j}$, and $\dfrac{\partial\theta}{\partial n}$ denotes the outer normal derivative of θ at x to $\partial\Omega$, g is the gravitational vector, and ρ(density), ν(kinematic viscosity), β(coefficient of volume expansion), χ(thermal diffusivity) are positive constants. $\xi(x,t)$ (resp.$\eta(x,t)$) is a function defined on $\Gamma_1 \times [0, \infty)$ (resp. $\Gamma_2 \times [0, \infty)$), and $a_0(x)$ (resp. $\tau_0(x)$) is a vector (resp. scalar) function defined on Ω.

This system of equations describes the motion of fluid of heat convection (Boussinesq approximation). The stationary problem corresponding to it is as follows:

$$\begin{cases} (u \cdot \nabla)u = -\dfrac{1}{\rho}\nabla p + \nu\Delta u + \beta g\theta \\ \mathrm{div}\, u = 0, \qquad\qquad \text{in } \Omega, \\ (u \cdot \nabla)\theta = \chi\Delta\theta, \end{cases} \tag{4}$$

$$\begin{cases} u = 0, \quad \theta = \xi, & \text{on } \Gamma_1, \\ u = 0, \quad \dfrac{\partial \theta}{\partial n} = \eta, & \text{on } \Gamma_2, \end{cases} \tag{5}$$

where $\xi = \xi(x)$ (resp. $\eta = \eta(x)$) is a given function defined on Γ_1 (resp. Γ_2).

We studied the stationary problem in [2], [3], the evolutional problem in [4]. The existence of periodic solution was also studied in [4]. In this paper, the asymptotic behavior of the weak solution of evolutional equation will be studied, and we show that, in 2 dimensional case, a stationary solution is asymptotically stable(Theorem 1 and its Corollary). We obtain the similar result for periodic problem(Theorem 2, 3).

In order to state our results, we introduce some

Function spaces ([2],[3],[4]).

$L^p(\Omega)$ and the Sobolev space $W_p^\ell(\Omega)$ are defined as usual. We also denote $H^\ell(\Omega) = W_2^\ell(\Omega)$. Whether the elements of the space are scalar or vector functions is understood from the contexts unless stated explicitly.

$D_\sigma = \{$vector function $\varphi \in C^\infty(\Omega)\,|\text{supp}\varphi \subset \Omega, \text{div}\varphi = 0 \text{ in } \Omega\}$,

$H = $ completion of D_σ under the $L^2(\Omega)$-norm,

$V = $ completion of D_σ under the $H^1(\Omega)$-norm,

$D_0 = \{$scalar function $\varphi \in C^\infty(\overline{\Omega})\,|\varphi \equiv 0 \text{ in a neighborhood of } \Gamma_1\}$,

$W = $ completion of D_0 under the $H^1(\Omega)$-norm,

V', W' are dual spaces of V, W.

Definition 1

$\{u, \theta\}$ is called a weak solution of evolutional problem (1), (2) if, for some function θ_0 such that
$$\theta_0 \in L^2(0, T : H^1(\Omega)), \quad \theta_0 = \xi \text{ on } \Gamma_1,$$

$\{u, \theta\}$ satisfies following conditions:

$$u \in L^2(0, T : V), \quad \theta - \theta_0 \in L^2(0, T : W),$$

$$\begin{cases} \dfrac{d}{dt}(u, v) + \nu(\nabla u, \nabla v) + ((u \cdot \nabla)u, v) - (\beta g\theta, v) = 0, & \forall v \in V, \\ \dfrac{d}{dt}(\theta, \tau) + \chi(\nabla\theta, \nabla\tau) + ((u \cdot \nabla)\theta, \tau) - \chi(\eta, \tau)_{\Gamma_2} = 0, & \forall \tau \in W, \end{cases} \tag{6}$$

where
$$(\eta, \tau)_{\Gamma_2} = \int_{\Gamma_2} \eta(x')\tau(x')d\sigma.$$

Definition 2

$\{u_\infty, \theta_\infty\}$ is called a weak solution of stationary problem (4),(5) if, for some function θ_0 in $H^1(\Omega)$ such that $\theta_0 = \xi$ on Γ_1, $\{u_\infty, \theta_\infty\}$ satisfies the following:

$$u_\infty \in V \text{ and } \theta_\infty - \theta_0 \in W,$$

$$\begin{cases} \nu(\nabla u_\infty, \nabla v) + ((u_\infty \cdot \nabla)u_\infty, v) - (\beta g\theta_\infty, v) = 0, \ \forall v \in V, \\ \chi(\nabla\theta_\infty, \nabla\tau) + ((u_\infty \cdot \nabla)\theta_\infty, \tau) - \chi(\eta, \tau)_{\Gamma_2} = 0, \ \ \forall\tau \in W. \end{cases} \quad (7)$$

For the proof of existence of weak solutions, we suppose the following condition.
Condition(H).
$\partial\Omega$ is of class C^1 and

$$\partial\Omega = \Gamma_1 \cup \Gamma_2, \Gamma_1 \cap \Gamma_2 = \phi, \text{ measure of } \Gamma_1 \neq 0,$$

and $\overline{\Gamma_1} \cap \overline{\Gamma_2}$ is an $n-2$ dimensinal C^1 manifold.

Remark 1([4])
(i) We suppose $2 \leq n \leq 4$, Condition(H), $\xi \in C^1(\overline{\Gamma_1} \times [0,T])$, $\eta \in L^2(\Gamma_2 \times (0,T))$, $g \in L^\infty(\Omega \times (0,T))$, $a_0 \in H$, $\tau_0 \in L^2(\Omega)$. Then there exists a weak solution $\{u,\theta\}$ of evolutional problem (1), (2) satisfying the initial condition (3). Furthermore

$$u \in L^\infty(0,T:H), \ \theta \in L^\infty(0,T:L^2(\Omega)).$$

(ii) For $n = 2$, the weak solution $\{u,\theta\}$ of evolutional problem (1), (2) satisfying the initial condition (3) is unique. Furthermore, $u \in C([0,T]:H)$ and $\theta \in C([0,T]:L^2(\Omega))$ holds.

Remark 2
For any $n \geq 2$, the existence of weak solutions of stationary problem (4), (5) is known ([2],[3]) under the condition (H).

Theorem 1
Let $2 \leq n \leq 4$. Let ξ, η, g depend only on x, $\{u,\theta\}$ be a weak solution of (1), (2), and $\{u_\infty, \theta_\infty\}$ be that of (4), (5). We suppose:

$$\frac{du}{dt} \in L^2(0,T:V'), \frac{d\theta}{dt} \in L^2(0,T;W') \text{ for } \forall T > 0, \quad (8)$$

$$c\|u_\infty\|_p + \frac{1}{4\chi}(c\|\theta_\infty\|_p + c'\beta g_\infty)^2 < \nu, \quad (9)$$

where $p > 2$ if $n = 2$, $p = n$ if $n = 3, 4$, c and c' are constants depending only on Ω, g_∞ is the L^∞-norm of the vector g. Then $\|u(t) - u_\infty\|^2 + \|\theta(t) - \theta_\infty\|^2$ decays exponentially as t tends to $+\infty$.

Remark 3
(i) For $n = 2$, (8) holds ([4]).
(ii) For any $n \geq 2$, the weak solution of (4), (5) satisfying the condition (9) is unique ([3]).

Corollary
Let $n = 2$, $\xi \in C^1(\overline{\Gamma_1})$, $\eta \in L^2(\Gamma_2)$, $g \in L^\infty(\Omega)$. Ω satisfies the Condition (H). Then the weak solution of (4), (5) satisfying (9) is asymptotically stable.

Definition 3
$\{u,\theta\}$ is called a periodic weak solution of (1), (2) with period T_0, if $\{u,\theta\}$ is a weak solution of (1), (2) for $T = T_0$ satisfying

$$u(x, T_0) = u(x, 0), \quad \theta(x, T_0) = \theta(x, 0). \tag{10}$$

Remark 4([4])

Let Ω be a bounded domain in R^2 satisfying Condition (H). Let $\xi(x,t), \eta(x,t), g(x,t)$ be periodic with respect to t with period T_0, satisfying $\xi \in C^1(\overline{\Gamma}_1 \times [0, T_0])$, $\eta \in L^2(\Gamma_2 \times (0, T_0))$ and $g \in L^\infty(\Omega \times (0, T_0))$. Set $g_\infty = \|g\|_{L^\infty(\Omega \times (0, T_0))}$. If $\dfrac{\beta g_\infty}{\sqrt{\nu \chi}}$ is sufficiently small, then there exists a periodic weak solution of (1), (2) with period T_0. Furthermore

$$u \in C([0, \infty) : H), \ \theta \in C([0, \infty) : L^2(\Omega)).$$

Theorem 2

Let $n = 2$, $\{u_\pi, \theta_\pi\}$ be a periodic weak solution of (1), (2) with period T_0 such that for some $p > 2$,

$$\operatorname*{ess.sup}_t \{c\|u_\pi(t)\|_p + \frac{1}{4\chi}(c\|\theta_\pi(t)\|_p + c'\beta g_\infty)^2\} < \nu \tag{11}$$

where c and c' are constants depending on Ω. If $\{u_\pi + u, \theta_\pi + \theta\}$ is a periodic weak solution of (1), (2) with period T_0, then $u = 0, \theta = 0$.

Let $n = 2$ and $\xi \in C^1(\overline{\Gamma}_1 \times [0, \infty))$, $\eta \in L^2(\Gamma_2 \times (0, \infty))$, $g \in L^\infty(\Omega \times (0, \infty))$, $a_0 \in H, \tau_0 \in L^2(\Omega)$. Let T be any positive number. Then there exists one and only one weak solution $\{u_T, \theta_T\}$ of (1),(2) satisfying (3). Therefore, for $T < T'$,

$$u_T(t) = u_{T'}(t), \quad \theta_T(t) = \theta_{T'}(t) \quad \text{for } \forall t \in (0, T)$$

hold, and we can omit T.

Theorem 3

Let $n = 2$ and ξ, η, g be periodic with respect to t with period T_0. Let $\{u_\pi, \theta_\pi\}$ be a periodic weak solution satisfying (11). Then $\{u_\pi, \theta_\pi\}$ is asymptotically stable.

Remark 5

(i) Since $u_\pi \in L^2(0, T : V) \cap C([0, T] : H)$, u_π belongs to the space $L^{2p/(p-2)}(0, T : \mathbf{L}^p(\Omega))$ for $\forall p > 2$. Similarly θ_π is in $L^{2p/(p-2)}(0, T : L^p(\Omega))$. The condition (11) is stronger than this one.

(ii) When (11) holds, such periodic solution is unique (Theorem 2).

Proof of Theorem 1

Let $w = u - u_\infty, \zeta = \theta - \theta_\infty$. According to our assumption,

$$\frac{dw}{dt} = \frac{du}{dt} \in L^2(0, T : V'), \quad \frac{d\zeta}{dt} = \frac{d\theta}{dt} \in L^2(0, T : W')$$

hold for $\forall T > 0$. Subtracting (7) from (6), we have:

$$\frac{d}{dt}(w, v) + \nu(\nabla w, \nabla v) + ((u \cdot \nabla)u - (u_\infty \cdot \nabla)u_\infty, v) = (\beta g \zeta, v), \quad \forall v \in V, \tag{12}$$

$$\frac{d}{dt}(\zeta, \tau) + \chi(\nabla \zeta, \nabla \tau) + ((u \cdot \nabla)\theta - (u_\infty \cdot \nabla)\theta_\infty, \tau) = 0, \quad \forall \tau \in W. \tag{13}$$

Since we suppose (8), $w(\text{resp.}\zeta)$ is continuous from $[0,T]$ to H (resp. $L^2(\Omega)$) and

$$\frac{d}{dt}\|w(t)\|^2 = 2 < \frac{dw}{dt}, w > \quad (\text{resp.}\frac{d}{dt}\|\zeta(t)\|^2 = 2 < \frac{d\zeta}{dt}, \zeta >)$$

holds where $< \cdot, \cdot >$ denotes the duality of V' and V (resp. W' and W) ([6]). Therefore we can take $v = w$ and $\tau = \zeta$ in (12), (13), and obtain:

$$\begin{aligned}
\frac{1}{2}\frac{d}{dt}&\|w(t)\|^2 + \nu\|\nabla w(t)\|^2 \\
&= -((w\cdot\nabla)u, w) - ((u_\infty\cdot\nabla)w, w) + (\beta g\zeta, w) \\
&= ((w\cdot\nabla)w, u) + (\beta g\zeta, w) \\
&= ((w\cdot\nabla)w, u_\infty) + (\beta g\zeta, w),
\end{aligned} \tag{14}$$

$$\begin{aligned}
\frac{1}{2}\frac{d}{dt}&\|\zeta(t)\|^2 + \chi\|\nabla\zeta(t)\|^2 \\
&= -((w\cdot\nabla)\theta, \zeta) = ((w\cdot\nabla)\zeta, \theta) = ((w\cdot\nabla)\zeta, \theta_\infty).
\end{aligned} \tag{15}$$

We prove the theorem for $n = 3, 4$. For $n = 2$, the proof is similar and is omitted. Let us estimate the right hand side of (14), (15), using Hölder's inequality, and Poincaré's inequality, and we have:

$$\begin{aligned}
\frac{1}{2}\frac{d}{dt}&\|w(t)\|^2 + \nu\|\nabla w(t)\|^2 \\
&\leq \|w\|_{2n/(n-2)}\|\nabla w\|\,\|u_\infty\|_n + \beta g_\infty\,\|\zeta\|\,\|w\| \\
&\leq c\|\nabla w\|^2\|u_\infty\|_n + \beta g_\infty c'\|\nabla\zeta\|\,\|\nabla w\|,
\end{aligned} \tag{16}$$

$$\begin{aligned}
\frac{1}{2}\frac{d}{dt}&\|\zeta(t)\|^2 + \chi\|\nabla\zeta(t)\|^2 \\
&\leq \|w\|_{2n/(n-2)}\|\nabla\zeta\|\,\|\theta_\infty\|_n \leq c\|\nabla w\|\,\|\nabla\zeta\|\,\|\theta_\infty\|_n,
\end{aligned} \tag{17}$$

where c and c' are constants depending on Ω. Suppose

$$\nu - c\|u_\infty\|_n > 0, \tag{18}$$

$$(\beta g_\infty c' + c\|\theta_\infty\|_n)^2 - 4\chi(\nu - c\|u_\infty\|_n) < 0 \tag{19}$$

hold. The following bilinear form

$$(\nu - c\|u_\infty\|_n)\|\nabla w\|^2 + \chi\|\nabla\zeta\|^2 - (\beta g_\infty c' + c\|\theta_\infty\|_n)\|\nabla w\|\|\nabla\zeta\| \tag{20}$$

is then positive definite, and we can find positive constants α and α^* such that

$$\text{bilinear form (20)} \geq \alpha^*\{\|\nabla w\|^2 + \|\nabla\zeta\|^2\} \geq \alpha\,\{\|w\|^2 + \|\zeta\|^2\}$$

holds. Thereby we have

$$\frac{1}{2}\frac{d}{dt}\{\|w\|^2 + \|\zeta\|^2\} + \alpha\{\|w\|^2 + \|\zeta\|^2\} \leq 0. \tag{21}$$

Integrating this inequality, we obtain the desired decay property. The condition (19) is equivalent to (9), and implies (18). The theorem is proved.

Proof of Corollary

Let $\{u_\infty, \theta_\infty\}$ be a weak solution of (4), (5) satisfying (9). Since $n = 2$, there exists a global weak solution $\{u, \theta\}$ of (1), (2) satisfying the initial condition (3), where $a_0 \in H$, $\tau_0 \in L^2(\Omega)$, and (8) holds (Remark 3). Then, by Theorem 1, we know there exists a positive constant α such that

$$\|u(t) - u_\infty\|^2 + \|\theta(t) - \theta_\infty\|^2 \leq e^{-2\alpha t}(\|a_0 - u_\infty\|^2 + \|\tau_0 - \tau_\infty\|^2), \quad t > 0.$$

Corollary is proved.

Proof of Theorem 2

Since $\{u_\pi, \theta_\pi\}$ $\{u_\pi + u, \theta_\pi + \theta\}$ are periodic weak solutions of (1) (2), $\{u, \theta\}$ satisfies

$$\begin{cases} \dfrac{d}{dt}(u, v) + (((u_\pi + u) \cdot \nabla)(u_\pi + u), v) - ((u_\pi \cdot \nabla)u_\pi, v) \\ \qquad = -\nu(\nabla u, \nabla v) + (\beta g\theta, v) \qquad\qquad \forall v \in V, \\ \dfrac{d}{dt}(\theta, \tau) + (((u_\pi + u) \cdot \nabla)(\theta_\pi + \theta), \tau) - ((u_\pi \cdot \nabla)\theta_\pi, \tau) \\ \qquad = -\chi(\nabla\theta, \nabla\tau) \qquad\qquad \forall \tau \in W, \end{cases}$$

Since $n = 2$, $\dfrac{du}{dt}$ is in $L^2(0, T : V')$ and $\dfrac{d\theta}{dt}$ in $L^2(0, T : W')$. Therefore

$$< \frac{d}{dt}u, u > = \frac{1}{2}\frac{d}{dt}\|u\|^2, \qquad < \frac{d}{dt}\theta, \theta > = \frac{1}{2}\frac{d}{dt}\|\theta\|^2$$

hold, where $< , >$ denotes the duality between V' and V (resp.W' and W). Since $u \in V$, $\theta \in W$, we have

$$\frac{1}{2}\frac{d}{dt}\|u\|^2 + ((u_\pi \cdot \nabla)u, u) + ((u \cdot \nabla)(u_\pi + u), u) = -\nu\|\nabla u\|^2 + (\beta g\theta, u)$$

$$\frac{1}{2}\frac{d}{dt}\|\theta\|^2 + ((u_\pi \cdot \nabla)\theta, \theta) + ((u \cdot \nabla)(\theta_\pi + \theta), \theta) = -\chi\|\nabla\theta\|^2.$$

Therefore

$$\frac{1}{2}\frac{d}{dt}\|u\|^2 + \nu\|\nabla u\|^2 = ((u \cdot \nabla)u, u_\pi) + (\beta g\theta, u)$$

$$\frac{1}{2}\frac{d}{dt}\|\theta\|^2 + \chi\|\nabla\theta\|^2 = ((u \cdot \nabla)\theta, \theta_\pi)$$

Summing each side of these equations, and using Sobolev's inequality and Poincaré's inequality, we have:

$$\frac{1}{2}\frac{d}{dt}(\|u\|^2 + \|\theta\|^2) + \nu\|\nabla u\|^2 + \chi\|\nabla\theta\|^2$$
$$\leq \|u_\pi\|_p \|u\|_{p'}\|\nabla u\| + \beta g_\infty\|\theta\| \|u\| + \|\theta_\pi\|_p\|u\|_{p'}\|\nabla\theta\|$$
$$\leq c\|u_\pi\|_p\|\nabla u\|^2 + c'\beta g_\infty\|\nabla\theta\| \|\nabla u\| + c\|\theta_\pi\|_p\|\nabla\theta\| \|\nabla u\|$$

where c and c' are constants depending only on Ω and $1/p + 1/p' = 1/2$. Therefore

$$\frac{1}{2}\frac{d}{dt}(\|u\|^2 + \|\theta\|^2) + (\nu - c\|u_\pi\|_p)\|\nabla u\|^2 + \chi\|\nabla\theta\|^2$$
$$-(c\|\theta_\pi\|_p + c'\beta g_\infty)\|\nabla\theta\|\,\|\nabla u\| \leq 0$$

Under the condition (11), the following matrix

$$\begin{bmatrix} \nu - c\|u_\pi(t)\|_p & -(c\|\theta_\pi(t)\|_p + c'\beta g_\infty)/2 \\ -(c\|\theta_\pi(t)\|_p + c'\beta g_\infty)/2 & \chi \end{bmatrix}$$

has two positive eigenvalues for a.e.t, that is, there exists a measurable function $\lambda(t)$ such that

$$\lambda(t) > 0 \text{ a.e. } t, \quad \lambda \in L^2(0, T_0),$$

and

$$\frac{1}{2}\frac{d}{dt}(\|u\|^2 + \|\theta\|^2) + \lambda(t)(\|\nabla u\|^2 + \|\nabla\theta\|^2) \leq 0.$$

hold. According to Poincaré's inequality, we can find an L^2 function $\lambda^*(t)$ such that

$$\lambda^*(t) > 0 \quad \text{a.e.}t$$

$$\frac{1}{2}\frac{d}{dt}(\|u(t)\|^2 + \|\theta(t)\|^2) + \lambda^*(t)(\|u(t)\|^2 + \|\theta(t)\|^2) \leq 0.$$

Integrating this inequality, we obtain:

$$\|u(t)\|^2 + \|\theta(t)\|^2 \leq e^{-2\int_0^t \lambda^*(s)ds}(\|u(0)\|^2 + \|\theta(0)\|^2).$$

Since $u(T_0) = u(0)$, $\theta(T_0) = \theta(0)$, and

$$e^{-2\int_0^{T_0} \lambda^*(s)ds} < 1,$$

we have $\|u(0)\| = 0$, $\|\theta(0)\| = 0$. Therefore $u(t) = 0$, $\theta(t) = 0$ for $\forall t$.

Proof of Theorem 3

Let $\{u, \theta\}$ be a weak solution of (1), (2) with (3). Put $w = u - u_\pi$, $\zeta = \theta - \theta_\pi$. Calculating in a similar way to Theorem 2, we have

$$\frac{1}{2}\frac{d}{dt}\|w(t)\|^2 + \nu\|\nabla w(t)\|^2 = ((w \cdot \nabla)w, u_\pi) + (\beta g\zeta, w)$$

$$\frac{1}{2}\frac{d}{dt}\|\zeta(t)\|^2 + \chi\|\nabla\zeta(t)\|^2. = ((w \cdot \nabla)\zeta, \theta_\pi).$$

Therefore,

$$\|w(t)\|^2 + \|\zeta(t)\|^2 \leq e^{-2\int_0^t \lambda^*(s)ds}(\|w(0)\|^2 + \|\zeta(0)\|^2).$$

holds. Put

$$\frac{1}{T_0} \int_0^{T_0} \lambda^*(s)ds = \alpha.$$

For sufficiently large t, we put $t = nT_0 + t'$, $n = $ positive integer, $0 \le t' < T_0$. Since $\lambda^*(t)$ is periodic in t with period T_0,

$$\int_0^t \lambda^*(s)ds = nT_0\alpha + \int_0^{t'} \lambda^*(s)ds \ge \alpha t - \alpha T_0.$$

Therefore the following estimate holds:

$$e^{-2\int_0^t \lambda^*(s)ds} \le e^{2\alpha T_0} e^{-2\alpha t},$$

and we obtain the desired result.

References

[1] Hishida, T., Existence and regularizing properties of solutions for the nonstationary convection problem, Funkcialaj Ekvacioj, **34** (1991), pp.449-474.

[2] Morimoto, H., On the existence of weak solutions of equations of natural convection, J. Fac. Sci. Univ. Tokyo, Sect.IA, **36** (1989), pp.87-102.

[3] Morimoto, H., On the existence and uniqueness of the stationary solution to the equation of natural convection, Tokyo J. Math., Vol.14(1991), pp.217-226.

[4] Morimoto, H., On non-stationary Boussinesq equations, Proc.Japan Acad., **67** Ser A (1991), pp.159-161, Nonstationary Boussinesq equations, J. Fac. Sci. Univ. Tokyo, Sect.IA, **39** (1992), pp.61-75.

[5] Ōeda, K., Weak and strong solutions of the heat convection equations in regions with moving boudaries, J. Fac. Sci. Univ. Tokyo, Sec. IA, **36**, No.3 (1989), pp.491-536.

[6] Temam, R., Navier-Stokes Equations, North-Holland, Amsterdam, 1979.

Uniform restricted parabolic Harnack inequality, separation principle, and ultracontractivity for parabolic equations

Minoru Murata
Department of Mathematics, Faculty of Science, Kumamoto University
Kumamoto, 860 Japan

§0. Introduction

A. Freire [F] made use of a parabolic Harnack inequality due to P. Li & S. T. Yau [LY] in establishing a beautiful splitting theorem for positive harmonic functions on a Riemannian product. Inspired by his work and a subsequent paper by J. C. Taylor [T], we introduce in this paper the notion of uniform restricted parabolic Harnack inequality (URPH) for nonnegative solutions of a linear second order parabolic equation

$$(\partial_t + L)u = 0$$

on a cylinder of R^{n+1}, where L is a time independent elliptic operator.

The purpose of this paper is to discuss several properties related to URPH.

In §1, we establish (see Theorem 1.3 below) that URPH is equivalent to the separation principle (which says, roughly speaking, that a nonnegative solution of a parabolic equation is minimal iff it is a product of $e^{\lambda t}$ and a minimal positive solution of an elliptic equation). We also show there (see Theorem 1.4 below) that URPH implies uniqueness of the positive Cauchy problem, and in §2 (see Theorem 2.2 below) that URPH yields a splitting theorem for positive solutions of an elliptic equation. In §3 we establish (see Theorem 3.1 below) that the intrinsic ultracontractivity (IU) of a Schrödinger semigroup (cf. [DS]) implies non-uniqueness of the positive Cauchy problem; therfore, if IU holds, then URPH does not hold. This non-uniqueness theorem is of independent interest. While we treat in §§1 ∼ 3 parabolic equations in $R^n \times I$, where I is an open interval, we do in §4 equations in general cylinders.

The results to be given in this paper are somehow related to the uniqueness problem of the positive Cauchy problem, which will be extensively investigated elsewhere.

Acknowledgment The author thanks Massimo A. Picardello for informing him of interesting papers [F], [Mo], and [T]; and Yoichi Oshima for [An]. He is indebted to the late professor Kôsaku Yosida who gave him encouraging and stimulating comments when he first gave a talk at RIMS, Kyoto University.

§1. Uniform restricted parabolic Harnack inequality and separation principle

We consider nonnegative solutions of a parabolic equation

$$\partial_t u = -Lu \equiv \left(\sum_{i,j=1}^{n} \partial_i a_{ij}(x)\partial_j - V(x) \right) u(x,t) \quad \text{in} \ \ R^n \times (-\infty, 0), \qquad (1)$$

where $\partial_t = \partial/\partial t$, $\partial_i = \partial/\partial x_i$, $(a_{ij}(x))_{i,j=1}^{n}$ is a positive definite symmetric matrix such that for any bounded set K of R^n there exists $\Lambda > 0$ with

$$\Lambda|\xi|^2 \le \sum_{i,j=1}^{n} a_{ij}(x)\xi_i\xi_j \le \Lambda^{-1}|\xi|^2, \qquad x \in K, \qquad \xi \in R^n,$$

and V is a real-valued function in $L_{p,loc}(R^n)$ with $p > n/2$ for $n > 1$ and $p = 1$ for $n = 1$. A solution of (1) means a function which belongs to $L_2(J, H^1(U)) \cap L_\infty(J, L_2(U))$ for any bounded interval $J \subset (-\infty, 0)$ and any bounded domain $U \subset R^n$, and satisfies (1) in the weak sense; by the regularity theorem, a solution may be assumed to be locally Hölder continuous (cf. [Ar]). We say that u is a *positive solution* of (1) if u is a solution of (1) which is continuous, nonnegative, and not identically zero on $R^n \times (-\infty, 0]$. We say that a positive solution u of (1) is *minimal* when for any positive solution v with $v \le u$ there exists a constant C such that $v = Cu$. We denote by $\mathcal{P}_1$ the set of all minimal positive solutions u with $u(0,0) = 1$ equipped with the topology induced by the uniform convergence on each compact set of $R^n \times (-\infty, 0)$.

Definition 1.1. We say that the uniform restricted parabolic Harnack inequality (which will be referred to as URPH) for (1) holds when the following conditions (a) and (b) are satisfied.

(a) There exists a positive solution of (1).

(b) For any $\delta > 0$ and $\tau > 0$ there exists a positive constant C such that

$$u(x, t - \tau) \le Cu(x,t), \qquad x \in R^n, \qquad t \le -\delta, \qquad (2)$$

for all positive solutions u of (1).

Remark We have allowed δ-dependece in (2) in order to similarly treat nonnegative solutions in a cylinder vanishing on the lateral boundary (see §4).

It is seen from the parabolic Harnack inequality in [Ar] that if

$$\sup\{\|V(x)\|_{L_p(\{x \in R^n; \ |x-y|<1\})} \ ; \ y \in R^n\} < \infty,$$

then URPH for (1) holds.

We put

$$\lambda_0(L) = -\inf\{Q_L(\varphi, \varphi); \ \varphi \in C_0^\infty(R^n), \ \|\varphi\|_{L_2(R^n)} = 1\}, \qquad (3)$$

$$Q_L(\varphi, \psi) = \int_{R^n} \left(\sum_{i,j=1}^{n} a_{ij}\partial_i\varphi \ \overline{\partial_j \psi} + V\varphi\overline{\psi} \right) dx.$$

Clearly, $\lambda_0(L)$ is a real number or ∞. It is well-known (see, for example, [Si] or [Mu2]) that for a real number λ the equation

$$(L + \lambda)u = 0 \quad \text{in} \quad R^n \tag{4}$$

has a positive solution if and only if $\lambda \geq \lambda_0(L)$.

If $\lambda_0(L) < \infty$, then $\{e^{-tH}\}_{t\geq 0}$ is a C_0-semigroup on $L_2(R^n)$ with H being the Friedrichs extension for L. Thus there exists a minimal fundamental solution for $\partial_t + L$ (cf. [AT], [Ar], and [Si]); and we can obtain a *parabolic Martin representation theorem* in the same way as in [F]. It is described as follows. A limiting procedure yields a compact metric space $\mathcal{P}$ and a continuous function $H(x, t, \rho)$ on $R^n \times (-\infty, 0] \times \mathcal{P}$ such that $\mathcal{P}_1$ is a G_δ-subset of $\mathcal{P}$, and for each ρ, $H(\cdot, \cdot, \rho)$ is a nonnegative solution of (1) and $H(x, t, \rho) = \rho(x, t)$ for ρ in $\mathcal{P}_1$. Then the parabolic Martin representation theorem is: For any positive solution u of (1) there exists a unique finite positive Borel measure μ on $\mathcal{P}$ such that $\mu(\mathcal{P}\backslash\mathcal{P}_1) = 0$ and

$$u(x, t) = \int_{\mathcal{P}_1} H(x, t, \rho) d\mu(\rho); \tag{5}$$

furthermore, for such a measure μ a function given by the right hand side of (5) is a nonnegative solution of (1).

Suggested by [T], we introduce the notion of separation principle.

Definition 1.2. We say that the separation principle (which will be referred to as SP) holds for (1) when there exists a positive solution of (1) and $u(x, t)$ is a minimal positive solution if and only if there exist a real number $\lambda \geq \lambda_0(L)$ and a minimal positive solution $v(x)$ of (4) (i.e. any positive solution $w(x)$ with $w \leq v$ is a constant multiple of v) such that

$$u(x, t) = e^{\lambda t} v(x) \quad \text{on} \quad R^n \times (-\infty, 0]. \tag{6}$$

Now we are ready to state a main result of this section.

Theorem 1.3. The following conditions are equivalent.
(i)   URPH for (1) holds.
(ii) There hold the condition (a) of Definition 1.1 and the following condition (b'): For any positive solution u of (1), $\delta > 0$, and $\tau > 0$ there exists a positive constant C such that (2) is valid.
(iii) SP for (1) holds.

Proof. Clearly, (i) implies (ii). Assume (ii). Let u be a minimal positive solution of (1) such that $u(0,0) = 1$. We will show (6) applying the argument of [Mo] (see also [KT], [F] and [P]). Since $u \neq 0$, there exists $\delta > 0$ such that $u(x, t) > 0$ for $t \geq -2\delta$. Let $\Gamma_R(x, y, s)$ be the fundamental solution for $\partial_t + L$ on $\{|x| < R\}$ with Dirichlet zero boundary condition. Then the maximum principle yields

$$u(0, s - \delta) \geq \int_{R^n} \Gamma_{\tilde{R}}(0, y, s)u(y, -\delta)dy$$

for $0 \leq s \leq \delta$, where $\Gamma_{\tilde{R}}(0, y, s) = \Gamma_R(0, y, s)$ for $|y| \leq R$ and $\Gamma_{\tilde{R}}(0, y, s) = 0$ otherwise; furthermore, $\Gamma_{\tilde{R}} \leq \Gamma_{\tilde{S}}$ for $R < S$. Thus

$$\infty > u(0, s - \delta) \geq \int \Gamma(0, y, s)u(y, -\delta)dy, \quad 0 \leq s \leq \delta, \tag{7}$$

where $\Gamma = \lim_{R \to \infty} \Gamma_R^{\sim}$ is the minimal fundamental solution. Put

$$\tilde{u}(x, t - \tau) = \int \Gamma(x, y, t + \delta) u(y, -\delta - \tau) dy \qquad \text{for } t > -\delta$$

$$= u(x, t - \tau) \qquad \text{for } t \leq -\delta.$$

Then, by (2) and (7), $\tilde{u}(x, t - \tau) \leq C u(x, t)$ for $x \in R^n$ and $t \leq 0$. Since u is minimal and $\tilde{u}(\cdot, \cdot, -\tau)$ is also a positive solution, there exists C_τ such that

$$\tilde{u}(x, t - \tau) = C_\tau u(x, t).$$

Then $C_\tau = u(0, -\delta - \tau)/u(0, -\delta)$ is continuous, $C_0 = 1$ and $C_{\tau + \sigma} = C_\tau C_\sigma$. Thus $C_\tau = e^{-\lambda \tau}$ for some $\lambda \in R^1$, which implies that

$$u(x, t) = e^{\lambda t} v(x) \equiv e^{\lambda t}[e^{\lambda \delta} u(x, -\delta)] > 0, \qquad t \leq -\delta.$$

Thus every minimal positive solution of (1) is everywhere positive on $R^n \times (-\infty, 0]$. This implies uniqueness of the positive Cauchy problem for $\partial_t + L$, since the existence of a nontrivial nonnegative solution of the Cauchy problem with zero initial data implies the existence of a positive solution of (1) which is zero on $R^n \times (-\infty, -T]$ for some $T > 0$ (cf. [KT]). On the other hand, v is a positive solution of (4), because $u(x, t) = e^{\lambda t} v(x)$ satisfies the equation $(\partial_t + L) u(x, t) = 0$ in $R^n \times (-\infty, -\delta)$. Since $u(x, t)$ and $e^{\lambda t} v(x)$ are both positive solutions of the Cauchy problem

$$(\partial_t + L) w(x, t) = 0 \quad \text{in} \quad R^n \times (-\delta, 0], \qquad w(x, -\delta) = u(x, -\delta),$$

the uniqueness mentioned above shows (6). Furthermore, v is minimal because $w \leq v$ implies $e^{\lambda t} w(x) \leq e^{\lambda t} v(x)$. By virtue of the Krein-Milman theorem, the existence of a positive solution of (1) implies that of a minimal solution. But we have shown that a minimal solution must be of the form (6). Thus there exists a positive solution of (4). Hence $\lambda_0(L) < \infty$, and we get the parabolic Martin representation theorem. Furthermore, since λ and $v(x)$ are determined uniquely by the minimal positive solution $u(x, t)$, there exists a continuous map Φ from $\mathcal{P}_1$ to the set

$$\mathcal{N} = \cup_{\lambda \geq \lambda_0(L)} \{\lambda\} \times \mathcal{M}_1^\lambda \tag{8}$$

which is considered to be a subspace of $[\lambda_0(L), \infty) \times C^0(R^n)$, where $\mathcal{M}_1^\lambda$ is the set of all minimal positive solutions $v(x)$ of (4) normalized as $v(0) = 1$. Writing $K_\lambda(x, \omega) = \omega(x)$ for $\omega \in \mathcal{M}_1^\lambda$ and $(\lambda(\rho), \omega(\rho)) = \Phi(\rho)$ for $\rho \in \mathcal{P}_1$, we obtain (5) with H replaced by the right hand side of

$$H(x, t, \rho) = e^{\lambda(\rho) t} K_{\lambda(\rho)}(x, \omega(\rho)). \tag{9}$$

Now let us show that if $a \geq \lambda_0(L)$ and v is a minimal positive solution of (4) with $\lambda = a$ such that $v(0) = 1$, then $u(x, t) = e^{at} v(x)$ is a minimal positive solution of (1). We have

$$u(x, t) = \int_{\mathcal{P}_1} e^{\lambda(\rho) t} K_{\lambda(\rho)}(x, \omega(\rho)) d\mu(\rho) \tag{10}$$

for some measure μ. Thus

$$1 = \int_{\mathcal{P}_1} e^{(\lambda(\rho) - a) t} d\mu(\rho).$$

Since $\lim_{t \to -\infty} \exp[(\lambda(\rho) - a)t]$ is equal to ∞ if $\lambda(\rho) < a$ and 0 if $\lambda(\rho) > a$, we have $\mu(\{\rho \in \mathcal{P}_1; \lambda(\rho) \neq a\}) = 0$. Then (10) becomes

$$v(x) \ = \ \int_{\mathcal{P}_1 \cap \{\lambda(\rho) = a\}} K_a(x, \omega(\rho)) d\mu(\rho).$$

Since v is minimal, this implies that

$$1 = \mu(\mathcal{P}_1) \ = \ \mu(\{\rho \in \mathcal{P}_1; \lambda(\rho) = a \quad \text{and} \quad \omega(\rho) = \omega_0\})$$

for some $\omega_0 \in \mathcal{M}_1^a$, which shows that $e^{at}v(x)$ is minimal. Note that we have also shown that Φ is homeomorphism from $\mathcal{P}_1$ onto $\mathcal{N}$.

It remains to prove that (iii) implies (i). Assume (iii). Let u be any positive solution of (1), $\tau > 0$, and $t \leq 0$. Then we have (10) in the same way as above. For $\lambda \geq \lambda_0 \equiv \lambda_0(L)$, $\lambda(t - \tau) \leq \lambda t - \lambda_0 \tau$. Thus

$$u(x, t - \tau) \ = \ \int e^{\lambda(\rho)(t-\tau)} K_{\lambda(\rho)}(x, \omega(\rho)) d\mu(\rho) \ \leq \ e^{-\lambda_0 \tau} \, u(x, t).$$

This shows (i).   Q.E.D.

We conclude this section with a simple consequence of URPH. Consider the Cauchy problem

$$\partial_t u(x, t) \ = \ Lu(x, t) \quad \text{in} \quad R^n \times (0, T), \qquad u(x, 0) = u_0(x), \tag{11}$$

where $T > 0$. A solution u is assumed, for simplicity, to be continuous on $R^n \times [0, T)$, although we can treat a solution of (11) which attains the initial value u_0 in a weak sense (cf. [Ar]). Since URPH for (1) implies that any positive solution is everywhere positive, as we have already mentioned in the proof of Theorem 1.3, the following theorem can be shown in the same way as in [KT].

Theorem 1.4. Suppose that URPH for (1) holds. Then a nonnegative solution of (11) is unique.

§2. Product decomposition and URPH

We consider the equation (1) in the case where

$$L \ = \ L_1 + L_2 \,, \qquad L_k \ = \ - \sum_{i,j=n_{k-1}+1}^{n_{k-1}+n_k} \partial_i a_{ij}(x_k)\partial_j \ + \ V_k(x_k), \tag{12}$$

$$k \ = \ 1, 2, \qquad n_0 \ = \ 0, \qquad x_1 \ = \ (x_1, \cdots, x_{n_1}), \qquad x \ = \ (x_1, x_2) \in R^{n_1} \times R^{n_2} \ = \ R^n,$$

together with two more equations

$$\partial_t u_k(x_k, t) \ = \ -L_k u_k(x_k, t) \quad \text{in} \quad R^{n_k} \times (-\infty, 0). \tag{$1)_k$}$$

Theorem 2.1. URPH for (1) holds if and only if URPH for both $(1)_1$ and $(1)_2$ hold.

Proof. Suppose that URPH for (1) hold. Since $\lambda_0(L) < \infty$, we see that $\lambda_0(L_k) < \infty$ for $k = 1, 2$. This implies the existence of a positive solution u_k of $(1)_k$ such that $u_k(x_k, t) = \exp(\lambda_k t)v_k(x_k)$. Let $u_1(x_1, t)$ be any positive solution of $(1)_1$. Choose $\lambda \geq \lambda_0(L_2)$ and a positive solution v_2 of the equation $(L_2 + \lambda)v_2 = 0$. Then $u(x, t) = u_1(x_1, t)e^{\lambda t}v_2(x_2)$ is a positive solution of (1). By (2),

$$u_1(x_1, t - \tau)e^{\lambda(t-\tau)}v_2(x_2) \leq Cu_1(x_1, t)e^{\lambda t}v_2(x_2), \quad (x_1, x_2) \in R^n, \quad t \leq -\delta.$$

Thus $u_1(x_1, t - \tau) \leq Ce^{\lambda \tau}u_1(x_1, t)$, $x_1 \in R^{n_1}$, $t \leq -\delta$, which shows that URPH holds for $(1)_1$. Similarly, URPH for $(1)_2$ holds.

Conversely, suppose that URPH holds for both $(1)_1$ and $(1)_2$. By virtue of Theorem 1.3 it suffices to show SP for (1). Let $\Gamma_k(x_k, y_k, t)$ be the minimal fundamental solution for $\partial_t + L_k$. Then

$$\Gamma(x, y, t) = \Gamma_1(x_1, y_1, t)\Gamma_2(x_2, y_2, t) \tag{13}$$

is the minimal fundamental solution for $\partial_t + L$. Put

$$\gamma(x, t; y, s) = \begin{cases} \Gamma(x, y, t - s), & s < t \leq 0 \\ \\ 0, & t \leq s \leq 0. \end{cases} \tag{14}$$

The parabolic Martin representation theorem says (cf. [F]) that for any $\rho \in \mathcal{P}_1$ the parabolic Martin kernel $H(x, t, \rho)$ is given by

$$H(x, t, \rho) = \lim_{j \to \infty} \gamma(x, t; y^j, s^j)/\gamma(0, 0; y^j, s^j) \tag{15}$$

for some sequence $\{(y^j, s^j)\}_{j=1}^{\infty}$ in $R^n \times (-\infty, 0)$. By the parabolic Harnack principle, choosing a subsequence if necessary, we may assume that there exist the limits

$$\lim_{j \to \infty} \gamma_k(x_k, t; y_k^j, s^j)/\gamma_k(0, 0; y_k^j, s^j) \equiv H_k(x_k, t, \rho_k), \quad k = 1, 2,$$

where γ_k is defined similarly from Γ_k. This together with (13) and (15) implies $H(x, t, \rho) = H_1(x_1, t, \rho_1)H_2(x_2, t, \rho_2)$. If H is minimal, then both H_k are minimal. Thus $H_k = \exp(\lambda_k t)v_k(x_k)$, which implies that $H = \exp[(\lambda_1 + \lambda_2)t]v_1(x_1)v_2(x_2)$. Q.E.D.

Theorem 2.2. Suppose that URPH holds for both $(1)_1$ and $(1)_2$. Then a positive solution $u(x)$ of (4) is minimal if and only if

$$u(x) = u_1(x_1)u_2(x_2), \tag{16}$$

where $\lambda_k \geq \lambda_0(L_k)$, $k = 1, 2$, $\lambda_1 + \lambda_2 = \lambda$, and each u_k is a minimal positive solution of

$$(L_k + \lambda_k)u_k = 0 \quad \text{in} \quad R^{n_k}. \tag{4}_k$$

Proof. The same argument as in the proof of Theorem 2.1 shows that if u is a minimal positive solution, then u is of the form (16). Conversely, let u be of the form (16). Then, making use of the representation (5) with (9) and uniqueness of the positive Cauchy problem (Theorem 1.4) we can show that u is minimal. (We omit the detail, because the author has learned his original proof is essentially same as that of B. Davies given in [An, pp 41 & 42].) Q.E.D.

We conclude this section with an application of Theorem 2.2.

Example 2.3. Let $n_1 = n_2 = R^3$, $V_1(x_1) = |x_1|^{-b} + a$ for some $0 < b < 1$ and $a \geq 0$, $V_2(x_2) = |x_2|^{-b}$, and $L_k = -\Delta_{x_k} + V_k(x_k)$. Then $\lambda_0(L_1) = a$, $\lambda_0(L_2) = 0$, and the minimal Martin boundary σ for (L, R^6) is homeomorphic to $S^2 \times S^2 \times [0, a]$ (cf. [Mu1]).

Remark Although Theorem 2.2 is powerful in constructing the minimal Martin boundary, it does not give any information on the non-minimal Martin boundary and Martin compactification; for which, see [Mu1 ∼ 3] and [GT], for example.

§3. Intrinsic ultracontractivity and non-uniqueness of the positive Cauchy problem

We have shown (Theorem 1.4) that a nonnegative solution of the Cauchy problem (11) is unique if URPH for (1) holds. In this section we shall show that if the semigroup associated with the equation (1) is intrinsically ultracontractive, then there exists a positive solution of (11) with initial value $u_0 = 0$.

Suppose that $\lambda_0(L) < \infty$ (for $\lambda_0(L)$, see (3) in §1), and let H be a self-adjoint operator on $L_2(R^n)$ associated with the closure of the sesqui-linear form Q_L with $domain(Q_L) = C_0^\infty(R^n)$. Let $B_R = \{x \in R^n; |x| < R\}$ for $R > 0$, and H_R be the self-adjoint operator on $L_2(B_R)$ defined by: $domain(H_R) = \{u \in H_0^1(B_R); Lu \in L_2(B_R)\}$ and $H_R u = Lu$. Choose $\lambda > \lambda_0(L)$. We see that the integral kernel of $(H_R + \lambda)^{-1}$ converges monotonically to that of $(H + \lambda)^{-1}$. This implies that the integral kernel of e^{-tH} is equal to the minimal fundamental solution $\Gamma(x, y, t)$ for $\partial_t + L$. Following [DS] we say that the semigroup e^{-tH} on $L_2(R^n)$ is *intrinsically ultracontractive* (which will be referred to as IU) when the following conditions (i) and (ii) are satisfied.

(i) $E_0 = -\lambda_0(L)$ is an eigenvalue of H with strictly positive eigenfunction $\varphi_0(x)$ normalized by $\|\varphi_0\|_2 = 1$.

(ii) The integral kernel $\Gamma(x, y, t)$ of e^{-tH} satisfies

$$\Gamma(x, y, t) \leq C_t \varphi_0(x) \varphi_0(y), \qquad (x, y) \in R^{2n}, \tag{17}$$

for each $t > 0$, where C_t is a positive constant depending only on t.

Theorem 3.1. Suppose that the semigroup e^{-tH} is IU. Then there exists a solution $u(x, t)$ of the equation

$$\partial_t u(x, t) = -Lu(x, t) \qquad \text{in} \quad R^n \times R^1 \tag{18}$$

such that

$$u(x, t) > 0 \qquad \text{for} \quad t > 0, \qquad u(x, t) = 0 \qquad \text{for} \quad t \leq 0. \tag{19}$$

Proof. Choose a sequence $\{y^k\}_{k=1}^{\infty}$ in R^n such that $|y^k| \to \infty$ as $k \to \infty$. In view of (17), put

$$A_k = \sum_{j=1}^{\infty} \Gamma(0, y^k, j)/2^j C_j \varphi_0(0). \tag{20}$$

Clearly, $A_k \leq \varphi_0(y_k)$. Define u_k by

$$u_k(x, t) = \begin{cases} \Gamma(x, y_k, t)/A_k, & t > 0 \\ 0, & t \leq 0. \end{cases} \tag{21}$$

Then u_k is a nonnegative solution of $(\partial_t + L)u_k = 0$ in $R^{n+1} \setminus \{(y^k, 0)\}$ satisfying $u_k(0, j) \leq 2^j C_j \varphi_0(0)$ for each positive integer j. By the usual parabolic Harnack inequality, there exists a subsequence $\{y^{k'}\}$ of $\{y^k\}$ such that u_k converges to a nonnegative solution u of (18) uniformly on each compact subset of R^{n+1}. By (21), $u(x, t) = 0$ for $t \leq 0$. On the other hand, by virtue of Theorem 3.2 in [DS], for any $t > 0$ there exists a positive constant B_t such that

$$\Gamma(x, y, t) \geq B_t \varphi_0(x) \varphi_0(y), \qquad (x, y) \in R^n. \tag{22}$$

Thus $u_k(x, t) \geq B_t \varphi_0(x) \varphi_0(y_k)/\varphi_0(y_k) = B_t \varphi_0(x)$ for $t > 0$. Hence $u(x, t) \geq B_t \varphi_0(x) > 0$ for $t > 0$. Q.E.D.

Example 3.2 ([Davies, Theorem 4.5.4, p.121]). Let $L = -\Delta + V$ with V satisfying

$$c_1 |x|^{a_1} - c_2 \leq V(x) - E_0 \leq c_3 |x|^{a_2} + c_4, \qquad x \in R^n,$$

where $c_i > 0$, $a_i > 2$, and $a_2 < 2a_1 - 2$. Let H be the self-adjoint operator associated with L. Then the semigroup e^{-tH} is IU.

§4. Parabolic equations in cylinders

Let D be a domain of R^n, and I an open interval. Consider the equation

$$\partial_t u(x, t) = -Lu(x, t) \qquad \text{in} \quad D \times I, \tag{23}$$

where L is the elliptic operator satisfying the condition in §1. The purpose of this section is to give results analogous to those in previous sections for positive solutions of (23) vanishing continuously on $\partial D \times I$.

Throughout the present section we assume the conditions (B) and (C) to be stated in what follows. For $T > 0$, let $D_T = D \times (0, T)$ and $\partial_p D_T = (\partial D \times (0, T)) \cup (\overline{D} \times \{0\})$. For $r > 0$ and $(Q, s) \in \partial D \times (0, T)$, put

$$\Psi_r(Q, s) = \{(x, t); 0 < t < T, |x - Q| < r, |t - s| < r^2\}$$

$$\Delta_r(Q, s) = \partial_p D_T \cap \overline{\Psi}_r(Q, s).$$

The following condition (B) (parabolic boundary Harnack inequality) is satisfied, for example, if D is a Lipshitz domain and $V = 0$ (cf. [FGS, Theorem 1.6]).

(B) For any $T > s > 0$ and $Q \in \partial D$, there exist $0 < R_0 < \min(\sqrt{s}/2, \sqrt{T-s}/2)$, $C > 0$, and one parameter family $\{P_r; 0 < r \leq R_0\}$ in D such that $|P_r - Q| \leq r$ and

$$u(x,t)/v(x,t) \leq Cu(P_r, s + 2r^2)/v(P_r, s - 2r^2) \tag{24}$$

for any $0 < r \leq R_0$, $(x,t) \in \Psi_{r/8}(Q,s) \cap D_T$, and any positive continuous solutions u, v of $(\partial_t + L)u = 0$ in $\Psi_{2r}(Q,s) \cap D_T$ vanishing continuously on $\Delta_{2r}(Q,s)$.

We denote the minimal fundamental solution for (23), if it exists, by $\Gamma_D(x,y,t)$. The following condition (C) is satisfied, for example, if D is a Lipshitz domain (cf. [B] and [DS]).

(C) There exists an increasing sequence $\{D_j\}_{j=1}^{\infty}$ of domains in D such that

$$\overline{D}_j \text{ is compact}, \qquad \cup_{j=1}^{\infty} D_j = D, \qquad \partial D_j \supset \partial D \cap \{|x| < j\},$$

and each fundamental solution $\Gamma_{D_j}(x,y,t)$ is continuous on $\overline{D}_j^2 \times (0,\infty)$ and vanishes on $\partial(\overline{D}_j^2) \times (0,\infty)$.

4.1. URPH and SP for cylinders Let $I = (-\infty, 0)$. We say that u is a positive solution vanishing on the lateral boundary when u is a solution of (1) which is continuous, nonnegative, not identically zero on $D \times \overline{I} \cup \partial D \times I$ and zero on $\partial D \times I$. Let $\mathcal{P}_1$ be the set of all minimal positive solutions u vanishing on the lateral boundary with $u(x_0, 0) = 1$ for a fixed point $x_0 \in D$. For (23), we define URPH, $\lambda_0(L)$, and SP in a way similar to that in §1. Note that if $\lambda_0(L) < \infty$, (B) and (C) are valid, then the minimal fundamental solution Γ_D vanishes continuously on the lateral boundary. We can obtain

Theorem 1.3.D. For (23), the result corresponding to Theorem 1.3 holds.

Example 4.1. Suppose that $V = 0$ and ∂D is uniformly Lipshitz continuous: There exist $r_0 > 0$ and $m > 0$ such that for any $Q \in \partial D$, in some coordinate system of R^n,

$$B_{r_0} \cap D = B_{r_0} \cap \{(x', x_n); x' \in R^{n-1}, \quad x_n > \varphi(x')\},$$

$$B_{r_0} \cap \partial D = B_{r_0} \cap \{(x', \varphi(x'); x' \in R^{n-1}\},$$

$$B_{r_0} = \{|x - Q| < r_0\}, \qquad |\varphi(x') - \varphi(y')| \leq m|x' - y'|.$$

Then Theorem 1.6 of [FGS] implies URPH for (23). In particular, if D is a bounded Lipshitz domain, then Theorem 1.3.D implies that any positive solution of (23) vanishing on the lateral boundary is a constant multiple of

$$\exp[\lambda_0(L)t]\varphi_0(x),$$

where φ_0 is a positive solution in $H_0^1(D)$ of the equation $(L + \lambda_0)u = 0$ in D. A similar uniqueness theorem has been shown by [N] for possibly time dependent domains.

4.2. Uniqueness theorem for cylinders We can obtain

Theorem 1.4.D. Suppose that URPH for (23) with $I = (-\infty, 0)$ holds. Then a nonnegative solution of (23) with $I = (0, T)$ which is continuous on $\partial_p D_T \cup D_T$ and vanishes on $\partial D \times (0, T)$ is uniquely determined by the initial data.

Example 4.2. Combining Example 4.1 and Theorem 1.4.D we get a uniqueness theorem of the positive Cauchy problem for a uniform Lipshitz domain D and L with $V = 0$. Similar results were obtained by [Su].

4.3. Product decomposition for cylinders Suppose that L is of the form (12) and $D = D_1 \times D_2$ for some domains D_k in R^{n_k}. Then we can obtain

Theorem 2.1.D. For (23), the result corresponding to Theorem 2.1 holds.

Consider positive solutions of

$$(L_k + \lambda)u = 0 \quad \text{in} \quad D_k \qquad (25)_k$$

vanishing continuously on ∂D_k, where $k = 1, 2, 3$, $D_3 = D$ and $L_3 = L$. Then we can obtain

Theorem 2.2.D. For $(25)_k$, the result corresponding to Theorem 2.2 holds.

4.4. IU for cylinders Defining a self-adjoint operator H_D on $L_2(D)$ and IU of e^{-tH_D} similarly, we have

Theorem 3.1.D. Suppose that D is unbounded and e^{-tH_D} is IU. Then there exists a solution u of (23) with $I = (-\infty, \infty)$ which is continuous on $\overline{D} \times I$, zero on $(\partial D \times I) \cup D \times (-\infty, 0]$, and positive on $D \times (0, \infty)$.

The proof of Theorem 3.1 works if we choose a sequence $\{y^k\}$ such that it has no accumulation point in $\overline{D}$.

Example 4.3. If $L = -\Delta$ and D is such an unbounded domain as in Theorem 9.6 of [DS] or Theorem 1 in [Davis], then e^{-tH_D} is IU. Thus there exists a positive solution of (23) with $I = (0, \infty)$ with initial and boundary data being zero. This is surprising in view of the fact that uniqueness of the positive Cauchy problem concerning the heat equation holds for $D = R^n$ or a bounded domain.

References

[An]   A. Ancona, Theorie du potentiel sur les graphes et les varietes, in: P. L. Hennekin ed., Lecture Notes in Math., Springer Verlag, **1427**(1990), 1-112.

[AT]   A. Ancona and J. C. Taylor, Some remarks on Widder's theorem and uniqueness of isolated singularities for parabolic equations, to appear in Proc. of Workshop on Partial Differential Equations, IMA and Univ. Chicago.

[Ar]   D. G. Aronson, Non-negative solutions of linear parabolic equations, Ann. Scuola Norm. Sup. Pisa, **22**(1968), 607-694.

[B]   R. Bañuelos, Intrinsic ultracontractivity and eigenfunction estimates for Schrödinger operators, J. Funct. Anal., **100** (1991), 181-206.

[Davies]   E. B. Davies, Heat kernels and spectral theory, Cambridge Univ. Press, Cambridge, 1989.

[Davis]   B. Davis, Intrinsic ultracontractivity and the Dirichlet Laplacian, J. Funct. Anal., **100**(1991), 162-180.

[DS]   E. B. Davies and B. Simon, Ultracontractivity and the heat kernel for Schrödinger operators and Dirichlet Laplacians, J. Funct. Anal., **59**(1984), 335-395.

[F]   A. Freire, On the Martin boundary of Riemannian products, J. Diff. Geometry, **33**(1991), 215-232.

[FGS]   E. B. Fabes, N. Garofalo and S. Salsa, A backward Harnack inequality and Fatou theorem for nonnegative solutions of parabolic equations, Illinois J. Math., **30**(1986), 536-565.

[GT]   Y. Guivarc'h and J. C. Taylor, The Martin compactification of the polydisc at the bottom of the positive spectrum, to appear in Colloquium Mathematicum.

[KT]   A. Koranyi and J. C. Taylor, Minimal solutions of the heat equation and uniqueness of the positive Cauchy problem on homogeneous spaces, Proc. Amer. Math. Soc., **94**(1985), 273-278.

[LY]   P. Li and S. T. Yau, On the parabolic kernel of the Schrödinger operator, Acta Math., **156**(1986), 153-201.

[Mo]   S. A. Molchanov, On Martin boundaries for the direct product of Markov chains, Theory Probab. Appl., **12**(1967), 307-310.

[MT]   B. Mair and J. C. Taylor, Integral representation of positive solutions of the heat equation, in: G. Mokobodzki and D. Pinchon eds, Lecture Notes in Math., Springer-Verlag, **1096**(1984), 419-433.

[Mu1]   M. Murata, Structure of positive solutions to $(-\Delta + V)u = 0$ in R^n, Duke Math. J., **53**(1986), 869-943.

[Mu2]   M. Murata, On construction of Martin boundaries for second order elliptic equations, Publ. RIMS, Kyoto Univ., **26**(1990), 585-627.

[Mu3]   M. Murata, Martin compactification and asymptotics of Green functions for Schrödinger operators with anisotropic potentials, Math. Ann., **288**(1990), 211-230.

[N]   M. Nishio, The uniqueness of positive solutions for parabolic equations of divergence form on an unbounded domain, Preprint, Osaka City Univ., 1991.

[P]   Y. Pinchover, Representation theorems for positive solutions of parabolic equations, Proc. Amer. Math. Soc., **104** (1988), 507-515.

[Si]   B. Simon, Schrödinger semigroups, Bull. Amer. Math. Soc., **7**(1982), 447-526.

288

[Su]   N. Suzuki, Huygens property of parabolic functions and a uniqueness theorem, Hiroshima Math. J., **19**(1989), 355-361.

[T]   J. C. Taylor, The product of minimal functions is minimal, Bull. London Math. Soc., **22**(1990), 499-504.

The Separable Quotient Problem for Barrelled Spaces

P. P. Narayanaswami*
Department of Mathematics, Memorial University of Newfoundland
St. John's, Newfoundland, Canada A1C 5S7

Abstract

In this survey, we discuss the classical open problem:

"Does every infinite-dimensional Banach space admit a quotient (by a closed subspace) which is infinite-dimensional and separable?"

We furnish several equivalent formulations of this famous unsolved problem, in terms of certain Baire-type covering and barrel properties of locally convex spaces, and also in terms of the structure theory of (metrizable, normable) (LF)-spaces. We solve the corresponding *"Separable Quotient Problem"* for the class of (LF)-spaces in the affirmative, by actually constructing the separable quotient. Based on the Baire-type covering properties, all (LF)-spaces are partitioned into three mutually disjoint and sufficiently rich classes, called $(LF)_1, (LF)_2$ and $(LF)_3$-spaces. These three classes are then characterized in terms of the sequence space φ. We show that every $(LF)_3$-space admits a separable quotient, that is a Fréchet space. Every $(LF)_2$ and $(LF)_3$ space (more generally, all non-strict (LF)-spaces) possesses a defining sequence, each of whose members has a separable quotient. The intimate interaction between the Separable Quotient Problem for Banach spaces, and the existence of metrizable, as well as normable (LF)-spaces will be studied, resulting in a rich supply of metrizable, as well as normable (LF)-spaces. Finally, we discuss *"Properly Separable"* quotients in the setting of barrelled spaces.

AMS Subject Classification: 46 A
Key words: Separable and properly separable quotients, Fréchet, Banach and Barrelled spaces, $(LF)_1, (LF)_2, (LF)_3$-spaces, the sequence space φ, Baire-like, Quasi-Baire, and (db)-spaces,

1 Introduction

The still open and classical *Separable Quotient Problem* for Banach spaces ([42, 33]) asks whether the following is true:

---
*Supported by NSERC Grant No. A-8772. Financial support from the organizers of the International Conference on Functional Analysis and Related Topics in memory of Professor Kôsaku Yosida, is gratefully acknowledged.

"Does every infinite-dimensional Banach space admit a quotient (by a closed subspace) which is infinite-dimensional and separable?" i.e., admit a continuous linear map onto a separable, infinite-dimensional Banach space?

Of course, such a map onto a finite-dimensional space always exists. An affirmative answer to this question would simplify the study of Banach spaces. For instance, a *normal sequence* for a Banach space E is a sequence $\{f_n\} \subset E'$ such that $f_n \to 0$ weak*, but $\|f_n\|$ does not converge to 0. In general, a normal sequence exists; easy proofs are available for separable, and reflexive spaces, and for Banach spaces having a separable quotient, normal sequences with some "nice" properties exist. Also a positive solution to this problem will throw some new insight into the structure theory of barrelled spaces, in particular, in the study of metrizable [normable] (LF)-spaces.

Perhaps, Banach himself, around 1932 has considered this problem, but it is not explicitly mentioned in literature earlier than 1969 (see H. P. Rosenthal [28], Remark 2, p. 188, where the problem has been stated in terms of the existence of quasi-complemented, separable subspaces.). Lacey [18] addressed this problem as *property (S)*, (which in turn implies *property (*)*, namely, the existence of a normal sequence), and proved that if X is compact Hausdorff, then the Banach space $C(X)$ has a separable quotient.

The answer is "yes", in the following cases:

- Separable Banach spaces (trivially);

- Reflexive Banach space;

- WCG (weakly compactly generated Banach spaces) (note that reflexive $\Longrightarrow$ WCG, and separable $\Longrightarrow$ WCG);

- ℓ_∞ (which is neither separable nor WCG);

- all known standard Banach spaces.

More generally, given any class $\mathcal{A}$ of locally convex spaces, one could phrase a Separable Quotient Problem for the class $\mathcal{A}$, and ask:

whether every member of $\mathcal{A}$ possesses a separable, infinite-dimensional quotient by a closed subspace.

The answer is "yes", again in the separable case - for example, the scalar sequence space ω (a non-normable Fréchet space) and the nuclear Fréchet space s.

In this exposition, we shall consider the cases where $\mathcal{A}$ is the class of all

- Banach spaces;

- non-normable Fréchet spaces;

- (LF)-spaces;

- Barrelled spaces.

We generally follow the terminology in Horváth [12] for all unexplained terminology.

2 Banach Space case

Consider the following eight properties which a Banach space $(E, \|.\|)$ may or may not have:

P_1: E has a separable (infinite-dimensional) quotient.

P_2: E has a dense, non-barrelled subspace.
  (Dense, barrelled subspaces are plentiful, however, for example see [32])

P_3: E has a dense, S_σ-subspace.
  (An S_σ-subspace is the union of a <u>strictly</u> increasing sequence of <u>closed</u> subspaces.

P_4: E has a dense BE-subspace. ([4])
  (A BE-subspace is a proper subspace of E which can be given a larger complete norm, i.e., it is a Banach space properly and continuously included in E. For example, ℓ_1 is a Bc_0-subspace of c_0.) (B stands for Banach!)

P_5: E has a dense locally convex FE-subspace.
  (in the preceding definition, replace "Banach" by "Fréchet')

P_6: There exists a normal sequence $\{f_n\} \subset E'$ with $\bigcup_{n=1}^\infty \bigcap_{i=n}^\infty f_i^\perp$ dense in E, and $\|f_n\| = 1$ for all n.

P_7: There exists closed subspaces A and B of E with $A + B$ a dense proper subset of E. ([28]) A and B can be found such that $A \cap B = \{0\}$, i.e., A and B are quasi-complementary.

P_8: There exists a dense subspace E_0 and a topology $\mathcal{T}_0$ on E_0 <u>finer than</u> the relative norm topology on E_0 such that $(E_0, \mathcal{T}_0)$ is a normable (LF)-space. ([34])

Theorem 2.1 (Saxon-Wilansky [35], Bennett-Kalton [4], Narayanaswami-Saxon [34])
For a Banach space E,

$$P_1 \iff P_2 \iff P_3 \iff P_4 \iff P_5 \iff P_6 \iff P_7 \iff P_8.$$

The equivalence of P_2, P_3 and P_5 was shown in [4]; and that of P_1 and P_3 was first given in [13]. Simplified proofs of the equivalence of P_1 through P_7 can be found in [35], and P_8 has been extensively discussed in [34]. It is still a wide open problem whether every infinite-dimensional Banach space enjoys any (hence all) of these equivalent properties.

Remark 2.2 A. Nissenzweig [23] and B. Josefson [14, 15] independently showed that every Banach space has a normal sequence. Josefson uses this powerful result in linear functional analysis, to prove the existence of entire functions of unbounded type in the context of Infinite dimensional holomorphy.

Consider the *property* (*): *every Banach space has a normal sequence.*
  Each of the properties $P_i (1 \le i \le 8)$ implies (*). But, the following example shows that (*) does not imply P_6.

Example 2.3 Let $bs := \{x = (x_n) : \sup_n \|x\| = |\sum_{k=1}^n x_k| < \infty\}$, and consider $E = bs \cap c_0$. If $f_n(x) = x_n \, (x \in E)$, then $\{f_n\}$ is (essentially) a normal sequence, but the set given in P_6 is not dense, since E is not separable. $\square$

Remark 2.4 In P_8, the phrase "finer than" can not *a priori* be relaxed.

3 Fréchet Space case

With some modifications of proofs, we can obtain the following Theorem analogous to Theorem 2.1, for the case of Fréchet spaces.

Theorem 3.1 (Narayanaswami-Saxon [33, 34]) *Let $(E, \mathcal{T})$ be a Fréchet space. The following statements are equivalent:*

(i) *E has a separable (infinite-dimensional) quotient (by a closed subspace).*

(ii) *E has a dense S_σ-subspace.*

(iii) *E has a dense non-barrelled subspace.*

(iv) *E has a dense non-(db)-space. (i.e. E has a dense subspace F, that can be covered by an increasing sequence of subspaces, none of which is both dense and barrelled.)*

(v) *E densely, properly and continuously includes a Fréchet space M. (M can be chosen so as to contain any specified countable subset of E).*

(vi) *There exists a dense subspace E_0 and a topology $\mathcal{T}_0$ on E_0 <u>finer than</u> $\mathcal{T}|_{E_0}$ such that $(E_0, \mathcal{T}_0)$ is a metrizable (LF)-space.*

Remark 3.2 In (vi), the phrase "finer than" is very crucial.

Since the proof employs several Baire-like "covering properties" of locally convex spaces, as well as some information from the structure theory of (LF)-spaces, we shall briefly digress to these topics. We return to the proofs of Theorems 2.1 and 3.1 in Sections 9 and 10.

4 Baire-type covering properties

The classical Baire Category Theorem states that if E is complete and metrizable, then $E \neq \cup_{n=1}^\infty E_n$, where $\{E_n\}$ is an increasing sequence of nowhere dense sets. Amemiya-Kōmura [1] improved this result by showing that if E is barrelled and metrizable, then $E \neq \cup_{n=1}^\infty E_n$, where $\{E_n\}$ is an *increasing* sequence of nowhere dense <u>absolutely convex</u> sets. (the current terminology for this (A-K) property is (B-L) or Baire-Like!

This result is the starting point for several new types of "Baire-like covering properties":

Definitions:
A locally convex space E is

- *Baire-like* if it is *not* the union of an <u>increasing</u> sequence of nowhere dense, absolutely convex sets. ([1, 29]

- *Unordered Baire-like* if it is *not* the union of a sequence of nowhere dense, absolutely convex sets. ([37])
  equivalently: E has *property (R-R)* [26]: if $E = \cup_{n=1}^{\infty} E_n$, E_n's subspaces $\Longrightarrow$ at least one of the E_n's is *both* dense and barrelled in E.

- *(db)-Space (Suprabarrelled)* space) if E is *not* the union of an <u>increasing</u> sequence of subspaces none of which is both dense and barrelled. ([27, 33, 40])

- *Quasi-Baire* if it is barrelled and is *not* the union of an increasing sequence of nowhere dense subspaces; ([22])

- E is *Barreled* $\Longrightarrow$ $E \neq \cup_{n=1}^{\infty} nB$, where B is a nowhere dense absolutely convex set.

$$\text{Baire} \Longrightarrow \text{Unordered Baire-like} \Longrightarrow \text{(db)} \Longrightarrow$$
$$\Longrightarrow \text{Baire-like} \Longrightarrow \text{quasi-Baire} \Longrightarrow \text{barrelled}$$

Remark 4.1 Non-locally convex analogues of some of these notions have been studied by Y. Terao, N. Kitzunezaki, M. Abe and K. Kera ([16, 36])

All these spaces (except Baire spaces) enjoy all "reasonable" permanence properties: i.e., stability under arbitrary products, quotients, countable-codimensional subspaces, to mention a few (see [22, 29, 37]).

The Wilansky-Klee Conjecture [37] asserts that every one-codimensional, dense subspace of a (locally convex) Baire space is Baire. This was disproved by Arias de Reyna [2] who showed (using Martin's axiom) that every infinite dimensional Banach space has a dense hyperplane of First Category.

Next, we ask when some of these classes coincide? The Amemiya-Kōmura result mentioned earlier shows that in the class of metrizable spaces, barrelled $\Longleftrightarrow$ Baire-Like.

The scalar sequence space $\varphi := \{x = (x_n) : x_n = 0$ eventually, with the finest locally convex topology$\}$ is invoked in this context. It is a strict, non-metrizable (LF)-space, and plays an important role in the further development. Saxon [29] generalized Amemiya-Kōmura result by showing: *a barrelled space that does not contain (an isomorphic copy of) φ must be Baire-like.* So, in the class of spaces that do *not* contain a copy of φ, barrelled $\Longrightarrow$ Baire-Like. In a still wider class, namely, in the class of barrelled spaces that do not contain a complemented copy of φ, *barrelled $\Longleftrightarrow$ Quasi-Baire* (Theorem 5.6). Valdivia [38] generalized the Amemiya-Kōmura result by showing that a Hausdorff barrelled space whose completion is a Baire space must itself be a Baire-like space. So it turns out that in the smallest "*variety*" of locally convex spaces [7], namely the variety of real Hausdorff locally convex spaces with their weak topology, the completion of any member is a product of reals, and hence a Baire-space. So, in the "*smallest variety*", all the concepts between Baire-like and barrelled inclusive coincide.

Next, we show that none of the implication arrows is reversible. There exist plenty of Unordered Baire-like spaces that are not Baire. Examples can be found in [6, 30, 37]. Examples of (db)-spaces that are not unordered Baire-like arise from the following result.

Theorem 4.2 (Saxon-Narayanaswami [33]) *Every infinite dimensional Fréchet space E has a dense subspace G which is a (db)-space, but not an Unordered Baire-Like space.*

(Choose a biorthogonal sequence $\{x_i, f_i\}$ in E, and set $A = \{x \in E :$ the real and imaginary parts of $f_i(x)$ are rational for each $i\}$, and let $G = span\ (A)$).

For the remaining implications

$$(db) \Longrightarrow Baire - like \Longrightarrow Quasi - Baire \Longrightarrow barrelled$$

we exploit the Baire-type Covering Properties of (generalized) (LF)-spaces.

5 A Classification of (LF)-spaces based on their Baire Covering Properties

Definition: Let $\{(E_n, \tau_n)\}$ be a sequence of Fréchet spaces such that for each n,

- $E_n \underset{\neq}{\subseteq} E_{n+1}$,

- $E_{n+1}|_{E_n} \leq E_n$ ($\leq$ means 'coarser than'),

- $E = \cup_{n=1}^\infty E_n$,

and let τ be the finest Hausdorff locally convex topology on E such that $\tau|_{E_n} \leq \tau_n$ for each n. Then, (E, τ) is an *inductive limit of Fréchet spaces*, or simply, an (LF)-space. We write

$$(E, \tau) = \lim \operatorname{ind}_n (E_n, \tau_n).$$

The sequence $\{(E_n, \tau_n)\}$ is a *defining sequence* of Fréchet spaces. (also called *steps*). If each (E_n, τ_n) is a Banach space, we have an (LB)-space.

An (LF)-space is *strict* if $\tau_{n+1}|_{E_n} = \tau_n$ for each n (which implies $\tau|_{E_n} = \tau_n$ for each n).

The spaces considered by Grothendieck, Dieudonne, and Schwartz [11, 8] are strict (LF)-spaces. They are never metrizable.

Remark 5.1 A strict (LF)-space can possess a non-strict defining sequence; an (LB)-space can have a non-Banach defining sequence.

Example 5.2 For each n, consider

$$E_n := \underbrace{\ell_1 \times \ell_1 \times \cdots \times \ell_1}_{n \text{ terms}} \times \{0\} \times \{0\} \times \cdots$$

with the product (Banach space) topology. $\{E_n\}$ is a strict defining sequence for the strict (LB)-space $\lim \operatorname{ind}_n E_n$. On the other hand, consider

$$F_n := \underbrace{\ell_1 \times \ell_1 \times \cdots \times \ell_1}_{n \text{ terms}} \times s \times \{0\} \times \{0\} \times \cdots$$

where s is the nuclear Fréchet space of all rapidly decreasing sequences. Since $s \hookrightarrow \ell_1$ continuously, we have $\lim \operatorname{ind}_n F_n = \lim \operatorname{ind}_n E_n$, and $\{F_n\}$ is a non-strict defining sequence of non-Banach spaces.

Replacing ℓ_1 by ℓ_2, and s by ℓ_1 in the above example, one obtains a strict (LB)-space with a non-strict defining sequence of Banach spaces. $\square$

- No (LF)-space is *both* complete and metrizable.

- Strict (LF)-spaces are always complete, hence non-metrizable.

- Some (LB)-spaces are complete: e.g. $\ell_{\overline{p}} = \lim \operatorname{ind}_n \ell_{p-\frac{1}{K+n}}$, (where $p - \frac{1}{K+1} > 1$).

- Some (LB)-spaces are *not* complete: e.g. Köthe's Text [17], Vol. I, page 437.

- No (LB)-space is metrizable.

- There exist non-strict (LF), non-(LB), non-metrizable (LF)-space: e.g. $\omega \times \ell_{\overline{p}}$ (non-metrizable, since it contains the (LB)-space $\ell_{\overline{p}}$, and non-strict (LF), since there is no coarser norm topology on ω (no Hausdorff topology strictly weaker than the product topology).

So, a natural question would be: *Are (LF)-spaces metrizable? (normable?); if so, when?*

The answer is "yes": Here is a quick and easy

Example 5.3 For each n, let

$$E_n := \underbrace{\omega \times \omega \times \cdots \times \omega}_{n \text{ terms}} \times \ell_1 \times \ell_1 \cdots$$

be the Fréchet space, with the product topology. The (LF)-space $E = \cup_{n=1}^\infty E_n$ is dense in the Fréchet space $\omega \times \omega \times \cdots$, which is isomorphic to ω, and hence metrizable. $\square$

Since an (LF)-space is "*covered*" by its steps, we are reminded of Baire-type coverings. We now pause to make several observations about (LF)-spaces that lead us to a nice classification scheme:

- all (LF)-spaces are barrelled.

- No (LF)-space is a (db)-space.

- No (LB)-space is Baire-like.

- No strict (LF)-space is quasi-Baire.

- Metrizable (LF)-spaces are neither strict (LF)-spaces, nor (LB)-spaces.

Theorem 5.4 (Narayanaswami-Saxon [22]) *An (LF)-space E is metrizable $\iff E$ is Baire-like $\iff E$ does not contain an isomorphic copy of φ.*

So if an (LF)-space E contains an isomorphic copy of φ, E is non-metrizable, and E is not Baire-like.

The next two theorems bring out the interaction of the sequence space φ with Baire-Like, and Quasi-Baire spaces. The role of φ in various other contexts is discussed in [21].

Theorem 5.5 (Saxon [29]) *Let E be a barrelled space which does not contain φ. Then, E is Baire-like.*

Since φ is not metrizable, it follows that a metrizable, barrelled space must be Baire-like. For quasi-Baire spaces, the relationship is much stronger, as shown by

Theorem 5.6 (Narayanaswami-Saxon [22]) *Let E be a barrelled space. The following statements are equivalent:*

a) *E is not quasi-Baire;*

b) *E contains a complemented copy of the space φ;*

c) *E contains a closed $\aleph_0$-codimensional subspace;*

d) *E is isomorphic to the product space $E \times \varphi$;*

e) *E is the strict inductive limit of a strictly increasing sequence of closed barrelled subspaces of E.*

We make the following useful observations.

- Every strict (LF)-space contains a complemented copy of φ.

- Some (LB)-spaces contain a complemented copy of φ.

- In the class of locally convex spaces not containing a complemented copy of φ, quasi-Baire $\Longleftrightarrow$ barrelled.

- an (LF)-space that is quasi-Baire, but not Baire-like contains φ, but *not* as a complemented subspace.

- E contains φ, but not φ complemented, $\Longrightarrow$ E is quasi-Baire but *not* Baire-like.

These observations lead to the following classification of all (LF)-spaces: (Narayanaswami - Saxon [22, 33, 34])
Definition: (Saxon-Narayanaswami)

An (LF)-space is called an

$(LF)_1$-*space* $\Longleftrightarrow$ there exists a defining sequence *none* of whose members is dense in E;

$(LF)_2$-*space* $\Longleftrightarrow$ it is non-metrizable and there exists a defining sequence *each* of whose members is dense in E;

$(LF)_3$-*space* $\Longleftrightarrow$ it is metrizable.

Note:: $(LB)_3$ is meaningless, since an (LB)-space is never metrizable.
Each of these (mutually disjoint) classes is sufficiently rich;

- Every strict (LF)-space is an $(LF)_1$-space;

- some (LB)-spaces are $(LF)_1$-spaces - for example, $\varphi \times \ell_{\overline{p}}$ is a non-strict $(LB)_1$-space;

- The example of Köthe [17] (Vol I. page 437) and $\ell_{\overline{p}}$ are $(LB)_2$-spaces;

- Metrizable/Normable (LF)-spaces exist in abundance [34] and they are all $(LF)_3$-spaces.

Theorem 5.7 (Characterization Theorem (Narayanaswami-Saxon [22])) *An* (LF)-*space E is an*

$\begin{aligned}
(LF)_1\text{-space} &\iff E \text{ contains a complemented copy of } \varphi \\
&\iff E \text{ is not quasi-Baire.} \\
(LF)_2\text{-space} &\iff E \text{ contains a copy of } \varphi \text{ but not a complemented copy} \\
&\iff E \text{ is quasi-Baire, but not Baire-like.} \\
(LF)_3\text{-space} &\iff E \text{ does not contains a copy of } \varphi \\
&\iff E \text{ is Baire-like, (but not } (db)).
\end{aligned}$

Summary:

	$(LF)_3$			$(LF)_2$			$(LF)_1$	
	$\not\Longleftarrow$			$\not\Longleftarrow$			$\not\Longleftarrow$	
(db)	$\Longrightarrow$	Baire-like		$\Longrightarrow$	Quasi-Baire		$\Longrightarrow$	barrelled
	$\not\supseteq \varphi$			$\supseteq \varphi$, but			$\supseteq \varphi$	
			$\not\supseteq \varphi$complement			complement		

6 Properties of (LF)-spaces

- The product of an $(LF)_i$ space with an $(LF)_j$-space is an $(LF)_k$-space where $k = \min\{i,j\}$.

- For all $i = 1, 2, 3$, the product $(LF)_i \times \varphi$ is an $(LF)_1$-space.

- The product of a Fréchet space with φ is an $(LF)_1$-space.

- The product of a metrizable (LF)-space with $\ell_{\overline{p}}$ is an $(LF)_2$-space.

- An infinite product of (LF)-spaces is never an (LF)-space.

- A Hausdorff inductive limits of an increasing sequence of (LF)-spaces is an (LF)-space.

- A finite codimensional subspace of an $(LF)_i$-space is an $(LF)_j$-space $\iff$ it is closed and $i = j$.

- A countable codimensional subspace F of an (LF)-space is an (LF)-space $\iff$ F is closed and not contained in any of the steps E_n.

- The completion of an $(LF)_3$-space is always a Fréchet space; it is never an (LF)-space.

- The completion of an $(LF)_1$ or $(LF)_2$ space may/may not be an (LF)-space.

Proofs of these statements can be found in [22]. The quotient structure of $(LF)_i$-spaces deserves special attention, and is studied in the next section.

7 Quotients of (LF)-spaces

Theorem 7.1 ([34]) *If E is $(LF)_i$-space, $(1 \leq i \leq 3)$, M a closed subspace , and if $E = E_n + M$ for some n, then E/M is a Fréchet space; otherwise, E/M is an $(LF)_j$-space for some $j \geq i$; that is*

$$(LF)_i/M = (LF)_j, \text{ for some } j \geq i \text{ or else a Fréchet space.}$$

In view of this theorem, it is tempting to regard Fréchet spaces as $(LF)_4$-spaces. The usage of the index 4, rather than 0 is obvious, since the index does not decrease when passing to quotients.

Thus, the product of a non-$(LF)_{1,2,3,4}$-space with φ is neither an (LF)-space, nor a Fréchet space.

Theorem 7.2 ([22]) *An $(LF)_3$-space always admits an $(LF)_4$ (i.e., Fréchet) quotient that is infinite dimensional and separable.*

Proof: (Sketch)

Suppose no such quotient exists. Let $(E, \tau) = \lim \text{ind}_n(E_n, \tau_n)$ be an $(LF)_3$-space, so it is Baire-like, and hence Quasi-Baire. Without loss of generality, assume E_1 is dense in E. By Open mapping Theorem, there is an absolutely convex τ_1-nbd U_1 of 0 such that $\overline{U_1}$ is not a τ-nbd of 0; i.e, span$(\overline{U_1})$ is proper in E.

Inductively define absolutely convex τ_n-nbd U_n of 0 such that $C_n = \overline{U_1 + U_2 + \cdots + U_n}$ is not a τ-nbd of 0. Now, $\{nC_n\}$ is an increasing sequence of nowhere dense absolutely convex sets whose union is $E = \bigcup_{n=1}^{\infty} E_n$, a contradiction to the fact that E is Baire-like. $\square$

Remark 7.3 The result is not necessarily true for $(LF)_1$ or $(LF)_2$-spaces, as seen from the example below.

Example 7.4 φ is an $(LF)_1$-space, but no quotient is Fréchet; $\ell_{\overline{p}}$ is an $(LF)_2$-space, but no quotient is Fréchet. On the other hand, if E is a Fréchet space, then $E \times \varphi \in (LF)_1$, and $E \times \ell_{\overline{p}} \in (LF)_2$, and both spaces have E as a Fréchet quotient.

Remark 7.5 Metrizable (LF)-spaces constitute a rich class of incomplete quotients of complete spaces [33]

8 Construction of Metrizable (LF)-spaces

Theorem 8.1 (Narayanaswami-Saxon [34]) *Let $(F, \mathcal{T})$ be a Fréchet space with a sequence $\{P_n\}_{n=1}^{\infty}$ of orthogonal projections such that each of the (necessarily closed) subspaces $P_n[F]$ has a separable (Hausdorff, infinite-dimensional) quotient. Then F contains a dense subspace F_0 which, with the relative topology, is a metrizable (LF)-space.*

(In short, a Fréchet space has a dense (LF)-subspace if it splits into infinitely many parts, each of which has a separable quotient.)

A Banach space E *splits into infinitely many parts* if there exists sequences $\{M_n\}, \{N_n\}$ of closed subspaces of E such that $E = M_1 \oplus N_1, N_1 = M_2 \oplus N_2, N_2 = M_3 \oplus N_3, \cdots$.

Sketch of the **Proof:**

Since each $P_n[F]$ has a separable quotient, by our equivalence (i) $\Longrightarrow$ (v) of Theorem 3.1, there exists a dense proper subspace G_n of $P_n[F]$ which, with a topology τ_n finer than the relativization of $\mathcal{T}$, is a Fréchet space. Set $F_n = P_n^{-1}[G_n]$, and $E_k = \bigcap_{n \geq k} F_n$ for all n, k. Then, one can show, E_k is a Fréchet space with a topology $\mathcal{T}_k$, having a basis of neighborhoods of 0, the set $\{E_k \cap U \cap (\bigcap_{n=k}^p P_n^{-1}[V_n]) : p \geq k$, and U a neighborhood of 0 in $(F, \mathcal{T})$ and V_n is a τ_n-neighborhood of 0 in G_n for $k \leq n \leq p\}$. Clearly, $E_k \subseteq E_{k+1}$, and $\mathcal{T}_{k+1}|_{E_k} \leq \mathcal{T}_k$.

Let $\mathcal{T}_0$ be the finest locally convex topology on $E_0 = \bigcup_{k=1}^\infty E_k$, such that $\mathcal{T}_0|_{E_k} \leq \mathcal{T}_k$, for each k.

Next, one shows

- E_k is *properly* contained in E_{k+1}; (there exists $x \in P_k[F] \setminus G_k$, and since P_n are orthogonal projections, $x \in E_{k+1} \setminus E_k$)

- E_1 is dense in F;

- $\mathcal{T}|_{E_0} \leq \mathcal{T}_0$ (easy!)

- $\mathcal{T}|_{E_0} \geq \mathcal{T}_0$ (some work!)

proving that $(E_0, \mathcal{T}_0)$ is the required dense $(LF)_3$-subspace of the given Fréchet space $E, \mathcal{T})$. $\square$

Corollary 8.2 *The familiar Banach spaces* $\ell_p(1 \leq p \leq \infty), c_0, C[0,1]$ *and* $L_p[0,1](p \geq 1)$ *and the familiar (nuclear) Fréchet spaces* s *and* ω *all have dense subspaces which, with relative topology, are* $(LF)_3$-*spaces.*

Indeed so do all Fréchet spaces with an unconditional basis.

Proof. If $\{x_i\}_{i=1}^\infty$ is an unconditional basis for a Fréchet space E, then letting $\{S_k\}_{k=1}^\infty$ be any partition of $\mathbb{N}$ into infinite, disjoint sets and, for each k, defining $P_k(\sum_{i=1}^\infty a_i x_i) = \sum_{j \in S_k} a_j x_j$ for each $x = \sum_{i=1}^\infty a_i x_i$ in E, we see that $\{P_k\}_{k=1}^\infty$ is a sequence of orthogonal projections, and each infinite-dimensional subspace $P_k[E]$ admits a separable quotient (by the trivial subspace $\{0\}$. Hence Theorem 8.1 applies.

For ℓ_∞, each $P_k[\ell_\infty]$ is isomorphic to ℓ_∞, which is known to have a separable quotient.

For $C[0,1]$, choose an infinite sequence $\{[a_n, b_n]\}_{n=1}^\infty$ of disjoint, non-degenerate subintervals of $[0,1]$ and choose $\{[c_n, d_n]\}_{n=1}^\infty$ such that $a_n < c_n < d_n < b_n$ for each n. Define projections $P_n : C[0,1] \longrightarrow C[0,1]$ by:

$$P_n(f)(t) = \begin{cases} f(t) & \text{for } c_n \leq t \leq d_n \text{ ,} \\ 0 & \text{for } t \notin (a_n, b_n), \\ \text{linear} & \text{on } [a_n, c_n] \text{ and } [d_n, b_n] \end{cases}$$

Each $P_n[C[0,1]]$ is isomorphic to $C[0,1]$, thus is infinite-dimensional and separable, and $\| P_n \| = 1$. Theorem 8.1 applies.

Remark 8.3 $C[0,1]$ has a basis, but not an unconditional basis.

The next Theorem, in conjunction with Theorem 8.1 shows that if a Fréchet space F has a separable quotient which splits into infinitely many parts, then F has a dense subspace which is an (LF)-space.

Theorem 8.4 ([34]) *Let $Q : F \longrightarrow G$ be a continuous, linear surjection of a Fréchet space $(F, \mathcal{T})$ onto a Fréchet space (G, τ). G has a dense subspace G_0 which, with the relative topology, is an (LF)-space if and only if F has a dense subspace F_0 which, with the relative topology, is an (LF)-space containing $Q^{-1}[0]$.*

We now have an approach to the Separable Quotient Problem for Banach spaces via $(LF)_3$-spaces.

We raise the question: *Does every Banach space contain a dense $(LF)_3$-subspace?*

Observe that (by Corollary 8.2) all familiar Banach and Fréchet spaces contain dense metrizable (LF)-subspaces. The following theorem readily follows from Theorems 8.1 and 8.4.

Theorem 8.5 (Saxon-Narayanaswami [34]) *A Fréchet space E contains a dense $(LF)_3$-subspace provided*

    either   (i) *E has separable quotient, which splits infinitely often*

    or   (ii) *E splits infinitely often, and each split admits a separable quotient.*

Note that condition (ii) is satisfied by every non-normable Fréchet space from [10], so it follows that every non-normable Fréchet space is the completion of some (normable) (LF)-space [41].

Open Question: *Does every infinite dimensional Banach space split infinitely often?*

Thus, if the Separable Quotient Problem, and the Splitting Problem have affirmative solutions in the class of Banach Spaces, then, every Banach space will be the completion of an $(LF)_3$-space.

9 Proof of the Theorem 3.1 for the Fréchet Space case

Since we have already developed the necessary tools to prove Theorem 3.1, we devote the next two sections toward this end.

Proof. (i) $\Longrightarrow$ (ii): If M is a closed subspace of E and E/M is infinite-dimensional and separable, then there exists a linearly independent sequence $\{x_n\}$ in E such that $M \cap sp(\{x_n\}) = \{0\}$ and $M + sp(\{x_n\})$ is dense in E. Then, $\cup_{n=1}^{\infty}[M + sp(\{x_1, x_2, \ldots, x_n\})]$ is a dense S_σ-subspace of E.

(ii) $\Longrightarrow$ (iii): clear, since by the Amemiya-Kōmura result, metrizable, barrelled spaces are Baire-like.

(iii) $\Longrightarrow$ (iv) follows trivially from the definition of a (db)-space.

(iv) $\Longrightarrow$ (iii): Suppose M is a dense subspace and M is the union of an increasing sequence $\{M_n\}$ of subspaces none of which is both dense in M and barrelled. Then, if M

is non-barrelled, we are through. If M is barrelled, then M is quasi-Baire, so that some M_n is dense in M, hence dense and non-barrelled in E. Thus (iii) and (iv) are equivalent.

(iii) $\implies$ (v): Suppose N is any dense non-barrelled subspace of E, and let C be any countable subset of E. Then, $sp\,(N \cup C)$ is non-barrelled [32] so there exists a closed, absolutely convex set V such that $M = sp\,(V) \supset N \cup C$ and V is not a $\mathcal{T}|_M$-neighborhood of 0. The collection $\{k^{-1}V \cap U : k$ is a positive integer and U is a closed neighborhood of 0 in $(E, \mathcal{T})\}$ forms a base of $\mathcal{T}$-complete neighborhoods of 0 for a metrizable topology $\mathcal{T}_0$ on M. By Horváth ([12] Prop. 5, page 207), $(M, \mathcal{T}_0)$ is complete, and thus is a Fréchet space continuously included in E. M is a dense proper subspace, since it contains N and since V is not a neighborhood of 0 in the barrelled space E.

Clearly, (v) $\implies$ (iii) by the Open Mapping Theorem.

(iii) $\implies$ (i): if E is Banach, the implication is given in [35]; if E is non-Banach, (i) holds by Eidelheit [10], since ω is separable. Thus (i) through (v) are equivalent. $\square$

The equivalence of [(i)–(v)] with (vi) is provided in the next section.

10 Non-Banach Fréchet Space case

For a non-normable Fréchet space, condition (vi) of Theorem 3.1 can be improved, as shown by

Theorem 10.1 (Eidelheit [10], Valdivia-Perez Carreras [41], Narayanaswami-Saxon [34]) *If E is a non-Banach Fréchet space, then, conditions (i)–(v) of Theorem 3.1 hold, and moreover E has a dense (metrizable) (LF)-subspace.*

Proof. By Eidelheit [10] , E has a quotient isomorphic to ω, so conditions (i)–(v) of Theorem 3.1 hold. Also, ω has a dense (LF)-subspace by Example 5.3 Thus, so does E. $\square$

Theorem 10.2 ([34]) *Let $(E, \mathcal{T})$ be a Banach space. The conditions (i)–(v) of Theorem 3.1 (for Fréchet space case) are equivalent to the following condition:*
(vi) $[(P_8)]$: there exists a dense subspace E_0 and a topology $\mathcal{T}_0$ on E_0 finer than $\mathcal{T}|_{E_0}$ such that $(E_0, \mathcal{T}_0)$ is a normable (LF)-space.

Proof. (vi) implies [(i) - (v)]: Suppose $(E_0, \mathcal{T}_0)$ exists as in (vi). Then there is a strictly increasing sequence $\{(E_n, \mathcal{T}_n)\}_{n=1}^{\infty}$ of Fréchet spaces such that $E_0 = \cup_{n=1}^{\infty} E_n$ where each $\mathcal{T}_n$ is finer than $\mathcal{T}_0|_{E_n}$, hence finer than $\mathcal{T}|_{E_n}$. Thus, either $\cup_{n=1}^{\infty} \bar{E}_n$ is a dense S_σ-subspace of E [(ii)] or one of the Fréchet spaces $(E_n, \mathcal{T}_n)$ is a dense (and necessarily proper) subspace of E [(v)].

(i) implies (vi) :

Assume M is a closed subspace of E such that E/M is a separable (infinite-dimensional) Banach space. One readily sees that E/M must densely and continuously include a copy of the Banach space ℓ_1. [Let $\{x_i, f_i\}_{i=1}^{\infty}$ be any biorthogonal sequence such that $sp\,(\{x_i\})$ is dense in E/M; identify the unit vectors in ℓ_1 with small multiples of the x_i's.] Let Q be the quotient map of E onto E/M. One shows that $F = Q^{-1}[\ell_1]$ is a Fréchet space with a topology μ finer than $\mathcal{T}|_F$, such that $Q|_F$ is μ-continuous onto ℓ_1. [In fact, (F, μ) is a Banach space.] Also, F is dense in E, trivially, since ℓ_1 is dense in E/M and Q is open. Clearly, ℓ_1 has a dense (LF)-subspace G_0 (Corollary 8.2). The desired conclusion follows from Theorem 8.4 on quotients. $\square$

11   The Separable Quotient Problem for (LF)-spaces

Theorem 11.1 (Saxon-Narayanaswami [22]) *Every (LF)-space has a separable quotient.*

Proof: If (E, τ) is not quasi-Baire, it contains φ complemented; φ is infinite dimensional and separable.

If $(E, \tau) = \lim \operatorname{ind}_n (E_n, \tau_n)$ is quasi-Baire, some E_k is dense in E. By Ptak's open mapping theorem, the identity map $I_k : (E_k, \tau_k) \longrightarrow (E, \tau)$ is not almost open since $E_k \subsetneqq E_{k+1} \subset E$, so for there exists an absolutely convex τ_k-neighborhood V of 0 such that $\bar{V}$ is not a neighborhood of 0 in (E, τ), although it is a τ-barrel. Let $F = sp\,(\bar{V})$, so F is dense in (E, τ) and $\bar{V}$ is not a τ-relative neighborhood of 0 in F. Then, F must have uncountable codimension in the barrelled space E. Hence for some $j, E_j \cap F$ must be infinite dimensional in E_j. Let $\{U_n\}$ be a sequence of absolutely convex (basic) neighborhoods of 0 in (E_j, τ_j) such that $U_n + U_n \subset U_{n-1}$ for $n \geq 2$. Inductively choose $\{V_n\}, \{x_n\}, \{f_n\}$ such that $V_1 = \bar{V}, V_{n+1} = V_n + \{ax_n : |a| \leq 1\}, x_n \in U_n \setminus sp\,(V_n), f_n \in V_n^\circ, f_n(x_i) = 1$ if $i = n$, and 0 if $i > n$.

Given $y \in V_1$, inductively define $a_1 = -f_1(y), a_n = -f_n(y + \sum_{i=1}^{n-1} a_i x_i)$ for $n \geq 2$. Observe that $|a_1| \leq 1$, and if $|a_i| \leq 1$ for $1 \leq i \leq p$, then $y + \sum_{i=1}^{p} a_i x_i$ is in V_{p+1}, so that $|a_{p+1}| \leq 1$. Thus, $a_i x_i \in U_i$ for each i, $\sum_{i=1}^{\infty} a_i x_i$ is (absolutely) convergent to some z in E_j. Using the defining property of (LF)-spaces we see that it is also τ-convergent in E. Next, a simple computation shows that for each n, $f_n(y + z) = 0$. Thus, $y + z$ belongs to (the τ-closed subspace) $N = \cap_{n=1}^{\infty} f_n^{-1}[0]$, and $\{y + z - \sum_{i=1}^{n} a_i x_i\}_{n=1}^{\infty}$ is a sequence in $N + sp\,(\{x_i\}_{i=1}^{\infty})$, which is τ-convergent to y. Therefore, the τ-closure of the subspace $N + sp\,(\{x_i\}_{i=1}^{\infty})$ contains V_1, hence F, and hence E; that is $N + sp\,(\{x_i\}_{i=1}^{\infty})$ is dense in (E, τ). We conclude that $sp\,(\{x_i + N\}_{i=1}^{\infty})$ is dense in the quotient space E/N of (E, τ), and therefore, E/N is infinite dimensional and separable. $\square$

Corollary 11.2 *Let (E, η) be a Hausdorff locally convex space, and suppose that $(E, \tau) = \lim \operatorname{ind}_n(E_n, \tau_n)$ for some topology τ with $\eta \leq \tau$. Then, (E, η) has a separable quotient.*

The following are some interesting results in this context.

- Every $(LF)_3$-space admits a quotient which is a separable, infinite-dimensional Fréchet space, i.e, an $(LF)_4$-space). (Theorem 7.2)

- Every $(LF)_2$ and $(LF)_3$ space (more generally non-strict (LF)-spaces) have a defining sequence each of whose members admit a separable quotient.

- There exists a defining sequence for an (LF)-space E each of whose members have a separable quotient if and only if E is not isomorphic to $F \times \varphi$, where F is a Fréchet space not having a separable quotient.

12   Properly Separable Quotients in Barreled Spaces

This section summarizes the work done in [25, 24]. Let us recall that an infinite dimensional locally convex space is separable if and only if there exists a dense $\aleph_0$-dimensional subspace.

Definition [25]: A locally convex space is *Properly Separable* if and only if there exists a *proper* dense $\aleph_0$-dimensional subspace.

The following are examples of properly separable spaces:

- A complete metrizable space (since it has dimension $> \aleph_0$)

- metrizable barrelled spaces (since φ is the only barrelled space of dimension $\aleph_0$)

If the dimension is $\aleph_0$, then, properly separable $\iff$ $E' \neq E^*$ (algebraic dual). If the dimension is $\aleph_0$, and the space has Mackey topology, then, properly separable $\iff$ it does *not* have the finest locally convex topology. An infinite dimensional separable metrizable space is properly separable (since it does *not* have the finest locally convex topology). Countable dimensional quotients of barrelled spaces are *not* properly separable (for, quotients of barrelled spaces are barrelled, so countable dimension implies finest locally convex topology), for example φ.

Question: *Does a barrelled Hausdorff locally convex space admit a properly separable quotient? (equivalently, a separable quotient of uncountable dimension?)*

For Banach spaces, this amounts to asking: *does there exist an infinite dimensional separable quotient?* (since no quotient can be countable dimensional)

We define three classes of locally convex spaces

$\mathcal{A}_0 = \{E \mid E \text{ has a separable quotient }\}$

$\mathcal{A} = \{E \mid E \text{ has a properly separable quotient }\}$

$\mathcal{B} = \{E \mid E \text{ has a dense non-barrelled subspace }\}$

For Banach spaces, the classes $\mathcal{A}_0$ and $\mathcal{A}$ coincide, and $E \in \mathcal{A} \iff E \in \mathcal{B}$ (Saxon-Wilansky [35]). For Fréchet spaces $E \in \mathcal{A} \iff E \in \mathcal{B}$ (Saxon-Narayanaswami [34]) For non-normable Fréchet spaces $E \in \mathcal{A}$ and $E \in \mathcal{B}$ (Eidelheit [10], Valdivia-Pérez Carreras [41]) For barrelled spaces $E \in \mathcal{A} \implies E \in \mathcal{B}$ (Robertson-Narayanaswami [25])

We do not know whether the reverse implication holds for barrelled spaces. But, there exists a non-barrelled space $E \in \mathcal{B} \setminus \mathcal{A}$ (an example is provided later).

It is easy to see that

- $E \in \mathcal{A},(\text{ or } \mathcal{B})$, F any locally convex space $\implies E \times F \in \mathcal{A}(\text{ or } \mathcal{B})$.

- If M is closed subspace of E, $E/M \in \mathcal{A}(\text{ or } \mathcal{B}) \implies E \in \mathcal{A}(\text{ or } \mathcal{B})$.

- $E \in \mathcal{A} (\text{or } \mathcal{B})$ does *not* imply $E/M \in \mathcal{A} (\text{or } \mathcal{B})$.

Example 12.1 Let E_0 have finest locally convex topology, and let $E = E_0 \times \omega$. Then, $E \in \mathcal{A}$, $E/\omega \cong E_0 \notin \mathcal{A}$. (same example works for $\mathcal{B}$.) □

- If $E \in \mathcal{A}$, a (closed, barrelled) subspace of E need not be in $\mathcal{A}$.

Example 12.2 Let E_0 be as in the above example, and let $E = \omega \times E_0$, $F = \{0\} \times E_0$. F is closed in E, both E and F are barrelled, $E \in \mathcal{A}$ (respectively in $\mathcal{B}$) $F \notin \mathcal{A}$ (respectively not in $\mathcal{B}$) □

Question: *If some (closed barrelled) subspace of a barrelled space E is in $\mathcal{A}$, does it follow that $E \in \mathcal{A}$?*

If the answer is "yes", then every Banach space would have a separable quotient, for, $D \subset E$, $\aleph_0$-dimensional implies $\bar{D} \in \mathcal{A}$.

- Every (LF)-space belongs to $\mathcal{A}_0$ (Saxon-Narayanaswami).

- $(LF)_2$ and $(LF)_3$-spaces belong to $\mathcal{A}$, hence to $\mathcal{B}$.

- φ is an $(LF)_1$-space, $\varphi \notin \mathcal{A}$.

- If $E \in (LF)_i (i = 2, 3)$ or a non-normable Fréchet space, and $F \in (LF)_1$, $E \times F$ is $(LF)_1$ in $\mathcal{A}$.

  Question: *is there an $(LF)_1$-space, other than φ not in $\mathcal{A}$ (not in $\mathcal{B}$)?*

- If there exists a Banach space F *without* a separable quotient, there exists an $(LB)_1$-space $E = F \times \varphi$ not in $\mathcal{B}$ (not in $\mathcal{A}$)

- A vector space in its finest locally convex topology is not in $\mathcal{A}$ (or $\mathcal{B}$). But there are others.

We recall (Eberhardt-Roelcke [9]) that a locally convex space E is a (GM)-space if each linear map from E into any locally convex metrizable space with a closed graph is continuous. Clearly, every (GM)-space is barrelled. Spaces with the finest locally convex topology are (GM)-spaces, but not conversely.

Also, every dense subspace of a (GM)-space is barrelled, so no (GM)-space is in $\mathcal{B}$ (in $\mathcal{A}$).

Definition: (Narayanaswami, Robertson) [25]: A locally convex space E is a (GS)-*space* if each linear map from E into any separable locally convex metrizable space with a closed graph is continuous. We say that E is a (QS)-*space* if every quotient of E with a coarser separable metrizable locally convex topology has its finest locally convex topology.

- In barrelled spaces, $(GS) \iff (QS)$.

- (QS)-spaces have no properly separable quotients.

- A (QS)-space (respectively, (GM)-space) E has a separable quotient if and only if $E \cong \varphi + M$ for some (QS)-space (respectively, (GM)-space) M.

- Thus a non-barrelled (QS)-space is an example in $\mathcal{B}$ but not in $\mathcal{A}$ (promised example). There exist plenty of non-barrelled (QS)-spaces (Köthe [17] §22.5 (5)).

- In barrelled (QS)-spaces, every dense subspace is barrelled, so they do not belong to $\mathcal{A}$ (to $\mathcal{B}$).

Remark 12.3   1. If every separable quotient of E is properly separable, then E can not contain φ complemented;

2. Conversely, if E has Mackey topology, and E does not contain φ complemented, then every separable quotient of E is properly separable. This condition covers all Fréchet spaces.

3. If E contains φ complemented, a separable quotient of E may or may not be properly separable: ($F \in \mathcal{A} \Rightarrow F \times \varphi \in \mathcal{A}$; but $\varphi \cong \varphi \times \varphi \notin \mathcal{A}$.)

The next theorem yields two interesting classes that admit properly separable quotients.

Theorem 12.4 (Robertson [24]) *Let E be a strict inductive limit of a strictly increasing sequence $\{E_n\}$ of Fréchet spaces (respectively, metrizable barrelled spaces), with at least one of the E_n's non-normable. Then E has properly separable quotient, that is, $E \in \mathcal{A}$.*

We conclude this section with the following

Theorem 12.5 (Robertson [24]) *Let E be a normed barrelled space. Then E has a separable quotient if and only if E has a dense non-barrelled subspace.*

13 Some Questions

In addition to the several questions raised throughout this exposition, we add the following open questions.

1. The Separable Quotient Problem for Banach spaces.

2. The Splitting Problem for Banach spaces.

3. For which classes of spaces, Separable Quotient Problem is equivalent to the existence of dense non-barrelled subspace?

4. If E is a strict inductive limit of a sequence of Banach spaces, (respectively, metrizable barrelled spaces) each with a separable quotient, must E have a properly separable quotient?

5. In the class of barrelled spaces, Quasi-baire $\iff \not\supseteq \varphi$ complemented. Characterize the class of barrelled spaces that just do not contain φ by means of a suitable Baire covering property.

6. Is it true that a Fréchet space E has a dense, Baire-like (equivalently barrelled), non-(db)-subspace if and only if E has a dense (metrizable) (LF)-subspace?

References

[1] I. AMEMIYA and Y. KŌMURA, *Über nicht-vollständige Montelräume*, Math. Ann. **177** (1968), 273—277.

[2] J. ARIAS de REYNA, *Dense hyperplanes of first category*, Math. Ann. **249** (1980), 111—114.

[3] J. ARIAS de REYNA, *Normed barely Baire spaces*, Israel J. Math. **42** (1982), 33—36.

[4] G. BENNETT and N.J. KALTON, *Inclusion theorems for FK-spaces* , Canadian J. Math. **25** (1973), 511—524.

[5] P. PÉREZ CARRERAS and J. BONET, *Barrelled Locally Convex Spaces*, North Holland Mathematics Studies, Vol. **131** (1987).

[6] P. DIEROLF, S. DIEROLF and L. DREWNOWSKI, *Remarks and examples concerning unordered Baire-like and ultrabarrelled spaces*, Colloq. Math. **39** (1978), 109—116.

[7] J. DIESTEL, S.A. MORRIS and S.A. SAXON, *Varieties of Linear Topological Spaces*, Trans. Amer. Math. Soc. **172** (1972), 207—230.

[8] J. DIEUDONNÉ and L. SCHWARTZ, *La dualité dans les espaces* (F) *et* (LF), Ann. Inst. Fourier (Grenoble) **1** (1950), 61—101.

[9] V. EBERHARDT and W. ROELCKE, *Über eine Graphensatz für lineare abbildungen mit metrisierbar Zielräum*, Manuscripta Math. **13** (1974), 53—78.

[10] M. EIDELHEIT, *Zur Theorie der Systeme linearer Gleichungen*, Studia Math. **6** (1936), 139—148.

[11] A. GROTHENDIECK, *Produits tensoriels topologiques et espaces nucleaires*, Memoirs Amer. Math. Soc. No. **16** (1959)

[12] J. HORVÁTH, *Topological Vector Spaces and Distributions*, Addison-Wesley 1966.

[13] W. B. JOHNSON and H. P. ROSENTHAL, *On* w^* *basic sequences*, Studia Math. **43** (1972), 77—95.

[14] B. JOSEFSON, *Weak sequential convergence in the dual of a Banach space does not imply norm convergence*, Bull. Amer. Math. Soc. **81** (1975) 166—168.

[15] B. JOSEFSON, *Weak sequential convergence in the dual of a Banach space does not imply norm convergence*, Arkiv für Math. **13**, (1975), 78—89.

[16] N. KITSUNEZAKI, K. KERA and Y. TERAO, *On Unordered Ultra Baire-Like Spaces*, TRU Math. **19-1** (1983), 89—92.

[17] G. KÖTHE, *Topological Vector Spaces I*, Die Grundlehren der Mathematischen Wissenschaften, **159**, Springer-Verlag, Berlin-Heidelberg-New York, 1969.

[18] H. E. LACEY, *Separable Quotients of Banach Spaces*, An. Acad. Brasil. Cienc., **44** 2 (1972) 185—189.

[19] J. LINDERSTRAUSS, *Some Aspects of the Theory of Banach Spaces*, Advances in Mathematics **5**, (1970) 159—180.

[20] P. P. NARAYANASWAMI, *Baire Properties of* (LF)-*spaces*, Note di Mathematica **VII**, 1—18 (1987)

[21] P. P. NARAYANASWAMI, *The Strongest Locally Convex Topology on an* $\aleph_0$-*dimensional Space*, International Conference on Functional Analysis and Related Topics, Hokkaido University, Sept. 1990 (to appear).

[22] P.P. NARAYANASWAMI and S.A. SAXON, (LF)-*spaces, Quasi-Baire spaces and the Strongest Locally Convex Topology*, Math. Ann. **274** (1986), 627—641.

[23] A. NISSENZWEIG, *On* w^* *sequential convergence*, Israel J. Math. **22**, (1975), 266—272.

[24] W. ROBERTSON, *On Properly Separable Quotients of Strict* (LF)-*spaces*, J. Austral. Math. Soc. (Series A) **47** (1989) 307—312.

[25] W. ROBERTSON and P. P. NARAYANASWAMI, *On Properly Separable Quotients and Barrelled Spaces*, University of Western Australia Research Reports No. **50** (1988), 1-18.

[26] A.P. ROBERTSON and W. ROBERTSON, *On the Closed Graph Theorem*, Proc. Glasgow. Math. Assoc. **3** (1956), 9—12.

[27] W.J. ROBERTSON, I. TWEDDLE and F.E. YEOMANS, *On the Stability of Barrelled Topologies III*, Bull. Austral. Math. Soc. **22** (1980), 99—112.

[28] H. P. ROSENTHAL, *On Quasi-Complemented Subspaces of Banach Spaces, with an Appendix on Compactness of Operators from* $L^p(\mu)$ *to* $L^r(\nu)$, J. Functional Analysis **4** (1969), 176—214.

[29] S.A. SAXON, *Nuclear and Product spaces, Baire-like spaces and the strongest Locally Convex Topology*, Math. Ann. **197** (1972), 87—106.

[30] S.A. SAXON, *Some Normed Barrelled Spaces which are not Baire*, Math. Ann. **209** (1974), 153—160.

[31] S.A. SAXON, *Two characterizations of linear Baire spaces*, Proc. Amer. Math. Soc. **45** (1974), 204—208.

[32] S. A. SAXON, and M. LEVIN, *Every countable-codimensional subspace of a barrelled space is barrelled*, Proc. Amer. Math. Soc. **29** No. 1 (1971), 91-96.

[33] S.A. SAXON and P.P. NARAYANASWAMI, *Metrizable* (LF)-*spaces, (db)-spaces and the Separable Quotient Problem*, Bull. Austral. Math. Soc. **23** (1981), 65—81.

[34] S.A. SAXON and P.P. NARAYANASWAMI, *Metrizable[Normable] (LF)-spaces and two classical problems in Fréchet [Banach] spaces*, Studia Math. **T. XCIII** (1989), 1—16.

[35] S.A. SAXON and A. WILANSKY, *The Equivalence of some Banach Space Problems*, Colloq. Math. **37** (1977), 217—226.

[36] Y. TERAO, N. KITSUNEZAKI, M. ABE and K. KERA, *On (dub)-spaces and Related Topics*, TRU Math. **21-1** (1985) 55—60.

[37] A. TODD and S.A. SAXON, *A Property of Locally Convex Baire Spaces*, Math. Ann. **206** (1973), 23—34.

[38] M. VALDIVIA, *Absolutely Convex Sets in Barrelled Spaces*, Ann. Inst. Fourier, Grenoble, **21** (1971), 3—13.

[39] M. VALDIVIA, *On Weak Compactness*, Studia Math. **49** (1973), 35—40.

[40] M. VALDIVIA, *On Suprabarrelled Spaces*, Functional Analysis, Holomorphy and Approximation Theory, Springer-Verlag, Lecture Notes in Mathematics, **843** (1981), 572—580.

[41] M. VALDIVIA and P. PERÉZ CARRERAS, *Sobre espacios (LF) metrizables*, Collectanea Math. **33** (1982), 299—303.

[42] A. WILANSKY, *Modern Methods in Topological Vector Spaces*, McGraw-Hill, Inc. (1978).

Department of Mathematics
Memorial University of Newfoundland
St. John's, Newfoundland
CANADA A1C 5S7

A computer-assisted analysis of the two dimensional Navier-Stokes equations

H. OKAMOTO[1], M. SHŌJI[2] AND M. KATSURADA[3]

Abstract. Complicated behaviour is common in solutions to the Navier-Stokes equations when the viscosity is small. The mechanism of the complexities is obscure even today. Here is a reason why numerical computations play an important role in making a qualitative picture of the Navier-Stokes flows. The purpose of this paper is to explain, through examples, necessity of numerical computations in the analysis of the Navier-Stokes equations. More specifically, we consider two dimensional freely decaying flow and the Kolmogorov problem.

§1. Freely decaying flow.

Let us begin with a simple example which clearly shows the necessity of numerical computations. The content of this section is a collection of arguments which are rather well known in physics. We include this as a mathematical introduction to the Kolmogorov problem, which will be discussed later in this paper. We consider the two dimensional Navier-Stokes equations:

$$(1.1) \qquad \frac{\partial u}{\partial t} + u\frac{\partial u}{\partial x} + v\frac{\partial u}{\partial y} = \nu\Delta u - \frac{1}{\rho}\frac{\partial p}{\partial x} + f_1(x,y),$$

$$(1.2) \qquad \frac{\partial v}{\partial t} + u\frac{\partial v}{\partial x} + v\frac{\partial v}{\partial y} = \nu\Delta v - \frac{1}{\rho}\frac{\partial p}{\partial y} + f_2(x,y),$$

$$(1.3) \qquad \frac{\partial u}{\partial x} + \frac{\partial v}{\partial y} = 0,$$

$$(1.4) \qquad u(0,x,y) = u_0(x,y), \quad v(0,x,y) = v_0(x,y)$$

where $(u,v), p, \rho, \nu$ are velocity vector, pressure, mass density, and the kinematic viscosity, respectively. In the vector notation, we write as

$$\vec{u} = (u,v), \quad \vec{u}_0 = (u_0,v_0), \quad \vec{f} = (f_1,f_2).$$

The force $\vec{f} = (f_1, f_2)$ is independent of the time t. The flow region is a rectangle $[-a,a] \times [-b,b]$ and the periodic boundary condition is imposed in both directions. Thus we consider the Navier-Stokes equations on a two dimensional flat torus, which is denoted by T.

As is usual, we define the following spaces of solenoidal vector fields:

$$H = \left\{ (u(x,y),v(x,y)) \;\middle|\; \frac{\partial u}{\partial x} + \frac{\partial v}{\partial y} = 0, \; \int_T (|u|^2 + |v|^2)dxdy < \infty \right\}$$

[1] Research Institute for Mathematical Sciences, Kyoto University, Kyoto 606-01 Japan
[2] Department of Mathematical Sciences, University of Tokyo, Tokyo 113 Japan
[3] Department of Mathematics, Meiji University, Tama-ku, Kawasaki, 214 Japan

$$X = \left\{ (u(x,y), v(x,y)) \,\middle|\, \frac{\partial u}{\partial x} + \frac{\partial v}{\partial y} = 0, \ u, v \in C^{1,\epsilon}(T) \right\}$$

where $C^{1,\epsilon}(T)$ is the Hölder space ($\epsilon > 0$).

As is well known, the following two propositions hold true:

(A) Assume that $\nu > 0$. Then the above system of evolution equations is well-posed in H for any time interval, under a mild condition on $\vec{f}$. For instance, $\vec{f} \in L^2(T)^2$ is sufficient ([4]).

(B) Assume that $\nu = 0$. Then the above system of evolution equations is well-posed in X for any time interval, under a certain condition on $\vec{f}$. For instance, $\vec{f} \in C^{\epsilon}(T)^2$ is sufficient ([2]). It is well-posed also in the Sobolov space $H^3(T)$ ([3]).

Thus, as far as the well-posedness is concerned, we can say that a basic theory has already been established in the two dimensional problem. However, there remains much room for the theory of two dimensional turbulence, which is quite independent of the well-posedness. This becomes clear if we note the following proposition.

(C) Assume that $\vec{f} \equiv 0$ and $\nu > 0$. Then $\int_T \vec{u}\,dx\,dy$ is constant in time. If $\int_T \vec{u}_0$ is zero, then the solution goes to zero exponentially (the energy inequality, [10]).

This proposition implies that the vector field approaches zero whatever the viscosity or the initial velocity may be. In particular, the universal attractor is trivial in $H/\mathbf{R}$, i.e., it equals $\{0\}$. This, however, should not be interpreted as that the flow is trivial if the outer force is absent. On the contrary, the flow can be complex. The complexity eventually disappears as is implied by (C). However, the duration of the complexity increases indefinitely as $\nu \to 0$. The flows described by (1.1-4) with $\vec{f} \equiv 0$ are, by definition, freely decaying flows. From here to the end of this section, we consider the freely decaying flows.

In order to see turbulent behaviour, we take a look at the vorticity distribution. The vorticity, which is a scalar function defined by

$$\omega = \frac{\partial v}{\partial x} - \frac{\partial u}{\partial y},$$

changes in a complex manner. Since we are assuming that $\vec{f} \equiv 0$, we have

$$\|\vec{u}(t)\| \le \|\vec{u}_0\| \quad \text{if} \quad \vec{u}_0 \in H, \quad \|\vec{\omega}(t)\| \le \|\vec{\omega}_0\| \quad \text{if} \quad \vec{u}_0 \in H \cap H^1(T)^2.$$

Furthermore,

$$\|\vec{u}(t)\| \equiv \|\vec{u}_0\|, \quad \|\vec{\omega}(t)\| \equiv \|\vec{\omega}_0\|$$

if $\nu = 0$ and $\vec{u}_0 \in X$. On the other hand, no estimate is available for the second or higher order derivatives of $\vec{u}$ as $t \to \infty$ when $\nu = 0$. Kida and Yamada [7] numerically computed the 2D Euler flows in the case where $\vec{f} \equiv 0$ and $\nu = 0$. They found that, when the initial conditions are appropriately chosen, the vorticity behaves

$$\|\nabla \omega\|_{L^2(T)} \sim e^{at} \qquad \text{as } t \to \infty,$$

for some positive constant a. The growth like this does not always happen: for instance, if the solution is stationary, then $\|\nabla\omega\|$ is constant. It is, however, a plausible assumption that exponential growth of $\|\nabla\omega\|$ is typical, i.e., the initial conditions which do not lead to the exponential growth are negligible in $X \cap C^2(T)$. This proposition makes us cautious in the long-time numerical integration of the Euler equations. In fact, for sufficiently large $t > 0$, the derivatives of ω become quite big somewhere in the torus T. If they are so big that a computer can not realize them, then we will see spurious " blow-up in finite time". However, the well-posedness in X prohibits the blow-up in finite time. In other words, C^∞ solutions look like $C^1 \setminus C^2$ solutions. Note, however, that not all exponential growth mean turbulent motions. We say complicated or turbulent only if the set where $|\nabla\omega|$ is large is a complicated subset of the plane. How to define complicatedness is ambiguous. But Majda' criterion ([8]) is a very charming one. From his argument, it seems to us that the following are true: The contours of ω, i.e., $\{x \; ; \; \omega(t,x) = c \,\}$ where c is a constant, are tamed curves while t is small. If ν is small, they get winded for sufficiently large t. Thus the length of the curve becomes indefinitely large and the curvature is not a bounded function of t. He conjectured as follows: If ω is piecewise constant and $\nu = 0$, then the initially smooth contours may form cusps in finite time. We will give a numerical evidence for the case of $\nu > 0$ in §3.

As we have seen, we have a very limited number of tools for mathematical analysis of two dimensional turbulence in freely decaying flows. Dynamical systems theory, which mostly neglects transient phenomena, does not seem to be able to play an important role. As of now, dynamical systems theory works well when the dimension of the attractor is relatively small but nonzero, say five or ten. Freely decaying flow has an attractor of dimension zero and is excluded. What is called fully developed turbulence would have an attractor of dimension more than hundreds of thousands. On the other hand, the Kolmogorov flow, which we will introduce in the next section, seems to be accessible from the dynamical systems theory.

§2. The Kolmogorov problem.

The Kolmogorov problem is to solve the Navier-Stokes equations in two dimensional flat tori under a special driving force. More precisely, we solve the Navier-Stokes equations (1.1-4) in T with $f_1(x,y) = \gamma \sin(\pi y/b)$ and $f_2(x,y) \equiv 0$, where γ is a positive constant.

We now introduce a stream function ψ. Thus the components of the velocity field are $u = \psi_y$ and $v = -\psi_x$. Here and hereafter the subscripts mean differentiations. Under the present force $\vec{f}$, the equations (1.1-3) are equivalent to :

$$(2.1) \qquad \frac{\partial}{\partial t}\Delta\psi - \nu\Delta^2\psi - J(\psi, \Delta\psi) = \frac{\gamma\pi}{b}\cos(\pi y/b),$$

where J is a bilinear form defined by

$$J(u,v) = u_x v_y - u_y v_x.$$

Our nondimensional form is obtained by the following transformation of variables:

$$(x',y') = \left(\frac{\pi x}{b}, \frac{\pi y}{b}\right), \qquad \psi'(x',y') = \frac{\nu\pi^3}{\gamma b^3}\psi(x,y), \qquad t' = \frac{\gamma b}{\nu\pi}t.$$

We then define the Reynolds number R by

$$R = \frac{\gamma b^3}{\nu^2 \pi^3}.$$

After dropping the primes, we have the following equation:

(2.2)
$$\frac{\partial}{\partial t}\triangle\psi - \frac{1}{R}\triangle^2\psi - J(\psi, \triangle\psi) = \frac{1}{R}\cos y.$$

This equation is considered in $[-\pi/\alpha, \pi/\alpha] \times [-\pi, \pi]$, where $\alpha = b/a$. We denote this domain by the symbol T_α and call it a two dimensional flat torus of aspect ratio α. If the right hand side of (2.2) is absent, then it represents freely decaying flows.

We first note that $\psi(t, x, y) \equiv -\cos y$ satisfies all the requirements for any $R > 0$. We call this a basic solution. The velocity field of the basic solution is given by $(u, v) = (\sin y, 0)$, which represents a shear flow parallel to the x-axis. We consider (2.2) in the following function space:

$$Y = \{\phi \in H^4(\mathsf{T}_\alpha)/\mathsf{R} \ ; \quad \phi(x, y) \equiv \phi(-x, -y) \},$$

where $H^4(\mathsf{T}_\alpha)$ is a Sobolev space. The symbol $/\mathsf{R}$ implies that only those functions with zero spatial mean are collected.

As $R \to +\infty$, (2.2) approaches the Euler equation. It is easy to prove the following

PROPOSITION 2.1. *For a given $T > 0$ and an initial condition, the solution to (2.2) converges, as $R \to +\infty$, to the solution of the Euler equation uniformly in $t \in [0, T]$.*

However, the behaviour of the solutions with finite R as $t \to \infty$ are different from those of the Euler equation. For instance, the following theorem of Iudovich is very remarkable:

THEOREM 2.1 (IUDOVICH [1]). *For any $1 \le \alpha < +\infty$, $0 < R < \infty$, and any initial value $\psi(0, x, y) \in H^4(\mathsf{T}_\alpha)/\mathsf{R}$, the solution to (2.2) approaches exponentially toward the basic solution as $t \to +\infty$. Namely the basic solution is globally and asymptotically stable for any Reynolds number if $\alpha \ge 1$.*

Since the Euler equation, as an infinite dimensional Hamiltonian system, does not allow asymptotically stable steady state, Iudovich's theorem prohibits the convergence uniform in $t \in [0, +\infty)$ of the Navier-Stokes flow to the Euler flow.

§3. Bifurcations of stationary solutions.

In this section we consider stationary solutions to (2.2) in X. Thus, we seek solutions to the following equations in Y:

(3.1)
$$\frac{1}{R}\triangle^2\psi + J(\psi, \triangle\psi) + \frac{\cos y}{R} = 0.$$

Iudovich's theorem implies, in particular, that there is one and only one stationary solution – the basic solution – when $\alpha \ge 1$. The point is that this is true *however large the Reynolds number may be*. On the other hand, as we will show later, there are stationary solutions other than the basic ones when $\alpha < 1$. It is easy to verify that if $(R, \psi(x,y))$ ($|x| < \pi/\alpha$, $|y| < \pi$) solves (3.1) in Y with aspect ratio α, then $(R, \psi(x,y))$ ($|x| < m\pi/\alpha$, $|y| < \pi$) satisfies (3.1) in Y with aspect ratio α/m for any positive integer m. But the converse is not true in general. Therefore the smaller α implies in general larger number of stationary solutions.

The dynamics of (2.2) is trivial from the dynamical systems viewpoint when $\alpha \geq 1$. The question now is to see what happens when $\alpha < 1$. Since there is one and only one stationary solution when $\alpha \geq 1$, the first question is " How many stationary solutions are there when $\alpha < 1$?" and " What are their stability?". As we will show below, these simplest questions are far from easy to answer. Although the details are presented later in this section, we here list some new results:

(1) When $1/2 \leq \alpha < 1$, there is one and only one bifurcation from the basic solution.

(2) The bifurcating branch has turning points if $\alpha < 1$ is sufficiently close to one.

(3) For some $\alpha < 1$, solutions different from the basic one can exist not only in the range $R > R^*(\alpha)$ but also in the range $R < R^*(\alpha)$, where $R^*(\alpha)$ is the critical Reynolds number.

(4) A Hopf bifurcation exists when $\alpha = 0.43$.

(5) Some bifurcating stationary solutions with cat's eyes pattern are unstable for small Reynolds numbers but recover stability for Reynolds numbers large enough.

We confirmed these results by numerical computations ([5,9]). The numerical method for this is described in these papers. We used Keller's path continuation method ([6]).

We are now going to explain what is the geometry of the set of stationary solutions. In the present paper we consider R as a bifurcation parameter and α as a supplementary splitting parameter. We first consider the case of $\alpha = 0.7$. In this case, the basic flow loses stability at $R = R^*(0.7) \approx 3.011193 \cdots$ and another steady state bifurcates (Figure 1). The bifurcation, which is a pitchfork, is *supercritical*. What is interesting is that we could not find any secondary bifurcation from this branch in Y nor did we find any turning point. Only one simple pitchfork is present in the whole range of Y and the Reynolds number. We, thereby, have Figure 2, where the R-axis consists of the basic solutions.

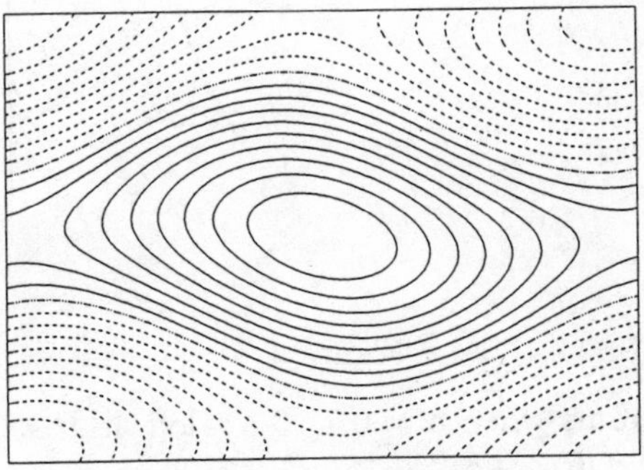

Figure 1

Stream lines when $\alpha = 0.7$, $R = 3.6$

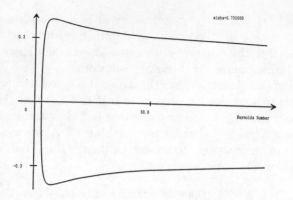

Figure 2

Bifurcation diagrams when $\alpha = 0.7$: $a(0,1)$ vs. the
Reynolds number R

Dynamics of (2.2) is presumably simple in Y when $\alpha = 0.7$. Our numerical computation seems to conjecture that

(1)   when $R < R^*(0.7)$ all the solutions approach the basic solution. The dynamics in the phase space is schematically shown in Figure 3-a.

(2)   when $R^*(0.7) < R < \infty$ all the solutions approach one of the three solutions. The basic solution, which is unstable now, has an unstable manifold of dimension one. The dynamics is schematically shown in Figure 3-b.

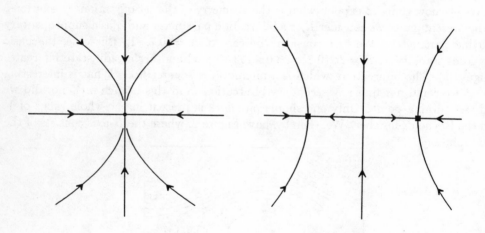

Figure 3

Phase diagrams when $\alpha = 0.7$. 3-a (left): $R < R^*$. 3-b (right): $R^* < R < +\infty$.

Again, we remark that the transient behaviour can be complex. Figure 4 shows contours of the vorticity, which clearly shows winding and stretching of the contours.

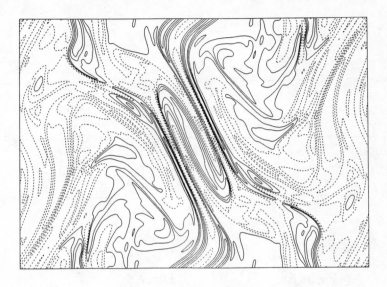

Figure 4

Contours of the vorticity $\omega(t, x, y)$. $\alpha = 0.7, R = 500.0$, t = 4.0. The vortex lines of positive vorticity are drawn by the broken lines. Those of negative vorticity are drawn by the solid lines. The initial vorticity is $\omega_0(x, y) = -10[\cos(y) + \cos(2y)] + 5\cos(2x) + 5\cos(2x + y) - 5\cos(x - y)$.

We carried out computations with other values of α. We found that *for* $0.5 \leq \alpha < 0.966 \cdots$, *the phase diagram is the same as that of* $\alpha = 0.7$.

In view of Theorem 2.1, one might naively imagine that larger α implies simpler structure of bifurcation. Actually, the branch is of different nature when α is close to one. When $0.966 \cdots < \alpha < 1.0$, the branch of the stationary solutions is folded and we have what is called a hysteresis. When $\alpha = 0.98$, we have Figure 5 as a bifurcation diagram. In Figure 5, the primary bifurcation is supercritical. But we can observe two turning points on the branch. As we decrease α, the turning points come closer. At $\alpha = 0.966 \cdots$, two pairs of turning points coalesce and make hysteresis points (Figure 6). As we increase α, the turning point on the right moves further to the right, thus we have deep dents. When $\alpha = 0.984$, we have a pitchfork below and another branch with a turning point, which is substantially separated from the primary branch (Figure 7). Figures 8 shows the case of $\alpha = 0.999$. In this case, the turning points lie in the left hand side of the pitchfork bifurcation point from the basic solution. This means existence of nontrivial solutions in the region $R < R^*(\alpha)$. This also implies that basic solution is stable but not globally stable if $0 < 1 - \alpha$ is sufficiently small and $R < R^*$ is sufficiently close to R^*.

These numerical solutions are found in [5,9]. For more details, see these papers.

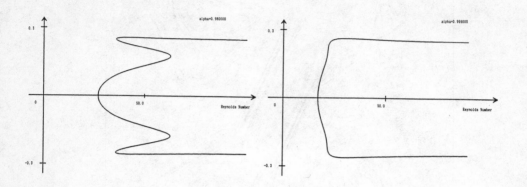

Figure 5

Bifurcation diagram. $\alpha = 0.98$.

Figure 6

Bifurcation diagram. $\alpha = 0.966$.

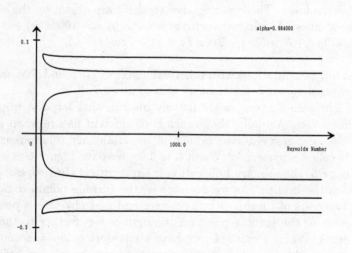

Figure 7

Bifurcation diagram. $\alpha = 0.984$.

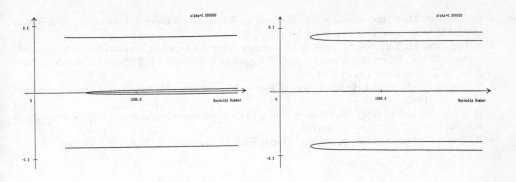

Figure 8

Bifurcation diagram. $\alpha = 0.999$. $a(1,0)$ vs. the Reynolds number (left). $a(0,1)$ vs. the Reynolds number (right).

§4. Other solutions.

As we have seen, there is one and only one bifurcation point along the branch of the trivial solutions when $1/2 \leq \alpha < 1$. It is known (Iudovich [1]) that there are n bifurcation points when $1/(n+1) \leq \alpha < 1/n \quad n = 1, 2, \cdots$. Thus the bifurcation diagrams are substantially different for $\alpha \gtrless 1/2$. In particular, the attractor has a dimension > 1. Its geometry is made complicated by the presence of secondary bifurcations and a Hopf bifurcation. More details in the case of $\alpha < 1/2$ are reported in [9].

Acknowledgement. The present work was partially supported by Ohbayashi Corporation. We computed nonstationry solutions with their supercomputer NEC SX-1.

REFERENCES

1. V.I. Iudovich, *Example of the generation of a secondary stationary or periodic flow when there is loss of stability of the laminar flow of a viscous incompressible fluid*, J. Appl. Math. Mech. **29** (1965), 527–544.
2. T. Kato, *On classical solutions of the two dimensional nonstationary Euler equation*, Arch. Rat. mech. Anal. **25** (1967), 188 – 200.
3. T. Kato and C. Y. Lai, *Nonlinear evolution equations and the Euler flow*, J. Func. Anal. **56** (1984), 15 – 28.
4. T. Kato and H. Fujita, *On the nonstationary Navier-Stokes system*, Rend. Sem. Mat. Padova **32** (1962), 243–260.
5. M. Katsurada, H. Okamoto, and M. Shōji, *Bifurcation of the stationary Navier-Stokes flows on a 2-D flat torus*, to appear in Proc. Katata workshop on "Nonlinear PDEs and Applications", eds. T. Nishida and M. Mimura.

318

6. H.B. Keller, "Lectures on Numerical Methods in Bifurcation Theory (Tata Institute of Fundamental Research No. 79)," Springer Verlag, 1987.

7. S. Kida and M. Yamada, *Singularity and energy spectrum in a two dimensional incompressible inviscid flow*, in " Turbulence and Chaotic Phenomena in Fluids", ed. T. Tatsumi, North-Holland, Proc. IUTAM Symposium in 1984.

8. A. Majda, *Vorticity and the mathematical theory of incompressible fluid flow*, Comm. Pure Appl. Math. **39** (1986).

9. H. Okamoto, and M. Shōji, *Bifurcation diagrams in Kolmogorov's problem of viscous incompressible fluid on 2-D flat tori.*, submitted.

10. R. Temam, "Navier-Stokes equations," North-Holland, Amsterdam, New York, Oxford, 1979.

A priori Estimates for Some Nonlinear Parabolic Equations via Lyapunov Functions

Mitsuharu ÔTANI

Department of Applied Physics
School of Science and Engineering, Waseda University
3-4-1, Okubo, Shinjuku-ku, Tokyo, Japan

1 Introduction

Since the pioneering work of Professor Kôsaku Yosida, the theory of semigroup has been regarded as a very important tool for solving partial differential equations. Especially, in the field of semilinear partial differential equations, the perturbation theory for semigroup has been very successful, e.g., Segal's theory [11] for nonlinear wave equations and Kato-Fujita's theory [5] for the Navier-Stokes equation, etc. The nonlinear version of Hille-Yosida theory, nonlinear semigroup theory, was established by Kōmura [7]; and Brézis [4] found out that the subdifferential operator, roughly speaking generalized Fréchet derivative of a convex functional on a real Hilbert space, forms a subclass of maximal monotone operators which generates a nonlinear semigroup with smoothing effect. Kōmura-Brézis theory itself has beautiful applications to nonlinear partial differential equations such as some nonlinear diffusion equations or porous media-type equations, etc. On the other hand, however, there are many parabolic equations which can not be treated in this theory. Therefore, as was in the theory of linear operators, it would be very important to study the perturbation problem within this framework aiming at much wider applicability. In fact, several attempts were made in this direction : Attouch-Damlamian [2], Biroli [3], Koi-Watanabe [6] and Ôtani [9] discussed the existence of local strong solutions for the following equation in a real Hilbert space H :

$$(E) \quad \begin{cases} \dfrac{d\,u}{d\,t}\,(t) \;+\; \partial\varphi\,(\,u\,(t)\,) \;+\; B\,(\,u\,(t)\,) \;=\; 0\,, \qquad t > 0, \\ u\,(0) \;=\; u_o\,, \end{cases}$$

where $\partial\varphi$ is the subdifferential of φ, defined by the following and $B(\cdot)$ is regarded as a perturbation for $\partial\varphi$:

$\varphi \in \Phi\,(H)\ =\ \{\varphi:\ H \to [0,+\infty];\ \varphi$ is lower semicontinuous convex, $D(\varphi) \neq \emptyset\}$,

$D\,(\,\varphi\,)\ =\ \{\,z \in H\ ;\ \varphi\,(z)\ <\ +\infty\ \ \}\ ,$

$\partial\,\varphi\,(z)\ =\ \{\,f \in H\ ;\ \varphi\,(v)\ -\ \varphi\,(z)\ \geq\ (\,f,\ v\ -\ z\,)_H\ \ {}^{\forall}v \in D\,(\varphi)\,\}\ ,$

$D\,(\partial\varphi)\ =\ \{\,z \in H\ ;\ \partial\varphi\,(z)\ \neq\ \emptyset\ \}.$

In general, $\partial\varphi$ and B could be multivalued operators in the following argument. For the sake of simplicity, however, we here always assume that they are both singlevalued. One of the most general results among the works cited above is given as follows (see [9]) :

Theorem 1.   *Let the following* (A.1)-(A.3) *be satisfied.*

(A.1) *For all* $L > 0$, *the set* $\{\,u \in H\ ;\ \varphi\,(u)\ +\ |u|_H\ \leq\ L\ \}$ *is compact in* H ,

(A.2) B *is* φ-*demiclosed ,i.e., If* $u_n\ \to\ u$ *in* $C([0,T];H)$; $\varphi\,(u_n)$ *is bounded in* $L^{\infty}(0,T)$;

$\partial\varphi(u_n) \rightharpoonup \partial\varphi(u),\ B(u_n) \rightharpoonup b$ *weakly in* $L^2(0,T;H)$, *then* $b\ =\ B(u)$ *follows,*

(A.3) ${}^{\exists}\ell \in \mathcal{L},\ {}^{\exists}k \in [0,1)$ *s.t.* $|B(u)|_H \leq k|\partial\varphi(u)|_H +\ \ell\,(\varphi(u)+\ |u|_H),\ {}^{\forall}u \in D(\partial\varphi).$

Then for all $u_o \in D(\varphi)$, *there exists a positive number* T_o *depending only on* $\varphi(u_o)$, $\ell(\cdot)$ *and* k *such that* (E) *has a strong solution* u *on* $[0,T_o]$ *satisfying :*

($\sharp$) $u \in C([0,T_o];H)$; $\varphi(u(t)) \in L^{\infty}(0,T_o);\ du(t)/dt,\ \partial\varphi(u)\ ,B(u) \in L^2(0,T_o;H).$

Here and throughout this paper, $\mathcal{L}$ *denotes the family of all monotone increasing continuous functions on* $[0,+\infty)$.

This type of framework can give a unified abstract treatment for the initial boundary value problems of some nonlinear parabolic equations in bounded domains $\Omega \subset R^N$ such as :

$$(P)\ \begin{cases} \dfrac{\partial\,u}{\partial\,t}\ (x,t)\ +\ A\,u(x,t)\ +\ b\,(x,u,\nabla u)\ =\ 0\ ,\ (x,t) \in \Omega \times (0,\infty)\ , \\[2mm] \text{with some boundary condition,} \\[2mm] u\,(x,0)\ =\ \ u_o(x)\ ,\ x \in \Omega\ , \end{cases}$$

where A is a (nonlinear) differential operator which has a subdifferential form. Since we must always work in real Hilbert spaces and the choice of spaces where we work depends crucially on the form of A (this is peculiar to the nonlinear problems), this framework has the following disadvantages :

(1) When the Hilbert space H is a large space such as $H^{-1}(\Omega) =$ the dual space of $H_o^1(\Omega)$, the verification of (A.1) or (A.2) is not always clear,

(2) Boundedness condition (A.3) requires some growth conditions on $b(x, u, p)$ with respect to u and p depending on the space dimension N, which comes from Sobolev's embedding theorem,

(3) For the case where A has some singularity or degeneracy, the regularity of solutions satisfying ($\sharp$) is not enough to assure the uniqueness.

On the other hand, these difficulties are often cleared off if we could treat (P) in some Banach spaces, say in L^∞. The main purpose of this paper is to give a **Hilbert-space** framework in which we can feel free from these restrictions. We state our main results and give sketches of proofs for them in the next section; and a couple of examples of applications are given in §3.

2 Main Results

In our framework, it is essential to choose a suitable convex function j which can help the leading term $\partial\varphi$ to control the singularity of the perturbation term $B(\cdot)$ and at the same time whose subdifferential ∂j becomes a good perturbation for $\partial\varphi$ as follows :

Theorem 2. *Assume that there exist a function* $j \in \Phi(H)$ *and* $\varepsilon_o > 0$ *s.t.*

(a.0) $^\forall \varepsilon \in (0, \varepsilon_o]$, $\partial\varphi + \varepsilon\partial j$ *become maximal monotone; and moreover* $^\exists \ell \in \mathcal{L}$ *s.t.*

$$|\partial\varphi(u)|_{L^2(0,T;H)} \leq \ell \left(|(\partial\varphi + \varepsilon\partial j)^\circ(u)|_{L^2(0,T;H)} + |j(u)|_{L^\infty(0,T)}\right), ^\forall \varepsilon \in (0, \varepsilon_o],$$

where $A^\circ(u)$ *denotes the nearest point of* $A(u)$ *from the origin.*

(a.1) $^\forall L > 0$, *the set* $\{u \in H; \varphi(u) + j(u) + |u|_H \leq L\}$ *is compact in* H,

(a.2) $B(\cdot)$ *is* φ_j-*demiclosed,i.e.,* $u_n \to u$ *in* $C(0, T; H)$; $\varphi(u_n)$ *and* $j(u_n)$ *are bounded in* $L^\infty(0, T)$; $\partial\varphi(u_n) \rightharpoonup \partial\varphi(u), B(u_n) \rightharpoonup b$ *weakly in* $L^2(0, T; H)$,

then $b = B(u)$ *follows,*

(a.3) $^\exists k_1, k_2 \in [0, 1)$ *with* $k_1 + k_2 < 1$, $^\exists \ell \in \mathcal{L}$ *s.t.*

(i) $|B(u)|_H \leq k_1 |(\partial\varphi + \varepsilon\partial j)^\circ(u)|_H + \ell \left(\varphi(u) + j(u) + |u|_H\right), ^\forall \varepsilon \in (0, \varepsilon_o], ^\forall u \in D(\partial\varphi) \cap D(\partial j)$,

(ii) $(-\partial\varphi(u) - B(u), f)_H \leq k_2 |\partial\varphi(u)|^2 + \ell \left(\varphi(u) + j(u) + |u|_H\right), ^\forall u \in D(\partial\varphi) \cap D(\partial j)$,

$$^\forall f \in \partial j(u).$$

Then for all $u_o \in D(\varphi) \cap D(j)$, *there exists a positive number* T_o *depending on* $\varphi(u_o) + j(u_o)$, *such that* (E) *has a strong solution* u *on* $[0, T_o]$ *satisfying :*

($\sharp$)' u *satisfies* ($\sharp$) *and* $j(u(t)) \in L^\infty(O, T_o)$.

This is our main result of general form. However, one may feel uncomfortable, since there appears an artificial parameter ε in the assumptions. On the other hand, it is often the case that the semigroup $S(t) = e^{-t\partial\varphi}$ generated by $\partial\varphi$ has **Lyapunov functions** $h(\cdot) \in \Phi(H)$, i.e.,

$$(L) \qquad h(S(t)\, z) \leq h(z) \qquad {}^\forall t \geq 0, \quad {}^\forall z \in \overline{D(\partial\varphi)}.$$

In this case, Lyapunov functions are good candidates for j in the above theorem, and our result can be stated more neatly as follows :

Corollary 3.   *Assume that there exists a function* $j \in \Phi(H)$ *satisfying* (a.1), (a.2) *and the following conditions* (a.0)$'$ *and* (a.3)$'$:

(a.0)$'$ $j((I + \lambda\partial\varphi)^{-1}z) \leq j(z)$ ${}^\forall\lambda > 0$, ${}^\forall z \in D(j)$,

(a.3)$'$ ${}^\exists k_1, k_2 \in [0, 1)$ *with* $k_1 + k_2 < 1$, ${}^\exists \ell \in \mathcal{L}$ *s.t.*

(i) $|B(u)|_H \leq k_1 |\partial\varphi(u)|_H + \ell\, (\varphi(u) + j(u) + |u|_H)$, ${}^\forall u \in D(\partial\varphi) \cap D(\partial j)$,

(ii) $(-\partial\varphi(u) - B(u), f)_H \leq k_2 |\partial\varphi(u)|^2 + \ell\, (\varphi(u) + j(u) + |u|_H)$, ${}^\forall u \in D(\partial\varphi) \cap D(\partial j)$,

$${}^\forall f \in \partial j(u).$$

Then the same assertion as in Theorem 2 *holds.*

Remark 1.   Condition (a.0)$'$ implies the relation :

(a.0)$''$ $(\,\partial\varphi(u), f\,)_H \geq 0$ ${}^\forall u \in D(\partial\varphi) \cap D(\partial j)$, ${}^\forall f \in \partial j(u)$.

Moreover (a.0)$''$ implies (L). Conversely, if $j\,(Proj_{\overline{D(\partial\varphi)}}\, z) \leq j(z)$ ${}^\forall z \in D(j)$, then (a.0)$'$, (a.0)$''$ and (L) are equivalent to each other (see Th.4.4. of [4]).

Proof of Cor. 3.   By Proposition 2.17 of [4], (a.0)$'$ assures that $\partial\varphi + \varepsilon\partial j$ becomes maximal monotone and furthermore the following estimate holds :

(1) $|\partial\varphi(z)|_H \leq |(\partial\varphi + \varepsilon\partial j)^o(z)|_H$ ${}^\forall z \in D(\partial\varphi + \varepsilon\partial j) = D(\partial\varphi) \cap D(\partial j)$.

Hence (a.0) is assured. Moreover (i) of (a.3) follows from (1) and (i) of (a.3)$'$.

[Q.E.D.]

Sketch of proof of Theorem 2.   First of all we introduce the following approximation equation for each $\varepsilon \in (0, \varepsilon_o]$:

$$(E)_\varepsilon \quad \begin{cases} du_\varepsilon(t)/dt + \varepsilon\, \partial j(u_\varepsilon(t)) + \partial\varphi(u_\varepsilon(t)) + B(u_\varepsilon(t)) \ni 0, & t > 0, \\ u_\varepsilon(0) = u_o. \end{cases}$$

The existence of local strong solution u_ε on $[0, T_1]$ is assured by Theorem 1, since assumptions (a.0),(a.1),(a.2), and (i) of (a.3) assure all the assumptions in Theorem 1 with φ replaced by $\varphi + \varepsilon\, j$. Moreover, the existence time T_1 is a monotone decreasing function of $\varphi(u_o) + \varepsilon\, j(u_o)$, so we can take T_1 independent of ε. Put $\overline{\varphi}(u) = \varphi(u) + |u|_H^2$,i.e., $\partial\overline{\varphi}(u) = \partial\varphi(u) + 2\, u$ and $\overline{B}(u) = B(u) - 2\, u$. Under this transformation, the equation is unchanged and conditions (a.0)-(a.3) are satisfied with φ and B replaced by $\overline{\varphi}$ and $\overline{B}$. Thus, without loss of generality, we may assume that

$$(2) \qquad \varphi(u) \geq |u|_H^2, \quad {}^\forall u \in D(\varphi).$$

Multiplying $(E)_\varepsilon$ by $\partial\varphi(u_\varepsilon) + \partial j(u_\varepsilon)$ and using (a.3), we can show that there exist $\delta > 0$, $\ell \in \mathcal{L}$ such that

$$\frac{d}{dt}\left(\, j(u_\varepsilon(t)) + \varphi(u_\varepsilon(t))\,\right) + \frac{1}{2}\,\varepsilon\,|\partial j(u_\varepsilon(t))|_H^2 + \delta|\partial\varphi(u_\varepsilon(t))|_H^2 \leq \ell\left(\, j(u_\varepsilon(t)) + \varphi(u_\varepsilon(t))\,\right).$$

Hence, by the standard argument, we can find a number $T_o \in (0, T_1)$ such that

$$(3) \qquad \max_{0 \leq t \leq T_o}\{j(u_\varepsilon(t)) + \varphi(u_\varepsilon(t))\} + \int_0^{T_o}(\varepsilon\,|\partial j(u_\varepsilon(t))|_H^2 + |\partial\varphi(u_\varepsilon(t))|_H^2)dt \leq Const.$$

Then, by using boundedness condition (i) of (a.3) and the fact that u_ε is a solution of $(E)_\varepsilon$, we get the boundedness of $B(u_\varepsilon(t))$ and $du_\varepsilon(t)/dt$ in $L^2(0, T_o; H)$. Hence $\{u_\varepsilon(t)\}_{\varepsilon \in (0, \varepsilon_o]}$ is equicontinuous on $[0, T_o]$. Moreover, (3) and (a.1) ensure that $\{u_\varepsilon(t)\}_{\varepsilon \in (0, \varepsilon_o]}$ forms a precompact set in H for each $t \in [0, T_o]$. Then, by Ascoli's theorem, (3) and (a.2) and the demiclosedness of $\partial\varphi$, we can extract a subsequence u_{ε_n} denoted by u_n such that

$$u_n \to u \qquad strongly\ in\ C([0, T_o]; H) \quad as\ n \to \infty,$$

$$\partial\varphi(u_n) \rightharpoonup \partial\varphi(u) \quad weakly\ in\ L^2([0, T_o]; H) \quad as\ n \to \infty,$$

$$B(u_n) \rightharpoonup B(u) \quad weakly\ in\ L^2([0, T_o]; H) \quad as\ n \to \infty,$$

$$du_n/dt \rightharpoonup du/dt \quad weakly\ in\ L^2([0, T_o]; H) \quad as\ n \to \infty,$$

$$\varepsilon_n\,\partial j(u_n) \to 0 \quad strongly\ in\ L^2([0, T_o]; H) \quad as\ n \to \infty.$$

Thus u is proved to be the desired strong solution of (E).   [Q.E.D.]

3 Applications

In order to illustrate the scope of our abstract treatment, we here give some typical examples of equations in bounded domain Ω in R^N:

(Ex.1) We first deal with the porous media-type equation :

$$(P_m) \quad \begin{cases} \partial u(x,t)/\partial t - \Delta\beta(u(x,t)) + b(x,u,\nabla u) = 0, & (x,t) \in \Omega \times (0,T), \\ u(x,t) = 0, & (x,t) \in \partial\Omega \times (0,T), \\ u(0,x) = u_o(x), & x \in \Omega. \end{cases}$$

Following the formulation in [4], we choose $H = H^{-1}(\Omega) = V^*$, $(V = H_o^1(\Omega))$ as the Hilbert space where we work and define the inner product by

$$(u,v)_H = {}_V<\Lambda^{-1}u, v>_{V^*}, \qquad \Lambda = -\Delta : \text{ isomorphism } V \to V^*$$

and as a convex function φ, we put

$$\varphi(u) = \begin{cases} \int_\Omega \gamma(u(x))dx & \text{if } u, \gamma(u) \in L^1(\Omega), \\ +\infty & \text{otherwise,} \end{cases}$$

where $\gamma \in \Phi(R^1)$ with $\beta = \partial\gamma$, $R(\beta) = R^1$. Then $\varphi \in \Phi(H^{-1}(\Omega))$ and $\partial\varphi(u) = -\Delta\,\beta(u)$; the typical example is $\gamma(u) = |u|^p/p$, $1 < p < +\infty$, $\beta(u) = |u|^{p-2}u$. We also assume that $b(x,p,q) = b(x,p) \in C(\overline{\Omega} \times R^1)$. As the function $j(\cdot)$, we can choose a Lyapunov function $j(u) = |u|_{L^\infty(\Omega)}$. In fact, let $w = (I + \lambda\partial\varphi)^{-1} z$, then w satisfies

$$w - \lambda\,\Delta\,\beta(w) = z, \quad \text{in } H^{-1}(\Omega).$$

Then, taking the inner product between this and $\Lambda(|w|^{n-2}w)$, we get

$$\int_\Omega |w|^n dx - \lambda \int_\Omega \Delta\beta(w)|w|^{n-2}w dx = \int_\Omega z\,|w|^{n-2}w dx \le |z|_{L^n}\,|w|_{L^n}^{n-1}.$$

By the standard argument with use of the monotonicity of Yosida approximation of β, we can easily show that the second term of the left side of the above equality is nonnegative. Hence we derive $|w|_{L^n} \le |z|_{L^n}$ for all n. Then letting n tend to ∞, we obtain $j(w) \le j(z)$, whence follows (a.0)'. Since $L^\infty(\Omega)$ is compactly embedded in $H^{-1}(\Omega)$, (a.1) is also easily verified. (Note that if the term j(u) is absent, the verification of (a.1) is delicate.) To get (i) of (a.3)', it suffices to note that L^∞ is continuously embedded in H^{-1} and $|b(\cdot,u)|_{L^\infty} \le \ell\,(j(u))$ for some $\ell \in \mathcal{L}$. To derive (ii) of (a.3)', we note, by the definition of subdifferential,

$$j(-B(u)) - j(u) \ge (\partial j(u), -B(u) - u), \qquad \text{and hence}$$

$$(-B(u), \partial j(u)) \le j(-B(u)) + (\partial j(u),\ u) - j(u)$$

$$\le j(B(u)) + (\partial j(u),\ 2u\ -\ u) - j(2u) + j(u) \le j(B(u)).$$

Then , recalling (a.0)″, we obtain (ii) of (a.3)′. Thus, by Cor.3, we conclude that for every $u_o \in L^\infty(\Omega)$, (P_m) has a local strong solution $u \in L^\infty(\Omega \times (0, T_o))$ such that u_t, $\Delta\beta(u) \in L^2(0, T_o; H^{-1}(\Omega))$. Furthermore, if b(x,p) is Lipschitz continuous with respect to p, then the solution is unique.

Remark 2. Several people already constructed a local solution (in a weaker class than ours) via vanishing viscosity method,(cf. [1, 8]). However, this method need some technical assumptions which we do not such as : β has the inverse β^{-1} and β should be locally absolutely continuous , etc.

(Ex.2) In general, the check of (ii) of (a.3) or (a.3)' requires delicate arguments when ∂j is multivalued as above. In the above example, however, $j(\cdot)$ has good approximations $j_n(\cdot)$, which are also Lyapunov functions. In such a case, we can give a variant of Corollary 3 which provides us a framework where the more minute estimate is feasible.

Corollary 4 *Assume that there exist a function* $j \in \Phi(H)$ *satisfying* (a.0)′, (a.1), (a.2) *and functions* $j_n \in \Phi(H)$ $(n \in \mathbf{N})$ *satisfying* :

(a.4) (i) $\quad j_n(z) \to j(z)$ as $n \to +\infty$ $\quad ^\forall z \in D(j)$, and $\quad D(\partial j_n) \supset D(\partial\varphi) \cap D(\partial j)$,

(ii) $\quad j_n((I + \lambda\partial\varphi)^{-1} z) \leq j_n(z)$ $\quad ^\forall z \in D(j)$, $\quad ^\forall \lambda > 0$, $\quad ^\forall n$,

(iii) $\quad j((I + \lambda\partial j_n)^{-1} z) \leq j(z)$ $\quad ^\forall z \in D(j)$, $\quad ^\forall \lambda > 0$, $\quad ^\forall n$,

(a.3)″ $\quad$ *condition* (a.3)′ *is satisfied with* (ii) *replaced by*

$$\limsup_{n\to\infty}(-\partial\varphi(u) - B(u), \partial j_n(u))_H \leq k_2|\partial\varphi(u)|^2 + \ell(\varphi(u) + j(u) + |u|_H),$$
$$^\forall u \in D(\partial\varphi) \cap D(\partial j).$$

Then the same assertion as in Theorem 2 *holds.*

Proof. Since (ii), (iii) of (a.4) imply $(\partial\varphi(u), \partial j_n(u))_H \geq 0$, $(\partial j(u), \partial j_n(u))_H \geq 0$, multiplying $(E)_\varepsilon$ by $\partial\varphi(u_\varepsilon) + \partial j_n(u_\varepsilon)$, using (a.3)″ and letting n tend to ∞, we can derive the same a priori estimate as (3). [Q.E.D.]

By applying this result, we can prove the same assertion as before for the following type of perturbations :

$$b(x, u, \nabla u) = \sum_{i=1}^{N} b^i(x, u) u_{x_i} + b^o(x, u), \quad \text{with} \quad b^o, b^i, b^i_{x_i} \in C(\overline{\Omega} \times R^1).$$

As is seen in the above example, $j(u) = |u|_{L^\infty(\Omega)}$, $j_n(u) = |u|_{L^n(\Omega)}$ satisfy (i),(ii) of (a.4). Since $w = (I + \lambda\partial j_n)^{-1} z$ satisfies $(1 + \lambda |w|_{L_n(\Omega)}^{1-n}) w = z$, (iii) of (a.4) is also satisfied. In order to verify (a.3)″, put $B^i(x, p) = \int_0^p b^i(x, s) \, ds$,

$B_n^i(x, p) = \int_0^p b^i(x, s) |s|^{n-2} s \, ds$. Then, by integration by parts, we get

$$\int_\Omega b^i(x,u)\, u_{x_i}\, h\, dx \;=\; \int_\Omega -B^i(x,u)\, h_{x_i}\, dx \;-\; \int_\Omega \int_0^{u(x)} b^i_{x_i}(x,s)\, h\, ds\, dx$$

$$\leq\; |h|_V\, \ell\,(j(u))$$

$$\int_\Omega b^i(x,u)\, u_{x_i}\, |u|^{n-2}u\, dx \;=\; \int_\Omega (B^i_n(x,u))_{x_i}\, dx \;-\; \int_\Omega \int_0^{u(x)} b^i_{x_i}(x,s)\, |s|^{n-2}s\, ds\, dx$$

$$\leq\; \ell\,(j(u))\, |u|_{L^n}^{n-1},$$

whence follows (a.3)$''$.

(Ex.3)   It is possible to give another example of $j(\cdot)$. Put

$$H \;=\; L^2(\Omega), \qquad\qquad \varphi(u) \;=\; \tfrac{1}{p} \int_\Omega |\nabla u|^p dx, \qquad\qquad D(\varphi) = W_o^{1,p}(\Omega),$$

$$b(x,u,\nabla u) = b_o(x,u)\,|\nabla u|^\alpha, \quad b_o(x,p) \in C(\overline{\Omega} \times R^1), \qquad\qquad b_o(x,0) = 0, \quad \alpha \geq 0,$$

$$j(u) = |u|_{L^\infty(\Omega)} + |\nabla u|_{L^\infty(\Omega)}, \quad j_n(u) = |u|_{L^n(\Omega)} + |\nabla u|_{L^n(\Omega)},$$

Then we can apply Cor.4 and find that for every $u_o \in W_o^{1,\infty}(\Omega)$,

$$\partial u(x,t)/\partial t \;-\; div(|\nabla u|^{p-2}\nabla u) \;+\; b(x,u,\nabla u) \;=\; 0, \;\; u(x,t)|_{\partial\Omega} \;=\; 0, \;\; u(x,0) \;=\; u_o(x),$$

has a local strong solution $u(\cdot,t) \in L^\infty(0,T_o; W_o^{1,\infty}(\Omega))$ such that $div(|\nabla|^{p-2}\nabla u)$,

$\partial u(\cdot,t)/\partial t \in L^2(0,T_o; H)$.

Remark 3.   We can extend our framework to much more general situation such as in [9, 10],i.e., (i) the initial data can be taken in the nonlinear interpolation classes between $D(\varphi)$ and $\overline{D(\varphi)}$. To do this, as a matter of course, boundedness conditions in (a.3),(a.3)$'$ and (a.3)$''$ should be replaced properly.
(ii) φ and B can depend on time t, by which we can discuss the existence of strong solutions of some nonlinear parabolic equations satisfying L^p or L^∞-type estimates in regions with moving boundaries.

References

[1] D. Aronson, M. G. Crandall and L. A. Peletier ; Stabilization of solutions of a degenerate nonlinear diffusion problem, Nonlinear Analysis TMA **6**, 1001-1022 (1982).

[2] H. Attouch and A. Damlamian ; On multivalued evolution equations in Hilbert spaces, Israel J. Math. **12**, 373-390 (1972).

[3] M. Birolli ; Sur un équation d'évolution multivoque et non monotone dans un espace de Hilbert, Rend. Sem. Mat. Univ. Padova **52**, 313-331 (1974).

[4] H. Brézis ; *Opérateurs Maximaux Monotones et Semi-Groupes de Contractions dans les Espaces de Hilbert*, Amsterdam: North-Holland Publ. Co., 1973.

[5] H. Fujita and T. Kato ; On the Navier-Stokes initial value problem, I, Arch. Rational Mech. Anal. **16**, 269-315 (1970).

[6] Y. Koi and J. Watanabe ; On nonlinear evolution equations with a difference term of subdifferentials, Proc. Japan Acad. **52**,413-416 (1976).

[7] Y. Kōmura ; Nonlinear semigroups in Hilbert space, J. Math. Soc. Japan **19**, 493-507 (1967).

[8] H. A. Levine and P. Sacks ; Some existence and nonexistence theorems for solutions of degenerate parabolic equations, J. Differential Equations **52**, 135-161 (1984).

[9] M. Ôtani ; Nonmonotone perturbations for nonlinear parabolic equations associated with subdifferential operators, Cauchy problems, J. Differential Equations **46**, 268-299 (1982).

[10] M. Ôtani ; Nonmonotone perturbations for nonlinear parabolic equations associated with subdifferential operators, Periodic problems, J. Differential Equations **54**, 248-273 (1984).

[11] I. Segal ; Nonlinear semigroups, Annals of Math. **8**, 339-364 (1963).

[12] M. Tsutsumi ; Existence and nonexistence of global solutions for nonlinear parabolic equations, Publ. RIMS Kyoto Univ. **8**, 211-229 (1972).

[13] K. Yosida ; *Functional Analysis*, Springer-Verlag, 1980.

REMARKS ON RECURRENCE CRITERIA FOR PROCESSES OF ORNSTEIN–UHLENBECK TYPE

KEN-ITI SATO AND MAKOTO YAMAZATO

College of General Education, Nagoya University
and
Nagoya Institute of Technology

The Ornstein-Uhlenbeck process on $\mathbf{R}^d$ is a Markov process generated by the differential operator

$$(1) \qquad \mathbf{G}_0 f(x) = \frac{1}{2}\Delta f - \sum_{j=1}^{d} x_j D_j f, \qquad D_j = \frac{\partial}{\partial x_j}.$$

It was introduced by G. E. Uhlenbeck and L. S. Ornstein [9]. The $\frac{1}{2}\Delta$ in (1) corresponds to the Brownian motion. So the Ornstein-Uhlenbeck process is Brownian motion undergoing a drift force, which is directed toward the origin and proportional to the distance with the origin. Since the Brownian motion is spatially homogeneous, its distribution diffuses as time passes. But the Ornstein-Uhlenbeck process has spatial inhomogeneity due to the restoring force; it has a limiting distribution, which is Gaussian with mean at the origin.

A generalization of the Brownian motion is a Lévy process, that is, a stochastically continuous process with stationary independent increments or, equivalently, a Markov process with spatial homogeneity. Replacing the Brownian motion part in (1) by a general Lévy process and making the drift force slightly more general, we get a process of Ornstein-Uhlenbeck type on $\mathbf{R}^d$. Namely, let $\mathbf{G}$ be an operator defined on $C_c^2(\mathbf{R}^d)$, the set of real-valued C^2 functions with compact supports, by

$$(2) \qquad \mathbf{G}f(x) = \frac{1}{2}\sum_{j,k=1}^{d} B_{jk} D_j D_k f(x)$$

$$+ \int_{\mathbf{R}^d} [f(x+y) - f(x) - \sum_{j=1}^{d} y_j D_j f(x) 1_{\{|y|<1\}}(y)]\rho(dy)$$

$$+ \sum_{j=1}^{d} b_j D_j f(x) - \sum_{j,k=1}^{d} Q_{jk} x_k D_j f(x).$$

Here $B = (B_{jk})$ is a symmetric nonnegative-definite constant matrix, ρ is a measure on $\mathbf{R}^d$ with $\rho(\{0\}) = 0$ and $\int(1 \wedge |y|^2)\rho(dy) < \infty$, $b = (b_j)$ is a constant vector,

$Q = (Q_{jk})$ is a constant matrix whose all eigenvalues have positive real parts, and $1_{\{|y|<1\}}(y)$ is the indicator function of the set $\{y : |y| < 1\}$. Consider the real Banach space $C_0(\mathbf{R}^d)$ of continuous functions vanishing at infinity with the supremum norm. As is proved in the authors' paper [5], the operator $\mathbf{G}$ is closable and the smallest closed extension $\bar{\mathbf{G}}$ is the infinitesimal generator (in the sense of Yosida's book [12]) of a strongly continuous semigroup of positive linear operator P_t with norm 1, so that a Markov process $\mathbf{X}$ is associated. As usual (see [1]) the Markov process $\mathbf{X}$ is represented by $(\Omega, \mathcal{F}, \mathcal{F}_t, X_t(\omega), \mathbf{P}^x)$ with $\mathbf{P}^x(X_0 = x) = 1$. Note that

$$\mathbf{E}^x(f(X_t)) = \int_\Omega f(X_t(\omega))\mathbf{P}^x(d\omega) = P_t f(x)$$

The process $\mathbf{X}$ is called *the process of Ornstein-Uhlenbeck type associated with* $\mathbf{G}$. The measure ρ , called *Lévy measure* of the process $\mathbf{X}$, controls size and frequency of jumps of $X_t(\omega)$ as a function of t. The operator $\mathbf{G}$ is written as

$$\mathbf{G} = \mathbf{H} - \sum_{j,k=1}^{d} Q_{jk} x_k D_j,$$

and the smallest closed extension $\bar{\mathbf{H}}$ of $\mathbf{H}$ is the infinitesimal generator of a Lévy process $\{Z_t\}$. So $\mathbf{X}$ is sometimes called the process of Ornstein-Uhlenbeck type associated with $\{Z_t\}$ and Q. We consider elements of $\mathbf{R}^d$ as d-column vectors, and denote the inner product and the norm by $\langle x, y \rangle = \sum_{j=1}^{d} x_j y_j$ and $|x| = \langle x, x \rangle^{1/2}$ for $x = (x_j)_{1 \leq j \leq d}$, $y = (y_j)_{1 \leq j \leq d}$. An equivalent definition of the process $\mathbf{X}$ is given by the unique solution $\{X_t\}$ of the equation

$$(3) \qquad X_t = Z_t - \int_0^t QX_s ds.$$

Limiting behaviors of the processes of Ornstein-Uhlenbeck type as $t \to \infty$ are of interest. Two groups of results have been obtained so far: limiting distributions and rercurrence-transience.

A probability measure μ on $\mathbf{R}^d$ is called the limiting distribution of $\mathbf{X}$ if, for every $f \in C_c^+(\mathbf{R}^d)$ and $x \in \mathbf{R}^d$,

$$\lim_{t \to \infty} P_t f(x) = \int f(y)\mu(dy).$$

Here $C_c^+(\mathbf{R}^d)$ is the set of nonnegative continuous functions with compact supports. In [5] it is shown that $\mathbf{X}$ has limiting distribution if and only if

$$(4) \qquad \int_{|x| \geq c} \log |x|\, \rho(dx) < \infty.$$

Here $c > 0$ is arbitrarily fixed. Special cases of this result are found by Çinlar and Pinsky [2] and Wolfe [11]. The condition (4) shows that only a property of ρ outside compact sets is relevant. No influence of the dimension d appears. The class of limiting distributions of the processes satisfying (4) with $Q = I =$ unit matrix coincides with the class L (sometimes called the class of self-decomposable distributions) introduced by P. Lévy. The correspondence between the processes and the class L is one-to-one and, moreover, continuous in some sense. The class L is an important subclass of the class of infinitely divisible distributions (see Gnedenko and Kolmogorov [3]). Characterization of the class of limiting distributions is known also in the case of general Q. It is the class of operator-selfdecomposable distributions, which is very close to the class studied by Urbanik [10]. See [5] for the proof.

The process $\mathbf{X}$ is said to be *recurrent* if there is a point y (called a recurrent point) such that, for every x,

$$\mathbf{P}^x \left(\liminf_{t \to \infty} |X_t - y| = 0 \right) = 1.$$

It is said to be *transient* if, for every x,

$$\mathbf{P}^x \left(\lim_{t \to \infty} |X_t| = \infty \right) = 1.$$

Shiga [8] shows that $\mathbf{X}$ is either recurrent or transient. Expressed by the transition operator P_t, the process $\mathbf{X}$ is recurrent if and only if

$$\int_0^\infty P_t f(x) dt = \infty$$

for every f in $C_c^+(\mathbf{R}^d)$ not identically zero and for every x; $\mathbf{X}$ is transient if and only if

$$\int_0^\infty P_t f(x) dt < \infty$$

for every $f \in C_c^+(\mathbf{R}^d)$ and x. Thus, if $\mathbf{X}$ has limiting distribution, then $\mathbf{X}$ is recurrent. The authors gave in [5] an example of $\mathbf{X}$ which is recurrent but does not have limiting distribution, and raised the problem to find a criterion of recurrence and transience for processes of Ornstein-Uhlenbeck type in terms of the quantities appearing in the expression (2) of the infinitesimal generators. In one-dimensional case ($d = 1$), Shiga [8] proves that $\mathbf{X}$ is recurrent if and only if

$$(5) \qquad \int_0^1 \frac{dv}{v} \exp \left[\int_v^1 \frac{du}{Qu} \int_{|x| \geq 1} (e^{-u|x|} - 1) \rho(dx) \right] = \infty,$$

where Q is identified with a positive number. Further he discusses some special cases in general dimensions. However, criteria for recurrence and transience in case of general d and Q are not known.

Here we make a short mention of the corresponding problem for Lévy processes. Lévy processes are objects of extensive study, but a satisfactory criterion of recurrence and transience in terms of the quantities in their infinitesimal generators is not known. Shepp [7] is suggestive in this respect. If $\{Z_t\}$ is a Lévy process generated by $\mathbf{H}$, then the logarithmic characteristic function $\psi(z)$ is defined by

$$\mathbf{E}(\exp(i\langle z, Z_{s+t} - Z_s \rangle)) = e^{t\psi(z)}$$

and we have

$$\psi(z) = -\frac{1}{2}\langle Bz, z\rangle + \int_{\mathbf{R}^d} [e^{i\langle z,y\rangle} - 1 - i\langle z, y\rangle 1_{\{|y|<1\}}(y)]\rho(dy) + i\langle b, z\rangle.$$

A deep result obtained by K. L. Chung and W. H. J. Fuchs, F. Spitzer, D. Ornstein, C. J. Stone, S. C. Port and C. J. Stone, and M. Itô (see [4] for references) is that $\{Z_t\}$ is recurrent if and only if

$$\int_{|z|<1} \mathrm{Re}\left(\frac{1}{-\psi(z)}\right) dz = \infty.$$

But the connection of this criterion with B, ρ, b is indirect.

In the following we discuss the problem of criteria for recurrence and transience of a process of Ornstein-Uhlenbeck type $\mathbf{X}$ under some additional assumptions on the matrix Q. We say that Q is diagonalizable with n eigenvalues, or satisfies Condition (D_n), if all eigenvalues of Q are real and positive, the eigenvectors of Q span the whole space $\mathbf{R}^d$, and the number of distinct eigenvalues is n. If Q satisfies Condition (D_n), we denote the distinct eigenvalues of Q by $\alpha_1, \ldots, \alpha_n$ and the eigenspace of α_j by V_j. Thus

$$\mathbf{R}^d = V_1 \oplus \cdots \oplus V_n.$$

We denote the projectors associated with this direct sum decomposition by $T_1, \ldots, T_n$, so that

$$x = T_1 x + \cdots + T_n x, \quad T_j x \in V_j \quad \text{for} \quad j = 1, \ldots, n.$$

We have arrived at the following

Conjecture. *Assume that Q satisfies Condition (D_n). Fix $c > 0$ arbitrarily. Then $\mathbf{X}$ is recurrent if and only if*

$$(6) \qquad \int_0^1 \frac{dv}{v} \exp\left[\int_v^1 \frac{du}{u} \int_{|x|\geq c} (\exp(-\sum_{j=1}^n u^{\alpha_j}|T_j x|) - 1)\rho(dx)\right] = \infty.$$

This conjecture is proved to be true for $n = 1$ and 2. Namely the following two theorems are given in [6].

Theorem 1. *Assume that Q satisfies Condition (D_1), that is, $Q = \alpha_1 I$. Then $\mathbf{X}$ is recurrent if and only if*

$$(7) \qquad \int_0^1 \frac{dv}{v} \exp\left[\int_v^1 \frac{du}{u} \int_{|x|\geq c} (\exp(-u^{\alpha_1}|x|) - 1)\rho(dx)\right] = \infty.$$

Theorem 2. *Assume that Q satisfies Condition (D_2). Then $\mathbf{X}$ is recurrent if and only if (6) holds with $n = 2$.*

In one dimension $(d = 1)$ Theorem 1 is none other than Shiga's criterion (5). His proof in [8] is simplified and extended by [6]. Proof of Proposition 6 of the present paper contains proof of Theorem 1 as a special case. Analytical manipulation in the proof of Theorem 2 in [6] is much more complicated than that of Theorem 1. We have not yet succeeded in extending the method of proof to the case $n \geq 3$.

We will give some remarks related to Theorems 1 and 2 and Conjecture.

1. For recurrence and transience of Lévy processes on $\mathbf{R}^d$ the dimension d and the symmetricity are often decisive. Thus, for $d \geq 3$, any non-degenerate Lévy process on $\mathbf{R}^d$ is transient; a stable process on $\mathbf{R}^1$ with index 1 is recurrent if and only if its Lévy measure is symmetric. See the book [4]. Theorem 1 shows that the situation for processes of Ornstein-Uhlenbeck type with $Q = \alpha_1 I$ is quite different. The criteria (7) is independent of the dimension d. Further (7) shows that the process is recurrent if and only if its symmetrization or isotropic modification is recurrent. More precisely, given the process $\mathbf{X}$ associated with $\mathbf{G}$ of (2) with $Q = \alpha_1 I$, $\alpha_1 > 0$, let ρ_0 and ρ_{00} be the symmetrization and the isotropic modification of ρ, respectively, that is,

$$\rho_0(E) = 2^{-1}(\rho(E) + \rho(-E)),$$

$$\rho_{00}(E) = \frac{1}{c_d} \int_S \sigma_0(d\xi) \int_0^\infty 1_E(r\xi)\tau(dr),$$

where σ_0 is the Euclidean surface measure of the unit sphere S in $\mathbf{R}^d$, $c_d = \sigma_0(S)$, and τ is a measure on $(0, \infty)$ such that $\tau((a, b]) = \int_{a < |x| \leq b} \rho(dx)$. Define $\mathbf{G}_0$ and $\mathbf{G}_{00}$ by (2) with ρ replaced by ρ_0 and ρ_{00}, respectively. Let $\mathbf{X}_0$ and $\mathbf{X}_{00}$ be the processes associated with $\mathbf{G}_0$ and $\mathbf{G}_{00}$, respectively. Then, recurrence of $\mathbf{X}$, recurrence of $\mathbf{X}_0$, and recurrence of $\mathbf{X}_{00}$ are equivalent.

2. Let $\mathbf{X}$ be a process of Ornstein-Uhlenbeck type on $\mathbf{R}^d$ associated with $\mathbf{G}$ of (2) with Q satisfying Condition (D_1) or (D_2). For $c > 0$ denote by ρ^c the restriction of ρ to the set $\{|x| \geq c\}$, and let $\{Z_t^c\}$ be the compound Poisson process with Lévy measure ρ^c. Let $\mathbf{X}^c$ be the process of Ornstein-Uhlenbeck type associated with $\{Z_t^c\}$ and Q. Theorems 1 and 2 imply that, $\mathbf{X}$ is recurrent if and only if $\mathbf{X}^c$ is so. This situation is also different from general Lévy processes. A one-dimensional Lévy process with finite mean is recurrent if and only if the mean is zero, which implies that the whole Lévy measure as well as drift term is relevant.

3. The recurrence criteria in Theorem 2 involves both the eigenvalues α_1, α_2 of Q. But, under a mild additional condition on the Lévy measure ρ, we can give a criterion which involves only the smaller of α_1 and α_2.

Proposition 3. *Assume that Q satisfies Condition (D_2) and that $\alpha_1 < \alpha_2$. Fix $c > 0$. Suppose that*

$$(8) \qquad \rho(\{x : |x| \geq c, T_1 x = 0\}) = 0,$$

$$(9) \qquad \int_{|x| \geq c} \log \frac{|x|}{|T_1 x|} \rho(dx) < \infty.$$

Then, $\mathbf{X}$ is recurrent if and only if (7) holds.

For proof of this proposition we need the following lemma. This is Lemma 4.1 of [6]. Proof is not difficult.

Lemma 4. *Let S_1, S_2 be matrices and α_1, α_2, c_1, c_2 be positive reals. Let U_1, U_2, R be invertible matrices. Then*

$$\int_0^1 \frac{dv}{v} \exp\left[\int_v^1 \frac{du}{u} \int_{|U_1 x| \geq c_1} (\exp(-u^{\alpha_1}|S_1 x| - u^{\alpha_2}|S_2 x|) - 1)\rho(dx)\right] = \infty$$

if and only if

$$\int_0^1 \frac{dv}{v} \exp\left[\int_v^1 \frac{du}{u} \int_{|U_2 x| \geq c_2} (\exp(-u^{\alpha_1}|RS_1 x| - u^{\alpha_2}|RS_2 x|) - 1)\rho(dx)\right] = \infty.$$

Proof of Proposition 3. Write $\alpha_1 = \alpha$, $\alpha_2 = \beta$. There exists an invertible matrix R such that $RQR^{-1} = D$ is diagonal. The diagonal entries of D consist of α and β. Assume (7). Then we claim that (6) holds with $n = 2$. Here we do not need the assumptions (8), (9). By Lemma 4 we have

$$(10) \qquad \int_0^1 \frac{dv}{v} \exp\left[\int_v^1 \frac{du}{u} \int_{|Rx| \geq c} (e^{-u^{\alpha}|Rx|} - 1)\rho(dx)\right] = \infty.$$

Since $\langle RT_1 x, RT_2 x \rangle = 0$, we have $|Rx| \geq 2^{-1/2}(|RT_1 x| + |RT_2 x|)$. Hence, using $\alpha < \beta$, we see that

$$\int_v^1 \frac{du}{u} \int_{|Rx| \geq c} (e^{-u^{\alpha}|Rx|} - 1)\rho(dx)$$

$$\leq \int_{2^{-1/(2\alpha)}v}^{2^{-1/(2\alpha)}} \frac{du}{u} \int_{|Rx| \geq c} (e^{-u^{\alpha}|RT_1 x| - u^{\beta}|RT_2 x|} - 1)\rho(dx).$$

Hence (10) implies that

$$(11) \qquad \int_0^1 \frac{dv}{v} \exp\left[\int_v^1 \frac{du}{u} \int_{|Rx| \geq c} (\exp(-u^{\alpha}|RT_1 x| - u^{\beta}|RT_2 x|) - 1)\rho(dx)\right] = \infty,$$

which is equivalent to (6) with $n = 2$ by Lemma 4.

Conversely, suppose that (6) with $n = 2$ holds. Then we have (11) with c replaced by arbitrary $c' > 0$. Let us prove (7). First note that

$$\int_v^1 \frac{du}{u} \int_{|Rx| \geq c'} (e^{-u^{\alpha}|RT_1 x| - u^{\beta}|RT_2 x|} - 1)\rho(dx)$$

$$\leq \int_v^1 \frac{du}{u} \int_{|Rx| \geq c'} (e^{-u^{\alpha}(|RT_1 x| + |RT_2 x|)} - 1)\rho(dx) + J,$$

where

$$J = \int_0^1 \frac{du}{u} \int_{|Rx| \geq c'} (e^{-u^{\alpha}|RT_1 x| - u^{\beta}|RT_2 x|} - e^{-u^{\alpha}(|RT_1 x| + |RT_2 x|)})\rho(dx).$$

Note that $a|x| \leq |Rx| \leq b|x|$ for some $0 < a \leq b$. If $c' \geq bc$, then, from the assumptions (8), (9), we see that J is finite. In fact,

$$0 \leq J \leq \int_0^1 \frac{du}{u} \int_{|Rx| \geq c'} (e^{-u^\alpha |RT_1 x|} - e^{-u^\alpha(|RT_1 x| + |RT_2 x|)}) \rho(dx)$$

$$\leq \frac{1}{\alpha} \int_{|Rx| \geq c', T_1 x \neq 0} \log \frac{|RT_1 x| + |RT_2 x|}{|RT_1 x|} \rho(dx),$$

because

$$(12) \qquad \int_0^1 (e^{-Au^\alpha} - e^{-Bu^\alpha}) \frac{du}{u} \leq \frac{1}{\alpha} \log \frac{B}{A}$$

for any $0 < A \leq B$, and hence

$$J \leq \frac{1}{\alpha} \int_{|x| \geq c'/b, T_1 x \neq 0} \log \frac{|x|}{|T_1 x|} \rho(dx) + \text{const}$$

$$\leq \frac{1}{\alpha} \int_{|x| \geq c, T_1 x \neq 0} \left(0 \vee \log \frac{|x|}{|T_1 x|} \right) \rho(dx) + \text{const}.$$

Therefore

$$(13) \qquad \int_0^1 \frac{dv}{v} \exp \left[\int_v^1 \frac{du}{u} \int_{|Rx| \geq c'} (e^{-u^\alpha(|RT_1 x| + |RT_2 x|)} - 1) \rho(dx) \right] = \infty,$$

if $c' \geq bc$. So, for every $c' > 0$, (13) holds with R deleted. This is again by Lemma 4. Now (7) follows, since $|T_1 x| + |T_2 x| \geq |x|$. $\square$

4. An important special case of Proposition 3 is the case where the spherical component of the Lévy measure ρ is uniform. The following result is related to Theorem 4.3 of Shiga [8].

Proposition 5. *Assume that Q satisfies Condition (D_2) and that $\alpha_1 < \alpha_2$. Fix $c > 0$. Suppose that ρ^c has the polar decomposition*

$$(14) \qquad \rho^c(E) = \int_S \sigma_0(d\xi) \int_{[c,\infty)} 1_E(r\xi) \tau_\xi(dr),$$

where σ_0 is the Euclidean surface measure of the unit sphere $S = \{\xi \in \mathbf{R}^d : |\xi| = 1\}$, τ_ξ is a measure on $[c, \infty)$ with total mass equal to $\rho^c(\mathbf{R}^d)/\sigma_0(S)$ for each ξ, and $\tau_\xi(F)$ is measurable in ξ for each Borel set F in $[c, \infty)$. Then $\mathbf{X}$ is recurrent if and only if (7) holds.

Proof. The condition (8) is evidently fulfilled. So it suffices to show that (9) is also satisfied. By (14) the condition (9) is equivalent to

$$(15) \qquad \int_S \log \frac{1}{|T_1 \xi|} \sigma_0(d\xi) < \infty.$$

But (15) is equivalent to

$$(16) \qquad \int_{|x|<1} \log \frac{|x|}{|T_1 x|} dx < \infty.$$

Use the matrix R in the proof of Proposition 3 with $a|x| \le |Rx| \le b|x|$, $0 < a \le b$. We have

$$\int_{|x|<1} \log \frac{|Rx|}{|RT_1 x|} dx \ge \int_{|x|<1} \log \frac{|x|}{|T_1 x|} dx + \text{const.}$$

On the other hand

$$\int_{|x|<1} \log \frac{|Rx|}{|RT_1 x|} dx = \text{const} \int_{|R^{-1}x|<1} \log \frac{|x|}{|RT_1 R^{-1} x|} dx$$

$$\le \text{const} \int_{|x|\le b} \log \frac{|x|}{|\tilde{T}_1 x|} dx$$

$$\le \text{const} \int_{|x|\le b} \log \frac{|x|}{|x_1|} dx,$$

where x_1 is the first coordinate of x, since $\tilde{T}_1 = RT_1 R^{-1}$ is an orthogonal projector and $|x|/|\tilde{T}_1 x|$ is dominated by that of the case where the range of $\tilde{T}_1$ is of one dimension. Using the polar coordinates in $\mathbf{R}^d$, we get

$$\int_{|x|\le b} \log \frac{|x|}{|x_1|} dx \le \text{const} \int_0^{\pi/2} \log \frac{1}{\cos \theta} d\theta < \infty.$$

Hence (16) holds. $\square$

5. We remark that Conjecture is true if, essentially, the process $\mathbf{X}$ is a direct product process. This is an extension of Theorem 4.2 of Shiga [8].

Proposition 6. *Let Q satisfy Condition (D_n) for some n. Suppose that, for some $c > 0$,*

$$(17) \qquad \rho^c \left(\mathbf{R}^d \setminus \bigcup_{j=1}^n V_j \right) = 0.$$

Then $\mathbf{X}$ is recurrent if and only if (6) holds.

Proof. If we use a natural extension of Lemma 4, proof of this proposition is reduced to the case that the subspaces $V_1, \ldots, V_n$ are orthogonal (see the second step in the proof of Theorem 2 given in [6]). So we assume that $V_1, \ldots, V_n$ are orthogonal.

Let $p_t(x, dy)$ and $p_t^a(x, dy)$ be the transition probabilities of $\mathbf{X}$ and $\mathbf{X}^a$ (defined in **2** with c in place of a), respectively. Let $\hat{p}_t(x, z)$ and $\hat{p}_t^a(x, z)$ be the characteristic functions of $p_t(x, \cdot)$ and $p_t^a(x, \cdot)$, respectively. We know that $\mathbf{X}$ is transient if and only if

$$(18) \qquad \int_0^\infty dt \int_{|x|<b} p_t(0, dx) < \infty \quad \text{for every} \quad b.$$

Assume that **X** is transient and let us prove that

(19)
$$\int_0^1 \frac{dv}{v} \exp\left[\int_v^1 \frac{du}{u} \int_{|x|\ge a} (\exp(-\sum_{j=1}^n u^{\alpha_j}|T_j x|)-1)\rho(dx)\right] < \infty.$$

for every $a > 0$. By [8], p.439, or Lemma 2.1 of [6], it follows from the transience of **X** that $\mathbf{X}^a$ is transient for every $a > 0$. Let

$$h(x) = \prod_{j=1}^d ((1-|x_j|)\vee 0),$$

$$\hat{h}(z) = \int e^{i\langle z,x\rangle} h(x)dx = \prod_{j=1}^d (4z_j^{-2}\sin^2(2^{-1}z_j)).$$

We have (18) with p_t^a in place of p_t. Hence, by a standard argument,

$$\int_0^\infty dt \int \hat{h}(z)\operatorname{Re}\hat{p}_t^a(0,z)dz < \infty$$

for every $a > 0$. Note that

$$\hat{p}_t^a(0,z) = \exp\left[\int_0^t ds \int (\exp(i\langle e^{-sQ'}z,x\rangle)-1)\rho^a(dx)\right],$$

where Q' is the transpose of Q by Theorem 3.1 of [5]. Let

$$F_a(t,z) = \int_0^t ds \int \sin\langle z,e^{-sQ}x\rangle\rho^a(dx),$$

$$G_a(t,z) = \int_0^t ds \int [\cos\langle z,e^{-sQ}x\rangle - 1]\rho^a(dx),$$

$$H_a(t,z) = \int_0^t ds \int [\cos\langle z,e^{-sQ}x\rangle - \exp(-|\langle z,e^{-sQ}x\rangle|)]\rho^a(dx).$$

Then
$$\operatorname{Re}\hat{p}_t^a(0,z) = (\cos F_a(t,z))\exp G_a(t,z).$$

Now, let $a \ge c$. Then, by the assumption (17) and by the orthogonality of $V_1, \ldots, V_n$,

$$F_a(t,z) = \sum_{j=1}^n \int_0^t ds \int_{V_j} \sin(e^{-\alpha_j s}\langle z,x\rangle)\rho^a(dx),$$

$$G_a(t,z) = \sum_{j=1}^n \int_0^t ds \int_{V_j} [\exp(-e^{-\alpha_j s}|\langle z,x\rangle|) - 1]\rho^a(dx) + H_a(t,z),$$

$$H_a(t,z) = \sum_{j=1}^n \int_0^t ds \int_{V_j} [\cos(e^{-\alpha_j s}\langle z,x\rangle) - \exp(-e^{-\alpha_j s}|\langle z,x\rangle|)]\rho^a(dx).$$

By change of variables we get

$$F_a(t, z) = \sum_{j=1}^n \int_{V_j} \rho^a(dx) \int_{u_j}^{v_j} \frac{\sin u}{\alpha_j u} du,$$

$$H_a(t, z) = \sum_{j=1}^n \int_{V_j} \rho^a(dx) \int_{u_j}^{v_j} \frac{\cos u - e^{-|u|}}{\alpha_j u} du,$$

where $u_j = e^{-\alpha_j t} \langle z, x \rangle$, $v_j = \langle z, x \rangle$. Hence

$$|F_a(t, z)| \le K \sum_{j=1}^n \rho^a(V_j)/\alpha_j,$$

$$|H_a(t, z)| \le K \sum_{j=1}^n \rho^a(V_j)/\alpha_j$$

with an absolute constant K. If a is large enough, then we have $|F_a(t, z)| < \pi/4$ and $\cos F_a(t, z) > 1/\sqrt{2}$, which implies that

$$\int_0^\infty dt \int \hat{h}(z) \exp G_a(t, z) dz < \infty.$$

Thus, for some z with $0 < |z| \le 1$, we get

$$\int_0^\infty \exp G_a(t, z) dt < \infty.$$

Hence we have, using the boundedness of $|H_a(t, z)|$,

$$\int_0^\infty dt \exp \left[\sum_{j=1}^n \int_0^t ds \int_{V_j} \left(\exp(-e^{-\alpha_j s} |\langle z, x \rangle|) - 1 \right) \rho^a(dx) \right] < \infty$$

for large a. Since $|\langle z, x \rangle| \le |x|$, we obtain the same inequality with $|x|$ in place of $|\langle z, x \rangle|$. Hence

$$(20) \qquad \int_0^\infty dt \exp \left[\int_0^t ds \int_{\mathbf{R}^d} (\exp(-\sum_{j=1}^n e^{-\alpha_j s} |T_j x|) - 1) \rho^a(dx) \right] < \infty,$$

which is equivalent to (19).

Conversely, assume that (19) holds for some $a > 0$. Then (19) or, equivalently, (20) holds for every $a > 0$. We claim that $\mathbf{X}$ is transient. By a standard argument we see that it suffices to show that

$$\int_0^\infty dt \int_{|z|<1} |\hat{p}_t(0, z)| dz < \infty.$$

Use

$$|\hat{p}_t(0, z)| \le |\hat{p}_t^a(0, z)| = \exp G_a(t, z)$$

and

$$\int_{|z|<1} \exp G_a(t, z)\,dz$$
$$= \int_{|z|<1} dz \exp\left[\int_0^t ds \int (\exp(-|e^{-sQ}x|) - 1)\rho^a(dx) + H_a(t, z) + I_a(t, z)\right],$$

where

$$I_a(t, z) = \int_0^t ds \int [\exp(-|\langle z, e^{-sQ}x\rangle|) - \exp(-|e^{-sQ}x|)]\rho^a(dx)$$
$$= \sum_{j=1}^n \int_0^t ds \int_{V_j} [\exp(-e^{-\alpha_j s}|\langle z, x\rangle|) - \exp(-e^{-\alpha_j s}|x|)]\rho^a(dx)$$
$$= \sum_{j=1}^n \int_{V_j} \rho^a(dx) \int_{e^{-t}}^1 [\exp(-u^{\alpha_j}|\langle z, x\rangle|) - \exp(-u^{\alpha_j}|x|)]\frac{du}{u}$$
$$\le \sum_{j=1}^n \frac{1}{\alpha_j} \int_{V_j} \log \frac{|x|}{|\langle z, x\rangle|}\rho^a(dx)$$

by (12). The term $|H_a(t, z)|$ is bounded in t and z, and

$$\int_{|z|<1} \sup_t (\exp I_a(t, z))\,dz < \infty$$

by virtue of Lemma 2.2 of [6], if $\sum_{j=1}^n (1/\alpha_j)\rho^a(V_j) < 1$. Thus

$$\int_0^\infty dt \int_{|z|<1} \exp G_a(t, z)\,dz < \infty$$

for large a, because (20) means

$$\int_0^\infty dt \exp\left[\int_0^t ds \int (\exp(-|e^{-sQ}x|) - 1)\rho^a(dx)\right] < \infty.$$

The proof is complete. $\square$

REFERENCES

1. R. M. Blumenthal and R. K. Getoor, *Markov processes and potential theory*, Academic Press, New York, 1968.
2. E. Çinlar and M. Pinsky, *A stochastic integral in storage theory*, Zeit. Wahrsch. verw. Gebiete **17** (1971), 227–240.
3. B. V. Gnedenko and A. N. Kolmogorov, *Limit distributions for sums of independent random variables*, 2nd ed., Addison-Wesley, Reading, Mass., 1968.

4. K. Sato, *Stochastic processes with stationary independent increments*, (Japanese), Kinokuniya, Tokyo, 1990.

5. K. Sato and M. Yamazato, *Operator-selfdecomposable distributions as limit distributions of processes of Ornstein-Uhlenbeck type*, Stoch. Proc. Appl. **17** (1984), 73-100.

6. K. Sato and M. Yamazato, *Recurrence conditions for multidimensional processes of Ornstein-Uhlenbeck type*, Preprint.

7. L. A. Shepp, *Recurrent random walks with arbitrary large steps*, Bull. Amer. Math. Soc. **70** (1964), 540–542.

8. T. Shiga, *A recurrence criterion for Markov processes of Ornstein-Uhlenbeck type*, Probab. Th. Rel. Fields **85** (1990), 425-447.

9. G. E. Uhlenbeck and L. S. Ornstein, *On the theory of the Brownian motion*, Phys. Rev. **36** (1930), 823–841.

10. K. Urbanik, *Lévy's probability measures on Euclidean spaces*, Studia Math. **44** (1972), 119-148.

11. S. J. Wolfe, *On a continuous analogue of the stochastic difference equation $X_n = \rho X_{n-1} + B_n$*, Stoch. Proc. Appl. **12** (1982), 301–312.

12. K. Yosida, *Functional analysis*, Springer, Berlin Heidelberg New York, 1965.

Some remarks about singular perturbed solutions for Emden-Fowler equation with exponential nonlinearity

Takashi SUZUKI

Department of Mathematics, Faculty of Science,
Tokyo Metropolitan University
Minamiohsawa 1-1, Hachiojishi, Tokyo 192-03, Japan

1 Introduction

Our purpose is to construct singular perturbed solutions for the nonlinear eigenvalue problem of Emden-Fowler type (P):

$$- \Delta u = \lambda e^u \quad (in \ \Omega) \tag{1}$$

with

$$u = 0 \quad (on \ \partial \Omega), \tag{2}$$

where λ is a positive constant and $\Omega \subset R^2$ is a bounded domain with smooth boundary $\partial \Omega$.

In the previous work [7] [14], we have proven the following theorems for classical solutions $\{(u, \lambda)\} \subset C^2(\Omega) \cap C^0(\overline{\Omega}) \times R_+$, where $R_+ = (0, \infty)$ and $\Sigma \equiv \int_\Omega \lambda e^u dx$.

Theorem 1.1. *As $\lambda \downarrow 0$, the values $\{\Sigma\}$ accumulate to $8\pi m$ for some $m = 0, 1, 2, ...,$ $+\infty$. The solutions $\{u\}$ behave as follows:*

(a) If $m = 0$, then $|u|_{L^\infty(\Omega)} \longrightarrow 0$.

(b) If $0 < m < +\infty$, there exists a set of m-points $\mathcal{S} = \{x_1^*, ..., x_m^*\} \subset \Omega$ such that $|u|_{L_{loc}^\infty(\overline{\Omega} \backslash \mathcal{S})} \in O(1)$ and $u |_{\mathcal{S}} \longrightarrow +\infty$.

(c) If $m = +\infty$, then $u(x) \longrightarrow +\infty \ (x \in \Omega)$.

Theorem 1.2. *If the domain is star-shaped, the case (c) does not occur. In the case of (b), the limit function of $\{u\}$, the singular limit, is given as*

$$u_0(x) = 8\pi \sum_{j=1}^m G(x, x_j^*) \tag{3}$$

and the blow-up points $\mathcal{S} = \{x_1^*, ..., x_m^*\}$ are located so that

$$\frac{1}{2} \nabla R(x_j^*) + \sum_{\ell \neq j} \nabla_x G(x_\ell^*, x_j^*) = 0 \quad (1 \leq j \leq m). \tag{4}$$

Here, $G(x, y)$ denotes the Green function of $-\Delta$ in Ω under the Dirichlet boundary condition and $R(x) = \left[G(x, y) + \frac{1}{2\pi} \log | x - y | \right]_{y=x}$ is the Robin function. In other words, $(x_1^*, ..., x_m^*) \in \Omega \times ... \times \Omega$ is a critical point of the function

$$K(x_1, ..., x_m) = \frac{1}{2} \sum_{j=1}^{m} R(x_j) + \sum_{\ell \neq \ell'} G(x_\ell, x_{\ell'}) \tag{5}$$

in $\Omega^m = \Omega \times ... \times \Omega$.

Theorem 1.3. *Suppose the simply-connectedness of Ω and the existence of classical solutions $\{(u, \lambda)\}$ such that the value $\Sigma = \int_\Omega \lambda e^u dx$ tends to 8π from below. Then the singular limit $u_0(x)$ in (3) with $m = 1$ is connected to the trivial solution $u = 0$ of $\lambda = 0$ in $\lambda - u$ plane, through a one-dimensional manifold T bending just once.*

Later, [15] improved Theorem 1.2 and showed that the case (c) in Theorem 1.1 does not occur when Ω is simply-connected. Furthermore, [11], [7], and [13] showed the existence of a positive constant $C_\varepsilon > 0$ for given $\varepsilon > 0$, such that $|u|_{L^\infty(\Omega)} \leq C_\varepsilon$ for any classical solution u of (P) with $\lambda \geq \varepsilon$. Therefore, from the theorem of [10], any component of the solution set is absorbed into the hyperplane $\lambda = 0$. In this way, the construction of singularly perturbed solutions, classical solutions close to $u_0(x)$ in (3) for $\lambda > 0$ small, becomes an important issue and that is the object of the present paper.

The work of [18], [5] has already succeeded in it for the case of $m = 1$. Refining the methods, we can reduce their assumptions to construct one-point blow-up solutions. Namely, for any *non-degenerate* critical point $x_1^* \in \Omega$ of the Robin function $R(x)$, there exists a family of classical solutions close to the singular limit $u_0(x) = 8\pi G(x, x_1^*)$ for $\lambda > 0$ sufficiently small, provided that Ω is simply-connected.

For that family, in the same way as in [12], the condition for Σ to tend to 8π from below can be given in terms of the conformal mapping $g : D \equiv \{| \zeta | < 1\} \longrightarrow \Omega$ with $g(0) = x_1^*$. Namely,

$$\frac{C}{\pi} \equiv - | a_1 |^2 + \sum_{k=3}^{\infty} \frac{k^2}{k - 2} | a_k |^2 < 0, \tag{6}$$

where $g(\zeta) = \sum_{k=0}^{\infty} a_k \zeta^k$. Thus we have the condition (6), a quantitative criterion about how Ω should be close to a ball, for the conclusion of Thorem 1.3 to hold.

The meaning of the non-degeneracy of x_1^* can be understood clearly if the perturbation of the domain is taken into account. Actually, the following phenomenon was observed by [2] and [9] for perturbing domains with two axile symmetries; Under those circumstances the degeneracy occurs just when the singular limit $u_0(x) = 8\pi G(x, x_1^*)$ is bifurcating.

The success of the use of complex function theory is due to pulling back the problem to the unit disc, the origin corresponding to the maximal point of the solution. This is actually efficient for one-point blow-up solutions by virtue of the flexibility of Möbius transformations. Unfortunately, it does not work for the multi-point blow-up, in spite that all possibilities $m = 0, 1, 2, ..., +\infty$ are expected in Theorem 1.1 when the domain is an annulus ([8], [3]). Thus, function theoretic construction of multi-point blow-up solutions seems to be still open.

There remain other problems. For example, we expect the local uniqueness around $\lambda = 0$ of the family $\{u\}$ converging to $u_0(x) = 8\pi G(x, x_1^*)$ when x_1^* is non-degenerate. In connection with this, we note that the existence of non-minimal solution is assured for any domain via the mountain pass lemma ([1]). If it holds that the mountain pass solutions always make one-point blow-up, the singular perturbation might be performed even when x_1^* is degenerate, in which case we may have the non-uniqueness of classical solutions close to $8\pi G(x, x_1^*)$ for any $\lambda > 0$ sufficiently small. Also, except for the annulus domain, the entire blow-up solution, i.e. the case (c) in Theorem 1.1, has not been constructed, where the domain can not be simply-connected.

From the complex function theoretic viewpoint, our study is to construct a family of holomorphic functios $\{f\}$ on $D \equiv \{|\zeta| < 1\} \subset C$ satisfying

$$\rho(f) \equiv \frac{|f'|}{1 + |f|^2} = \left(\frac{\lambda}{8}\right)^{1/2} |p| \quad (on \ \partial D), \tag{7}$$

where p is a given holomorphic function satisfying $p'(0) = 0$ and $|p''(0)/p(0)| \neq 2$. Therefore, it is related to harmonic or minimal surfaces ([16]).

2 One-point blow-up solutions

Henceforth, the domain $\Omega \subset R^2$ is supposed to be simply-connected.

As is described in §1, one-point blow-up solutions have been constructed. The procedure for construction is divided into two parts.

In the first part asymptotic solutions are constructed ([18]). Thus we have, for each $n = 1, 2, ...$, a family $\{v\}$ of classical solutions for

$$-\Delta v = \lambda e^v \quad (in \ \Omega) \tag{8}$$

satisfying

$$|v|_{L^\infty(\partial\Omega)} \in O(\lambda^n) \tag{9}$$

and

$$v(x) \longrightarrow u_0(x) = 8\pi G(x, x_1^*) \tag{10}$$

as $\lambda \downarrow 0$ by a complex function theoretic way. Actually, discovered by [4], the equation (8) has a complex structure. Later, [5] refined the procedure to reduce the assumptions about the blow-up point $x_1^* \in \Omega$.

Genuine solutions are constructed by a modified Newton's iteration starting from the n-th asymptotic solution. By means of a delicate study on the associated Legendre equation, [18] showed that the inverses of the linearized operators are dominated by $constant \times \lambda^{-1}$ in the operator norm in $L^\infty(\Omega)$, under some additional assumption on x_1^*. From this fact, it is concluded that the iteration converges if $\lambda > 0$ is sufficiently small and $n \geq 3$ in that case.

Weston's additional assumption was examined to hold when Ω is convex and has two axile symmetries by [17]. In this case, Moseley's criterion for constructing asymptotic solutions also holds and hence the existence of one-point blow-up solutions arises. However, Weston's additional condition looks very comlicated and difficult to examine for general domains.

What we shall show in the present section is that Moseley's asymptotic solutions $\{v\}$ actually converge as $n \longrightarrow +\infty$, $\lambda > 0$ being fixed, to produce genuine solutions $\{u\}$ without imposing any other conditions. Furthermore, Meseley's criterion for the existence of asymptotic solutions is nothing but the non-degenaracy of the critical point $x_1^* \in \Omega$ of $R(x)$. Thus we can prove the following theorem:

Theorem 2.1. *Let Ω be simply-connected and $x_1^* \in \Omega$ be a non-degenerate critical point of the Robin function $R(x)$. Then, there exists a family $\{(u, \lambda)\}$ of classical solutions of (P) with $\lambda > 0$ sufficiently small such that*

$$u(x) \longrightarrow 8\pi G(x, x_1^*) \tag{11}$$

for each $x \in \overline{\Omega} \setminus \{x_1^*\}$ as $\lambda \downarrow 0$.

As is described in §1, we can take a conformal mapping $g : D \equiv \{|\zeta| < 1\} \longrightarrow \Omega$ so that $g(0) = x_1^* \in \Omega$, the blow-up point. The Green function $G(x, y)$ on Ω is expressed in terms of the inverse mapping $h = g^{-1} : \Omega \longrightarrow D$. That is,

$$G(x, y) = -\frac{1}{2\pi} \log \left| \frac{h(z) - h(w)}{1 - \overline{h(w)}h(z)} \right|, \tag{12}$$

where $z = x_1 + \imath x_2$ and $w = y_1 + \imath y_2$ for $x = (x_1, x_2)$ and $y = (y_1, y_2)$.

In particular we have

$$G(x, y) = -\frac{1}{2\pi} \log |z - w| + \frac{1}{2\pi} \log \frac{1 - |h(w)|^2}{|h'(w)|} + O(|z - w|)$$

as $w \to z$ so that

$$R(x) = \frac{1}{4\pi} \log \left\{ |g'(\zeta)|^2 (1 - |\zeta|^2)^2 \right\} \tag{13}$$

for $x = g(\zeta)$ with $\zeta \in D$. From this expression, we have

$$4\pi \frac{\partial}{\partial \zeta} R(x) = \frac{g''(\zeta)}{g'(\zeta)} - \frac{2\overline{\zeta}}{1 - |\zeta|^2} = 4\pi \overline{\frac{\partial}{\partial \zeta} R(x)}, \tag{14}$$

$$4\pi \frac{\partial^2}{\partial \zeta^2} R(x) = \frac{g'''(\zeta)}{g'(\zeta)} - \left\{ \frac{g''(\zeta)}{g'(\zeta)} \right\}^2 - \frac{2\overline{\zeta}^2}{(1 - |\zeta|^2)^2} = 4\pi \overline{\frac{\partial^2}{\partial \zeta^2} R(x)}, \tag{15}$$

and

$$4\pi \frac{\partial^2}{\partial \zeta \partial \overline{\zeta}} R(x) = -\frac{2}{(1 - |\zeta|^2)^2}. \tag{16}$$

Therefore, $x_1^* = g(0) \in \Omega$ is a non-degenerate critical point of $R(x)$ if and only if

$$g''(0) = 0 \tag{17}$$

and

$$|g'''(0)/g'(0)| \neq 2. \tag{18}$$

The Liouville integral [4] is formulated as follows ([7] and [13]), where $\omega(\psi_1, \psi_2) = \psi_1 \psi_2' - \psi_1' \psi_2$ denotes the Wronskian of two holomorphic functions $\psi_1(z)$ and $\psi_2(z)$.

Lemma 2.2. *Let $\Omega \subset R^2$ be a simply-connected domain, and the C^2 function u solve*

$$-\Delta u = \lambda e^u \quad (in \ \Omega). \tag{19}$$

Then there exists a pair of holomorphic functions $\{\psi_1, \psi_2\}$ satisfying

$$\omega(\psi_1, \psi_2) = 1 \quad (in \ \Omega) \tag{20}$$

and

$$w \equiv e^{-u/2} = |\psi_1|^2 + \frac{\lambda}{8} |\psi_2|^2 . \tag{21}$$

Proof Introducing the complex variables $z = x_1 + \imath x_2$ and $\bar{z} = x_1 - \imath x_2$ for $x = (x_1, x_2) \in \Omega$, we put

$$s = u_{zz} - \frac{1}{2} u_z^2. \tag{22}$$

Because of $u_{z\bar{z}} = -\frac{\lambda}{4} e^u$ we have

$$s_{\bar{z}} = u_{zz\bar{z}} - u_z u_{z\bar{z}} = 0$$

and hence $s = s(z)$ is a holomorphic function.

The relation (22) can be regarded as a Riccati equation so that is to be linearized. Thus the function $\phi = e^{-u/2}$ solves

$$\phi_{zz} + \frac{1}{2} s\phi = 0. \tag{23}$$

We introduce a fundamental sysytem $\{\phi_1, \phi_2\}$ of solutions, which are holomorphic functions of z. Then we have

$$\phi \equiv e^{-u/2} = \overline{f_1}(\bar{z})\phi_1(z) + \overline{f_2}(\bar{z})\phi_2(z) \tag{24}$$

with some functions $\overline{f_1}(\bar{z})$ and $\overline{f_2}(\bar{z})$ independent of z.

Let $x^* = (x_1^*, x_2^*) \in \Omega$ be a maximal point of u. We can assume that

$$\phi_1(z^*) = \phi_2'(z^*) = 1, \ \phi_1'(z^*) = \phi_2(z^*) = 0 \tag{25}$$

for $z^* = x_1^* + \imath x_2^*$. Then it holds that

$$\omega(\phi_1, \phi_2) \equiv \phi_1 \phi_2' - \phi_1' \phi_2 = 1.$$

Therefore, we have

$$\overline{f_1}(\bar{z}) = \omega(\phi, \phi_2) = \phi \phi_2' - \phi_z \phi_2$$

and

$$\overline{f_2}(\bar{z}) = -\omega(\phi, \phi_1) = -\phi \phi_1' + \phi_z \phi_1$$

for $\phi(z, \bar{z}) = e^{-u/2}$. Here, the left-hand sides are independent of z. We put $z = z^*$ to obtain

$$\overline{f_1}(\bar{z}) = \phi(z^*, \bar{z}) \tag{26}$$

and

$$\overline{f_2}(\bar{z}) = \phi_z(z^*, \bar{z}). \tag{27}$$

The real-valued function $\phi = e^{-u/2}$ solves (23) with $s_{\bar{z}} = 0$. Therefore, so does $f_1(z) = \overline{f_1}(\bar{z})$. In other words, those functions f_1 and f_2 are linear combinations of ϕ_1 and ϕ_2. By virtue of $\nabla u(x^*) = 0$ we have

$$\overline{f_1}|_{\bar{z}=\bar{z}^*} = \phi(z^*, \bar{z}^*) = e^{-u/2}|_{x=x^*} = C_1 > 0,$$

$$\frac{\partial}{\partial \bar{z}}\overline{f_1}\mid_{\bar{z}=z^*}= \phi_{z\bar{z}}(z^*,\overline{z^*}) = -\frac{1}{2}e^{-u/2}u_{\bar{z}}\mid_{x=x^*}= 0,$$

$$\overline{f_2}\mid_{\bar{z}=\overline{z^*}}= \phi_z(z^*,\overline{z^*}) = -\frac{1}{2}e^{-u/2}u_z\mid_{x=x^*}= 0,$$

and

$$\frac{\partial}{\partial \bar{z}}\overline{f_2}\mid_{\bar{z}=\overline{z^*}}= \phi_{z\bar{z}}(z^*,\overline{z^*}) = -\frac{1}{2}e^{-u/2}u_{z\bar{z}}\mid_{x=x^*}= \frac{\lambda}{8}e^{u/2}\mid_{x=x^*}= \frac{\lambda}{8}C_1^{-1}.$$

Therefore, we get

$$f_1 = C_1\phi_1 \tag{28}$$

and

$$f_2 = \frac{\lambda}{8}C_1^{-1}\phi_2. \tag{29}$$

Substituting those relations into (24) we obtain the expression (21) with (20), where $\psi_1 = C_1^{1/2}\phi_1$ and $\psi_2 = C_1^{-1/2}\phi_2$. **Q.E.D.**

The expression (21) is important. We can write the singular limit $u_0(x) = 8\pi G(x, x_1^*)$ in this form. In fact, in terms of the conformal mapping $h = g^{-1} : \Omega \longrightarrow D$ we obtain the equality

$$w_0(x) \equiv e^{-u_0(x)/2} =\mid h(x)\mid^2$$

by (12). The relation is to be pulled-back to $\overline{D}$. Thus the function $W = g^*w \equiv w \circ g$ is to satisfy

$$W \equiv \mid \Psi_1 \mid^2 +\frac{\lambda}{8}\mid \Psi_2 \mid^2 \tag{30}$$

with

$$\omega(\Psi_1, \Psi_2) \equiv \Psi_1\Psi_2' - \Psi_1'\Psi_2 = g' \quad (in\ D) \tag{31}$$

$$W = 1 \quad (on\ \partial D) \tag{32}$$

and

$$W(\zeta) \longrightarrow W_0(\zeta) \equiv\mid \zeta \mid^2, \tag{33}$$

where $\Psi_1(\zeta)$ and $\Psi_2(\zeta)$ are holomorphic functions on D.

From the proof of Theorem 1.1 in §1 ([7]), the asymptotics (33) splits in

$$\Psi_1(\zeta) \longrightarrow \zeta \tag{34}$$

and

$$\{\Psi_2\} : bounded \tag{35}$$

uniformly on $\overline{D}$, when one-point blow-up actually occurs. Taking this into account, we put

$$\Psi_1 = \zeta/G \tag{36}$$

and

$$\Psi_2 = M/G. \tag{37}$$

Noting the equality

$$\omega(\zeta/G, M/G) = \omega(\zeta, M)/G^2,$$

we shall construct a family $\{G, M\}$ of pairs of holomorphic functions such that

$$\omega(\zeta, M) = g'G^2 \quad (in\ D), \tag{38}$$

$$| G |^2 = 1 + \frac{\lambda}{8} | M |^2 \quad (on\ \partial D), \tag{39}$$

and

$$|G - 1|_{L^\infty(D)} \to 0, \quad |M|_{L^\infty(D)} \in O(1) \tag{40}$$

as $\lambda \downarrow 0$. The important thing here is that from those pairs $\{G, M\}$, conversely, a family $\{u\}$ of classical solution of (P) is recovered with the relation (11) as $\lambda \downarrow 0$, as we shall see below.

To this end we deduce the equations for $w = e^{-u/2}$ to staisfy, where u denotes the classical solution of (P). In fact,

$$w\Delta w - | \nabla w |^2 = \frac{\lambda}{2}, \ w > 0 \quad (in\ \Omega) \tag{41}$$

with

$$w = 1 \quad (on\ \partial\Omega). \tag{42}$$

Thus the function $W = g^* w$ is to satisfy

$$W\Delta W - | \nabla W |^2 = \frac{\lambda}{2} | g' |^2, \ W > 0 \quad (in\ D) \tag{43}$$

with

$$W = 1 \quad (on\ \partial D). \tag{44}$$

Furthermore, the asymptotics (11) will follow from (33).

Take a family $\{G, M\}$ of paris of holomorphic functions satisfying (38), (39), and (40). Then the function

$$W = | \zeta/G |^2 + \frac{\lambda}{8} | M/G |^2 \tag{45}$$

obviously satisfies (44) and (33). In order to verify (43), we introduce the bilinear form

$$\{V, W\} = \frac{1}{2}(V\Delta W + W\Delta V) - \nabla V \cdot \nabla W \tag{46}$$

for real-valued C^2 functions V and W on $\overline{D}$. The equation (43) is written as

$$\{W\} = \frac{\lambda}{2} | g' |^2, \ W > 0 \quad (in\ D), \tag{47}$$

where $\{W\}$ stands for $\{W, W\}$. We note the following lemma.

Lemma 2.3. *For holomorphic functions $p(\zeta)$ and $q(\zeta)$ of $\zeta \in D$, it holds that*

$$\{| p |^2, | q |^2\} = 2 | \omega(p, q) |^2 . \tag{48}$$

In particular, we have

$$\{| p |^2\} = 0. \tag{49}$$

Proof Introducing the complex variable $\zeta = \xi_1 + \imath\xi_2$ and $\overline{\zeta} = \xi_1 - \imath\xi_2$ for $\xi = (\xi_1, \xi_2) \in D$, we obtain

$$\{V, W\} = 2(VW_{\zeta\overline{\zeta}} + V_{\zeta\overline{\zeta}}W - V_\zeta W_{\overline{\zeta}} - V_{\overline{\zeta}}W_\zeta).$$

Therefore, we get

$$\{|\,p\,|^2, |\,q\,|^2\} = 2(|\,p\,|^2|\,q'\,|^2 + |\,p'\,|^2|\,q\,|^2 - \bar{p}p'\overline{q'}q - p\bar{p}'q'\bar{q})$$

$$= 2\,|\,pq' - p'q\,|^2\,.$$

Q.E.D.

Now, for the function W given in (45) we can compute as

$$\{W\} = \{W, W\} = \frac{\lambda}{4}\{|\,\zeta/G\,|^2, |\,M/G\,|^2\} = \frac{\lambda}{2}\,|\,\omega(\zeta/G, M/G)\,|^2$$

$$= \frac{\lambda}{2}\,|\,\omega(\zeta, M)/G^2\,|^2 = \frac{\lambda}{2}\,|\,g'\,|^2\,.$$

Hence the equality (47) follows.

In this way, Theroem 2.1 has been reduced to constructing a family of pairs $\{G, M\}$ of holomorphic functions satisfying (38), (39), and (40).

The following lemma is elementary:

Lemma 2.4. *Given a holomorphic function $K(\zeta)$, there exists a holomorphic function $M(\zeta)$ such that*

$$\omega(\zeta, M) \equiv (\zeta\frac{d}{d\zeta} - 1)M = K \quad (in\ D) \tag{50}$$

if and only if $K'(0) = 0$. The solution is given as

$$M = \alpha(K) + a\zeta, \tag{51}$$

where

$$\alpha(K)(\zeta) = \sum_{n\neq 1}\frac{k_n}{n-1}\zeta^n \tag{52}$$

for $K(\zeta) = \sum_{n=0}^{+\infty} k_n\zeta^n$ and $a \in C$ is arbitrary.

We introduce some function spaces. Let HL be the Hardy-Lebesgue space ([19] e.g.):

$$HL = \{H(\zeta):\ holomorphic\ in\ D\ |$$

$$|H|^2_{HL} \equiv \sup_{0\leq r<1}\frac{1}{2\pi}\int_0^{2\pi}\,|\,H(re^{i\theta})\,|^2\,d\theta < +\infty\}. \tag{53}$$

We put

$$HL_0 = \{H \in HL \mid H'(0) = 0\} \subset HL \tag{54}$$

and

$$HL_0^1 = \{H \in HL_0 \mid H' \in HL\}. \tag{55}$$

The functional α defined in (52) is an isomorphism from HL_0 onto HL_0^1, because

$$|H|^2_{HL} = \sum_{n=0}^{+\infty}\,|\,h_n\,|^2 \tag{56}$$

for $H(\zeta) = \sum_{n=0}^{+\infty} h_n\zeta^n$.

We prepare the following Banach space:

$$X = A_0(D) \equiv \{H(\zeta) : \text{ holomorphic in } D, \text{ continuous on } \overline{D} \mid H'(0) = 0\}. \tag{57}$$

Obviously, $X = A_0(D)$ is continuously imbedded into HL_0. We have the following lemma:

Lemma 2.5. *The space HL_0^1 is continuously imbedded into X.*

Proof We have for $H(\zeta) = \sum_{n=0}^{+\infty} h_n \zeta^n$ that

$$|H|_X = |H|_{L^\infty(D)} \leq \sum_{n=0}^{+\infty} |h_n| \leq |h_0| + (\sum_{n\geq 1} n^2 |h_n|^2)^{1/2}(\sum_{n\geq 1} \frac{1}{n^2})^{1/2}$$

$$\leq |H|_{HL} + \text{constant} \times |H'|_{HL}$$

Q.E.D.

Therefore, α in (52) can be regarded as a bounded linear operator on X. The space X becomes an algebra. Each element in X has the continuous boundary value.

The smoothness of $\partial\Omega$ is supposed so that $g' \in X$. Note that $g''(0) = 0$. Hence the equation (38) with (39) is reduced to

$$|G|^2 = 1 + \frac{\lambda}{8} |\alpha(g'G^2) + a\zeta|^2 \quad (\text{on } \partial D) \tag{58}$$

with

$$M =' \alpha(g'G^2) + a\zeta \tag{59}$$

for $(G, a) \in X \times C$. We shall construct a family $\{G, a\}$ satisfying (58) for $\lambda > 0$ small as well as

$$|G - 1|_{L^\infty(D)} \in o(1), \quad |a| \in O(1). \tag{60}$$

Then the asymptotics (40) will be realized. Thus, Theorem 2.1 has been reduced to solving the equation (58) for $(G, a) \in X \times C$ with the asymptotics (60).

Setting $G = 1 + \lambda H$, we get the equation

$$H + \overline{H} = \frac{1}{8} |\alpha(g') + a\zeta|^2 + \lambda\Phi(H, a, \lambda) \quad (\text{on } \partial D), \tag{61}$$

where

$$\Phi(H, a, \lambda) = -|H|^2 + \frac{1}{8}\{(\alpha(g') + a\zeta)(\overline{2\alpha(g'H) + \lambda\alpha(g'H^2)})$$

$$+ \overline{(\alpha(g') + a\zeta)}(2\alpha(g'H) + \lambda\alpha(g'H^2)) + |2\alpha(g'H) + \lambda\alpha(g'H^2)|^2\}. \tag{62}$$

The asymptotics (60) will follow from

$$|H|_X, \quad |a| \in O(1). \tag{63}$$

The operator $\Phi = \Phi(H, a, \lambda) : X \times C \times R \longrightarrow X$ is bounded locally Lipschitz (i.e., bounded and Lipschitz continuous on every bounded set in $X \times C \times R$).

Here we define the holomophic function $I_0(\zeta)$ and the constant c_0 through

$$\alpha(g') = c_0 + \zeta^2 I_0(\zeta). \tag{64}$$

Then we have $c_0 = -g'(0)$ and $I_0(0) = \frac{1}{2}g'''(0)$. Furthermore,

$$|\alpha(g') + a\zeta|^2 = |a|^2 + 2\Re\{(a\overline{c_0} + \overline{a}I_0(\zeta))\zeta\} + |\alpha(g')|^2$$

if $|\zeta| = 1$. Therefore, (61) is written as

$$2\Re H = \frac{|a|^2}{8} + \frac{1}{4}\Re\{(a\overline{c_0} + \overline{a}I_0(\zeta))\zeta\} + \frac{1}{8}|\alpha(g')|^2 + \lambda\Phi(H, a, \lambda) \quad (on\ \partial D). \qquad (65)$$

We can solve the boundary value problem about the harmonic function $\Re H$. Thus, Schwarz's formula reduces (65) to

$$H(\zeta) = \frac{|a|^2}{16} + \frac{1}{8}(a\overline{c_0} + \overline{a}I_0(\zeta))\zeta$$

$$+ \frac{1}{2\pi\imath}\int_{\partial D}\{\frac{|\alpha(g')|^2}{8} + \lambda\Phi(H, a, \lambda)\}(\tilde{\zeta})\{\frac{1}{\tilde{\zeta} - \zeta} - \frac{1}{2\tilde{\zeta}}\}d\tilde{\zeta} \quad (\zeta \in D). \qquad (66)$$

We utilize the constant $a \in C$ for the condition $H'(0) = 0$ to hold in (66). Namely,

$$a\overline{c_0} + \overline{a}I_0(0) = \phi(a, H, \lambda), \qquad (67)$$

where

$$\phi(a, H, \lambda) = \frac{-1}{2\pi\imath}\int_{\partial D}\{|\alpha(g')|^2 + 8\lambda\Phi(H, a, \lambda)\}(\tilde{\zeta})\frac{d\tilde{\zeta}}{\tilde{\zeta}^2}. \qquad (68)$$

If (67) is satisfied, the right-hand side of (66) belongs to X.

Take a constant $R > \frac{1}{2\pi}\int_{\partial D}|\alpha(g')|^2|d\tilde{\zeta}|$. Given $H \in X$ there exists a constant $\lambda^* = \lambda^*(|H|_X)$ such that the mapping $a \longrightarrow \phi(a, H, \lambda)$ is a contraction on $\overline{B_R}(0) = \{|a| \leq R\} \subset C$ provided that $0 < \lambda < \lambda^*$. Therefore, the equation (67) concerning $a \in C$ is solvable for $0 < \lambda \ll 1$ if the linear part, the left-hand side, is invertible, which is equivalent to $|I_0(0)/c_0| \neq 1$, i.e., $|g'''(0)/g'(0)| \neq 2$.

Actually, let $a_0 \in C$ be the unique solution of the linear algebraic equation

$$a_0\overline{c_0} + \overline{a_0}I_0(0) = -\frac{1}{2\pi\imath}\int_{\partial D}|\alpha(g')|^2(\tilde{\zeta})\frac{d\tilde{\zeta}}{\tilde{\zeta}^2}. \qquad (69)$$

Then the solution $a \in C$ of (67) is written like

$$a(H, \lambda) = a_0 + \lambda a_1(H, \lambda), \qquad (70)$$

where $a_1 = a_1(H, \lambda)$ is a bounded locally Lipschitz operator on $\mathcal{M} = \{(H, \lambda) \mid H \in X, 0 < \lambda < \lambda^*(|H|_X)\}$ into C.

Substituting the relation into the right-hand side of (66), we obatain a fixed point equation

$$H = \mathcal{N}(H, \lambda) \equiv H_0 + \lambda\mathcal{N}_1(H, \lambda), \qquad (71)$$

where

$$H_0(\zeta) = \frac{|a_0|^2}{16} + \frac{1}{8}\{(a_0\overline{c_0} + \overline{a_0}I_0(\zeta))\zeta\}$$

$$+ \frac{1}{2\pi\imath}\int_{\partial D}\frac{1}{8}|\alpha(g')(\tilde{\zeta})|^2\{\frac{1}{\tilde{\zeta} - \zeta} - \frac{1}{2\tilde{\zeta}}\}d\tilde{\zeta} \in X \qquad (72)$$

and $\mathcal{N}_1 : \mathcal{M} \longrightarrow X$ is a bounded locally Lipschitz operator.

The fixed point equation (71) is solvable for sufficiently small $\lambda > 0$, where the ball $B \equiv \{H \in X \mid |H - H_0|_X \leq 1\}$ is prepared. In fact, then the mapping $H \longmapsto \mathcal{N}(H, \lambda)$ becomes a contraction on B. Thus the proof of Theorem 2.1 has been completed.

3 Remarks

1. We can extend Theorem 2.1 to the inhomogeneous problem

$$- \Delta u = \lambda e^u \quad (in \ \Omega) \tag{73}$$

with

$$u = \phi \quad (on \ \partial\Omega) \tag{74}$$

by the method of [5], where $\phi \in C^0(\partial\Omega)$.

In fact, let the function $u_0 = \mathcal{P}_\Omega \phi$ solve

$$- \Delta u_0 = 0 \quad (in \ \Omega) \tag{75}$$

and

$$u_0 = \phi \quad (on \ \partial\Omega), \tag{76}$$

and v_0 be a conjugate harmonic function of u_0. Then, the holomorphic function $h = u_0 + \imath v_0$ satisfies

$$e^{u_0} = \mid \exp(\frac{1}{2}h) \mid^2 . \tag{77}$$

Therefore, the function $u_1 = u - u_0$ is to solve

$$- \Delta u_1 = \lambda \mid \exp(\frac{1}{2}h) \mid^2 e^{u_1} \quad (in \ \Omega) \tag{78}$$

with

$$u_1 = 0 \quad (on \ \partial\Omega). \tag{79}$$

Introducing the conformal mapping $g : D \equiv \{\mid \zeta \mid < 1\} \longrightarrow \Omega$, the function $U = g^* u_1$ on D is to solve

$$- \Delta U = \lambda \mid p \mid^2 e^U \quad (in \ D) \tag{80}$$

with

$$U = 0 \quad (on \ \partial\Omega), \tag{81}$$

where $p = g' \exp(\frac{1}{2}g^* h)$ is a holomorphic function on D.

Through the procedure described in the previous section, we can construct a family of solutions $\{(U, \lambda)\}$ for $0 < \lambda \ll 1$ satisfying

$$e^{-U(\zeta)/2} \longrightarrow \mid \zeta \mid^2 \tag{82}$$

as $\lambda \downarrow 0$, provided that

$$p'(0) = 0 \tag{83}$$

and

$$\mid p''(0)/p(0) \mid \neq 2. \tag{84}$$

For $x_1^* = g(0) \in \Omega$, the relation (82) means that $u_1(x) \to 8\pi G(x, x_1^*)$ and hence

$$u(x) \to 8\pi G(x, x_1^*) + \mathcal{P}_\Omega \phi(x) \tag{85}$$

for each $x \in \overline{\Omega} \setminus \{x_1^*\}$ as $\lambda \downarrow 0$. On the other hand, (83) and (84) are equivalent to

$$g''(0) + \frac{1}{2}g'(0)(g^*h)'(0) = 0 \qquad (86)$$

and

$$|\,(\frac{g''}{g'})'(0) + \frac{1}{2}(g^*h)''(0)\,| \neq 2, \qquad (87)$$

respectively.

Since $g : D \to \Omega$ is conformal, we have

$$(g^*h)' = \psi_{\xi_1} - \imath\psi_{\xi_2} \qquad (88)$$

for $\zeta = \xi_1 + \imath\xi_2 \in D$, where $\psi = \mathcal{P}_D g^*\phi$ solves

$$-\Delta\psi = 0 \quad (in\ D) \qquad (89)$$

with

$$\psi = g^*\phi \quad (on\ \partial D). \qquad (90)$$

As we have seen in the previous section, it holds that

$$R(x) = S(\xi(x)) \qquad (91)$$

for $x = (x_1, x_2) \in \Omega$ corresponding to $\xi = (\xi_1, \xi_2) \in D$, where

$$S(\xi) = \frac{1}{4\pi} \log |\,g'(\zeta)\,|^2 + \frac{1}{2\pi} \log(1 - |\,\zeta\,|^2) \quad (\zeta = \xi_1 + \imath\xi_2). \qquad (92)$$

Hence

$$\frac{\partial S}{\partial \xi_1}(0) = \frac{1}{2\pi}\Re\frac{g''}{g'}(0), \quad \frac{\partial S}{\partial \xi_2}(0) = -\frac{1}{2\pi}\Im\frac{g''}{g'}(0). \qquad (93)$$

Thus, (86) is equivalent to

$$\nabla_\xi S(0) + \frac{1}{4\pi}\nabla_\xi\psi(0) = 0. \qquad (94)$$

or

$$\nabla_x R(x_1^*) + \frac{1}{4\pi}\nabla_x(\mathcal{P}_\Omega\phi)(x_1^*) = 0 \qquad (95)$$

for $x_1^* = g(0) \in \Omega$.

By virtue of (88) we have

$$(g^*h)'' = \psi_{\xi_1\xi_1} - \imath\psi_{\xi_1\xi_2}, \qquad (96)$$

where (92) deduces

$$S_{\xi_1\xi_1}(0) = \frac{1}{2\pi}\Re(\frac{g''}{g'})'(0) - \frac{1}{\pi} \equiv \frac{1}{2\pi}a - \frac{1}{\pi} \qquad (97)$$

$$S_{\xi_1\xi_2}(0) = -\frac{1}{2\pi}\Im(\frac{g''}{g'})'(0) \equiv -\frac{1}{2\pi}b \qquad (98)$$

and

$$S_{\xi_2\xi_2}(0) = -\frac{1}{2\pi}\Re(\frac{g''}{g'})'(0) - \frac{1}{\pi} = -\frac{1}{2\pi}a - \frac{1}{\pi}. \qquad (99)$$

Therefore, for the function $q(\xi) = Q(x(\xi))$ with

$$Q = R + \frac{1}{4\pi}\mathcal{P}_\Omega\phi, \tag{100}$$

the condition $Hess\ q(0) \neq 0$ is equivalent to

$$-(a - 2 + \frac{1}{2}\psi_{\xi_1\xi_1}(0))(a + 2 - \frac{1}{2}\psi_{\xi_2\xi_2}(0)) - (-b + \frac{1}{2}\psi_{\xi_1\xi_2}(0))^2 \neq 0,$$

or

$$(a + \frac{1}{2}\psi_{\xi_1\xi_1}(0))^2 + (b - \frac{1}{2}\psi_{\xi_1\xi_2}(0))^2 \neq 4 \tag{101}$$

by $\psi_{\xi_2\xi_2} = -\psi_{\xi_1\xi_1}$. That is, (87) is nothing but

$$Hess\ Q(x_1^*) \neq 0. \tag{102}$$

Theorem 3.1. *For each non-degenerate critical point $x_1^* \in \Omega$ of Q in (100), there exists a family of solutions $\{(u, \lambda)\}$ of (73) with (74) for $\lambda > 0$ sufficiently small, satisfying (85).*

2. Let us take the perturbed problem

$$-\Delta u = \lambda f(x, e^u) \quad (in\ \Omega) \tag{103}$$

with

$$u = \phi \quad (on\ \partial\Omega), \tag{104}$$

where $f = f(x, t) : \overline{\Omega} \times [0, +\infty) \longrightarrow R$ is continuous and satisfies

$$f(x, t) = t + o(t) \tag{105}$$

as $t \to +\infty$ uniformly in x.

As before we put $u_0 = \mathcal{P}_\Omega\phi$, take a conjugate harmonic function v_0, and put $h = u_0 + \imath v_0$. Then the function $U = g^* u_1$ with $u_1 = u - u_0$ is to satisfy

$$-\Delta U = \lambda \mid p \mid^2 \{e^U + j(\xi, e^U)\} \quad (in\ D) \tag{106}$$

with

$$U = 0 \quad (on\ \partial D), \tag{107}$$

where $g : D \equiv \{\mid \zeta \mid < 1\} \longrightarrow \Omega$ is a conformal mapping, $p = g'e^{\frac{1}{2}g^*h}$ and $j = j(\xi, t) : \overline{D} \times [0, +\infty) \longrightarrow R$ is continuous and satisfies

$$j(\xi, t) = o(t) \tag{108}$$

as $t \to +\infty$ uniformly in ξ.

Under the assumptions (83) and (84) we have a family $\{U_e\}$ of classical solutions for

$$-\Delta U_e = \lambda \mid p \mid^2 e^{U_e} \quad (in\ D) \tag{109}$$

with

$$U_e = 0 \quad (on\ \partial D), \tag{110}$$

$\lambda > 0$ being sufficiently small so that

$$e^{-U_e/2} \to |\zeta|^2 \tag{111}$$

as $\lambda \downarrow 0$. Writing $U = U_e + \lambda H$, we arrive at the equation with respect to H,

$$-\Delta H = \lambda |p|^2 \{e^{U_e}(e^{\lambda H} - 1) + J_\lambda(\xi, e^{\lambda H})\} \quad (in\ D) \tag{112}$$

with

$$H = 0 \quad (on\ \partial D), \tag{113}$$

where

$$J_\lambda(\xi, t) = j(\xi, e^{U_e}t). \tag{114}$$

Unfortunately, the problem is still a singular perturbation because

$$e^{U_e} \to 1/|\zeta|^4. \tag{115}$$

However, [6] constucted the asymptotic solutions $\{u\}_{0<\lambda\ll1}$ such that

$$|-\Delta u - \lambda f(x, e^u)|_{L^\infty(D)} \in O(\lambda^n) \tag{116}$$

with

$$|u - \phi|_{L^\infty(\partial D)} \in O(\lambda^n) \tag{117}$$

as $\lambda \downarrow 0$ for $n = 1, 2, ...$, in way of the equation for H in case of

$$f(x, e^u) = e^u + \nu(x)e^{-u}, \tag{118}$$

$\nu(x)$ being a real analytic function. Then, following the modified Newton method by [17] starting from the fourth asymptotic solution, he proved the generic existence of a family of classical solutions $\{(u, \lambda)\}$ for (103) with (104) satisfying (85).

4 Toward the multi-point blow-up solutions

In terms of the notation $\{V\} = \{V, V\}$ of (46), the construction of blow-up solutions for (73) with (74) is reduced to that for the equation

$$\{W\} = \frac{\lambda}{2} |p|^2, \ W > 0 \quad (in\ D) \tag{119}$$

with

$$W = 1 \quad (on\ \partial D), \tag{120}$$

where $p = g'e^{\frac{1}{2}g^*h}$ is a holomorphic function given in the previous section. Let $\{x_1^*, ..., x_m^*\} \subset \Omega$ denote the blow-up points and put $\delta_j = g^{-1}(x_j^*)$ $(1 \leq j \leq m)$.

The limiting function $u_0(x)$ of $\{u(x)\}$ can be expected to be of the form

$$u_0(x) = 8\pi \sum_{j=1}^m G(x, x_j^*) + \mathcal{P}_\Omega \phi(x), \tag{121}$$

so that we have to make the family $\{W\}$ satisfy

$$W \to |w|^2 \tag{122}$$

as $\lambda \downarrow 0$, where $w = \prod_{j=1}^{m} \frac{\zeta - \delta_j}{1 - \delta_j \zeta}$ denotes the finite Blaschke product.

Following the procedure of §2, it may look valid to put

$$W = |w/G|^2 + \frac{\lambda}{8} |M/G|^2, \tag{123}$$

where $G(\zeta)$ and $M(\zeta)$ are holomorphic functions of $\zeta \in D$. However, zero points of $\Psi_1 = w/G$ are suspected to correspond to the local maximal points of u under the mapping $g : D \to \Omega$. Thus, the one-point blow-up solutions in §2 are expected to have common maximal points. We do not think that this is the case for $m > 1$.

Actually, from the formula (48) we see that the equalities (119) and (120) are reduced to

$$\omega(w, M) \equiv (w \frac{d}{d\zeta} - w')M = pG^2 \quad (in \ D) \tag{124}$$

and

$$|G|^2 = 1 + \frac{\lambda}{8} |M|^2 \quad (on \ \partial D), \tag{125}$$

respectively. The relation (122) will be realized by

$$|G - 1|_{L^\infty(D)} \to 0, \quad |M|_{L^\infty(D)} \in O(1) \tag{126}$$

as $\lambda \downarrow 0$.

The problem arises in the linear equation (124). In fact, its homogeneous equation admits only the solution of the form $constant \times w$, while its solvability requires m-conditions for the right-hand side, local around the zero points of W, that is, $\{\delta_j\}_{j=1}^{m}$. The gap causes an obstruction for balancing (124) and (125).

We may be able to construct some sort of "multi-valued solutions", regarding ζ as an m-valued algebraic function of $w = \prod_{j=1}^{m} \frac{\zeta - \delta_j}{1 - \delta_j \zeta}$. Then the conformal mapping $g : D \to \Omega$ induces an m-valued function on D, denoted by $\hat{g}$, through the equality

$$\hat{g}(w) = g(\zeta). \tag{127}$$

Let us take a formal argument. If the conditions

$$\hat{p}'(0) = 0 \tag{128}$$

and

$$|\hat{p}''(0)/\hat{p}(0)| \neq 2 \tag{129}$$

are satisfied for

$$\hat{p}(w) = \hat{g}'(w) \exp(\frac{1}{2} h(\hat{g}(w))), \tag{130}$$

then m-valued analytic functions $\hat{G}(w), \hat{M}(w)$ may be constructed as

$$\hat{\omega}(w, \hat{M}) = (w \frac{d}{dw} - 1)\hat{M} = \hat{p}\hat{G}^2 \quad (in \ D) \tag{131}$$

$$| \hat{G} |^2 = 1 + \frac{\lambda}{8} | \hat{M} |^2 \quad (on \ \partial D) \tag{132}$$

and

$$|\hat{G} - 1|_{L^\infty(D)} \to 0, \ |\hat{M}|_{L^\infty(D)} \in O(1). \tag{133}$$

The family $\{G, M\}$ with $G(\zeta) = \hat{G}(w(\zeta))$ and $M(\zeta) = \hat{M}(w(\zeta))$ will produce the desired solutions, which may be, however, multi-valued because so is true for G or M.

Concluding the section, we shall examine the conditions (128) and (129).

They are nothing but

$$(\hat{g}''/\hat{g}')(0) + \frac{1}{2}(\hat{g}^*h)'(0) = 0 \tag{134}$$

and

$$| (\hat{g}''/\hat{g}')'(0) + \frac{1}{2}(\hat{g}^*h)''(0) | \neq 2, \tag{135}$$

respectively. From the chain rule we have

$$g' = \hat{g}'w', \ g'' = \hat{g}''w'^2 + \hat{g}'w'', \ and$$

$$g''' = \hat{g}'''w'^3 + 3\hat{g}'w'w'' + \hat{g}'w''' \tag{136}$$

so that

$$\hat{g}' = g'/w', \ \hat{g}'' = \{g''w' - g'w''\}/w'^3, \ and$$

$$\hat{g}''' = g'''/w'^3 - 3g''w''/w'^4 + \{-w'''w' + 3w''^2\}g'/w'^5. \tag{137}$$

Therefore, it holds that

$$\hat{g}''/\hat{g}' + \frac{1}{2}(\hat{g}^*h)' = \{g''/g' - w''/w' + \frac{1}{2}(g^*h)'\}/w' \tag{138}$$

because

$$(\hat{g}^*h)' = \frac{d}{dw}h(\hat{g}(w)) = \frac{d}{dw}h(g(\zeta))/\frac{dw}{d\zeta}.$$

Thus the condition (128) is equivalent to

$$\frac{g''}{g'}(\delta_j) - \frac{w''}{w'}(\delta_j) + \frac{1}{2}(g^*h)'(\delta_j) = 0 \quad (1 \leq j \leq m). \tag{139}$$

Similarly, we have

$$(\hat{g}''/\hat{g}')' = \hat{g}'''/\hat{g}' - (\hat{g}''/\hat{g}')^2 = \{(g''/g')' - (w''/w')(g''/g') - (w'''/w') + 2(w''/w')^2\}/w'^2$$

and

$$(\hat{g}^*h)'' = \frac{d^2}{dw^2}h(\hat{g}(w)) = \frac{d^2}{dw^2}h(g(\zeta)) = \{(g^*g)''w' - (g^*h)'w''\}/w'^3.$$

Hence

$$(\hat{g}''/\hat{g}')' + \frac{1}{2}(\hat{g}^*h)'' = \{(g''/g')' - (w''/w')(g''/g') - w'''/w' + 2(w''/w')^2$$

$$+ \frac{1}{2}(g^*h)'' - \frac{1}{2}(w''/w')(g^*h)'\}/w'^2. \tag{140}$$

Therefore, the condition (129) is equivalent to

$$| (\frac{g''}{g'})'(\delta_j) - (\frac{w'''}{w'})^2(\delta_j) + \frac{1}{2}(g^*h)''(\delta_j) |$$

$$\neq 2 \mid w'(\delta_j) \mid^2 \quad (1 \le j \le m) \tag{141}$$

by (139).

Here, the condition (139) means that $(x_1^*, ..., x_m^*)$ is a critical point of $K(x_1, ..., x_m)$, where $x_j^* = g(\delta_j)$ $(1 \le j \le m)$ and

$$K(x_1, ..., x_m) = \frac{1}{2}\sum_{j=1}^{m}\{R(x_j) + \frac{1}{4\pi}\mathcal{P}\phi(x_j)\} + \sum_{\ell \neq \ell'}G(x_\ell, x_{\ell'}). \tag{142}$$

In fact, let us recall that

$$w(\zeta) = \prod_{j=1}^{m}\frac{\zeta - \delta_j}{1 - \overline{\delta_j}\zeta}.$$

Since

$$(\frac{\zeta - \delta_j}{1 - \overline{\delta_j}\zeta})' = \frac{1 - \mid \delta_j \mid^2}{(1 - \overline{\delta_j}\zeta)^2},$$

we get the relation

$$w' = \psi w \tag{143}$$

with

$$\psi(\zeta) = \sum_{\ell=1}^{m}(1 - \mid \delta_\ell \mid^2)/(1 - \overline{\delta_\ell}\zeta)(\zeta - \delta_\ell) = \sum_{\ell=1}^{m}\{\frac{\overline{\delta_\ell}}{1 - \overline{\delta_\ell}\zeta} + \frac{1}{\zeta - \delta_\ell}\}. \tag{144}$$

We note the identity

$$\psi(\zeta) = \frac{1}{\zeta - \delta_j} + \psi_j(\zeta) \quad (1 \le j \le m), \tag{145}$$

where

$$\psi_j(\zeta) = \frac{\overline{\delta_j}}{1 - \overline{\delta_j}\zeta} + \sum_{\ell \neq j}(1 - \mid \delta_\ell \mid^2)/(1 - \overline{\delta_\ell}\zeta)(\zeta - \delta_\ell). \tag{146}$$

Then, (143) implies that

$$\frac{w''}{w'} = \psi + \frac{\psi'}{\psi} = \frac{2\psi_j + (\zeta - \delta_j)(\psi_j^2 + \psi_j')}{1 + (\zeta - \delta_j)\psi_j}. \tag{147}$$

Hence we have

$$\frac{w''}{w'}(\delta_j) = 2\psi_j(\delta_j) = \frac{2\overline{\delta_j}}{1 - \mid \delta_j \mid^2} + 2\sum_{\ell \neq j}\frac{1 - \mid \delta_\ell \mid^2}{(1 - \overline{\delta_\ell}\delta_j)(\delta_j - \delta_\ell)}. \tag{148}$$

Similarly, it holds that

$$(\frac{w''}{w'})'(\delta_j) = 3\psi_j'(\delta_j) - \psi_j(\delta_j)^2. \tag{149}$$

We now recall the relations (12) and (13):

$$R(x) = \frac{1}{4\pi} \log | g'(\zeta) |^2 (1- | \zeta |^2)^2, \qquad (150)$$

for $x = g(\zeta)$,

$$G(x,y) = -\frac{1}{2\pi} \log \left| \frac{\zeta - \eta}{1 - \overline{\eta}\zeta} \right| \qquad (151)$$

for $x = g(\zeta)$ and $y = g(\eta)$. Introducing the variables ζ_j $(1 \leq j \leq m)$ by $x_j = g(\zeta_j)$, we get

$$K(x_1, ..., x_m) = k(\zeta_1, ..., \zeta_m)$$
$$\equiv \frac{1}{8\pi} \sum_{j=1}^m \{\log | g'(\zeta_j) |^2 + \mathcal{P}_D\Phi(\zeta_j)\} + \frac{1}{4\pi}\psi(\zeta_1, ..., \zeta_m), \qquad (152)$$

where $\Phi = g^*\phi$, $\mathcal{P}_D$ denotes the Poisson operator on D, and

$$Q(\zeta_1, ..., \zeta_m) = \sum_{j=1}^m \log(1- | \zeta_j |^2) - \sum_{\ell < \ell'} \log \left| \frac{\zeta_\ell - \zeta_{\ell'}}{1 - \overline{\zeta_\ell}\zeta_{\ell'}} \right|^2 . \qquad (153)$$

Here, we note that

$$\frac{\partial Q}{\partial \zeta_j} = -\frac{\overline{\zeta_j}}{1- | \zeta_j |^2} + \sum_{\ell \neq j} \frac{1- | \zeta_\ell |^2}{(\zeta_\ell - \zeta_j)(1 - \overline{\zeta_\ell}\zeta_j)} \qquad (154)$$

and hence

$$\frac{\partial Q}{\partial \zeta_j}(\delta_1, ..., \delta_m) = -\frac{1}{2}\frac{w''}{w'}(\delta_j). \qquad (155)$$

On the other hand we have

$$\frac{\partial}{\partial \zeta_j}\mathcal{P}_D\Phi = \frac{1}{2}(\frac{\partial}{\partial \xi^1} - \imath\frac{\partial}{\partial \xi^2})\mathcal{P}_D\Phi = \frac{1}{2}(g^*h)', \qquad (156)$$

where $\zeta_j = \xi^1 + \imath\xi^2$.

Thus we obtain

$$\frac{\partial}{\partial \zeta_j a}k(\delta_1, ..., \delta_m) = \frac{1}{8\pi}\{\frac{g''}{g'}(\delta_j) - \frac{w''}{w'}(\delta_j) + \frac{1}{2}(g^*h)'(\delta_j)\}. \qquad (157)$$

The functioin k is real-valued so that the condition (139) is equivalent to $\nabla_x K(x_1^*, ..., k_m^*) = 0$ for $x_j^* = g(\delta_j)$.

References

[1] Crandall. M.G., Rabinowitz, P.H., Some continuation and variational methods for positive solutions for nonlinear elliptic eigenvalue problems, Arch. Rat. Mech. Anal., 58 (1975) 207-218.

[2] Gustafsson, B., On the motion of a vortex in two-dimensional flow of an ideal fluid in simply and multiply connected domains, technical report, Department of Math. Royal Institute of Technology, Stockholm 1979.

[3] Lin. S.S., On non-radially symmetric bifurcation in the annulus, J. Differential Equations 80 (1989) 251-279.

[4] Liouville, J., Sur l'équation aux différences partielles $\partial^2 \log \lambda / \partial u \partial v \pm \lambda / 2a^2 = 0$, J. de Math. 18 (1853) 71-72.

[5] Moseley, J.L., Asymptotic solutions for a Dirichlet problem with an exponential nonlinearity, SIAM J. Math. Anal. 14 (1983) 934-946.

[6] Moseley, J.L., A two-dimensional Dirichlet problem with an exponential nonlinearity, SIAM J. Math. Anal. 14 (1983) 934-946.

[7] Nagasaki, K., Suzuki, T., Asymptotic analysis for two-dimensional elliptic eigenvalue problems with exponetially-dominated nonlinearities, Asymptotic Analysis 3 (1990) 173-188.

[8] Nagasaki, K., Suzuki, T., Radial and nonradial solutions for the nonlinear eigenvalue problem $\Delta u + \lambda e^u = 0$ on annului in R^2, J. Differential Equations 87 (1990) 144-168.

[9] Nakane, S., A bifurcation phenomenon for a Dirichlet problem with an exponential nonlinearity, J. Math. Anal. Appl. 161 (1991) 227-240.

[10] Rabinowitz, P., Some aspects of nonlinear eigenvalue problems, Rocky Mountain J. Math. 3 (1973) 161-202.

[11] Spruck. J., The elliptic Sinh-Gordon equation and the construction of toroidal soap bubbles, In; Hildebrandt, S., Kinderlehrer, D., Miranda, M., (eds.), Calculus of Variations and Partial Differential Equations, Lecture Note in Math. 1340, Springer, Berlin et. al., pp 275-301, 1988.

[12] Suzuki, T., Nagasaki, K., On the nonlinear eigenvalue problem $\Delta u + \lambda e^u = 0$, Trans. Amer. Math. Soc. 309 (1988) 591-608.

[13] Suzuki, T., Two dimensional Emden-Fowler equation with exponential nonlinearity, In; Lloyd, N.G., Ni, W.M., Peletier, L.A., Serrin, J., (eds.), Nonlinear Diffusion Equations and their Equibrium States, 3, Gregynog 1989, Birkhäuser, Boston et.al., pp 493-512, 1992.

[14] Suzuki, T., Global analysis for a two-dimensional elliptic eigenvalue problem with the exponential nonlinearity, to appear in Ann. Inst. H. Poincaré, Analyse NonLinéaire.

[15] Suzuki, T., Harnack principle for spherically subharmonic functions, preprint 1991.

[16] Talenti, G., A note on the Gauss curvature of harmonic and minimal surfaces, Pacific J. Math. 101 (1982) 477-492.

[17] Wente, H., Counter example to a conjecture of H. Hopf, Pacific J. Math. 121 (1986) 193-243.

[18] Weston, V.H., On the asymptotic solution of a partial differential equation with an exponential nonlinearity, SIAM J. Math. Anal. 9 (1978) 1030-1053.

[19] Yosida, K., Functional Analysis, Springer, Berlin, 1964.

Fully Discrete Approximation of a Second Order Linear Evolution Equation Related to the Water Wave Problem

USHIJIMA, Teruo and MATSUKI, Mihoko

Department of Computer Science and Information Mathematics,
The University of Electro-Communications
1-5-1, Chofugaoka, Chofu-shi, Tokyo, 182, JAPAN

November 1991

Introduction

Consider the following initial value problem for the ordinary differential equation with values in a Hilbert space X:

$$(E) \quad \begin{cases} \dfrac{d^2\phi}{dt^2} + A\phi = \psi, \quad t > 0, \\[2mm] \phi(0) = \phi^1, \quad \dfrac{d\phi}{dt}(0) = \phi^0, \end{cases}$$

where the operator valued coefficient A is a positive definite selfadjoint operator acting in X. A linear problem, derived from the non linear water wave problem under the assumption of the infinitesimal amplitude of the wave, can be represented as (E), in which the operator A is a realization of a first order pseudo-differential operator. The problem is said to be the linear water wave problem in this paper.

Introduce a family of approximate problems $(E_{h,\tau})$ for the problem (E) depending on a parameter $h \in (0, \bar{h}]$ and a time mesh parameter $\tau \in (0, \infty)$. The problem $(E_{h,\tau})$ is obtained by replacing A with a positive definite selfadjoint operator acting on a finite dimensional subspace X_h in the problem (E) and by adopting the Newmark-β method as the way of discretization of the time variable.

Suppose that the lower bounds of the spectrums of A_h are uniformly bounded below by a positive constant. Assuming an appropriate growth rate of $\|A_h\|$ as h tends to 0, we have a stability criterion for the family of the approximate problems.

Assuming further appropriate conditions on the rate of convergence of the resolvents, we deduce error estimates for the evolution problem correspondingly to the imposed conditions on the rate of convergence of the resolvents.

Applying our abstract result to the finite element approximation of the evolution linear water wave equation, we can give L^2 error estimates for the approximate solutions of the problem. Incidentally the first author felt a substantial difficulty applying standard error estimation methods in the analysis of the finite element approximation, such as the energy method and the semi-group theoretical method, to the linear water wave problem. To overcome the difficulty, the present study has been developed.

As for the existence and the uniqueness of solutions of the problem (E), the theory of semi-groups of linear operaters due to Yosida and others is a strong tool to prove them (see Yosida [20]). A standard way is the conversion of (E) to the following first order problem :

$$\frac{d}{dt}\begin{pmatrix} u \\ v \end{pmatrix} = \begin{pmatrix} 0 & 1 \\ -A & 0 \end{pmatrix}\begin{pmatrix} u \\ v \end{pmatrix} + \begin{pmatrix} 0 \\ \psi \end{pmatrix},$$

which is an evolution problem with the value $(u, v) \in D(A^{1/2}) \times X$, where u, and v, correspond to ϕ and ϕ_t, respectively. The approximation problem of the first order linear evolution problem is strongly connected with the approximation theory for semi-groups of linear operaters, the origin of which goes back to the Trotter–Kato theory (see, for example, Kato[9]). In this direction, there have been many works, among them the works of Fujita–Mizutani[6], Fujita–Suzuki[7], and Ushijima[15][16] are quoted in the References of this paper. In [7], fairly satisfactory references are found. In the proper finite element method, the problems of approximation of the heat equation and the wave equation, or some other similar kinds of equations have been often treated by, so-called, the energy method (see, for example, H. Fujii[4], Strang-Fix[14] and Raviart–Thomas [12]).

The organization of this paper is as follows. In §1, the settings of continuous problems and its approximate problems are mentioned. In §2, a stability criterion for the Newmark-β method is shown in Theorem 1. In §3, under a condition, an error estimate is shown in Theorem 2 with an outline of its proof. The condition is a sufficient condition for the stability condition mentioned in Theorem 1. In §4, the conditions assumed in §1 are checked in the case of the linear water wave problem.

We express our sincere thanks to Professor H. Komatsu and the referee of the paper for their valuable comments on the work.

We are very grateful to Mr. I. Takanami for his assistance in the preparation of the machine readable manuscript of the paper.

§1. Abstract setting of the problem.

Fix a complex Hilbert space X. Its inner product, and norm, are denoted by $(\, , \,)$ and $\| \, \|$, respectively. Suppose that there is given a positive definite selfadjoint operator A acting in X. Assume that there exists a positive number $\underline{\alpha}$ such that the spectrum $\sigma(A)$ of A is contained in the interval $[\underline{\alpha}, \infty)$.

Hereafter $\bar{h}$ is a positive constant which should be determined appropriately in concrete problems. Suppose that, for any $h \in (0, \bar{h}]$, there are a finite dimensional subspace X_h of X and a positive definite bounded selfadjoint operator A_h acting on X_h. Assume that, for any $h \in (0, \bar{h}]$, there are positive numbers $\underline{\alpha}_h$ and $\bar{\alpha}_h$ with $\underline{\alpha}_h < \bar{\alpha}_h$ such that the spectrum $\sigma(A_h)$ is contained in the interval $[\underline{\alpha}_h, \bar{\alpha}_h]$. The following condition (1) is assumed to be satisfied.

$$(1) \qquad\qquad \underline{\alpha} \leq \underline{\alpha}_h.$$

In the above, the subspace X_h itself is considered to be a Hilbert space with the inner product $(\, , \,)$, and A_h to be an operator from X_h onto X_h. Let P_h be the orthogonal projection from X onto X_h. Assume that there is a positively valued function $\varepsilon(h)$ defined on the interval $(0, \bar{h}]$ satisfying the following two conditions $(\varepsilon\text{-}1)$ and $(\varepsilon\text{-}2)$.

$(\varepsilon\text{-}1) \qquad\qquad \lim_{h \to 0} \varepsilon(h) = 0.$

$(\varepsilon\text{-}2) \qquad\qquad \begin{cases} \textit{There is a positive constant } \alpha \textit{ independent of} \\ h \in (0, \bar{h}] \textit{ with the property that} \\ \qquad \|A_h\| \leq \dfrac{\alpha}{\varepsilon(h)}. \end{cases}$

Let s be a real number. We say that *condition* $(A_{\epsilon,s})$ *holds if there is a constant* C *independent of* h *such that the following inequality holds for any* $h \in (0, \bar{h}]$.

$$\left\| \left(A_h^{-1} P_h - A^{-1} \right) \phi \right\| \leq C\epsilon(h)^{1+s} \left\| A^s \phi \right\|, \qquad \phi \in D(A^s).$$

We say that *condition* (ε_s) *holds if both condition* $(A_{\epsilon,0})$ *and condition* $(A_{\epsilon,s})$ *hold*.

Let $C^m([0,\infty) : X)$ be the totality of m times continuously differentiable X-valued functions defined on the interval $[0,\infty)$. Analogously we use the notation $C^m([0,\infty) : D(A^s))$, where $D(A^s)$ is considered to be a Banach space with the graph topology. Introduce an operator L acting on $C^2([0,\infty) : X) \cap C([0,\infty) : D(A))$ through

$$L = \frac{d^2}{dt^2} + A.$$

Fix nonnegative numbers β and δ independent of $h \in (0, \bar{h}]$. For a positive number τ, define a difference operator $L_{h,\tau}^{\delta,\beta}$ acting on the sequence $\{\phi_m : m = 0, 1, 2, \cdots\}$ of elements of X_h through

$$L_{h,\tau}^{\delta,\beta} = \left(1 + \beta\tau^2 A_h\right) D_{\tau\bar{\tau}} + A_h + \delta\tau A_h D_{\bar{\tau}},$$

where

$$D_{\bar{\tau}}\phi_m = \frac{\phi_m - \phi_l}{\tau}, \quad D_\tau\phi_m = \frac{\phi_n - \phi_m}{\tau}, \quad D_{\tau\bar{\tau}}\phi_m = \frac{\phi_n - 2\phi_m + \phi_l}{\tau^2}.$$

Here and hereafter we use freely the following correspondence of the suffices:

$$(l, \ m, \ n) = (m-1, \ m, \ m+1).$$

Using above notations, our continuous problem $(E : \phi^1, \ \phi^0, \ \psi)$ and fully discrete approximate problem $(E_{h,\tau} : \phi_{h,\tau}^1, \ \phi_{h,\tau}^0, \ \psi_{h,\tau,m})$ are given in the following fashion.

$(E : \phi^1, \ \phi^0, \ \psi)$

$$\begin{cases} L\phi = \psi, & t > 0, \\ \phi(0) = \phi^1, & \dfrac{d\phi}{dt}(0) = \phi^0. \end{cases}$$

$(E_{h,\tau} : \phi_{h,\tau}^1, \ \phi_{h,\tau}^0, \ \psi_{h,\tau,m})$

$$\begin{cases} L_{h,\tau}^{\delta,\beta}\phi_{h,\tau,m} = \psi_{h,\tau,m}, & m = 1, 2, \cdots, \\ \phi_{h,\tau,0} = \phi_{h,\tau}^1, & D_\tau\phi_{h,\tau,0} = \phi_{h,\tau}^0. \end{cases}$$

In these problems, ϕ^i $(i = 0, 1)$, and $\phi_{h,\tau}^i$ $(i = 0, 1)$, are given initial data of elements of X, and X_h, respectively, and ψ is an inhomogeneous data of X-valued function of t, and $\{\psi_{h,\tau,m} : m = 1, 2, \cdots\}$ is an inhomogeneous data of X_h-valued sequence. The function ϕ of t, and the sequence $\{\phi_{h,\tau,m} : m = 0, 1, 2, \cdots, \}$, are unknowns which should be determined, in $(E : \phi^1, \ \phi^0, \ \psi)$, and in $(E_{h,\tau} : \phi_{h,\tau}^1, \ \phi_{h,\tau}^0, \ \psi_{h,\tau,m})$, respectively.

For simplicity of notations, sometimes problems $(E : \phi^1, \ \phi^0, \ \psi)$, and $(E_{h,\tau} : \phi_{h,\tau}^1, \ \phi_{h,\tau}^0, \ \psi_{h,\tau,m})$ are expressed as (E), and $(E_{h,\tau})$, respectively if their data are clearly recognized.

We followed Raviart–Thomas [12] concerning the formulation of the problem $(E_{h,\tau})$. The problem is obtained through the method which is called the Newmark-β method by many engineers.

§2. Stability criterion for the Newmark-β method.

Introduce a bounded selfadjoint operator $A_{h,\tau}$ through

$$A_{h,\tau} = \left(1 + \beta\tau^2 A_h\right)^{-1} A_h,$$

where the fixed parameter β is dropped in the symbol $A_{h,\tau}$ for simplicity of notation. Fix γ independently of $h \in (0, \bar{h}]$ satisfying

(2) $$0 < \gamma < 1.$$

THEOREM 1. *For any* $h \in (0, \bar{h}]$, *choose* τ *so as to satisfy the following stability condition* $(S^\delta_{2\gamma})$:

$(S^\delta_{2\gamma})$ $$\tau^2 (1+\delta)^2 \|A_{h,\tau}\| \le (2\gamma)^2.$$

Then the solution $\{\phi_{h,\tau,n} : n = 0, 1, 2, \cdots\}$ *of* $(E_{h,\tau})$ *is estimated as follows.*

(3)
$$
\begin{aligned}
&\|\phi_{h,\tau,n}\| \\
&\le \left(1 + \frac{\gamma}{\sqrt{1-\gamma^2}}\right) \|\phi^1_{h,\tau}\| + \frac{1}{\sqrt{1-\gamma^2}} \left\| A_{h,\tau}^{-\frac{1}{2}} \phi^0_{h,\tau} \right\| \\
&\quad + \frac{m\tau}{\sqrt{1-\gamma^2}} \cdot \max_{1 \le j \le m} \left\| \left\{ \left(1 + \beta\tau^2 A_h\right) A_h \right\}^{-\frac{1}{2}} \psi_{h,\tau,j} \right\|, \\
&\qquad\qquad n = 2, 3, \cdots, \qquad m = n - 1.
\end{aligned}
$$

Before giving the proof of Theorem 1, we remark a sufficient condition for $(S^\delta_{2\gamma})$ in terms of A_h instead of $A_{h,\tau}$. If

(i) $$\beta \left\{ \frac{2\gamma}{1+\delta} \right\}^2 < 1,$$

then set

$$\rho = \left\{ \frac{2\gamma}{1+\delta} \right\}^2 \frac{1}{1 - \beta \left\{ \dfrac{2\gamma}{1+\delta} \right\}^2},$$

and if

(ii) $$\beta \left\{ \frac{2\gamma}{1+\delta} \right\}^2 \ge 1,$$

then set ρ as an arbitrary positive number. Then the following condition $(T_{2\gamma}^{\delta})$ implies condition $(S_{2\gamma}^{\delta})$.

$(T_{2\gamma}^{\delta})$ $$\tau^2 \|A_h\| \leq \rho.$$

Proof of Theorem 1. An essential point to derive estimate (3) is in the consideration of the Newmark-β scheme with $\beta = 0$ applied to the scalar equation of single harmonic motion. The scheme is the following problem $(e_\tau^{\delta,0})$:

$(e_\tau^{\delta,0})$ $$\begin{cases} D_{\tau\bar{\tau}}a_m + \omega^2 a_m + \delta\tau\omega^2 D_{\bar{\tau}}a_m = b_m, & m = 1,2,3,\cdots, \\ a_0 = a^1, \quad D_\tau a_0 = a^0, \end{cases}$$

where the positive number ω is the angular frequency of the harmonic motion.

Let $\Omega = (\tau\omega)^2$. Suppose

(4) $$\left(\frac{1+\delta}{2}\right)^2 \Omega < 1.$$

It is to be noted that condition (4) implies $1 - \delta\Omega > 0$ since (4) assures the negativity of the discriminant of the following quadratic equation:

$$z^2 - \{2 - (1+\delta)\,\Omega\}\,z + (1 - \delta\Omega) = 0.$$

Introduce the following quantities:

$$r = (1 - \delta\Omega)^{\frac{1}{2}}, \qquad \omega_\tau = \frac{1}{\tau}\,\cos^{-1}\left(\frac{1 - \frac{1+\delta}{2}\Omega}{r}\right),$$
$$c_m = r^m\,\cos(m\tau\omega_\tau), \quad s_m = r^m\,\sin(m\tau\omega_\tau).$$

Then the solution a_m of $(e_\tau^{\delta,0})$ is represented through the following formula:

(5) $$a_m = \frac{\Omega^{\frac{1}{2}}}{s_1}\sum_{j=1}^{l} s_{m-j}\frac{b_j}{\omega}\tau + \left\{c_m + \frac{1-c_1}{s_1}s_m\right\}a^1 + \frac{\Omega^{\frac{1}{2}}}{s_1}s_m\frac{a^0}{\omega},$$

$$m = 0,1,2,\cdots, \qquad l = m-1,$$

under the notational convention:

$$\sum_{j=1}^{l} = 0 \qquad \text{if } l \leq 0.$$

Assume further

(6) $$\left(\frac{1+\delta}{2}\right)^2 \Omega \leq \gamma^2.$$

for some $\gamma \in (0,1)$. Then we have

$$\frac{\Omega^{\frac{1}{2}}}{s_1} \leq \frac{1}{\sqrt{1-\gamma^2}} \qquad \text{and} \qquad \frac{1-c_1}{s_1} \leq \frac{\gamma}{\sqrt{1-\gamma^2}}.$$

Hence representation (5) yields that under condition (6) we have

$$|a_m| \leq \left(1 + \frac{\gamma}{\sqrt{1-\gamma^2}}\right)|a^1| + \frac{1}{\sqrt{1-\gamma^2}}\left|\frac{a^0}{\omega}\right|$$

(7)

$$+ \frac{m\tau}{\sqrt{1-\gamma^2}} \max_{1 \leq j \leq m}\left|\frac{b_j}{\omega}\right|,$$

$$m = 2, 3, \cdots.$$

Using the spectral representation of $A_{h,\tau}$, we can transfer estimate (7) to estimate (3). ∎

§3. Error estimate for the approximate solution.

Set the following condition (D_s) with real s for the data (ϕ^1, ϕ^0, ψ) of problem (E) as follows.

(D_s)
$$\begin{cases} \phi^i \in D\left(A^{1+\frac{i}{2}+s}\right), & i = 0, 1, \\ \psi \in C^2\left([0,\infty) : X\right), \\ \psi^{(i)}(0) \in D\left(A^{\frac{1-i}{2}+s}\right), & i = 0, 1, \\ \psi(t),\ \psi^{(2)}(t) \in D(A^s), & t \geq 0, \\ A^s\psi(t),\ A^s\psi^{(2)}(t) \in C([0,\infty) : X). \end{cases}$$

THEOREM 2. *Suppose condition $\left(T^\delta_{2\gamma}\right)$ holds for some $\gamma \in (0,1)$. Assume condition (ε_s) and (D_s) with a fixed $s \geq -1/2$. Then the difference between the solution ϕ of problem $\left(E : \phi^1, \phi^0, \psi\right)$ and the solution $\phi_{h,\tau,m}$ of problem $\left(E_{h,\tau} : \phi^1_{h,\tau}, \phi^0_{h,\tau}, \psi_{h,\tau,m}\right)$ is estimated as follows.*

$$\|\phi(m\tau) - \phi_{h,\tau,m}\|$$
$$\leq \varepsilon(h)^{1+s} \cdot C_1 \cdot \{(1 + \underline{\alpha}^{-1/2} + m\tau) \cdot E(m\tau;\ s)$$

$$+ \left(1 + \frac{\gamma}{\sqrt{1-\gamma^2}}\right) \cdot E(0;\ s - \frac{1}{2})\}$$

$$+ \tau^{2\{1+\min(0,s)\}} \cdot (1+m\tau) \cdot \frac{\sqrt{\alpha^{-1} + \beta\tau^2} + 1}{\sqrt{1-\gamma^2}} \cdot (1 + \beta\rho^2)^{-\min(0,s)}.$$

(8)

$$\cdot C_2 \cdot E(m\tau;\ \min(0, s))$$

$$+ \beta\,\tau^2 \cdot \frac{\gamma}{\sqrt{1-\gamma^2}} \cdot \frac{1}{1+\delta} \cdot D(0)$$

$$+ \beta\,\tau^{2\{1+\min(0,s)\}} \cdot m\tau \cdot C_3 \cdot E((m-1)\tau;\ \min(0, s))$$

$$+ \delta\,\tau \cdot m\tau \cdot \frac{1}{\sqrt{1-\gamma^2}} \cdot C_4 \cdot D((m-1)\tau)$$

$$+ \|\varepsilon^I_{h,\tau,m}\|, \qquad\qquad m = 2, 3, \cdots,$$

where constants C_i $(i = 1, 2, 3, 4)$ are independent of parameters β, δ, γ, $\underline{\alpha}$, τ and function $\varepsilon(h)$, and ρ is the constant appeared in $\left(T_{2\gamma}^{\delta}\right)$. In (8), quantities $D(t)$, and $E(t; s)$, are defined by (9), and (10), below, respectively, and $\left\|\varepsilon_{h,\tau,m}^{I}\right\|$ is estimated as in (11) below.

$$
(9) \qquad D(t) = \left\|A\phi^1\right\| + \left\|A^{1/2}\phi^0\right\| + \|\psi(0)\| + t \cdot \max_{0 \le r \le t} \left\|\psi^{(1)}(r)\right\|.
$$

$$
(10) \qquad
\begin{aligned}
E(t; s) =& \left\|A^{3/2+s}\phi^1\right\| + \left\|A^{1+s}\phi^0\right\| + \left\|A^{1/2+s}\psi(0)\right\| + \left\|A^s\psi^{(1)}(0)\right\| \\
&+ \max_{0 \le r \le t} \|A^s\psi(r)\| + \left(t + \underline{\alpha}^{-1/2}\right) \cdot \max_{0 \le r \le t} \left\|A^s\psi^{(2)}(r)\right\|.
\end{aligned}
$$

$$
(11) \qquad
\begin{aligned}
\left\|\varepsilon_{h,\tau,m}^{I}\right\| \le& \left(1 + \frac{\gamma}{\sqrt{1-\gamma^2}}\right) \left\|P_h\phi^1 - \phi_{h,\tau}^1\right\| \\
&+ \left(\frac{\alpha + \beta\tau^2}{1-\gamma^2}\right)^{1/2} \left\{ \left\|P_h\left\{\phi^0 + \frac{\tau}{2}\left[\psi(0) - A\phi^1\right]\right\} - \phi_{h,\tau}^0\right\| \right. \\
&\left. + (m-1)\tau \cdot \max_{1 \le j \le m-1} \|P_h\psi(j\tau) - \psi_{h,\tau,j}\| \right\}, \\
& m = 2, 3, \cdots.
\end{aligned}
$$

Outline of the proof. Our guiding principles are summarized in the following two points.

(I) Setting five intermediate problems in X_h, we telescope the difference using the solutions of these problems.

(II) The differences of solutions of two consecutive problems in the telescoping of (I) are firstly estimated with the norms of the data of the relevant problems corresponding to the solutions. Then the norms are estimated with the data of the continuous problem.

To describe our process, introduce operators L_h, and $\tilde{L}_h$, through

$$
L_h = \frac{d^2}{dt^2} + A_h, \qquad \text{and} \qquad \tilde{L}_h = \frac{d^2}{dt^2} + A_{h,\tau},
$$

respectively. Our semi-discrete approximate problem $\left(\mathrm{E_h} : \phi_h^1,\ \phi_h^0,\ \psi_h\right)$ is given as follows.

$$
(\mathrm{E_h} : \phi_h^1,\ \phi_h^0,\ \psi_h) \qquad
\begin{cases}
L_h\phi_h = \psi_h, & t > 0, \\
\phi_h(0) = \phi_h^1, & \dfrac{d\phi_h}{dt}(0) = \phi_h^0.
\end{cases}
$$

Auxiliarily we use the following intermediate semi-discrete approximate problem $\left(\tilde{\mathrm{E}}_h : \tilde{\phi}_h^1,\ \tilde{\phi}_h^0,\ \tilde{\psi}_h\right)$.

$$
(\tilde{\mathrm{E}}_h : \tilde{\phi}_h^1,\ \tilde{\phi}_h^0,\ \tilde{\psi}_h) \qquad
\begin{cases}
\tilde{L}_h\tilde{\phi}_h = \tilde{\psi}_h, & t > 0, \\
\tilde{\phi}_h(0) = \tilde{\phi}_h^1, & \dfrac{d\tilde{\phi}_h}{dt}(0) = \tilde{\phi}_h^0.
\end{cases}
$$

Now we set the five intermediate problems as follows.

Semi-discrete approximate problem with projected data:

$$\left(\mathbf{E}_h : P_h\phi^1,\ P_h\phi^0,\ P_h\psi\right).$$

Semi-discrete approximate problem with modified data:

$$\left(\mathbf{E}_h : A_h^{-1}P_hA\phi^1,\ P_h\phi^0,\ P_h\psi\right).$$

Intermediate semi-discrete approximate problem:

$$\left(\tilde{\mathbf{E}}_h : A_h^{-1}P_hA\phi^1,\ P_h\phi^0,\ \left(1+\beta\tau^2A_h\right)^{-1}P_h\psi\right).$$

Full-discrete approximate problem with adjusted data:

$$\left(\mathbf{E}_{h,\tau} : A_h^{-1}P_hA\phi^1,\right.$$
$$P_h\phi^0 + \frac{\tau}{2}\left(-A_{h,\tau}A_h^{-1}P_hA\phi^1 + \left(1+\beta\tau^2A_h\right)^{-1}P_h\psi(0)\right),$$
$$\left.P_h\psi(m\tau)\right).$$

Full-discrete approximate problem with ideal data:

$$\left(\mathbf{E}_{h,\tau} : P_h\phi^1,\ P_h\left\{\phi^0 + \frac{\tau}{2}\left(-A\phi^1 + \psi(0)\right)\right\},\ P_h\psi(m\tau)\right).$$

Denote the solutions of above five problems by ϕ_h, ϕ_h^M, $\tilde{\phi}_h$, $\phi_{h,\tau,m}^A$, and $\phi_{h,\tau,m}^I$, successively. Then we have the following telescoping:

$$\phi(m\tau) - \phi_{h,\tau,m}$$

(12)
$$= (\phi(m\tau) - \phi_h(m\tau)) + \left(\phi_h(m\tau) - \phi_h^M(m\tau)\right) + \left(\phi_h^M(m\tau) - \tilde{\phi}_h(m\tau)\right)$$
$$+ \left(\tilde{\phi}_h(m\tau) - \phi_{h,\tau,m}^A\right) + \left(\phi_{h,\tau,m}^A - \phi_{h,\tau,m}^I\right) + \left(\phi_{h,\tau,m}^I - \phi_{h,\tau,m}\right)$$
$$= I + II + III + IV + V + VI.$$

According to the guiding principle II above, the first five terms of the right hand side of (12) can be treated. And we obtain

$$\begin{cases} |I| = O\left(\varepsilon(h)^{1+s}\right), & |II| = O\left(\varepsilon(h)^{1+s}\right), \\ |III| = O\left(\beta\tau^2\right), & |IV| = O\left(\tau^2\right) + O(\delta\tau), \\ |V| = O\left(\varepsilon(h)^{1+s}\right) + O\left(\beta\tau^2\right). \end{cases}$$

A rearrangement of the sum of right hand sides of these five estimates yields the first four terms of the right hand side of estimate (8). Applying Theorem 1 to the sixth term of (12), we get the last term of the right hand side of estimate (8). ∎

§4. Linear Water Wave Problem.

1. Setting of Problem. When we analyze the motion of the water in a vessel under the assumption of infinitesimal amplitude, we encounter the following linear initial-boundary value problem (LWW).

$$(\text{LWW}) \quad \begin{cases} -\Delta\Phi = 0 & \text{in} \quad \Omega, \\ \Phi_{tt} + g\dfrac{\partial\Phi}{\partial n} = F_t & \text{on} \quad \Gamma_0, \\ \dfrac{\partial\Phi}{\partial n} = 0 & \text{on} \quad \Gamma_1, \\ \Phi(0, x) = \Phi^1(x) & \text{on} \quad \Gamma_0, \\ \Phi_t(0, x) = \Phi^0(x) & \text{on} \quad \Gamma_0. \end{cases}$$

Details of the derivation of (LWW) from the fundamental laws of the fluid dynamics are shown in Chapter 1 of Stoker's book [13], for example. Explanations of notations in (LWW) are as follows. The open set Ω is either the region of water at rest in the 3 dimensional Euclidean space, or the cross section of this region when it is uniform in a horizontal direction perpendicular to the considered cross section. The surface of the water at rest, and the rigid wall of the vessel in contact with the water at rest, are denoted by Γ_0, and Γ_1, respectively. The scalar function Φ is the potential function of the velocity of water. Its exterior normal derivative of Φ at the boundary of Ω is denoted by $\frac{\partial\Phi}{\partial n}$. The symbol g means the acceleration of the gravity. The scalar function F represents the additional external force per unit surface area affecting the water surface, and F_t is the time derivative of F.

Let $\mathbb{R}^N_-$ be the lower half space of $\mathbb{R}^N$ defined by

$$\mathbb{R}^N_- = \{x = (x_1, x_2, \cdots, x_N) : x_N < 0\}.$$

Denote the hyperplane with $x_N = 0$ in $\mathbb{R}^N$ by $\mathbb{R}^N_0$.

Here the water region Ω is assumed to be a bounded domain in $\mathbb{R}^N_-(N = 2 \text{ or } 3)$ with Lipshitz continuous boundary Γ in the sense of Nečas[11]. The boundary portion Γ_0 is the intersection of Γ and $\mathbb{R}^N_0$. Assume that Γ_0 is a bounded domain in $\mathbb{R}^{N-1}$ with Lipshitz continuous boundary having positive $N - 1$ dimensional Lebesgue measure, $m_{N-1}(\Gamma_0)$.

Set

$$X = L^2(\Gamma_0), \qquad Y = H^{\frac{1}{2}}(\Gamma_0), \qquad V = H^1(\Omega).$$

Define Hermitian bilinear forms $a(u, v)$, $b(u, v)$, $\langle\langle u, v \rangle\rangle$ and the norm $\|v\|$ on the space V as follows.

$$a(u, v) = \int_\Omega \nabla u \nabla\bar{v} \, dx, \qquad b(u, v) = \int_{\Gamma_0} u\bar{v} \, d\Gamma,$$
$$\langle\langle u, v \rangle\rangle = a(u, v) + b(u, v), \qquad \|v\| = \langle\langle v, v \rangle\rangle^{\frac{1}{2}}.$$

Let γ_0 be the trace operator from V into X. The operator γ_0 is completely continuous. The norm $\|v\|$ on V is equivalent to the original $H^1(\Omega)$-norm. (For the proof, consult [11], Chapitre 1, see also [17], Chapter 1). Set

$$V_0 = \{v \in V : v = 0 \text{ on } \Gamma_0\}.$$

Let V_1 be the orthogonal complement of V_0 in the space V with respect to the inner product $\langle\langle u, v\rangle\rangle$.

2. Evolution Problem in $\mathbf{L}^2(\Gamma_0)$. For any element Φ of Y, there is a unique solution u of the following problem (H_Φ), which belongs to V_1.

$$(H_\Phi) \qquad \begin{cases} a(u, v) = 0, & v \in V_0, \\ u = \Phi & \text{on } \Gamma_0, \\ u \in V. \end{cases}$$

Define an operator $H \in L(Y, V_1)$ through the relation

$$u = H\Phi,$$

where u is the solution of problem (H_Φ). The operator H is an isomorphism from Y onto V_1. Define the non-negative Hermitian form a_0 acting in X with form domain Y by the following relation:

$$a_0(\Phi, \Psi) = a(H\Phi, H\Psi), \quad \Phi, \Psi \in D(a_0) = Y.$$

Denote the inner product, and the norm, of X by (Φ, Ψ), and $|\Phi|$, respectively. Namely, we have

$$(\Phi, \Psi) = \int_{\Gamma_0} \Phi\bar{\Psi}\, d\Gamma, \qquad |\Phi| = \left(\int_{\Gamma_0} |\Phi|^2\, d\Gamma\right)^{\frac{1}{2}}$$
$$\text{for } \Phi, \Psi \in X = L^2(\Gamma_0).$$

Consulting [11] Chapitre 1, we see that the form domain Y is dense in X, and that the form a_0 is closed in the sense defined in §1.3 of Chapter 6 of Kato [9].

Since the bilinear form a_0 is a densely defined, closed non-negative Hermitian form, Theorems 2.1 and 2.23 of Chapter 6 of [9] assert that there exists uniquely the non-negative selfadjoint operator A acting in X with the densely defined domain $D(A)$, which satisfies the following properties from (13) to (15).

(13) $\qquad (A\Phi, \Psi) = a_0(\Phi, \Psi), \quad \text{for } \Phi \in D(A), \quad \Psi \in D(a_0).$

(14) $\qquad (A^{\frac{1}{2}}\Phi, A^{\frac{1}{2}}\Psi) = a_0(\Phi, \Psi), \quad \text{for } \Phi, \Psi \in D(A^{\frac{1}{2}}) = D(a_0).$

(15) $\qquad \begin{aligned} D(A) = \{\Phi \in Y : &\text{ There is a constant } C \\ &\text{such that } |a_0(\Phi, \Psi)| \leq C|\Psi| \text{ is} \\ &\text{valid for any } \Psi \in Y\}. \end{aligned}$

We see that an element Φ of X satisfies $A\Phi = 0$ if and only if Φ is constant on Γ_0. The complete continuity of operator γ_0 implies that $(A + 1)^{-1}$ is a completely continuous operator acting on X. The spectrum $\sigma(A)$ of A consists of non-negative eigenvalues λ_n $(n = 0, 1, 2, \cdots)$ such that

(16) $\qquad \sigma(A) = \{\lambda_0 = 0 < \lambda_1 \leq \lambda_2 \leq \cdots \to \infty\},$

where the multiple eigenvalues are repeatedly ordered according to their multiplicities.

Now we back to (LWW). Hereafter F_t is denoted by Ψ. Suppose the existence of smooth solution $\phi(t, x)$ of (LWW), which is denoted by ϕ below. Then we see that $\Phi = \gamma_0 \phi$ is the solution of the following evolution problem (ALWW).

(ALWW)
$$\begin{cases} \dfrac{d^2\Phi}{dt^2} + gA\Phi = \Psi(t), & t > 0, \\ \Phi(0) = \Phi^1, \quad \dfrac{d\Phi}{dt}(0) = \Phi^0. \end{cases}$$

We adopt this problem (ALWW) as the setting of continuous problem.

Set
$$X^0 = \{\Phi \in X : (\Phi, 1) = 0\} \qquad \text{and} \qquad Y^0 = Y \cap X^0.$$

Let P^0 be the orthogonal projection from X onto X^0. Then P^0 reduces A. Namely we have
$$AP^0 = P^0 A P^0.$$

Denote AP^0 by A^0. The operator A^0 is a positive definite selfadjoint operator acting in X^0.

3. Finite Element Approximation. Suppose there is given a finite dimensional subspace V_h of V containing constant functions. Let X_h be the image $\gamma_0(V_h)$. Regard X_h as a Hilbert space with the inner product induced from the space X. The space X_h is considered to be an approximation space of X. It is noted that $X_h \subset Y$.

Set
$$V_{0h} = V_h \cap V_0,$$

and notice the orthogonal decomposition of the space V_h:
$$V_h = V_{0h} \oplus V_{1h}$$

with respect to the inner product $\langle\!\langle\, , \,\rangle\!\rangle$.

Consider the following problem $(H_{h\Phi_h})$ which is the Galerkin type approximation problem of (H_Φ).

$(H_{h\Phi_h})$
$$\begin{cases} a(u_h, v_h) = 0, & v_h \in V_{0h}, \\ u_h = \Phi_h & \text{on } \Gamma_0, \\ u_h \in V_h. \end{cases}$$

There exists uniquely the solution u_h of $(H_{h\Phi_h})$ for any $\Phi_h \in X_h$. We have also the fact that $u_h \in V_{1h}$. Define an operator $H_h \in L(X_h, V_{1h})$ through the formula:
$$u_h = H_h \Phi_h \qquad \text{for } \Phi_h \in X_h,$$

where u_h is the solution of problem $(H_{h\Phi_h})$. The operator H_h is an isomorphism from X_h onto V_{1h}. Define the non-negative Hermitian form a_h defined on X_h by the following relation:
$$a_h(\Phi_h, \Psi_h) = a(H_h\Phi_h, H_h\Psi_h) \qquad \text{for } \Phi_h, \Psi_h \in X_h.$$

There exists uniquely the bounded selfadjoint operator A_h acting on X_h which satisfies

(17) $$(A_h\Phi_h, \Psi_h) = a_h(\Phi_h, \Psi_h) \qquad \text{for } \Phi_h, \Psi_h \in X_h.$$

Since we assumed that the constant function belongs to V_h, we see that

(18) $$\begin{cases} \text{an element } \Phi_h \text{ of } X_h \text{ satisfies } A_h\Phi_h = 0 \\ \text{if and only if } \Phi_h \text{ is constant on } \Gamma_0. \end{cases}$$

Since $N_{1h} = \dim X_h$ is finite, the non-negativity of A_h together with (18) asserts that the spectrum $\sigma(A_h)$ of A_h consists of non-negative eigenvalues λ_{hn} $(n = 0, 1, 2, \cdots, N_{1h} - 1)$ such that

(19) $$\sigma(A_h) = \{\lambda_{h0} = 0 < \lambda_{h1} \le \lambda_{h2} \le \cdots \le \lambda_{hN_{1h}-1}\}.$$

In (19) the multiple eigenvalues are repeatedly ordered according to their multiplicities. Set

$$X_h^0 = X_h \cap X^0.$$

Let P_h^0 be the orthogonal projection from X onto X_h^0. Then P_h^0 can be considered as the orthogonal projection from X_h onto X_h^0, and reduces A_h. Namely we have

$$A_h P_h^0 = P_h^0 A_h P_h^0.$$

Denote $A_h P_h^0$ by A_h^0. The operator A_h^0 is a bounded positive definite selfadjoint operator acting on X_h^0.

Since $H\Phi_h - H_h\Phi_h \in V_0$ for $\Phi_h \in X_h$, we have

(20) $$a(H\Phi_h, H\Phi_h - H_h\Phi_h) = 0$$

by definition of problem (H_{Φ_h}). By definition of a_h and a_0, (20) yields

$$a_h(\Phi_h, \Phi_h) - a_0(\Phi_h, \Phi_h) = a(H_h\Phi_h - H\Phi_h, H_h\Phi_h - H\Phi_h).$$

Hence it holds

$$a_h(\Phi_h, \Phi_h) \ge a_0(\Phi_h, \Phi_h) \qquad \text{for } \Phi_h \in X_h.$$

Therefore we have

(21) $$\lambda_{h1} \ge \lambda_1 > 0$$

from the following representations of λ_1 and λ_{h1}:

$$\lambda_1 = \inf_{\Phi \in Y^0} \frac{a_0(\Phi, \Phi)}{|\Phi|^2}, \qquad \lambda_{h1} = \inf_{\Phi_h \in X_h^0} \frac{a_h(\Phi_h, \Phi_h)}{|\Phi_h|^2}.$$

Through the weak formulation of (LWW), the following Galerkin approximation problem $(\mathcal{P}_h)$ is naturally derived.

$$(\mathcal{P}_h) \quad \begin{cases} ga(\phi_h(t), v_h) = b\left(\Psi(t) - \dfrac{\partial^2 \phi_h}{\partial t^2}(t), v_h\right), & t > 0, \ v_h \in V_h, \\[2mm] \phi_h(t) \in V_h, & t > 0, \\[1mm] b(\phi_h(0), v_h) = b(\Phi^1, v_h), & v_h \in V_h, \\[2mm] b\left(\dfrac{\partial \phi_h}{\partial t}(0), v_h\right) = b(\Phi^0, v_h), & v_h \in V_h. \end{cases}$$

Then we see that

$$\Phi_h = \gamma_0 \phi_h$$

is the solution of the following X_h-valued evolution problem (ALWW_h).

$$(\mathrm{ALWW}_h) \quad \begin{cases} \dfrac{d^2 \Phi_h}{dt^2} + gA_h \Phi_h = \Psi_h(t), & t > 0, \\[2mm] \Phi_h(0) = \Phi_h^1, \quad \dfrac{d\Phi_h}{dt}(0) = \Phi_h^0. \end{cases}$$

The data in $(\mathcal{P}_h)$ correspond to those in (ALWW_h) as follows.

$$\Phi_h^i = P_h \Phi^i \ (i = 0, 1), \quad \Psi_h = P_h \Psi,$$

where P_h is the orthogonal projection from X onto X_h.

Let $\bar{h}$ be a fixed positive constant. Hereafter we consider a family of approximate problems (ALWW_h) depending on the parameter $h \in (0, \bar{h}]$. To state the setting of problems more precisely, suppose there is given a family of finite dimensional subspaces V_h of V, $\{V_h : 0 < h \leq \bar{h}\}$. Assume that each V_h contains constant functions. Then the notions of X_h, V_{ih} $(i = 0, 1)$, H_h, a_h, A_h, X_h^0 are introduced as above. And we have the problem (ALWW_h) for each $h \in (0, \bar{h}]$.

As for the ability of the family $\{V_h : 0 < h \leq \bar{h}\}$ concerning the approximation of V, the following condition (h-1) is considered in the sequel.

$$(\text{h-1}) \quad \begin{cases} \text{For any } v \in H^2(\Omega), \text{ there is an element } v_h \in V_h \text{ such that} \\ \quad \|v - v_h\| \leq Ch\|v\|_{H^2(\Omega)}, \\ \text{where } C \text{ is independent of } h \text{ and } v. \end{cases}$$

4. Check of Condition $(A_{\varepsilon,\cdot})$ with $\varepsilon(h) = h$.

Firstly we prepare the following spaces.

$$V^0 = \{v \in V : b(v, 1) = 0\}, \quad V_1^0 = V_1 \cap V^0, \quad V_{1h}^0 = V_{1h} \cap V^0.$$

We see that $X_h^0 = \gamma_0(V_{1h}^0)$.

Consider the following problem (G_{Φ^0}) in the weak formulation for a given $\Phi^0 \in X^0$.

$$(\mathrm{G}_{\Phi^0}) \quad \begin{cases} a(u^0, v) = \int_{\Gamma_0} \Phi^0 \bar{v} \, d\Gamma, & v \in V, \\ b(u^0, 1) = 0, \\ u^0 \in V. \end{cases}$$

Then there exists uniquely the solution u^0 of (G_{Φ^0}) for any $\Phi^0 \in X^0$. Let us define an operator $G \in L\left(X^0, V_1^0\right)$ through the formula:

$$u^0 = G\Phi^0 \quad \text{for } \Phi^0 \in X^0,$$

where u^0 is the solution of problem (G_{Φ^0}).

We say that the property (R) is satisfied if the following condition (R) holds good.

(R) $\qquad\qquad G\Phi^0 \in H^2(\Omega) \qquad \text{for } \Phi^0 \in Y^0.$

If Ω is a convex polygon in the plane, the property (R) is satisfied. This fact is derived from Theorems 1.5.2.8 and 3.2.1.3 of Grisvard [8]. The author owes this derivation to Watanabe [19].

Following Definition 2.1 in Chapter 1 of Lions-Margenes [10], the domain $D\left(\left(A^0\right)^{\frac{1-\theta}{2}}\right)$ can be regarded as the Hilbertian interpolation space $[Y^0, X^0]_\theta$ for $\theta \in [0, 1]$. For $s \in \left[0, \frac{1}{2}\right]$, define the Hilbert space $\mathcal{X}^s$ by

$$\mathcal{X}^s = \left[Y^0, X^0\right]_{1-2s}.$$

The space $\mathcal{X}^s$ coincides with $D\left(\left(A^0\right)^s\right)$ with the inner product

$$(\Phi, \Psi)_s = (\Phi, \Psi) + (A^s\Phi, A^s\Psi)$$

$$\text{for } \Phi, \Psi \in \mathcal{X}^s, \quad 0 \leq s \leq \frac{1}{2}.$$

Denote the anti-linear dual space of $\mathcal{X}^s$ by $\mathcal{X}^{-s}$ for $s \in \left[0, \frac{1}{2}\right]$. Here we regard $\Psi^0 \in \mathcal{X}^0$ as an element of $\mathcal{X}^{-s}$ according to the identification:

$$\Psi^0\left(\Phi^0\right) = \left(\Psi^0, \Phi^0\right) \qquad \text{for } \Phi^0 \in \mathcal{X}^s.$$

Introduce the following notations.

$$\mathcal{X} = X^0, \quad \mathcal{X}_h = X_h^0, \quad \mathcal{A} = A^0, \quad \mathcal{A}_h = A_h^0.$$

Denote P_h^0 by $\mathcal{P}_h$, which is regarded as the orthogonal projection from $\mathcal{X}$ onto $\mathcal{X}_h$. We have the following

THEOREM 3. *Suppose that the property (R) is satisfied. Assume condition $(h-1)$. Let s and t be real numbers such that $-\frac{1}{2} \leq s, t \leq \frac{1}{2}$. Then the operator $\{\mathcal{A}^{-1} - \mathcal{A}_h^{-1}\mathcal{P}_h\}$, defined on $\mathcal{X}^{\frac{1}{2}}$, has the unique extension as a bounded operator from $\mathcal{X}^s$ to $\mathcal{X}^t$, which satisfies*

$$\|\mathcal{A}^{-1} - \mathcal{A}_h^{-1}\mathcal{P}_h\|_{L(\mathcal{X}^s, \mathcal{X}^t)} \leq Ch^{s-t+1}$$

with a constant independent of $s, t \in \left[-\frac{1}{2}, \frac{1}{2}\right]$ and $h \in (0, \bar{h}]$.

Consulting Bramble–Osborn[1], we obtained some crucial parts of the proof of Theorem 3. Here we write the main points of the proof. The detailed discussion is written in [17]. Firstly we introduce the following problem $(G_{h\Phi_h^0})$ for $\Phi_h^0 \in X_h^0$ which approximates the problem (G_{Φ^0}).

$$(G_{h\Phi_h^0}) \qquad \begin{cases} a(u_h^0, v_h) = \displaystyle\int_{\Gamma_0} \Phi_h^0 \bar{v}_h d\Gamma, & v_h \in V_h, \\ b(u_h^0, 1) = 0, \\ u_h^0 \in V_h. \end{cases}$$

Then there exists uniquely the solution u_h^0 of $(G_{h\Phi_h^0})$ for any $\Phi_h^0 \in X_h^0$. Let us define an operator $G_h \in L(X_h^0, V_{1h}^0)$ through the formula :

$$u_h^0 = G_h \Phi_h^0 \quad \text{for} \quad \Phi_h^0 \in X_h^0.$$

Let R_h be the orthogonal projection from V onto V_h with respect to the inner product $\langle\langle u, v \rangle\rangle$. Introduce the quantity

$$\|v\| = \left(a(v,v) + \left| \int_{\Gamma_0} v \, d\Gamma \right|^2 \right)^{\frac{1}{2}}$$

for $v \in V$. The quantity $\|v\|$ is a norm on V equivalent to $\|v\|$ (see Lemma 1.1.2 of [17]). We have the following 3 Propositions.

PROPOSITION 1. For any $\varphi, \psi \in \mathcal{X}$, we have

$$((\mathcal{A}^{-1} - \mathcal{A}_h^{-1}\mathcal{P}_h)\varphi, \psi) = a((G - G_h\mathcal{P}_h)\varphi, (1 - R_h)G\psi).$$

PROPOSITION 2. For any $\varphi \in \mathcal{X}$, we have

$$\|(G - G_h\mathcal{P}_h)\varphi\| \le \|(1 - R_h)G\varphi\|.$$

PROPOSITION 3. Suppose that the property (R) is satisfied. Assume condition $(h-1)$. Let s and t be real numbers such that $-\frac{1}{2} \le s,t \le \frac{1}{2}$. Then there exists a constant C independent of $h \in (0, \bar{h}]$ and $s \in [-\frac{1}{2}, \frac{1}{2}]$ such that

$$\|(1 - R_h)G\varphi\| \le Ch^{s+1/2}|\varphi|_s, \varphi \in \mathcal{X}^s, -\frac{1}{2} \le s \le \frac{1}{2}.$$

Proof of Theorem 3. Since $(G - G_h\mathcal{P}_h)\varphi \in V^0$, applying Schwarz inequality to the identity in Proposition 1, we have

$$|((\mathcal{A}^{-1} - \mathcal{A}_h^{-1}\mathcal{P}_h)\varphi, \psi)| \le \|(G - G_h\mathcal{P}_h)\varphi\|\|(1 - R_h)G\psi\|.$$

Proposition 2 together with the equivalence of the norm $\|\ \|$ with $\|\ \|$ yields

$$|((\mathcal{A}^{-1} - \mathcal{A}_h^{-1}\mathcal{P}_h)\varphi, \psi)| \le C\|(1 - R_h)G\varphi\|\|(1 - R_h)G\psi\|.$$

Hence Proposition 3 implies

$$|((\mathcal{A}^{-1} - \mathcal{A}_h^{-1}\mathcal{P}_h)\varphi, \psi)| \leq Ch^{s+1/2}|\varphi|_s h^{-t+1/2}|\psi|_{-t},$$
$$-\tfrac{1}{2} \leq s, t \leq \tfrac{1}{2}, \quad \varphi, \psi \in \mathcal{X}^{1/2}.$$

Since $\mathcal{X}^{1/2}$ is dense in $\mathcal{X}^s$ if $s \leq \tfrac{1}{2}$, we have

$$|(\mathcal{A}^{-1} - \mathcal{A}_h^{-1}\mathcal{P}_h)\varphi|_t \leq Ch^{s-t+1}|\varphi|_s,$$
$$-\tfrac{1}{2} \leq s, t \leq \tfrac{1}{2}, \quad \varphi \in \mathcal{X}^s.$$

Hence we have the conclusion. ∎

Proposition 1 corresponds to equality (4.13) in [1], and we skip its proof. It is noted that operators $\mathcal{A}^{-1}$, and $\mathcal{A}_h^{-1}$, correspond to T and T_h in [1], respectively. The correspondences in the present context are shown in Theorems 1.3.1 and 2.2.1 in [17], respectively.

Proof of Proposition 2. Since $(G - G_h\mathcal{P}_h)\varphi \in V^0$, we have

$$\begin{aligned}
&\| (G - G_h\mathcal{P}_h)\,\varphi \|^2 \\
&= a\left((G - G_h\mathcal{P}_h)\,\varphi, (G - G_h\mathcal{P}_h)\,\varphi\right) \\
&= a\left((G - G_h\mathcal{P}_h)\,\varphi, (G - R_h G)\,\varphi\right) + a\left((G - G_h\mathcal{P}_h)\,\varphi, (R_h G - G_h\mathcal{P}_h)\,\varphi\right).
\end{aligned}$$

Let $v_h = (R_h G - G_h\mathcal{P}_h)\varphi$. Then $v_h \in V_h$ and

$$\begin{aligned}
&\text{the second term of the right most hand} \\
&= a\left((G - G_h\mathcal{P}_h)\,\varphi, v_h\right) \\
&= a\left(G\varphi, v_h\right) - a\left(G_h\mathcal{P}_h\varphi, v_h\right) \\
&= \int_{\Gamma_0} \varphi \bar{v}_h \, d\Gamma - \int_{\Gamma_0} \mathcal{P}_h \varphi \bar{v}_h \, d\Gamma \\
&= 0.
\end{aligned}$$

Therefore we have

$$\| (G - G_h\mathcal{P}_h)\,\varphi \|^2 = a\left((G - G_h\mathcal{P}_h)\,\varphi, G\varphi - R_h G\varphi\right).$$

Schwarz inequality yields

$$\| (G - G_h\mathcal{P}_h)\,\varphi \|^2 \leq \| (G - G_h\mathcal{P}_h)\,\varphi \| \cdot \| (1 - R_h)\,G\varphi \|.$$

Hence we have the conclusion. ∎

Proof of Proposition 3. We have that $G \in L\left(\mathcal{X}^{-\frac{1}{2}}, H^1(\Omega)\right)$. The satisfaction of the property (R) implies

$$G \in L\left(\mathcal{X}^{\frac{1}{2}}, H^2(\Omega)\right).$$

The interpolation theory assures that there is a constant C independent of $s \in \left[-\frac{1}{2}, \frac{1}{2}\right]$ such that

$$(22) \qquad \|G\|_{L\left(\mathcal{X}^s, H^\sigma(\Omega)\right)} \leq C, \qquad \sigma = s + \frac{3}{2}, \quad s \in \left[-\frac{1}{2}, \frac{1}{2}\right]$$

(see Lions–Magenes [10]). The theory together with condition (h-1) also assures the estimate:

$$(23) \qquad \|(1 - R_h)\|_{L\left(H^{1+s}(\Omega), V\right)} \leq C h^s, \qquad s \in [0,1], \quad h \in (0, \bar{h}]$$

with a constant C independent of s and h. Combining above two estimates (22) and (23), we obtain the conclusion. ∎

5. Reduction to Problem (E) and (E_h). By definition, $\mathcal{X}^0$ coincides with $\mathcal{X}$. Define the constant function Φ_0 such that

$$\Phi_0 = (m_{N-1}(\Gamma_0))^{-1/2}.$$

Then we have an orthogonal decomposition:

$$X \ni \Phi = a\Phi_0 + P^0\Phi, \qquad a \in \mathbb{C},$$

which is represented as

$$\Phi = a\Phi_0 + \phi$$

in the sequel. As the h-version of the above representation, we adopt

$$X_h \ni \Phi_h = a_h\Phi_0 + P_h^0\Phi_h = a_h\Phi_0 + \phi_h.$$

For $i = 0, 1$, set

$$\phi^i = P^0\Phi^i, \quad \phi_h^i = P_h^0\Phi_h^i.$$

Due to (16), (19) and (21), it holds

$$\mathcal{A} \geq \lambda_1 > 0, \quad \text{and} \quad \mathcal{A}_h \geq \lambda_1 > 0.$$

Corresponding to the orthogonal decomposition of Φ, decompose $\Phi(t)$ and $\Psi(t)$ in problem (ALWW) as

$$\Phi(t) = a(t)\Phi_0 + \phi(t), \quad \Psi(t) = b(t)\Phi_0 + \psi(t).$$

The function $\phi(t)$ is the solution of the following evolution problem $(\mathcal{A})$.

$$(\mathcal{A}) \qquad \begin{cases} \dfrac{d^2\phi}{dt^2} + g\mathcal{A}\phi = \psi(t), \quad t > 0, \\ \phi(0) = \phi^1, \quad \phi^{(1)}(0) = \phi^0. \end{cases}$$

Define the h-version of the above operators, functions and problem as follows.

$$\Phi_h(t) = a_h(t)\Phi_0 + \phi_h(t), \quad \Psi_h(t) = b_h(t)\Phi_0 + \psi_h(t).$$

The function $\phi_h(t)$ is the solution of the following initial value problem $(\mathcal{A}_h)$ of ordinary differential equation.

$(\mathcal{A}_h)$
$$\begin{cases} \dfrac{d^2\phi_h}{dt^2} + \mathcal{A}_h\phi_h = \psi_h(t), \quad t > 0, \\ \phi_h(0) = \phi_h^1, \quad \phi_h^{(1)}(0) = \phi_h^0. \end{cases}$$

Now we see that the problems $(\mathcal{A})$, and $(\mathcal{A}_h)$, can be regarded as examples of problems (E), and (E_h), of §1 with the correspondences:

$$(X, A) = (\mathcal{X}, g\mathcal{A}) \quad \text{and} \quad (X_h, A_h, P_h) = (\mathcal{X}_h, g\mathcal{A}_h, \mathcal{P}_h).$$

6. Check of Condition $(\varepsilon - 2)$. For simplicity of consideration, we adopt the finite element space V_h which is generated by Lagrangian interpolation polynomials on each element simplex. More precisely, we require the following two assumptions (Ω) and (T).

(Ω) $\qquad\qquad$ Ω is a polygon if $N = 2$, and a polyhedron if $N = 3$.

(T) $\quad$
$$\begin{cases} \text{There is given a family } \{\mathcal{T}_h : 0 < h \leq \overline{h}\} \text{ of simplicial} \\ \text{decomposition of } \Omega \text{ which is regular and satisfies an inverse} \\ \text{condition in the sense described in p.147 of [3].} \end{cases}$$

The suffix h means the representative length of the diameters h_T of simplices T belonging to $\mathcal{T}_h$. Actually we impose

$$h_T \leq h \quad \text{for } T \in \mathcal{T}_h.$$

Fix $k \in \mathbb{N}$. Let $\mathbf{P}_k$ be the totality of polynomials of N variables with degree at most k. Set

$$V_h = \{v_h \in C(\overline{\Omega}) : v_h|_T \in \mathbf{P}_k, \forall T \in \mathcal{T}_h\}.$$

Then the family of spaces $\{V_h : 0 < h \leq \overline{h}\}$ satisfies condition (h-1) (see [3] Chapter 3).

THEOREM 4. *In the above setting, there is a positive constant α independent of h such that*

$$\|A_h\|_{L(X_h)} \leq \frac{\alpha}{h}, \quad h \in (0, \overline{h}],$$

where A_h is defined through (17).

An outline of the proof of Theorem 4 is shown in [18]. Its detail is written in [17].

7. An Error Estimate for the Solutions of the Fully Discrete Approximate Problem. $\qquad$ Throughout this subsection, we assume the conditions (Ω) and (T).

The property (R) is also assumed to be satisfied. Let V_h be defined as in the previous subsection for some integer $k \geq 1$. Now we can give the following fully discrete approximation problem (ALWW$_{h,\tau}$).

$$(\text{ALWW}_{h,\tau}) \quad \begin{cases} (1 + \beta\tau^2 g\mathcal{A}_h)D_{\tau\bar{\tau}}\Phi_{h,\tau,m} + g\mathcal{A}_h\Phi_{h,\tau,m} \\ \quad + \delta\tau g\mathcal{A}_h D_{\bar{\tau}}\Phi_{h,\tau,m} = \Psi_{h,\tau,m}, \quad m = 1, 2, \cdots, \\ \Phi_{h,\tau,0} = \Phi_{h,\tau}^1, \quad D_\tau\Phi_{h,\tau,0} = \Phi_{h,\tau}^0. \end{cases}$$

Fix $\gamma \in (0,1)$. Set ρ as in the remark before the proof of Theorem 1. Fix a positive number $\tau_{\max}$. For any h, choose τ so as to satisfy

$$\tau^2 \leq (\rho/\alpha)h$$

and $\tau \leq \tau_{\max}$, where α is the constant mentioned in the statement of Theorem 4.

For simplicity of the pesentation, we consider only the following case :

$$\Phi^1 = 0, \quad \int_{\Gamma_0} \Phi^0 d\Gamma = 0, \quad \Psi(t) = 0 \quad \text{for } t \geq 0,$$

$$\Phi_{h,\tau}^1 = 0, \quad \Phi_{h,\tau}^0 = P_h\Phi^0, \quad \Psi_{h,\tau,m} = 0 \quad \text{for } m = 1, 2, \cdots.$$

Under these settings, we have the following Theorem.

THEOREM 5. *Let $\Phi(t)$, and $\Phi_{h,m,\tau}$, $m = 1, 2, \cdots$, be the solutions of (ALWW), and (ALWW$_{h,\tau}$), respectively.*

If $\Phi^0 \in D(A^{1+s})$ for $s \in [0, 1/2]$, it holds for $m = 2, 3, \cdots$,

$$|\Phi(m\tau) - \Phi_{h,\tau,m}| \leq C(1 + m\tau)|A^{1+s}\Phi^0|\{h^{1+s} + \tau^2 + \delta\tau\}$$

with the constant C dependent on λ_1, γ, β and $\tau_{\max}$, independent of $\rho, \delta, h \in (0, \bar{h}], s \in [0, 1/2]$ and Φ^0.

If $\Phi^0 \in D(A^{1-s})$ for $s \in [0, 1/2]$, it holds for $m = 2, 3, \cdots$,

$$|\Phi(m\tau) - \Phi_{h,\tau,m}| \leq C(1 + m\tau)|A^{1-s}\Phi^0|\{h^{1-s} + \tau^{2(1-s)} + \delta\tau\}$$

with the constant C dependent on $\lambda_1, \gamma, \beta, \tau_{\max}$ and ρ, independent of $\delta, h \in (0, \bar{h}], s \in [0, 1/2]$ and Φ^0.

To see the validity of Theorem 5, we note that $\Phi(t)$ is the solution of the problem $(\mathcal{A})$, and that $\Phi_{h,\tau,m}$ is the solution of $(\mathcal{A}_{h,\tau})$, where $(\mathcal{A}_{h,\tau})$ is the problem $(\text{E}_{h,\tau})$ with $(X_h, A_h, P_h) = (\mathcal{X}_h, g\mathcal{A}_h, \mathcal{P}_h)$. The inequality $\tau^2 \leq (\rho/\alpha)h$ together with Theorem 4 implies condition $(T_{2\gamma}^\delta)$ for $(\mathcal{A}_{h,\tau})$. Theorem 2 implies Theorem 5 as a special case since Theorem 3 and Theorem 4 assure the validity of condition (ε_s) for $s \in [-1/2, 1/2]$ with $\varepsilon(h) = h$.

References

[1] J. H. Bramble and J. E. Osborn, Approximation of Steklov eigenvalues of non-selfadjoint second order elliptic operators, in: K. Aziz, Ed., *The Mathematical Foundation of the Finite Element Method with Applications in the Partial Differential Equations* (Academic Press, New York, 1972) 387–408.

[2] J. H. Bramble and J. E. Osborn, Rate of convergence estimates for nonselfadjoint eigenvalue approximations, *Math. Comp.* **27** (1973) 525–549.

[3] Ph. G. Ciarlet, *The Finite Element Method for Elliptic Problems* (North-Holland, Amsterdam, 1978).

[4] H. Fujii, Finite element method for mixed initial-boundary value problems in elasticity theory, Doctoral thesis, Kyoto Univ., 1973.

[5] H. Fujita, On the semi-discrete finite element approximation for the evolution equation $u_t + A(t)u = 0$ of parabolic type, in: J. J. H. Millar, Ed., *Topics in Numerical Analysis* **3** (Academic Press, London, 1977) 143–157.

[6] H. Fujita and A. Mizutani, On the finite element methods for parabolic equations I ; approximation of holomorphic semi-groups, *J. Math. Soc. Japan* **28** (1976) 749–771.

[7] H. Fujita and T. Suzuki, Evolution Problems, in: Ph. G. Ciarlet and J. L. Lions, Eds., *Handbook of Numerical Analysis, Vol.2 Finite Element Methods (Part 1)* (Elsevier, Amsterdam, 1991) 789-928.

[8] P. Grisvard, *Elliptic Problems in Nonsmooth Domains* (Pitman, Boston-London-Melbourne, 1985).

[9] T. Kato, *Perturbation Theory for Linear Operators* (Springer, Berlin, 1966).

[10] J. L. Lions and E. Magenes, *Non-Homogeneous Boundary Value Problems and Applications I* (Springer, Berlin-Heidelberg-New York, 1972).

[11] J. Nečas, *Les Méthodes Directes en Théorie des Équations Elliptiques* (Masson et Cie, Paris, Academia, Prague, 1967).

[12] P. A. Raviart and J. M. Thomas, *Introduction à L'analyse Numérique des Équations aux Dérivées Partielles* (Masson, Paris, 1983).

[13] J. J. Stoker, *Water Waves* (Interscience Publishers, New York, 1957).

[14] G. Strang and G. J. Fix, *An Analysis of the Finite Element Method* (Prentice-Hall, Englewood Cliffs, N. J., 1973).

[15] T. Ushijima, Approximation theory for semi-groups of linear operators and its applications to approximation of wave equations, *Japan J. Math. NS.* **1** (1975) 185–224.

[16] T. Ushijima, On the finite element type approximation of semi-groups of linear operators, in: H. Fujii et al., Eds., *Numerical Analysis of Evolution Equation* (Lecture Notes in Numerical and Applied Analysis **1**, Kinokuniya, Tokyo, 1979) 1–24.

[17] T. Ushijima, Finite element analysis of linear water wave problem, Report CSIM, No.90-12, Dept. of Computer Science and Information Mathematics, The University of Electro-Communicatins, 1990.

[18] T. Ushijima, *Computational aspect of linear water wave problem*, Journal of Computational and Applied Mathematics, 38 (1991), 425-445.

[19] T. Watanabe, On the regularity of the Neumann problem in a planer domain with angles (in Japanese), private communication, December 10, 1985.

[20] K. Yosida, Functional Analysis, (Springer, Berlin, 1965).

A Counterexample Concerning Imaginary Powers of Linear Operators

Alberto Venni

Dipartimento di Matematica, Università di Bologna

Piazza di Porta S.Donato, 5

I-40127 Bologna, Italy

Dedicated to the memory
of the late Professor K. Yosida

Summary

An example is exhibited of a linear operator A such that some imaginary powers of A are bounded and some else are unbounded.

1 Introduction

Recently some attention has been devoted to the boundedness of the imaginary powers of a linear operator in a Banach space. For instance, it was shown in [2] and in [10] that such boundedness influences the regularity of the solutions of abstract linear differential equations of parabolic type in L^p spaces, and consequently of the solutions of parabolic initial-boundary value problems. However, though it is known that some realizations of concrete linear differential operators have bounded imaginary powers (see e.g. [5], [6], [11]), it seems that to give general sufficient conditions for an operator to have bounded imaginary powers is a very hard task. Therefore it may be useful to give counterexamples, i.e. examples of operators with unbounded imaginary powers, to understand in a deeper way what are the 'natural bounds' of the problem.

The oldest counterexample can be found in the first paper that Komatsu devoted to the subject twenty-five years ago: see [8, sect. 14]. However in that example the Banach space is c_0 (the Banach space of scalar sequences convergent to 0), which is not reflexive; moreover the operator is not the negative of the infinitesimal generator of an analytic semigroup. At least in some simple cases of differential operators, a method to prove the boundedness of the imaginary powers makes use of a variant of the Mihlin multiplier theorem for functions with values in a Banach space (see e.g. [2, theorem 3.1] and [3, theorem 3.2]), which is known to hold when the

Banach space has a suitable geometric property, ζ-convexity, which is stronger than reflexivity and holds in a Hilbert space. Therefore the examples that were given very recently by McIntosh–Yagi [9] and Baillon–Clément [1] are more shocking: in both cases an operator A is exhibited in a Hilbert space, such that $-A$ is the infinitesimal generator of an analytic semigroup and A^{it} is unbounded $\forall t \in \mathbf{R} \setminus \{0\}$. As much as I know, no example is known of an operator A such that for suitable $t, s \in \mathbf{R} \setminus \{0\}$ A^{it} is bounded and A^{is} is unbounded.

The aim of this note is to give an example of the last mentioned type: more precisely it is proved that in every Banach space with a basis (and hence in every separable Hilbert space) there is an operator A such that $-A$ is the infinitesimal generator of an analytic semigroup and $A^{ik\pi}$ is unbounded if k is an odd integer, while it is the identity operator if k is an even integer.

I wish to thank Giovanni Dore for useful discussions on the subject.

2 Preliminaries

In the sequel, X is a complex Banach space. We denote by $\mathcal{L}(X)$ the Banach space of the bounded linear operators from X to X, with the usual norm. Whenever A is a linear operator 'acting in X', i.e. with both the domain $\mathcal{D}(A)$ and the range contained in X, we denote by $\rho(A)$ its resolvent set, i.e. the set $\{\lambda \in \mathbf{C}; \lambda - A$ is one-to-one and $(\lambda - A)^{-1} \in \mathcal{L}(X)\}$.

Definition 2.1 *Let A be a linear operator acting in X. We say that A is of positive type if* $] - \infty, 0] \subseteq \rho(A)$ *and* $\exists M \in \mathbf{R}^+$ *such that* $\forall \lambda \in] - \infty, 0]$

$$\|(\lambda - A)^{-1}\| \leq \frac{M}{|\lambda| + 1}.$$

For an operator A of positive type (and in fact even for more general operators) the complex powers A^z are defined $\forall z \in \mathbf{C}$. We refer to [13] for the definitions and basic properties.

When $a = (a_n)_{n \in \mathbf{N}}$ is a bounded sequence in a Banach space, we set $\|a\|_\infty = \sup_{n \in \mathbf{N}} \|a_n\|$. Moreover we set $V(a) = \sum_{n=0}^\infty \|a_n - a_{n+1}\|$, and we say that a is of bounded variation if $V(a) < +\infty$. We call BV the space of scalar sequences of bounded variation.

Every sequence of bounded variation in a Banach space is obviously convergent.

Lemma 2.2 *Let $a \in BV$ and let $(y_n)_{n \in \mathbf{N}}$ be an arbitrary sequence in X. We set $s_n = \sum_{k=0}^n y_k$, $t_n = \sum_{k=0}^n a_k y_k$.*

(a) *If $(s_n)_{n \in \mathbf{N}}$ is bounded, then also $(t_n)_{n \in \mathbf{N}}$ is bounded, with $\|t\|_\infty \leq \|s\|_\infty (\|a\|_\infty + V(a))$.*

(b) *If $(s_n)_{n \in \mathbf{N}}$ is convergent, then also $(t_n)_{n \in \mathbf{N}}$ is convergent.*

Proof. By partial summation we get $t_n = \sum_{k=0}^{n-1} (a_k - a_{k+1}) s_k + a_n s_n$. Thus (a) is immediate. If $(s_n)_{n \in \mathbf{N}}$ is convergent, then $(a_n s_n)_{n \in \mathbf{N}}$ is convergent too, while the series $\sum_{k=0}^\infty (a_k - a_{k+1}) s_k$ is absolutely convergent. This proves (b).

From now on, we suppose that X has a basis $(x_n)_{n \in \mathbf{N}}$. This means that $\forall x \in X$ $x = \sum_{n=0}^\infty \alpha_n(x) x_n$, for a unique sequence $(\alpha_n(x))_{n \in \mathbf{N}}$ of complex numbers. Clearly, $\alpha_n(x_k) = 0$ for $n \neq k$ and $\alpha_n(x_n) = 1$.

$\forall n \in \mathbf{N}$ let P_n be the projection operator of rank one associated to the vector x_n. This means that $P_n(x) = \alpha_n(x)x_n$; in particular $P_n(x_k) = 0$ for $n \neq k$, $P_n(x_n) = x_n$ and $x = \sum_{n=0}^{\infty} P_n(x)$ $\forall x \in X$. Moreover it is known that P_n is bounded, see [12, Chapter I, Theorem 3.1]. Therefore, by the uniform boundedness principle, $M_0 \equiv \sup_{n \in \mathbf{N}} \|\sum_{k=0}^{n} P_k\| < +\infty$.

Definition 2.3 *If $a = (a_n)_{n \in \mathbf{N}}$ is any scalar sequence, the linear operator A associated to a is defined as the operator whose domain $\mathcal{D}(A)$ is $\{x \in X; \sum_{n=0}^{\infty} a_n P_n(x) \text{ is convergent}\}$, and $Ax = \sum_{n=0}^{\infty} a_n P_n(x)$.*

Lemma 2.4 *Let a be a scalar sequence. Then the associated operator A is densely defined and closed. Moreover, if $a \in BV$, then $A \in \mathcal{L}(X)$, with $\|A\| \leq M_0(\|a\|_\infty + V(a))$.*

Proof. It is obvious that $\mathcal{D}(A)$ contains the linear subspace X_0 of the vectors x such that $\{n \in \mathbf{N}; P_n(x) \neq 0\}$ is finite, and owing to the definition of a basis, X_0 is dense in X. Now suppose that $(y_k)_{k \in \mathbf{N}}$ is a sequence in $\mathcal{D}(A)$ such that $y_k \to y$ and $Ay_k \to z$ as $k \to +\infty$. Hence $\forall n \in \mathbf{N}$ $P_n(y_k) \to P_n(y)$ and $a_n P_n(y_k) = P_n(Ay_k) \to P_n(z)$. Thus $P_n(z) = a_n P_n(y)$; in particular $\sum_{n=0}^{\infty} a_n P_n(y) = z$; otherwise stated $y \in \mathcal{D}(A)$ and $Ay = z$. Finally suppose that $a \in BV$. Then by lemma 2.2, $\forall x \in X$ the series $\sum_{n=0}^{\infty} a_n P_n(x)$ is convergent, so that $x \in \mathcal{D}(A)$, with

$$\|Ax\| = \|\sum_{n=0}^{\infty} a_n P_n(x)\| \leq \sup_{n \in \mathbf{N}} \|\sum_{k=0}^{n} P_k(x)\|(\|a\|_\infty + V(a)) \leq M_0 \|x\|(\|a\|_\infty + V(a)).$$

Remark that Lemma 2.4 admits a partial converse: if $A \in \mathcal{L}(X)$, then a is a bounded sequence. Indeed the uniform boundedness principle yields $\sup_{n \in \mathbf{N}} \|\sum_{k=0}^{n} a_k P_k\| < +\infty$ and $\forall n \in \mathbf{N}$

$$|a_n| = \frac{\|a_n x_n\|}{\|x_n\|} = \frac{1}{\|x_n\|}\|\sum_{k=0}^{n} a_k P_k(x_n)\| \leq \|\sum_{k=0}^{n} a_k P_k\|.$$

Another trivial remark is that the operator associated to the constant sequence with value $\lambda \in \mathbf{C}$ is λI. Moreover, if $a_n \neq 0$ $\forall n$, then the operator A is one-to-one and A^{-1} is the operator associated to the sequence $(a_n^{-1})_{n \in \mathbf{N}}$. Putting together these remarks we get:

Lemma 2.5 *Let A be the operator associated to the complex sequence $(a_n)_{n \in \mathbf{N}}$ and let $\lambda \in \mathbf{C} \setminus \{a_n; n \in \mathbf{N}\}$. Then $\lambda - A$ is one-to-one and $(\lambda - A)^{-1}$ is the operator associated to the sequence $((\lambda - a_n)^{-1})_{n \in \mathbf{N}}$. In particular $\lambda \in \rho(A)$ if and only if $\forall x \in X$ the series $\sum_{n=0}^{\infty} (\lambda - a_n)^{-1} P_n(x)$ is convergent.*

3 The Example

Lemma 3.1 *Let f be a non-negative function of a non-negative variable. Suppose that $\exists x_0 \geq 0$ such that f is increasing on $[0, x_0]$ and decreasing on $[x_0, +\infty[$. Let $(a_n)_{n \in \mathbf{N}}$ be an increasing sequence of non-negative real numbers, and $b = (f(a_n))_{n \in \mathbf{N}}$. Then $V(b) \leq 2f(x_0)$.*

The proof is straightforward, and is omitted.

Theorem 3.2 *Let $a = (a_n)_{n \in \mathbb{N}}$ be a decreasing sequence of real numbers and let A be the operator associated to a. Then A is the infinitesimal generator of an analytic semigroup $t \mapsto e^{tA}$, strongly continuous at 0, and $\forall t \geq 0$ e^{tA} is the operator associated to the sequence $(e^{ta_n})_{n \in \mathbb{N}}$.*

Proof. If $\lambda \in]a_0, +\infty[$, then $(\lambda - a_n)^{-1}$ is a decreasing sequence of positive real numbers, so that the operator associated to it belongs to $\mathcal{L}(X)$ by lemma 2.4; moreover by lemma 2.5 we have that $\lambda \in \rho(A)$ and $(\lambda - A)^{-1}$ is the operator associated to the sequence $(\lambda - a_n)^{-1}$. Then, by induction on k, it is readily seen that, for each positive integer k, $(\lambda - A)^{-k}$ is the operator associated to the sequence $(\lambda - a_n)^{-k}$, which is a decreasing sequence of positive real numbers. Thus, by lemma 2.4,

$$\|(\lambda - A)^{-k}\| \leq 2 M_0 \sup_{n \in \mathbb{N}} (\lambda - a_n)^{-k} = 2 M_0 (\lambda - a_0)^{-k}.$$

Since we already know that $\mathcal{D}(A)$ is dense in X, the Hille-Yosida theorem yields that A is the infinitesimal generator of a C_0-semigroup. Moreover, by formula E_9 of [7, section 11.8], for $t > 0$ we have

$$e^{tA} x = \lim_{k \to +\infty} t^{-k} k^k \left(\frac{k}{t} - A \right)^{-k} x;$$

therefore

$$P_n(e^{tA} x) = \lim_{k \to +\infty} t^{-k} k^k P_n \left((\frac{k}{t} - A)^{-k} x \right) =$$

$$\lim_{k \to +\infty} t^{-k} k^k \left(\frac{k}{t} - a_n \right)^{-k} P_n(x) = e^{ta_n} P_n(x)$$

which proves that for $t > 0$ e^{tA} is the operator associated to the sequence $(e^{ta_n})_{n \in \mathbb{N}}$ (and this is obvious for $t = 0$). In order to prove that the semigroup is analytic, it is enough to show that $\forall x \in X \ \forall t > 0 \ e^{tA} x \in \mathcal{D}(A)$, and that $\limsup_{t \to 0} \|t A e^{tA}\| < +\infty$. For both purposes we need an estimate of the BV-norm of the sequence $\varphi(t) = (t a_n e^{ta_n})_{n \in \mathbb{N}}$ for $t > 0$: if we find that $\varphi(t) \in BV$ and that $\limsup_{t \to 0} (\|\varphi(t)\|_\infty + V(\varphi(t))) < +\infty$, the proof will be over.

Remark that we can suppose that $a_0 \leq 0$: indeed if this is not true we can replace a_n by $a_n - a_0$, which corresponds to replace A by $A - a_0 I$ and the semigroup e^{tA} by $e^{-ta_0} e^{tA}$, which is analytic if and only if e^{tA} is analytic. Then if we set $b_n = t|a_n|$, we have $\sup_{n \in \mathbb{N}} b_n e^{-b_n} \leq \frac{1}{e}$, while (by lemma 3.1, with $f(x) = xe^{-x}$) the total variation of $(b_n e^{-b_n})_{n \in \mathbb{N}}$ is $\leq \frac{2}{e}$. This concludes the proof.

We now consider the operator A associated to an increasing sequence $(a_n)_{n \in \mathbb{N}}$ with $a_0 > 0$. From the proof of theorem 3.2 it follows that $] -\infty, a_0 [\subseteq \rho(A)$ and that $\forall \lambda \in] -\infty, a_0 [$ $\|(\lambda - A)^{-1}\| \leq \frac{2 M_0}{a_0 - \lambda}$. In particular A is an operator of positive type, hence it admits complex powers A^z, $z \in \mathbb{C}$, and for $\operatorname{Re} z < 0$ $A^z \in \mathcal{L}(X)$. When $\operatorname{Re} z > 0$, A^z is bounded if and only if A is bounded, which happens if and only if the sequence $(a_n)_{n \in \mathbb{N}}$ is bounded, as follows from lemma 2.4 and the remark after it. Our next aim is to characterize the complex powers A^z.

Lemma 3.3 *Let S, T be linear operators acting in X. Suppose that T is closed and that there exists a linear subspace Y of $\mathcal{D}(S) \cap \mathcal{D}(T)$, dense in $\mathcal{D}(S)$ with respect to the graph norm of S, such that $\forall y \in Y \ Sy = Ty$. Then $S \subseteq T$.*

The proof is a trivial exercise.

Lemma 3.4 $\forall m \in \mathbf{N} \ A^m$ *is the operator associated to the sequence $(a_n^m)_{n \in \mathbf{N}}$.*

Proof. Suppose that $m \in \mathbf{N}$ and that A^m is the operator associated to the sequence $(a_n^m)_{n \in \mathbf{N}}$. If $x \in \mathcal{D}(A^{m+1})$, then $Ax \in \mathcal{D}(A^m)$; therefore

$$A^{m+1}x = A^m A x = \sum_{n=0}^{\infty} a_n^m P_n(Ax) =$$

$$\sum_{n=0}^{\infty} a_n^m P_n(\sum_{k=0}^{\infty} a_k P_k(x)) = \sum_{n=0}^{\infty} a_n^{m+1} P_n(x).$$

Hence A^{m+1} is a restriction of the operator associated to the sequence $(a_n^{m+1})_{n \in \mathbf{N}}$. On the other hand, if the series $\sum_{n=0}^{\infty} a_n^{m+1} P_n(x)$ converges, then also $\sum_{n=0}^{\infty} a_n^m P_n(x)$ converges by lemma 2.2, as $(a_n^{-1})_{n \in \mathbf{N}}$ is a decreasing sequence of positive real numbers. Thus $x \in \mathcal{D}(A^m)$, with $P_n(A^m x) = P_n(\sum_{k=0}^{\infty} a_k^m P_k(x)) = a_n^m P_n(x)$. Therefore $A^m x \in \mathcal{D}(A)$ and hence $x \in \mathcal{D}(A^{m+1})$.

Theorem 3.5 $\forall z \in \mathbf{C} \ A^z$ *is the operator associated to the sequence $(a_n^z)_{n \in \mathbf{N}}$.*

Proof. We call A_z the operator associated to the sequence $(a_n^z)_{n \in \mathbf{N}}$. Then both A^z and A_z are closed operators, so that, by lemma 3.3, our task reduces to find a linear subspace Y_1 of $\mathcal{D}(A^z)$, dense in $\mathcal{D}(A^z)$ (with respect to the graph norm), and a linear subspace Y_2 of $\mathcal{D}(A_z)$, dense in $\mathcal{D}(A_z)$ (with respect to the graph norm), such that $\forall y \in Y_1 \cup Y_2 \ A^z y = A_z y$.

Having fixed $z \in \mathbf{C}$, we know that if $n, m \in \mathbf{N}$, with $0 < n + \operatorname{Re} z < m$, then $\forall x \in \mathcal{D}(A^m)$

$$A^z x = \frac{\Gamma(m)}{\Gamma(n+z)\Gamma(m-n-z)} \int_0^{\infty} \lambda^{z+n-1} (A(\lambda + A)^{-1})^m A^{-n} x \, d\lambda$$

(see [13, 1.15.1]) and $\mathcal{D}(A^m)$ is dense in $\mathcal{D}(A^z)$ with respect to the graph norm. Hence for $x \in \mathcal{D}(A^m)$,

$$P_k(A^z x) = \frac{\Gamma(m)}{\Gamma(n+z)\Gamma(m-n-z)} \int_0^{\infty} \lambda^{z+n-1} (a_k(\lambda + a_k)^{-1})^m a_k^{-n} P_k(x) \, d\lambda = a_k^z P_k(x).$$

This proves that $x \in \mathcal{D}(A_z)$, with $A_z x = A^z x$.

Next we consider the space X_0 of those x such that $\{n \in \mathbf{N}; P_n(x) \neq 0\}$ is finite. It follows at once from lemma 3.4 that $X_0 \subseteq \bigcap_{m \in \mathbf{N}} \mathcal{D}(A^m) = \bigcap_{z \in \mathbf{C}} \mathcal{D}(A^z)$; moreover it is obvious that $X_0 \subseteq \mathcal{D}(A_z)$, and from the preceding part of the proof it follows that A^z and A_z coincide on X_0. Finally, if $x \in \mathcal{D}(A_z)$, then we have $\sum_{k=0}^n P_k(x) \in X_0$, $\lim_{n \to +\infty} \sum_{k=0}^n P_k(x) = x$,

$A_z(\sum_{k=0}^n P_k(x)) = \sum_{k=0}^n a_k^z P_k(x) \to A_z x$, and this proves that X_0 is dense in $\mathcal{D}(A_z)$ with respect to the graph norm of A_z.

Now we specialize further the sequence $(a_n)_{n\in\mathbb{N}}$ (which up to now was assumed to be positive and increasing) in the following way: let us fix a sequence of 'signs' $(\varepsilon_n)_{n\in\mathbb{N}}$, where $\varepsilon_n \in \{-1,1\}$ $\forall n \in \mathbb{N}$. We set $\varepsilon_{-1} = -1$ and

$$b_n = \frac{1}{2} \sum_{k=0}^n |\varepsilon_k - \varepsilon_{k-1}|.$$

Thus $b_0 \in \{0,1\}$ and $b_{n+1} = b_n + \frac{1}{2}|\varepsilon_{n+1} - \varepsilon_n|$ which equals either b_n or $b_n + 1$, so that $(b_n)_{n\in\mathbb{N}}$ is an increasing sequence of non-negative integers. Finally, we set $a_n = e^{b_n}$, so that $(a_n)_{n\in\mathbb{N}}$ is an increasing sequence of positive real numbers. Since $a_n^{2\pi ki} = e^{2\pi k b_n i} = 1$, an immediate consequence of theorem 3.5 is:

Theorem 3.6 $\forall k \in \mathbb{Z}$ $A^{2\pi ki} = I$ *(the identity operator).*

What about $A^{(2k+1)\pi i}$? By theorem 3.5 this is the linear operator associated to the sequence $(a_n^{(2k+1)\pi i})_{n\in\mathbb{N}}$. Now $a_n^{(2k+1)\pi i} = e^{(2k+1)\pi i b_n} = e^{i\pi b_n}$. By induction on n we can see immediately that $e^{i\pi b_n} = -\varepsilon_n$. Thus $A^{(2k+1)\pi i}$ is the linear operator associated to the sequence $(-\varepsilon_n)_{n\in\mathbb{N}}$.

It is known that in every infinite-dimensional Banach space with a basis, there is always a *conditional* basis (see [12, Chapter II, Theorem 23.2]). A necessary and sufficient condition for a basis $(x_n)_{n\in\mathbb{N}}$ to be conditional is that there exist $\tilde{x} \in X$ and a sequence of signs $(\varepsilon_n)_{n\in\mathbb{N}}$ such that

$$\sup_{n\in\mathbb{N}} \left\| \sum_{k=0}^n \varepsilon_k P_k(\tilde{x}) \right\| = +\infty$$

(see [12, Chapter II, Theorem 16.1, equivalence between statements 1 and 8]; see also formula (16.7) thereof). Therefore, if we choose such a sequence $(\varepsilon_n)_{n\in\mathbb{N}}$, since the sequence $(-\varepsilon_n)_{n\in\mathbb{N}}$ has obviously the same property, $\tilde{x}$ *does not belong* to the domain of the operator associated to the sequence $(-\varepsilon_n)_{n\in\mathbb{N}}$, that is $A^{(2k+1)\pi i}$. Therefore:

Theorem 3.7 $\forall k \in \mathbb{Z}$ $A^{(2k+1)\pi i}$ *is unbounded.*

Final remarks. As our example shows, the boundedness of a sequence a is not sufficient for the boundedness of the operator associated to a: in fact in the example $|a_n^{i\pi k}| = 1$ $\forall k$. The example also shows that the spectrum of A may be much larger than the set $\{a_n; n \in \mathbb{N}\}$. Indeed for an odd integer k, $A^{ki\pi}$ is a closed operator, and $A^{2ki\pi} = I$ is the closure of $(A^{ki\pi})^2$. But the domain of $(A^{ki\pi})^2$ is contained in the domain of $A^{ki\pi}$, which is different from $X = \mathcal{D}(A^{2ki\pi})$. Hence $(A^{ki\pi})^2 \neq I$, so that $(A^{ki\pi})^2$ is not closed. This gives an example of a closed operator whose square is not closed, and proves, by [4, Theorem VII.9.7], that the resolvent set of $A^{ki\pi}$ is empty.

References

[1] J.B. BAILLON - Ph. CLEMENT: *Examples of unbounded imaginary powers of operators;* J. Funct. Anal. **100** (1991), 419–434.

[2] G. DORE - A. VENNI: *On the closedness of the sum of two closed operators;* Math. Z. **196** (1987), 189–201.

[3] G. DORE - A. VENNI: *An operational method to solve a Dirichlet problem for the Laplace operator in a plane sector;* Differential and Integral Equations 3 (1990), 323–334.

[4] N. DUNFORD - J.T. SCHWARTZ: *Linear Operators. Part I.* Interscience, New York, 1958.

[5] Y. GIGA: *Domains of fractional powers of the Stokes operator in L_r spaces;* Arch. Rat. Mech. Anal. **89** (1985), 251–265.

[6] Y. GIGA - H. SOHR: *On the Stokes operator in exterior domains;* J. Fac. Sci. Univ. Tokyo Sect. IA, Math. **36** (1989), 103–130.

[7] E. HILLE - R.S. PHILLIPS: *Functional Analysis and Semi-Groups.* American Mathematical Society, Providence R.I., 1957.

[8] H. KOMATSU: *Fractional powers of operators;* Pacific J. Math. **19** (1966), 285–346.

[9] A. McINTOSH - A. YAGI: *Operators of type ω without a bounded H_∞-functional calculus;* in: *Miniconference on Operators in Analysis, 1989. Proceedings of the Centre for Mathematical Analysis, ANU, Canberra,* **24** (1990), 159–172.

[10] J. PRÜSS - H. SOHR: *On operators with bounded imaginary powers in Banach spaces;* Math. Z. **203** (1990), 429–452.

[11] R. SEELEY: *Norms and domains of the complex powers $A_B{}^z$;* Amer. J. Math. **93** (1971), 299–309.

[12] I. SINGER: *Bases in Banach spaces I.* Spinger-Verlag, Berlin Heidelberg New York, 1970.

[13] H. TRIEBEL: *Interpolation Theory, Function Spaces, Differential Operators.* North-Holland, Amsterdam New York Oxford, 1978.

Global Solution to Some Quasilinear Parabolic Problem in Mathematical Biology

ATSUSHI YAGI

Department of mathematics, Himeji Institute of Technology,
Himeji, Hyogo 671-22, Japan

1. Introduction.

We study a strongly coupled parabolic system:

$$
(P.S)\begin{cases}
\dfrac{\partial u}{\partial t} = \dfrac{\partial}{\partial x}[\dfrac{\partial}{\partial x}(a_1 u + \alpha_{11}u^2 + \alpha_{12}uv) + b_1\Phi'(x)u] \\
\qquad\qquad\qquad\qquad + c_1 u - \gamma_{11}u^2 - \gamma_{12}uv \quad \text{in} \quad \Omega \times (0,\infty), \\
\dfrac{\partial v}{\partial t} = \dfrac{\partial}{\partial x}[\dfrac{\partial}{\partial x}(a_2 v + \alpha_{21}uv + \alpha_{22}v^2) + b_2\Phi'(x)v] \\
\qquad\qquad\qquad\qquad + c_2 v - \gamma_{21}uv - \gamma_{22}v^2 \quad \text{in} \quad \Omega \times (0,\infty), \\
\dfrac{\partial}{\partial x}(a_1 u + \alpha_{11}u^2 + \alpha_{12}uv) + b_1\Phi'(x)u = 0 \quad \text{on} \quad \partial\Omega \times (0,\infty), \\
\dfrac{\partial}{\partial x}(a_2 v + \alpha_{21}uv + \alpha_{22}v^2) + b_2\Phi'(x)v = 0 \quad \text{on} \quad \partial\Omega \times (0,\infty), \\
u(0,x) = u_0(x) \quad \text{and} \quad v(0,x) = v_0(x) \quad \text{in} \quad \Omega.
\end{cases}
$$

Here, $\Omega = (0,\ell)$ is a bounded interval in $\mathbb{R}$. $a_i > 0(i=1,2)$ are positive constants; α_{ij}, b_i, c_i and $\gamma_{ij}(i,j=1,2) \geq 0$ are non negative constans; $\Phi \in C^2(\bar{\Omega})$ is a given real function on $\bar{\Omega}$. u_0 and v_0 are initial functions which are assumed to satisfy:

$$(I.F) \qquad u_0, v_0 \in H^{\frac{1}{2}+\varepsilon}(\Omega) \quad \text{with some} \quad \varepsilon > 0 \quad \text{and} \quad u_0, v_0 \geq 0 \quad \text{in} \quad \Omega.$$

And, $u = u(t,x)$ and $v = v(t,x)$ are possibly non negative unknown functions.

This system has been introduced by Shigesada et al. [7] to describe the population dynamics in Ω of two competitive species which move under the influence of population pressure and of environmental potential $\Phi(x)$. The unknown functions $u = u(t,x)$ and $v = v(t,x)$ denote population densities of the two species. The boundary conditions show that the flow of individuals is 0 at the boundary $x = 0$ and $x = \ell$.

In this Note we are concerned with the question of global existence of L^2 valued C^1 solution to (P.S). We shall in fact show that, if one of the following Conditions is satisfied:

$(\alpha.\text{i}) \qquad 0 < \alpha_{21} < 8\alpha_{11} \quad \text{and} \quad 0 < \alpha_{12} < 8\alpha_{22};$

$(\alpha.\text{ii}) \qquad \alpha_{11} > 0 \quad \text{and} \quad \alpha_{21} = \alpha_{22} = 0;$

$(\alpha.\text{iii}) \qquad \alpha_{11} = \alpha_{22} = 0, \alpha_{12} > 0, \alpha_{21} > 0 \quad \text{and} \quad a_1 = a_2, b_1 = b_2;$

This research was partly supported by Grant-in-Aid for Scientific Reseach (No. 03640176), Ministry of Education, Science and Culture.

then there exists a unique global non negative $\mathcal{C}^1$ solution for any initial functions satisfying (I.F). In particular cases the global existence had been already obtained by several authors. When $\Phi = 0$, the case (α.ii) was considered by Masuda and Mimura [5](more strictly, they allowed the case $\alpha_{11} = 0$, our result can be also generalized to cover this critical case provided $\Phi'(0) \leq 0$ and $\Phi'(\ell) \geq 0$, see Remark 4.3), the case (α.iii) was handled by Kim [2]. On the other hand, when similarly $\Phi = 0$, Deuring [1] proved that (P.S) always possesses a global solution without any condition like (α.i,ii,iii) if the initial functions are suitably small. His result seems to be true even for the case $\Phi \neq 0$. We shall therefore seek Conditions on the coefficients which provide the global solution for every scale of initial function.

Some results in this Note can be generalized to more spatial dimensional cases. Analogous global existence of solution is true in the two dimensional case, see [11,12]. To the contrary, it seems to be more difficult to establish the global existence in the three dimensional case (We remark however that Deuring's result [1] is the case independently of the spatial dimension). Uniqueness and local existence of non negative solution to (P.S) is true even in the three dimensional space.

The next problem to be studied which is important from the viewpoint of Mathematical Biology is investigating the asymptotic behavior of the global solution obtained in the cases (α.i,ii,iii). Especially it is very important to show when the solution is coexistence solution, that is, when $\liminf_{t\to\infty} \int_0^\ell u\,dx$ and $\liminf_{t\to\infty} \int_0^\ell v\,dx > 0$ is the case. And ultimately we have to show how the density dependent diffusion together with the environmental potential really provides the more possibility of coexistence and the spatial segregation which were verified in [7] by numerical analysis. At this moment it is far away to give a satisfactory answer. But the author believe that some techniques in this Note are available also in the next stage of research of this system.

In the case when $\alpha_{11} = \alpha_{22} = 0$ and $\Phi = 0$, there are many interesting results concerning the asymptotic behavior and the stationary solution. The existence and stability of stationary solution was studied in various cases by Mimura and Kawasaki [4], by Mimura [5] and by Mimura et al. [6]. The asymptotic behavior and the stbility were studied by Matano and Mimura [3]. But the situation seems to be different much from the case when $\Phi \neq 0$; for example, if the initial functions u_0 and v_0 are constant with $\Phi = 0$, then the solution u, v is also spatially homogeneous for every t by the uniqueness of solution.

This Note consists of four sections. In Section 2, we shall first verify that any real solution to (P.S) must be non negative by the truncation method. In Section 3, the existence and uniqueness of local solution will be obtained (without assuming (α.i), (α.ii) or (α.iii)) by applying the theory of abstract parabolic equations in Banach spaces (see [9]). In Section 4, a priori estimates for the solution u, v will be established under (α.i), (α.ii) or (α.iii) to obtain the global existence of solution.

2. Positivity of Solutions.

We shall observe that every real strict solution to (P.S) is non negative. To this end let us first consider an auxiliary problem:

(2.1)
$$\begin{cases} \dfrac{\partial u}{\partial t} = \dfrac{\partial}{\partial x}[\dfrac{\partial}{\partial x}(a(t,x)u) + \Phi'(x)u] + c(t,x)u \quad \text{in} \quad \Omega \times (0,T], \\[2mm] \dfrac{\partial}{\partial x}(a(t,x)u) + \Phi'(x)u = 0 \quad \text{on} \quad \partial\Omega \times (0,T], \\[2mm] u(0,x) = u_0(x) \quad \text{in} \quad \Omega, \end{cases}$$

in Ω. Here, $a(t,x) \geq \delta$ is a given positive function on $[0,T] \times \bar{\Omega}$ satisfying:

(2.2)
$$\begin{cases} a \in \mathcal{C}([0,T]; H^{\frac{1}{2}+\varepsilon_1}(\Omega)) \cap \mathcal{C}((0,T]; H^2(\Omega)), \\[2mm] \|a(t)\|_{H^2} \leq At^{\frac{2\varepsilon_2-3}{4}} \quad \text{near} \quad t = 0 \end{cases}$$

with some $\varepsilon_1, \varepsilon_2 > 0$ and a constant A. $\Phi \in \mathcal{C}^2(\bar{\Omega})$ is a given real function. And $c \in \mathcal{C}([0,T]; L^\infty(\Omega))$ is a given real valued function on $\Omega \times [0,T]$. Then,

PROPOSITION 2.1. *Let $u_0 \in L^2(\Omega)$ be non negative (≥ 0). If u is a real valued strict solution to (2.1) such that*

(2.3)
$$u \in \mathcal{C}([0,T]; L^2(\Omega)) \cap \mathcal{C}((0,T]; H^2(\Omega)) \cap \mathcal{C}^1((0,T]; L^2(\Omega)),$$

then $u(t)$ must be non negative for every $0 \leq t \leq T$.

PROOF: We use the truncation method. Set a decreasing function H_0 on $\mathbb{R}$ such that $H_0(\sigma) = 0$ for $\sigma \geq 0, H_0(\sigma) = \sigma^4$ for $-\frac{1}{2} \leq \sigma \leq 0, H_0(\sigma) = 1$ for $\sigma \leq -1$, and that $H_0 \in \mathcal{C}^\infty((-\infty,0); \mathbb{R})$. And set $H_1(\sigma) = \int_0^\sigma H_0(\sigma)d\sigma$ and $H(\sigma) = H_2(\sigma) = \int_0^\sigma H_1(\sigma)d\sigma$. Clearly, H is a decreasing function such that $H(\sigma) > 0$ if $\sigma < 0$ and that $H(\sigma) = 0$ if $\sigma \geq 0$. We then consider a function $\phi(t)$ defined by

$$\phi(t) = \int_\Omega H(u(t,x))dx \quad \text{for} \quad 0 \leq t \leq T.$$

From (2.3), $\phi \in \mathcal{C}([0,T]; \mathbb{R}_+) \cap \mathcal{C}^1((0,T]; \mathbb{R})$. And, indeed, $\phi'(t)$ is given by

$$\phi'(t) = -\int_\Omega aH''(u)|\frac{\partial u}{\partial x}|^2 dx - \int_\Omega H''(u)u\frac{\partial(a+\Phi)}{\partial x}\frac{\partial u}{\partial x}dx + \int_\Omega cH'(u)u\,dx.$$

So that, noting that $H'(\sigma)\sigma, H''(\sigma)\sigma^2 \leq Const.H(\sigma), \sigma \in \mathbb{R}$, we obtain that

$$\phi'(t) \leq -\frac{\delta}{2}\int_\Omega H''(u)|\frac{\partial u}{\partial x}|^2 dx + \int_\Omega |\frac{\partial a}{\partial x}|H''(u)|u|\,|\frac{\partial u}{\partial x}|dx$$
$$+ Const.\{\delta^{-1}\|\Phi'\|_{\mathcal{C}}^2 + \|c\|_{L^\infty}\}\phi(t).$$

Further, since $|\frac{dH''^{\frac{1}{2}}}{d\sigma}(\sigma)\sigma| \leq Const.H''(\sigma)^{\frac{1}{2}}, \sigma \in \mathbb{R}$, it is observed that

$$\int_\Omega |\frac{\partial a}{\partial x}|H''(u)|u|\,|\frac{\partial u}{\partial x}|dx \leq \frac{\delta}{2}\|H''(u)^{\frac{1}{2}}\frac{\partial u}{\partial x}\|_{L^2}^2$$
$$+ Const.\{\delta^{-3}\|\frac{\partial a}{\partial x}\|_{L^2}^4 + \delta^{-1}\|\frac{\partial a}{\partial x}\|_{L^2}^2\}\|H''(u)^{\frac{1}{2}}u\|_{L^2}^2,$$

(note that $H''(u)^{\frac{1}{2}}u \in H^1(\Omega)$.) Therefore, $\phi'(t) \leq h(t)\phi(t), 0 < t \leq T$, where

$$h(t) = Const.\{\delta^{-3}\|a(t)\|_{H^2}^{\frac{4}{3}}\|a(t)\|_{H^{\frac{1}{2}}}^{\frac{8}{3}} + \delta^{-1}\|\Phi'\|_{\mathcal{C}}^2 + \|c(t)\|_{L^\infty}\}.$$

(2.2) then implies that h is in $L^1(0,T; \mathbb{R})$. Hence, $\phi(t) \leq \phi(0)\exp(\int_0^t h(\tau)d\tau) = 0$. ∎

Now we can state:

THEOREM 2.2. *Let u_0 and v_0 satisfy (I.F). Assume that a real solution u, v to (P.S) on an interval $[0, T]$ satisfy:*

$$(2.4) \quad \begin{cases} u, v \in \mathcal{C}([0, T]; H^{\frac{1}{2}+\varepsilon_1}(\Omega)) \cap \mathcal{C}((0, T]; H^2(\Omega)) \cap \mathcal{C}^1((0, T]; L^2(\Omega)), \\[2mm] \|u(t)\|_{H^2}, \|v(t)\|_{H^2} \leq At^{\frac{2\varepsilon_2 - 3}{4}} \quad \text{near} \quad t = 0 \end{cases}$$

with some $\varepsilon_1, \varepsilon_2 > 0$ and a constant A. Then $u(t)$ and $v(t)$ must be non negative for every $0 \leq t \leq T$.

PROOF: We use the contradictory. Let $T^* = \sup\{t \in [0, T]; u(s)$ and $v(s)$ are ≥ 0 for all $0 \leq s \leq t\}$ and suppose that $T^* < T$. Then, the Proposition 2.1 (with $a = a_1 + \alpha_{11}u + \alpha_{12}v$ and with $a = a_2 + \alpha_{21}u + \alpha_{22}v$) yields immediately a contradiction. ∎

3. Local Solutions.

Let u_0, v_0 satisfy (I.F). We shall show that there exists a unique local solution to (P.S). We have first, however, to rewrite (P.S) into another system which is almost symmetric near u_0, v_0 introducing suitable scale functions. Indeed, let $\varphi, \psi \in \mathcal{C}^2(\bar{\Omega})(\bar{\Omega} = [0, \ell])$ such that

$$0 < \varphi, \psi < \infty \quad \text{on} \quad \bar{\Omega} \quad \text{and} \quad \varphi'(x) = \psi'(x) = 0 \quad \text{at} \quad x = 0, \ell$$

satisfy:

$$(3.1) \quad |\alpha_{12}\varphi(x)\psi(x)^{-1}u_0(x) - \alpha_{21}\varphi(x)^{-1}\psi(x)v_0(x)| < 2\sqrt{a_1 a_2} \quad \text{on} \quad \bar{\Omega}.$$

And let us change unknown functions from u, v to $u_\varphi = \varphi(x)u(t, x), v_\psi = \psi(x)v(t, x)$ respectively. Then, (P.S) becomes:

$$(P.S)' \quad \begin{cases} \dfrac{\partial u_\varphi}{\partial t} = \dfrac{\partial}{\partial x}[a_{11}(u_\varphi, v_\psi)\dfrac{\partial u_\varphi}{\partial x} + a_{12}(u_\varphi)\dfrac{\partial v_\psi}{\partial x}] + b_{11}(u_\varphi, v_\psi)\dfrac{\partial u_\varphi}{\partial x} \\[3mm] \qquad\qquad\qquad + b_{12}(u_\varphi, v_\psi)\dfrac{\partial v_\psi}{\partial x} + f(u_\varphi, v_\psi) \quad \text{in} \quad \Omega \times (0, \infty), \\[3mm] \dfrac{\partial v_\psi}{\partial t} = \dfrac{\partial}{\partial x}[a_{21}(v_\psi)\dfrac{\partial u_\varphi}{\partial x} + a_{22}(u_\varphi, v_\psi)\dfrac{\partial v_\psi}{\partial x}] + b_{21}(u_\varphi, v_\psi)\dfrac{\partial u_\varphi}{\partial x} \\[3mm] \qquad\qquad\qquad + b_{22}(u_\varphi, v_\psi)\dfrac{\partial v_\psi}{\partial x} + g(u_\varphi, v_\psi) \quad \text{in} \quad \Omega \times (0, \infty), \\[3mm] a_{11}(u_\varphi, v_\psi)\dfrac{\partial u_\varphi}{\partial x} + a_{12}(u_\varphi)\dfrac{\partial v_\psi}{\partial x} + h_1 u_\varphi = 0 \quad \text{on} \quad \partial\Omega \times (0, \infty), \\[3mm] a_{21}(v_\psi)\dfrac{\partial u_\varphi}{\partial x} + a_{22}(u_\varphi, v_\psi)\dfrac{\partial v_\psi}{\partial x} + h_2 v_\psi = 0 \quad \text{on} \quad \partial\Omega \times (0, \infty), \\[3mm] u_\varphi(0, x) = \varphi(x)u_0(x) \quad \text{and} \quad v_\psi(0, x) = \psi(x)v_0(x) \quad \text{in} \quad \Omega. \end{cases}$$

Here,

$$a_{11}(u_\varphi, v_\psi) = a_1 + 2\alpha_{11}\varphi^{-1}u_\varphi + \alpha_{12}\psi^{-1}v_\psi, \quad a_{12}(u_\varphi) = \alpha_{12}\psi^{-1}u_\varphi$$

$$a_{21}(v_\psi) = \alpha_{21}\varphi^{-1}v_\psi, \quad a_{22}(u_\varphi, v_\psi) = a_2 + \alpha_{21}\varphi^{-1}u_\varphi + 2\alpha_{22}\psi^{-1}v_\psi.$$

$b_{ij}(u_\varphi, v_\psi), 1 \leq i, j \leq 2$, are similarly affine functions with respect to u_φ and v_ψ; $f(u_\varphi, v_\psi)$ and $g(u_\varphi, v_\psi)$ are square functions with respect to u_φ and v_ψ; and $h_i = b_i \Phi', i = 1, 2$. All the coefficients of these functions are in $\mathcal{C}(\bar{\Omega})$, which are determined by φ, ψ and Φ.

We then formulate (P.S)′, and therefore (P.S), as an abstract quasilinear equation:

$$(3.2) \qquad \begin{cases} \dfrac{dU}{dt} + A(U)U = F(U), 0 < t < \infty, \\ U(0) = U_0 \end{cases}$$

in the product $H^{1'}$-space $X = \{H^1(\Omega)'\}^2$. Set also two other product spaces: $Z = \{H^{\frac{1}{2}+\varepsilon_1}(\Omega)\}^2$ and $W = \{H^{\frac{1}{2}+\varepsilon_2}(\Omega)\}^2$ with arbitrarily fixed $0 < \varepsilon_1 < \varepsilon_2 < \min\{\varepsilon, \frac{1}{2}\}$. Then (I.F) implies that the initial value $U_0 = \begin{pmatrix} \varphi u_0 \\ \psi v_0 \end{pmatrix}$ is in $W \subset Z$. For $U = \begin{pmatrix} u \\ v \end{pmatrix} \in K = \{U \in Z; \|U - U_0\|_Z < R\}, R > 0, A(U)$ denote linear operators acting in X defined by the sesqilinear forms

$$a(U; \tilde{U}_1, \tilde{U}_2) = \int_\Omega \{[a_{11}(\Re u, \Re v)\frac{d\tilde{u}_1}{dx} + a_{12}(\Re u)\frac{d\tilde{v}_1}{dx}]\frac{d\bar{\tilde{u}}_2}{dx}$$

$$+ [a_{21}(\Re v)\frac{d\tilde{u}_1}{dx} + a_{22}(\Re u, \Re v)\frac{d\tilde{v}_1}{dx}]\frac{d\bar{\tilde{v}}_2}{dx}\}dx$$

$$+ \int_\Omega \{[b_{11}(u,v)\frac{d\tilde{u}_1}{dx} + b_{12}(u,v)\frac{d\tilde{v}_1}{dx}]\bar{\tilde{u}}_2 + [b_{21}(u,v)\frac{d\tilde{u}_1}{dx} + b_{22}(u,v)\frac{d\tilde{v}_1}{dx}]\bar{\tilde{v}}_2\}dx$$

$$+ \int_\Omega c\{\tilde{u}_1\bar{\tilde{u}}_2 + \tilde{v}_1\bar{\tilde{v}}_2\}dx + [h_1\tilde{u}_1\bar{\tilde{u}}_2]_{x=0}^{x=\ell} + [h_2\tilde{v}_1\bar{\tilde{v}}_2]_{x=0}^{x=\ell}$$

on $\{H^1(\Omega)\}^2$, c being a sufficiently large constant specified below. $F(U) = cU + \begin{pmatrix} f(u,v) \\ g(u,v) \end{pmatrix}$ is a function on K with values in X. Finally, $U = \begin{pmatrix} u_\varphi(t) \\ v_\psi(t) \end{pmatrix}, 0 \leq t < \infty$, is an unknown function.

We then apply the result on abstract parabolic equations in Banach spaces (actually in Hilbert spaces). In fact, according to [9], it is known that (3.2) admits a unique local strict solution provided that the following Conditions on $A(U), W \subset Z \subset X, F(U)$ and on U_0 are fulfilled.

(A.i) For $U \in K, \rho(A(U))$ (the resolvent sets of $A(U)$) $\subset \Sigma = \{\lambda \in \mathbb{C}; |\arg \lambda| \geq \theta_0\}, 0 < \theta_0 < \frac{\pi}{2}$; and an estimate

$$\|(\lambda - A(U))^{-1}\|_{\mathcal{L}(X)} \leq \frac{M}{(|\lambda| + 1)}, \quad \lambda \in \Sigma, U \in K,$$

holds with some constant M.

(A.ii) A Hölder condition

$$\|A(U)(\lambda - A(U))^{-1}\{A(U)^{-1} - A(V)^{-1}\}\|_{\mathcal{L}(X)} \leq \frac{N\|U - V\|_Z}{|\lambda| + 1}, \quad \lambda \in \Sigma, U, V \in K,$$

holds with some constant N.

(S.i) For some $0 < \gamma < 1, \|\cdot\|_Z \le D\|\cdot\|_W^\gamma \cdot \|\cdot\|_X^{1-\gamma}$ on W with some constant D.

(S.ii) For some $0 < \beta < 1, \mathcal{D}(A(U)^\beta)$ (the domains of fractional powers $A(U)^\beta) \subset W, U \in K$, with continuous embedding $\|\cdot\|_W \le D_\beta\|A(U)^\beta\cdot\|_X$ with some constant D_β.

(F) $\|F(U) - F(V)\|_X \le L\|U - V\|_Z, U, V \in K$, with some constant L.

(I) $U_0 \in \mathcal{D}(A(U_0)^\beta)(\subset W)$.

So that, let us check all these conditions. Fix $R > 0$ sufficiently small so that, on account of (3.1), every $U = \begin{pmatrix} u \\ v \end{pmatrix} \in K$, satisfies:

$$(3.3) \qquad \{a_{12}(\Re u) + a_{21}(\Re v)\}^2 < 4a_{11}(\Re u, \Re v)a_{22}(\Re u, \Re v) \quad \text{on} \quad \bar{\Omega}.$$

Then we verify:

PROPOSITION 3.1. $A(U), U \in K$, satisfy (A.i) if c is fixed sufficiently large.

PROOF: Since (3.3) yields that

$$a_{11}(\Re u, \Re v)\xi^2 + \{a_{12}(\Re u) + a_{21}(\Re v)\}\xi\eta + a_{22}(\Re u, \Re v)\eta^2 \ge \delta(\xi^2 + \eta^2), \xi, \eta \in \mathbb{R},$$

with some $\delta > 0$, it follows that

$$\Re a(U; \tilde{U}, \tilde{U}) \ge \frac{\delta}{2}\|\tilde{U}\|_{H^1}^2 + (c - Const.)\|\tilde{U}\|_{L^2}^2 \quad \text{on} \quad \{H^1(\Omega)\}^2.$$

Then, by virtue of the theory of sesquilinear forms, the linear operators $A(U)$ determined by $a(U; \cdot, \cdot)$ satisfy (A.i) with some uniform Σ and M provided c is sufficiently large. ∎

PROPOSITION 3.2. $A(U), U \in K$, satisfy (A.ii).

PROOF: Let $U, V \in K$, and $\lambda \in \Sigma$. For any $\tilde{F} \in X$ and $\tilde{G} \in \{H^1(\Omega)\}^2$, we have:

$$< A(U)(\lambda - A(U))^{-1}\{A(U)^{-1} - A(V)^{-1}\}\tilde{F}, \tilde{G} >_{H^1(\Omega)' \times H^1(\Omega)}$$
$$= a(V; A(V)^{-1}\tilde{F}, (\bar{\lambda} - A(U)^*)^{-1}\tilde{G}) - a(U; A(V)^{-1}\tilde{F}, (\bar{\lambda} - A(U)^*)^{-1}\tilde{G}).$$

In view of the definition of $a(U; \cdot, \cdot)$, it is estimated by

$$| < A(U)(\lambda - A(U))^{-1}\{A(U)^{-1} - A(V)^{-1}\}\tilde{F}, \tilde{G} > |$$
$$\le Const.\|U - V\|_{L^\infty}\|A(V)^{-1}\tilde{F}\|_{H^1}\|(\bar{\lambda} - A(U)^*)^{-1}\tilde{G}\|_{H^1}.$$

Since $A(U), U \in K$, are isomorhisms from $\{H^1(\Omega)\}^2$ to $\{H^1(\Omega)'\}^2$, and since the parts of $A(U)^*$ in $\{H^1(\Omega)\}^2, U \in K$, also satisfy the Condition (A.i) in $\mathcal{L}(\{H^1(\Omega)\}^2)$, we obtain that

$$\le Const.\|U - V\|_Z\|\tilde{F}\|_X\|\tilde{G}\|_{H^1}(|\lambda| + 1)^{-1},$$

this shows the result to be verified. ∎

In addition,

PROPOSITION 3.3. For any $\frac{1}{2} \le \theta_1 < \theta_2 < \theta_3 \le 1$,

$$\{H^{2\theta_3-1}(\Omega)\}^2 \subset \mathcal{D}(A(U)^{\theta_2}) \subset \{H^{2\theta_1-1}(\Omega)\}^2, \quad U \in K,$$

with continuous embedding.

PROOF: 1) Let $\tilde{F} \in \{H^{2\theta_3-1}(\Omega)\}^2$ and $\tilde{G} \in \{H^1(\Omega)\}^2$. Then, for $\sigma \ge 0$,

$$|< A(U)(\sigma + A(U))^{-1}\tilde{F}, \tilde{G} >| = |< \tilde{F}, A(U)^*(\sigma + A(U)^*)^{-1}\tilde{G} >|$$

$$\le \|\tilde{F}\|_{H^{2\theta_3-1}} \|(\sigma + A(U)^*)^{-1}A(U)^*\tilde{G}\|_{H^{2\theta_3-1'}}$$

$$\le Const.\|\tilde{F}\|_{H^{2\theta_3-1}} \|(\sigma + A(U)^*)^{-1}A(U)^*\tilde{G}\|_{H^{1'}}^{2\theta_3-1} \|(\sigma + A(U)^*)^{-1}A(U)^*\tilde{G}\|_{L^2}^{2(1-\theta_3)}$$

$$\le Const.\sigma^{-\theta_3} \|\tilde{F}\|_{H^{2\theta_3-1}} \|\tilde{G}\|_{H^1}.$$

So that, $\|A(U)(\sigma + A(U))^{-1}\tilde{F}\|_{H^{1'}} \le Const.\sigma^{-\theta_3}\|\tilde{F}\|_{H^{2\theta_3-1}}$; this then implies the first incusion to be shown.

2) Let now $\tilde{F} \in \mathcal{D}(A(U)^{\theta_2})$ and $A(U)^{\theta_2}\tilde{F} = \tilde{G}$. Then,

$$\|\tilde{F}\|_{H^{2\theta_1-1}} = \|A(U)^{-\theta_2}\tilde{G}\|_{H^{2\theta_1-1}} \le Const. \int_0^\infty \sigma^{-\theta_2} \|(\sigma + A(U))^{-1}\tilde{G}\|_{H^{2\theta_1-1}} d\sigma$$

$$\le Const. \int_0^\infty \sigma^{-\theta_2} \|(\sigma + A(U))^{-1}\tilde{G}\|_{H^1}^{2\theta_1-1} \|(\sigma + A(U))^{-1}\tilde{G}\|_{L^2}^{2(1-\theta_1)} d\sigma$$

$$\le Const. \int_0^\infty \sigma^{-\theta_2}(\sigma+1)^{\theta_1-1} \|\tilde{G}\|_{H^{1'}} d\sigma \le Const.\|A(U)^{\theta_2}\tilde{F}\|_{H^{1'}}. \blacksquare$$

Other conditions are verified as follows. (S.i) is valid with $\gamma = \frac{2\varepsilon_1+3}{2\varepsilon_2+3}$; because,

$$H^{\frac{1}{2}+\varepsilon_1}(\Omega) = \left[H^{\frac{1}{2}+\varepsilon_2}(\Omega), L^2(\Omega)\right]_{\frac{2(\varepsilon_2-\varepsilon_1)}{2\varepsilon_2+1}}, \quad L^2(\Omega) = \left[H^{\frac{1}{2}+\varepsilon_2}(\Omega), H^{\frac{1}{2}+\varepsilon_2}(\Omega)'\right]_{\frac{1}{2}}$$

$$\text{and} \quad H^{\frac{1}{2}+\varepsilon_2}(\Omega)' = \left[L^2(\Omega), H^1(\Omega)'\right]_{\frac{1}{2}+\varepsilon_2}.$$

From Proposition 3.3, $\{H^{\frac{1}{2}+\varepsilon}(\Omega)\}^2 \subset \mathcal{D}(A(U)^\beta) \subset \{H^{\frac{1}{2}+\varepsilon_2}(\Omega)\}^2$ for $U \in K$ if $\frac{2\varepsilon_2+3}{4} < \beta < \frac{2\min\{\varepsilon,\frac{1}{2}\}+3}{4}$; therefore, (S.ii) and (I) can take place. Finally (F) is obvious since $F(U)$ is a square function. In this way we have concluded that

THEOREM 3.4. Let u_0, v_0 satisfy (I.F). Let $\varepsilon_1, \varepsilon_2, \gamma$ and β be as above and take η arbitrary in $0 < \eta < \beta(1-\gamma)$. Then, in the function space: $\mathcal{C}^\eta([0,\infty); H^{\frac{1}{2}+\varepsilon_1}(\Omega))$, there exists a unique local real solution u, v to (P.S) on an interval $[0, T], T > 0$, such that

$$(3.4) \qquad u, v \in \mathcal{C}((0,T]; H^1(\Omega)) \cap \mathcal{C}^1((0,T]; H^1(\Omega)').$$

According to the regularity property of abstract parabolic equation (e.g.[10]), the obtained local solution U to (3.2) is shown to satisfy: $U \in \mathcal{C}^\mu((0,T]; \{H^1(\Omega)\}^2)$ with some $\mu > 0$. Then it is verified that the linear equation:

$$\begin{cases} \dfrac{dV}{dt} + A(U(t))V = F(U(t)), 0 < t \le T, \\ V(0) = U_0 \end{cases}$$

admits a strict solution V in $\{L^2(\Omega)\}^2$. Since $U = V$ of course, we observe that u and v are actually more regular than (3.4).

COROLLARY 3.5. *The solution* u, v *obtained above satisfy:*

$$u, v \in \mathcal{C}((0,T]; H^2(\Omega)) \cap \mathcal{C}^1((0,T]; L^2(\Omega)).$$

In particular, by virtue of Theorem 2.2, u *and* v *are non negative.*

4. Global Solutions.

In the three favorable cases that $(\alpha.\mathrm{i})$, $(\alpha.\mathrm{ii})$ or $(\alpha.\mathrm{iii})$ holds, we shall establish a priori estimates and shall show the global existence of solution.

In this section we use the notations:

$$P(u,v) = a_1 u + \alpha_{11} u^2 + \alpha_{12} uv, \quad Q(u,v) = a_2 v + \alpha_{21} uv + \alpha_{22} v^2$$
$$f(u,v) = c_1 u - \gamma_{11} u^2 - \gamma_{12} uv, \quad g(u,v) = c_2 v - \gamma_{21} uv - \gamma_{22} v^2.$$

Then,

(P.S)
$$\begin{cases} \dfrac{\partial u}{\partial t} = \dfrac{\partial}{\partial x}[\dfrac{\partial P}{\partial x} + b_1 \Phi'(x) u] + f \quad \text{in} \quad \Omega \times (0,\infty), \\[2mm] \dfrac{\partial v}{\partial t} = \dfrac{\partial}{\partial x}[\dfrac{\partial Q}{\partial x} + b_2 \Phi'(x) v] + g \quad \text{in} \quad \Omega \times (0,\infty), \\[2mm] \dfrac{\partial P}{\partial x} + b_1 \Phi'(x) u = 0 \quad \text{on} \quad \partial\Omega \times (0,\infty), \\[2mm] \dfrac{\partial Q}{\partial x} + b_2 \Phi'(x) v = 0 \quad \text{on} \quad \partial\Omega \times (0,\infty), \\[2mm] u(0,x) = u_0(x) \quad \text{and} \quad v(0,x) = v_0(x) \quad \text{in} \quad \Omega. \end{cases}$$

We first prove:

PROPOSITION 4.1. *Let* $(\alpha.\mathrm{i})$ *be satisfied. For* $0 < T < \infty$, *assume that there exists a non negative strict solution* u, v *to (P.S) on* $[0, T)$ *such that*

$$u, v \in \mathcal{C}([0,T); L^2(\Omega)) \cap \mathcal{C}((0,T); H^2(\Omega)) \cap \mathcal{C}^1((0,T); L^2(\Omega)).$$

Then, the norms $\|u(t)\|_{H^1(\Omega)}$ *and* $\|v(t)\|_{H^1(\Omega)}$ *remain being bounded as* $t \to T$.

PROOF: Without loss of generality we can assume (in addition to $(\alpha.\mathrm{i})$) that $\alpha_{12} = \alpha_{21}$. Because, if we change the unknown functions from u, v to $u_\varphi = \varphi u(t, x), v_\psi = \psi v(t, x)$ with two positive scale constants φ, ψ like (P.S)$'$, then it is easily observed that the Condition $(\alpha.\mathrm{i})$ is invariable by this change and that one can take some suitable φ and ψ so that $\alpha_{12} = \alpha_{21}$ in a reduced new system. Our a priori estimates consist of two steps.

Step 1. Multiply the first (resp. second) equation in (P.S) by u (resp. v) and integrate the product in $\Omega \times (0, t), 0 < t < T$. And add the two equalities. Then, after some estimations,

$$\frac{1}{2} \int_\Omega \{u(t)^2 + v(t)^2\} dx + \int_0^t \int_\Omega (a_1 |\frac{\partial u}{\partial x}|^2 + a_2 |\frac{\partial v}{\partial x}|^2) dx dt$$
$$+ \int_0^t \int_\Omega \{(2\alpha_{11} u + \alpha_{12} v)|\frac{\partial u}{\partial x}|^2 + (\alpha_{12} u + \alpha_{21} v)\frac{\partial u}{\partial x}\frac{\partial v}{\partial x} + (\alpha_{21} u + 2\alpha_{22} v)|\frac{\partial v}{\partial x}|^2\} dx dt$$
$$\leq \varepsilon \int_0^t \int_\Omega (|\frac{\partial u}{\partial x}|^2 + |\frac{\partial v}{\partial x}|^2) dx dt + \varepsilon^{-1} Const. \left\{1 + \int_0^t \int_\Omega (u^2 + v^2) dx dt\right\}$$

with an arbitrary number $\varepsilon > 0$. While, $(\alpha.i)$ and $\alpha_{12} = \alpha_{21}$ imply here that

$$(4.1) \quad (2\alpha_{11}u + \alpha_{12}v)\xi^2 + (\alpha_{12}u + \alpha_{21}v)\xi\eta + (\alpha_{21}u + 2\alpha_{22}v)\eta^2$$
$$\geq \delta(u+v)(\xi^2 + \eta^2) \quad \text{for} \quad \xi, \eta \in \mathbb{R}; u, v \geq 0,$$

with some $\delta > 0$. Hence, for any $0 \leq t < T$,

$$(4.2) \quad \int_\Omega \{u(t)^2 + v(t)^2\} dx \quad \text{and} \quad \int_0^t \int_\Omega (1 + u + v)(|\frac{\partial u}{\partial x}|^2 + |\frac{\partial v}{\partial x}|^2) dx\, dt \leq Const..$$

Step 2. Multiply next the first (resp. second) equation in (P.S) by $\frac{\partial P}{\partial t}$ (resp. $\frac{\partial Q}{\partial t}$) and integrate the product in $\Omega \times (\tau, t)$ with a fixed τ such that $0 < \tau < T$. And add the two equalities. Then, after some estimation of using (4.1), it is observed that

$$\int_\Omega \{|\frac{\partial P}{\partial x}(t)|^2 + |\frac{\partial Q}{\partial x}(t)|^2\} dx + \int_\tau^t \int_\Omega (1 + u + v)(u_t^2 + v_t^2) dx\, dt$$
$$\leq Const. \left\{ 1 + \int_\tau^t \int_\Omega (|\frac{\partial P}{\partial x}|^2 + |\frac{\partial Q}{\partial x}|^2) dx\, dt + \int_\tau^t \int_\Omega (u^5 + v^5) dx\, dt \right\}.$$

While, since $|\frac{\partial u}{\partial x}| + |\frac{\partial v}{\partial x}| \leq Const.(|\frac{\partial P}{\partial x}| + |\frac{\partial Q}{\partial x}|)$ in Ω, it follows from (4.2) that

$$\int_\Omega (u^5 + v^5) dx \leq Const. \left\{ 1 + \int_\Omega (|\frac{\partial P}{\partial x}|^2 + |\frac{\partial Q}{\partial x}|^2) dx \right\}.$$

Hence, for any $\tau \leq t < T$,

$$\int_\Omega \{|\frac{\partial u}{\partial x}(t)|^2 + |\frac{\partial v}{\partial x}(t)|^2\} dx \quad \text{and} \quad \int_\Omega \{|\frac{\partial P}{\partial x}(t)|^2 + |\frac{\partial Q}{\partial x}(t)|^2\} dx \leq Const.. \quad \blacksquare$$

We now handle the second case:

PROPOSITION 4.2. *Let $(\alpha.ii)$ be satisfied. Then the same assertion as in the Proposition 4.1 is true.*

PROOF: The proof consists of four steps.

Step 1. Multiply the second equation in (P.S) by $pv^{p-1}, 2 \leq p < \infty$, and integrate the product in $\Omega \times (0, t), 0 < t < T$. Then,

$$\int_\Omega v(t)^p dx + p(p-1) \int_0^t \int_\Omega a_2 v^{p-2} |\frac{\partial v}{\partial x}|^2 dx\, dt$$
$$\leq \varepsilon \int_0^t \int_\Omega v^{p-2} |\frac{\partial v}{\partial x}|^2 dx\, dt + \varepsilon^{-1} Const.\{1 + \int_0^t \int_\Omega v^p dx\, dt\}$$

with any $\varepsilon > 0$. So that, for any $0 \leq t < T$,

$$(4.3) \quad \int_\Omega v(t)^p dx \quad \text{and} \quad \int_0^t \int_\Omega v^{p-2} |\frac{\partial v}{\partial x}|^2 dx\, dt \leq Const..$$

Step 2. Integration of the first equation in (P.S) in $\Omega \times (0, t)$ yields that

$$\int_\Omega u(t) dx \leq \int_\Omega u_0 dx + c_1 \int_0^t \int_\Omega u \, dx \, dt$$

So that, for any $0 \leq t < T$,

(4.4)
$$\int_\Omega u(t) dx \leq Const..$$

Step 3. Multiply the first (resp. second) equation in (P.S) by u (resp. $\frac{\partial v}{\partial t}$) and integrate the product in $\Omega \times (\tau, t), 0 < \tau < T$. And add the two equalities. Then, after some estimation of using (4.3),

$$\int_\Omega \{u(t)^2 + |\frac{\partial v}{\partial x}(t)|^2\} dx + \int_\tau^t \int_\Omega (P_u|\frac{\partial u}{\partial x}|^2 + v_t^2) dx \, dt \leq \varepsilon \left\{ \int_\Omega |\frac{\partial v}{\partial x}(t)|^2 dx \right.$$
$$\left. + \int_\tau^t \int_\Omega (u|\frac{\partial u}{\partial x}|^2 + v_t^2) dx \, dt \right\} + \varepsilon^{-1} Const. \left\{ 1 + \int_\tau^t \int_\Omega (u^3 + |\frac{\partial v}{\partial x}|^3) dx \, dt \right\}$$

with an arbitrary $\varepsilon > 0$. While,

$$\int_\Omega |\frac{\partial v}{\partial x}|^3 dx \leq \varepsilon^2 \|v_t\|_{L^2}^2 + \varepsilon^{-2} Const.(1 + \|v\|_{H^1}^2) + Const.\|u\|_{L^3}^{\frac{3}{2}}.$$

In addition, from (4.4)

$$\int_\Omega u^3 dx \leq Const.\|u\|_{H^1}^{\frac{4}{3}} \leq \varepsilon^2 \|u\|_{H^1}^2 + \varepsilon^{-2} Const.(1 + \|u\|_{L^2}^2).$$

Therefore, for any $\tau \leq t < T$,

(4.5)
$$\int_\Omega \{u(t)^2 + |\frac{\partial v}{\partial x}(t)|^2\} dx \quad \text{and} \quad \int_\tau^t \int_\Omega \{(1+u)|\frac{\partial u}{\partial x}|^2 + v_t^2\} dx \, dt \leq Const..$$

Step 4. Multiply the first equation in (P.S) by $\frac{\partial P}{\partial t}$ and inegrate the product in $\Omega \times (\tau, t), 0 < \tau < T$. Some estimation yields that

$$\int_\Omega |\frac{\partial P}{\partial x}(t)|^2 dx + \int_\tau^t \int_\Omega P_u u_t^2 dx \, dt$$
$$\leq \varepsilon \int_\tau^t \int_\Omega u_t^2 dx \, dt + \varepsilon^{-1} Const. \left\{ 1 + \int_\tau^t \int_\Omega (|\frac{\partial P}{\partial x}|^2 + u^6 + u^2 v_t^2) dx \, dt \right\}$$

with an arbitrary $\varepsilon > 0$. While, from (4.5)

$$\int_\Omega u^6 dx \leq Const.\|u\|_{L^2}^4 \|u\|_{H^1}^2 \leq Const. \int_\Omega |\frac{\partial P}{\partial x}|^2 dx,$$
$$\int_\Omega u^2 v_t^2 dx \leq \|u\|_{L^\infty}^2 \|v_t\|_{L^2}^2 \leq Const.\|v_t\|_{L^2}^2 \int_\Omega |\frac{\partial P}{\partial x}|^2 dx.$$

Therefore we obtain in view of (4.5) that, for any $\tau \leq t < T$,

$$\int_\Omega |\frac{\partial P}{\partial x}(t)|^2 dx \quad \text{and} \quad \int_\tau^t \int_\Omega (1+u)u_t^2 dx dt \leq Const..$$

This jointed with (4.3) then shows the result. ∎

Remark 4.3. In Proposition 4.2, the positivity of α_{11} can be replaced by some condition on Φ. Indeed, it is possible to verify the same result under the assumption:

$(\alpha.\text{ii})'$ $\qquad\qquad \alpha_{21} = \alpha_{22} = 0 \quad \text{and} \quad \Phi'(0) \leq 0, \quad \Phi'(\ell) \geq 0.$

Because this condition yields an estimate that, for any $0 \leq t < T$, $\|v(t)\|_{C(\bar\Omega)} \leq Const.$ in Step 1(for detail, see [12,Proposition 3.1]).

Finally we prove:

PROPOSITION 4.4. *Let $(\alpha.\text{iii})$ be satisfied. Then the same assertion as in Proposition 4.1 is true.*

PROOF: The proof is divided in three steps.

Step 1. Integrate the first and second equations in (P.S) in $\Omega \times (0, t), 0 < t < T$. Then we easily obtain that, for any $0 \leq t < T$,

$$(4.6) \qquad \int_\Omega \{u(t) + v(t)\} dx \leq Const..$$

Step 2. As was noticed above, we can assume, without loss of generality, that $\alpha_{12} = \alpha_{21}$ (in addition to $(\alpha.\text{iii})$). Set: $w = u - v$ and set:

$$\tilde P(u, w) = au + \alpha u^2 - \alpha uw, \quad \tilde Q(v, w) = av + \alpha v^2 + \alpha vw,$$

where $a_1 = a_2 = a$ and $\alpha_{12} = \alpha_{21} = \alpha > 0$. Clearly we have:

$$(4.7) \quad \begin{cases} \dfrac{\partial u}{\partial t} = \dfrac{\partial}{\partial x}[\dfrac{\partial \tilde P}{\partial x} + b\Phi'u] + f & \text{in} \quad \Omega \times (0, T), \\[2mm] \dfrac{\partial v}{\partial t} = \dfrac{\partial}{\partial x}[\dfrac{\partial \tilde Q}{\partial x} + b\Phi'v] + g & \text{in} \quad \Omega \times (0, T), \\[2mm] \dfrac{\partial w}{\partial t} = \dfrac{\partial}{\partial x}[a\dfrac{\partial w}{\partial x} + b\Phi'w] + f - g & \text{in} \quad \Omega \times (0, T), \\[2mm] \dfrac{\partial \tilde P}{\partial x} + b\Phi'u = 0 & \text{on} \quad \partial\Omega \times (0, T), \\[2mm] \dfrac{\partial \tilde Q}{\partial x} + b\Phi'v = 0 & \text{on} \quad \partial\Omega \times (0, T), \\[2mm] a\dfrac{\partial w}{\partial x} + b\Phi'w = 0 & \text{on} \quad \partial\Omega \times (0, T), \end{cases}$$

where $b_1 = b_2 = b$. Multiply the third equation in (4.7) by w and integrate the product in $\Omega \times (\tau, t)$ with some fixed $\tau, 0 < \tau < T$. Similarly, multiply the same equation by $\frac{\partial w}{\partial t}$

and integrate the product in $\Omega \times (\tau, t)$. And add the two equalities. Then, after some estimation,

$$\int_\Omega (w^2 + |\frac{\partial w}{\partial x}|^2)dx + \int_\tau^t \int_\Omega w_t^2 dx dt \leq Const. \left\{1 + \int_\tau^t \int_\Omega (w^2 + u^4 + v^4)dx dt\right\}.$$

On the other hand, multiply the first (resp. second) equation in (4.7) by u (resp. v) and integrate the product in $\Omega \times (\tau, t)$. And add the two equalities. After some estimation it follows that

$$\int_\Omega (u^2 + v^2)dx + \int_\tau^t \int_\Omega (1 + u + v)\{|\frac{\partial u}{\partial x}|^2 + |\frac{\partial v}{\partial x}|^2\}dx dt$$

$$\leq Const. \left[1 + \int_\tau^t \int_\Omega \{(u+v)|\frac{\partial w}{\partial x}|^2 + u^2 + v^2\}dx dt\right].$$

While, from (4.6)

$$\int_\Omega (u^4 + v^4)dx \leq Const.(\|u\|_{L^1}^2 \|u\|_{H^1}^2 + \|v\|_{L^1}^2 \|v\|_{H^1}^2) \leq Const.(\|u\|_{H^1}^2 + \|v\|_{H^1}^2),$$

$$\int_\Omega (u+v)|\frac{\partial w}{\partial x}|^2 dx \leq \|\frac{\partial w}{\partial x}\|_{L^\infty}^2 \|u+v\|_{L^1} \leq Const.\|w\|_{H^2}^{\frac{3}{2}}\|w\|_{L^2}^{\frac{1}{2}}$$

$$\leq \varepsilon\|w\|_{H^2}^2 + \varepsilon^{-3}Const.\|w\|_{L^2}^2$$

with an arbitrary $\varepsilon > 0$. Therefore we conclude that, for any $\tau \leq t < T$,

(4.8) $\quad \int_\Omega (u^2 + v^2 + w^2 + |\frac{\partial w}{\partial x}|^2)dx \quad$ and

$$\int_\tau^t \int_\Omega [(1 + u + v)\{|\frac{\partial u}{\partial x}|^2 + |\frac{\partial v}{\partial x}|^2\} + w_t^2]dx dt \leq Const..$$

Step 3. Multiply the first (resp. second) equation in (4.7) by $\frac{\partial \tilde{P}}{\partial t}$ (resp. $\frac{\partial \tilde{Q}}{\partial t}$) and integrate the product in $\Omega \times (\tau, t), 0 < \tau < T$. And add the two equalities. After some estimation,

$$\int_\Omega \{|\frac{\partial \tilde{P}}{\partial x}|^2 + |\frac{\partial \tilde{Q}}{\partial x}|^2\}dx + \int_\tau^t \int_\Omega (\tilde{P}_u u_t^2 + \tilde{Q}_v v_t^2)dx dt$$

$$\leq Const. \left[1 + \int_\tau^t \int_\Omega \{|\frac{\partial \tilde{P}}{\partial x}|^2 + |\frac{\partial \tilde{Q}}{\partial x}|^2 + (u+v)w_t^2 + (u+v)^3|w_t| + u^5 + v^5\}dx dt\right].$$

While, since $|\frac{\partial u}{\partial x}| + |\frac{\partial v}{\partial x}| \leq Const.(|\frac{\partial \tilde{P}}{\partial x}| + |\frac{\partial \tilde{Q}}{\partial x}|)$, it follows from (4.8) that

$$\int_\Omega \{(u+v)w_t^2 + (u+v)^3|w_t|\}dx \leq \|u+v\|_{L^\infty}\|w_t\|_{L^2}^2 + \|u+v\|_{L^\infty}^3\|u+v\|_{L^2}\|w_t\|_{L^2}$$

$$\leq Const.(\|u+v\|_{H^1}^{\frac{1}{2}}\|w_t\|_{L^2}^2 + \|u+v\|_{H^1}\|w_t\|_{L^2})$$

$$\leq Const. \left[1 + \|w_t\|_{L^2}^2 \int_\Omega \{|\frac{\partial \tilde{P}}{\partial x}|^2 + |\frac{\partial \tilde{Q}}{\partial x}|^2\}dx\right].$$

In addition,

$$\int_\Omega (u^5 + v^5)dx \le Const.(\|u\|_{H^1}^{\frac{3}{2}} + \|v\|_{H^1}^{\frac{3}{2}}) \le Const. \left[1 + \int_\Omega \{|\frac{\partial \tilde{P}}{\partial x}|^2 + |\frac{\partial \tilde{Q}}{\partial x}|^2\}dx\right].$$

Therefore, for any $\tau \le t < T$,

$$\int_\Omega \{|\frac{\partial \tilde{P}}{\partial x}|^2 + |\frac{\partial \tilde{Q}}{\partial x}|^2\}dx \quad \text{and} \quad \int_\tau^t \int_\Omega (1 + u + v)(u_t^2 + v_t^2)dxdt \le Const.. \quad \blacksquare$$

It is not difficult to verify that, if the norms $\|u(t)\|_{H^1(\Omega)}, \|v(t)\|_{H^1(\Omega)}$ of a non negative strict solution to (P.S) on an interval $[0, T)$ do not blow up as $t \to T$, then the solution can be extended beyond the extrem point T. Thus we obtain:

THEOREM 4.5. *Let $(\alpha.i)$, (α,ii) or $(\alpha.iii)$ be satisfied. Then, for the initial functions u_0, v_0 satisfying (I.F), there exists a global non negative strict solution to (P.S) such that $u, v \in C^1((0, \infty); L^2(\Omega)) \cap C((0, \infty); H^2(\Omega))$.*

REFERENCES

1. P. Deuring, *An initial-boundary value problem for certain density-dependent diffusion system*, Math. Z. **194** (1987), 375–396.
2. J. U. Kim, *Smooth solutions to a quasi-linear system of diffusion equations for a certain population model*, Nonlinear Analysis **8** (1984), 1121–1144.
3. H. Matano and M. Mimura, *Pattern formation in competition-diffusion systems in nonconvex domains*, Publ. RIMS Kyoto Univ. **19** (1983), 1049–1079.
4. M. Mimura and K. Kawasaki, *Spatial segregation in competitive interraction-diffusion equations*, J. Math. Biology **9** (1980), 49–64.
5. M. Mimura, *Stationary pattern of some density-dependent diffusion system with competitive dynamics*, Hiroshima Math. J. **11** (1981), 621–635.
6. M. Mimura, Y. Nishiura, A. Tesei and T. Tsujikawa, *Coexistence problem for two competing species models with density-dependent diffusion*, Hiroshima Math. J. **14** (1984), 425–449.
7. N. Shigesada, K. Kawasaki and E. Teramoto, *Spatial segregation of interacting species*, J. theor. Biol. **79** (1979), 83–99.
8. A. Yagi, *Fractional powers of operators and evolution equations of parabolic type*, Proc. Japan Acad. **64 A** (1988), 227–230.
9. A. Yagi, *Abstract quasilinear evolution equations of parabolic type in Banach spaces*, Bollettino U. M. I. **5-B** (1991), 341–368.
10. A. Yagi, *Maximal regularity of abstract parabolic evolution equation, an alternative approach of using evolution operator*, Reports Fac. Sci. Himeji Insti. Tech. **2** (1991).
11. A. Yagi, *Global solution to some quasilinear parabolic system in population dynamics*, Preprint.
12. A. Yagi, *A priori estimates for some quasilinear parabolic system in population dynamics*, Preprint.
13. H. Tanabe, "Equation of Evolution," Pitman, London, 1979.

Quasilinear Geometric Optics Approximation

Atsushi Yoshikawa
Department of Applied Science
Faculty of Engineering
Kyushu University

1. Introduction

We will discuss an asymptotic expansion of a solution to the initial value problem for a (class of) quasilinear hyperbolic system of partial differential equations. The system is (symbolically) given by:

$$A_0(u)\frac{\partial}{\partial t}u + \sum_1^n A_j(u)\frac{\partial}{\partial x_j}u + B(u)u = 0 \tag{1.1}$$

for an m-vector valued function $u = u(x,t)$. Hyperbolicity of the system (1.1) means that $A_0(u), \ldots, A_n(u)$ are $m \times m$ symmetric matrices, and $A_0(u)$ is positive definite. For further technical requirements, see below (§2).

The initial data are supposed to take the form

$$u(x,0) = g(x,\lambda) = \lambda^{-1}a_0(\lambda S_0(x), x)r_1(S_{0,x}), \tag{1.2}$$

containing a large parameter $\lambda > 0$. Here $a_0(\sigma, x)$ is a scalar function, and r_1 is an m-vector, directed towards an eigendirection. $S_0(x)$ is the initial phase function (eikonal). Recall that, in the linear theory of hyperbolic systems, it is basic to consider the oscillatory initial data of the form

$$g(x,\lambda) = e^{i\lambda S_0(x)}\lambda^{-1}a_0(x)r_1(S_{0,x}).$$

However, in the quasilinear case, we cannot prescribe the shape of the initial data so that the phase and the amplitude evolve separately.

$a_0(\sigma, x)$ is supposed to be compactly supported with respect to x (i.e., $a_0(\sigma, x) = 0$ for large $|x|$). Furthermore, we require that $a_0(\sigma, x)$ be either compactly supported, together with the vanishing mean, with respect to σ, or periodic (of period 2π) with respect to σ. So one may think of the initial data either a "pulse" or "oscillation" in the phase $S_0(x)$, emphasized to the extreme as $\lambda \to \infty$. In other words, the initial data (1.2) represent a certain state defined through $\lambda \to \infty$.

Suppose that the space dimension is greater than 1. Observe that the Sobolev norm of the initial data behaves as

$$\|g\|_s \sim C\lambda^{s-1}, \quad \lambda \to \infty \quad (s \geq 0).$$

Then general theories of hyperbolic partial differential equations predict validity of solutions to the problem (1.1) (1.2) only in time intervals shrinking to zero as λ grows, however smooth the initial functions $a_0(\sigma, x)$ and $S_0(x)$ may be (See, e.g., Kato [10],[11], Klainerman [12], Majda [14]). Furthermore, existence (and uniqueness) proof behind such results does not provide any information on how the solutions evolve starting from the initial data of the form as given above.

On the other hand, formal solutions have been constructed in situations similar to the present one by many authors in analogy to linear cases (See Choquet-Bruhat [1], Hunter and Keller [5], Majda [15]). Recall that in linear cases formal solutions are obtained starting from the initial data of the form (1.2) by solving successively ordinary differential equations more or less of the same type, after separating the eikonal (See Courant-Hilbert [2], Chapter 6). In geometric language, these ordinary differential equations are analogous to those governing rays of the light in geometric optics. Therefore, such formal solutions are in some sense an analytical version of geometric optics. It should also be mentioned that one may choose $a_0(\sigma, x)$ discontinuous at $\sigma = 0$ and discuss propagation of singularities. It is the characteristic of the linear theory that the propagation of oscillations and that of singularities are governed by the common principle.

In quasilinear cases, one cannot separate the eikonal, and one is obliged to solve partial differential equations of very simple type instead of solving ordinary differential equations (See, in particular, [1] or [5]). We still call geometric optics approximation such formal solutions obtained through solving auxiliary partial differential equations.

Such an asymptotic solution has a merit of clarifying how nonlinearity of the system and the form of the initial data interact. In fact, there are at least two factors involved. One is the very nonlinearity of the system, such as the genuine nonlinearity assumption leading to a formation of shocks , and another is appearance of caustics in the eikonal. However, to handle these delicate problems more thoroughly, our formation is yet to be improved.

Actually, it is not very difficult to propose a formal asymptotic solution to the system (1.1) with the initial data of the form (1.2). We can even discuss weak formal solutions (See [16]). However, we will soon have to face multi-phase interactions, which may cause very complicated phenomena (See Joly and Rauch [7]). This means that a proposal of a formal solution is closely interrelated to a verification of its asymptotic nature relative to a genuine solution. Such justifications are so far carried out for the space dimension 1 (See e.g., Majda [15], Hunter, Majda and Rosales [6]), semi-linear cases (See Delort [3], Joly, Métivier and Rauch [9]) or with a quite restricted class of formal solutions (See Guès [4], Joly and Rauch [8]. See also [9]).

Now let us specify that the initial phase function is planar:

$$S_0(x) = x \cdot \eta, \quad \eta \in \mathbf{R}^n, \quad \eta \neq 0. \tag{1.3}$$

Here $\cdot$ stands for the scalar product in the Euclid space. Our contribution in the present study is to show that a geometric optics approximation is actually asymptotic to the genuine solution of the problem (1.1) (1.2), which turns out to be valid in a time interval independent of λ provided the space dimension is 2 or 3. This can basically be done by carrying out the computation of geometric optics a little bit further than customary. In fact, this is an advantage of the simplest choice (1.3), which excludes occurrence of caustics though. Since estimation of remainder terms is quite technical, we concentrate here on the geometric optics construction (See [17] [18] for details).

2. Assumptions

An important algebraic assumption on the system other than those already stated in §1 is strict hyperbolicity: for any non-vanishing n-vector $\xi = (\xi_1, \ldots, \xi_n) \in \mathbf{R}^n, \xi \neq 0$, the equation

$$\det\left(-\tau A_0(u) + \sum_1^n \xi_j A_j(u)\right) = 0$$

has m distinct real roots $\tau = p_1(u, \xi), \ldots, p_m(u, \xi)$. Thus, there are m right eigenvectors $r_1(u, \xi), \ldots, r_m(u, \xi)$ and left eigenvectors $r_1^*(u, \xi), \ldots, r_m^*(u, \xi)$ such that

$$(-p_k(u, \xi)A_0(u) + \sum_1^n \xi_j A_j(u))r_k(u, \xi) = 0$$

or

$$r_k^*(u, \xi)(-p_k(u, \xi)A_0(u) + \sum_1^n \xi_j A_j(u)) = 0,$$

$k = 1, \ldots, m$, and

$$r_i^*(u, \xi) \cdot r_j(u, \xi) = \delta_{ij}, \quad i, j = 1, \ldots, m.$$

We assume that the coefficient matrices are smooth with respect to u. Thus, $p_k(u, \xi)$, $r_k(u, \xi)$ or $r_k^*(u, \xi)$ are smooth in u and in ξ (at least locally). Actually we need some requirements on the behaviors of $A_0(u), \ldots, B(u)$ for large u, but since they are not restrictive in our present situation, we do not mention them explicitly (See [17] or [18]).

Let $X_k = {}^t r_k(u, \xi) \cdot \nabla_u$ in the sense

$$X_k f(u) = \nabla f(u) \cdot r_k(u, \xi)$$

for any smooth function $f(u)$. Thus, X_k is the k-th characteristic vector field (its first integrals are called k-th Riemann invariants). If $X_k p_k(u, \xi) \neq 0$ holds, then the k-th characteristic field is said to be *genuinely non-linear*. We require that at least one, say, the first, characteristic field is genuinely non-linear, although our discussions at the present level are not quite sensitive to such assumptions.

For simplicity, when we consider the periodic initial data, we require that either $m = 2$ or $m > 2$ and for each pair k, j the commutator $[X_k, X_j]$ be a linear combination of X_k and $X_j, k, j = 1, \ldots, m$. This requirement will later be referred to as *Hypothesis* (H).

Remark. For the initial data $a_0(\sigma, x)$ with the vanishing mean with respect to σ, Hypothesis (H) is not required.

Example. Equations for the 2-dimensional isentropic compressible flow ($m = 3, n = 2$) satisfy Hypothesis (H). The 2-D isentropic compressible fluid flow is described by the following system of equations for the density ρ (> 0) and the momentum vector (μ_1, μ_2):

$$\frac{\partial}{\partial t}\rho + \frac{\partial}{\partial x_1}\mu_1 + \frac{\partial}{\partial x_2}\mu_2 = 0,$$

$$\frac{\partial}{\partial t}\mu_1 + \frac{\partial}{\partial x_1}\left\{\frac{\mu_1^2}{\rho} + P(\rho)\right\} + \frac{\partial}{\partial x_2}\left(\frac{\mu_1\mu_2}{\rho}\right) = 0,$$

$$\frac{\partial}{\partial t}\mu_2 + \frac{\partial}{\partial x_1}\left(\frac{\mu_1\mu_2}{\rho}\right) + \frac{\partial}{\partial x_2}\left\{\frac{\mu_2^2}{\rho} + P(\rho)\right\} = 0.$$

Here $P(\rho)$ stands for the pressure, and we assume

$$P'(\rho) > 0, \quad P''(\rho) \neq 0, \qquad \rho > 0.$$

After a straightforward computation, the above system can be rewritten in the form (1.1): with $m = 3, n = 2$ and $u = {}^t(\rho, \mu_1, \mu_2)$ (^{t} for the transpose),

$$A_0(u) = \begin{pmatrix} \dfrac{\mu_1^2 + \mu_2^2}{\rho^2} + P'(\rho) & -\dfrac{\mu_1}{\rho} & -\dfrac{\mu_2}{\rho} \\[2mm] -\dfrac{\mu_1}{\rho} & 1 & 0 \\[2mm] -\dfrac{\mu_2}{\rho} & 0 & 1 \end{pmatrix},$$

$$A_1(u) = \begin{pmatrix} -\dfrac{P'(\rho)}{\rho}\mu_1 + \dfrac{\mu_1}{\rho^3}(\mu_1^2 + \mu_2^2) & P'(\rho) - \dfrac{\mu_1^2}{\rho^2} & -\dfrac{\mu_1\mu_2}{\rho^2} \\[2mm] P'(\rho) - \dfrac{\mu_1^2}{\rho^2} & \dfrac{\mu_1}{\rho} & 0 \\[2mm] -\dfrac{\mu_1\mu_2}{\rho^2} & 0 & \dfrac{\mu_1}{\rho} \end{pmatrix},$$

$$A_2(u) = \begin{pmatrix} -\dfrac{P'(\rho)}{\rho}\mu_2 + \dfrac{\mu_2}{\rho^3}(\mu_1^2 + \mu_2^2) & -\dfrac{\mu_1\mu_2}{\rho^2} & P'(\rho) - \dfrac{\mu_2^2}{\rho^2} \\[2mm] -\dfrac{\mu_1\mu_2}{\rho^2} & \dfrac{\mu_2}{\rho} & 0 \\[2mm] P'(\rho) - \dfrac{\mu_2^2}{\rho^2} & 0 & \dfrac{\mu_2}{\rho} \end{pmatrix}.$$

Now the eigenvalues $p_k(u, \xi)$, the right and left eigenvectors $r_1(u, \xi), \cdots, r_3^*(u, \xi)$ are given as follows:

$$p_1(u, \xi) = \frac{1}{\rho}(\mu_1\xi_1 + \mu_2\xi_2),$$

$$p_j(u, \xi) = \frac{1}{\rho}(\mu_1\xi_1 + \mu_2\xi_2) + (-1)^j\sqrt{P'(\rho)}|\xi|, \quad j = 2, 3,$$

$$r_1(u, \xi) = \begin{pmatrix} 0 \\ -\xi_2 \\ \xi_1 \end{pmatrix}, \quad r_j(u, \xi) = \begin{pmatrix} 1 \\ \dfrac{\mu_1}{\rho} + (-1)^j\sqrt{P'(\rho)}\dfrac{\xi_1}{|\xi|} \\ \dfrac{\mu_2}{\rho} + (-1)^j\sqrt{P'(\rho)}\dfrac{\xi_2}{|\xi|} \end{pmatrix}, \quad j = 2, 3,$$

and

$$r_1^*(u, \xi) = \frac{1}{|\xi|^2}\left(\frac{\mu_1\xi_2 - \mu_2\xi_1}{\rho}, -\xi_2, \xi_1\right)$$

$$r_j^*(u, \xi) = (-1)^j\frac{1}{2\sqrt{P'(\rho)}|\xi|}\left(-\frac{\mu_1\xi_1 + \mu_2\xi_2}{\rho} + (-1)^j\sqrt{P'(\rho)}|\xi|, \xi_1, \xi_2\right),$$

$j = 2, 3$. Here $|\xi| = \sqrt{\xi_1^2 + \xi_2^2}$. Therefore, the vector fields X_1, X_2, X_3 are given by

$$
\begin{aligned}
X_1 &= -\xi_2 \frac{\partial}{\partial \mu_1} + \xi_1 \frac{\partial}{\partial \mu_2}, \\
X_j &= \frac{\partial}{\partial \rho} + \left\{ \frac{\mu_1}{\rho} + (-1)^j \sqrt{P'(\rho)} \frac{\xi_1}{|\xi|} \right\} \frac{\partial}{\partial \mu_1} + \left\{ \frac{\mu_2}{\rho} + (-1)^j \sqrt{P'(\rho)} \frac{\xi_2}{|\xi|} \right\} \frac{\partial}{\partial \mu_2},
\end{aligned}
$$

$j = 2, 3$. It then follows that the system satisfies Hypothesis (H) since

$$
[X_1, X_j] = \frac{1}{\rho} X_1, \quad j = 2, 3,
$$

and

$$
[X_2, X_3] = \left(\frac{1}{\rho} - \frac{P''(\rho)}{2P'(\rho)} \right) (X_2 - X_3)
$$

for $\rho > 0, \mu_1, \mu_2 \in \mathbf{R}$.

3. Results

Let $S_0(x) = x \cdot \eta, \eta \in \mathbf{R}^n, \eta \neq 0$. Let $a_0(\sigma, x)$ vanish for large x:

$$
a_0(\sigma, x) = 0, \quad |x| \geq R \quad \text{large enough}.
$$

Suppose $a_0(\sigma, x)$ is either periodic with respect to σ:

$$
a_0(\sigma + 2\pi, x) = a_0(\sigma, x), \tag{3.1}
$$

or compactly supported with respect to σ with the vanishing mean:

$$
\int a_0(\sigma, x) d\sigma = 0. \tag{3.2}
$$

We will refer to (3.1) as the *periodic* case and to (3.2) as the *single pulse* case.

Let $r_1(S_{0,x}) = r_1(0, \eta)$. Then we have a formal (geometric optics) solution $U(x, t, \lambda)$ of the problem (1.1)(1.2) which is valid in an interval independent of λ and smooth there. The formal solution will be constructed in the form:

$$
U(x, t, \lambda) = \lambda^{-1} u_1(x, t, \lambda) + \lambda^{-2} u_2(x, t, \lambda) + \lambda^{-3} u_3(x, t, \lambda). \tag{3.3}
$$

Here u_1, u_2, u_3 behave rather nicely and are explicitly computed. Detailed structures of u_1, u_2, u_3 are different according to the cases (3.1) or (3.2).

It then follows that

$$
A_0(U) \frac{\partial}{\partial t} U + \sum_1^n A_j(U) \frac{\partial}{\partial x_j} U + B(U)U = O(\lambda^{-3})
$$

and

$$
U(x, 0, \lambda) - g(x, \lambda) = O(\lambda^{-3})
$$

hold. Note that $O(\lambda^{-3})$ should be interpreted in auxiliary Sobolev spaces (See §4). This is enough to absorb the remainder terms $O(\lambda^{-3})$ to get a genuine solution $u(x, t, \lambda)$ of

(1.1)(1.2) in a time interval *independent* of λ. $U(x,t,\lambda)$ is then asymptotic to $u(x,t,\lambda)$ as $\lambda \to \infty$ in this time interval:

$$u(x,t,\lambda) - U(x,t,\lambda) = O(\lambda^{-3}).$$

Observe that the term $\lambda^{-3}u_3(x,t,\lambda)$ is to be mixed up in the remainder $O(\lambda^{-3})$.

The periodic case: In the periodic case, we require Hypothesis (H), and then $u_1(x,t,\lambda)$, $u_2(x,t,\lambda)$, $u_3(x,t,\lambda)$ are given as follows:

$$u_1(x,t,\lambda) = a_1(\lambda S_1(x,t),x,t)r_1(0,\eta) + \sum_1^m \overline{a}_k(x,t)r_k(0,\eta) \tag{3.4}$$

$$u_2(x,t,\lambda) = \sum_1^m b_{kk}(\lambda S_k(x,t),x,t)r_k(0,\eta) \tag{3.5}$$
$$+ \sum_2^m b_{1k}(\lambda S_1(x,t),x,t)r_k(0,\eta) + \sum_1^m \overline{b}_k(x,t)r_k(0,\eta),$$

$$u_3(x,t,\lambda) = \sum_{\substack{k,l=1 \\ k \neq l}}^m c_{kl}(\lambda S_k(x,t),x,t)r_l(0,\eta) \tag{3.6}$$
$$+ \sum_{k=2}^m \sum_{l=1}^m c_{1kl}(\lambda S_1(x,t),\lambda S_k(x,t)x,t)r_l(0,\eta).$$

Here

$$S_j(x,t) = -p_j(0,\eta)t + x \cdot \eta, \quad j = 1,\ldots,m, \tag{3.7}$$

are the j-th phase functions, respectively.

Derivations of the equations satisfied by $a_1, \overline{a}_k$, etc., are first carried out by choosing the coefficients of $\lambda^0, \lambda^{-1}, \lambda^{-2}$ vanish when $U(x,t,\lambda)$ is substituted in (1.1). Then we appeal to a standard technique used by many authors ([1], [5], [15]).

The functions $a_1, \overline{a}_k$ are determined as follows. $\overline{a} = \sum_1^m \overline{a}_k(x,t)r_k(0,\eta)$ is the solution to the linear hyperbolic system (the linearized system of (1.1) at $u = 0$):

$$A_0(0)\frac{\partial}{\partial t}\overline{a} + \sum_1^n A_j(0)\frac{\partial}{\partial x_j}\overline{a} + B(0)\overline{a} = 0, \tag{3.8}$$

satisfying the initial condition

$$\overline{a}(x,0) = \frac{1}{2\pi}\int_0^{2\pi} a_0(\sigma,x)d\sigma r_1(0,\eta).$$

$a_1(\sigma,x,t)$ solves a first order (quasilinear) partial differential equation (essentially of Burgers' type):

$$\frac{\partial}{\partial t}a_1 + \sum_1^n p_1^{(j)}(0,\eta)\frac{\partial}{\partial x_j}a_1 + \frac{1}{2}\alpha_1(\eta)\frac{\partial}{\partial \sigma_1}a_1^2 \tag{3.9}$$
$$+ \beta_1(x,t,\eta)\frac{\partial}{\partial \sigma_1}a_1 + \gamma(\eta)a_1 = 0$$

with the initial data

$$a_1(\sigma,x,0) = a_0(\sigma,x) - \frac{1}{2\pi}\int_0^{2\pi} a_0(\sigma,x)d\sigma.$$

Here $p_1^{(j)}(0,\eta) = \partial p_1(0,\eta)/\partial \eta_j$,

$$\alpha_1(\eta) = \nabla_w p_1(w,\eta) \cdot r_1(w,\eta)\Big|_{w=0} = X_1 p_1(w,\eta)\Big|_{w=0},$$

$$\beta_1(x,t,\eta) = r_1^*(0,\eta) \cdot \sum_1^n \eta_j (\nabla_w (A_0(w)^{-1} A_j(w)))\Big|_{w=0} \cdot \overline{a}(x,t)) r_1(0,\eta)$$

$$\gamma(\eta) = r_1^*(0,\eta) \cdot A_0(0)^{-1} B(0) r_1(0,\eta).$$

Recall that $\alpha_1(\eta) \neq 0$ if the first characteristic field is genuinely non-linear.

The coefficients $b_{1k}(\sigma_1, x, t), k = 2, \ldots, m$, are determined from the equations

$$(-p_1(0,\eta) + p_k(0,\eta)) \frac{\partial}{\partial \sigma_1} b_{1k} + \alpha_{k1}(\eta) \frac{\partial}{\partial \sigma_1} a_1^2 \qquad (3.10)$$

$$+ \beta_{k1}(x,t,\eta) \frac{\partial}{\partial \sigma_1} a_1 + \gamma_{k1}(\eta) a_1 + \sum_1^n \kappa_{kj}(\eta) \frac{\partial}{\partial x_j} a_1 = 0$$

with

$$\int_0^{2\pi} b_{1k}(\sigma_1, x, t) d\sigma_1 = 0$$

in order to assure periodicity in σ_1 in a subsequent step. Here

$$\alpha_{k1}(\eta) = r_k^*(0,\eta) \cdot \sum_1^n \eta_j (\nabla_w (A_0(w)^{-1} A_j(w)))\Big|_{w=0} \cdot r_1(0,\eta)) r_1(0,\eta),$$

$$\beta_{k1}(x,t,\eta) = r_k^*(0,\eta) \cdot (\sum_1^n \eta_j \nabla_w (A_0(w)^{-1} A_j(w)))\Big|_{w=0} \cdot \overline{a}(x,t)) r_1(0,\eta),$$

$$\gamma_{k1}(\eta) = r_k^*(0,\eta) \cdot A_0(0)^{-1} B(0) r_1(0,\eta),$$

$$\kappa_{kj}(\eta) = r_k^*(0,\eta) \cdot A_0(0)^{-1} A_j(0) r_1(0,\eta).$$

Let $R(\sigma_1, x, t)$ be an extra term determined by $a_1, \overline{a}, b_{1k}$:

$$R(\sigma_1, x, t) = \frac{\partial}{\partial t} \sum_k b_{1k} r_k$$

$$+ \sum_j A_0(0)^{-1} A_j(0) \frac{\partial}{\partial x_j} \sum_k b_{1k} r_k + A_0(0)^{-1} B(0) \sum_k b_{1k} r_k$$

$$+ \sum_j (\nabla_w (A_0(w)^{-1} A_j(w)))\Big|_{w=0} \cdot (\overline{a} + a_1 r_1)) \frac{\partial}{\partial x_j} (\overline{a} + a_1 r_1)$$

$$+ (\nabla_w (A_0(w)^{-1} B(w)))\Big|_{w=0} \cdot (\overline{a} + a_1 r_1))(\overline{a} + a_1 r_1)$$

$$+ \sum_j (\eta_j \nabla_w (A_0(w)^{-1} A_j(w)))\Big|_{w=0} \cdot (\overline{a} + a_1 r_1)) \frac{\partial}{\partial \sigma_1} \sum_k b_{k1} r_k$$

$$+ \sum_k \sum_j (\eta_j \nabla_w (A_0(w)^{-1} A_j(w)))\Big|_{w=0} \cdot r_k) r_1 (\frac{\partial}{\partial \sigma_1} a_1) b_{1k}$$

$$+ \frac{1}{2} \sum_j (\eta_j \nabla_w^2 (A_0(w)^{-1} A_j(w)))\Big|_{w=0} \cdot (\overline{a} + a_1 r_1)^2) \cdot r_1 \frac{\partial}{\partial \sigma_1} a_1.$$

Here $r_1 = r_1(0, \eta), r_k = r_k(0, \eta)$.

Then $\overline{b}(x, t) = \sum_1^m \overline{b}_k(x, t) r_k(0, \eta)$ satisfies the following linear hyperbolic system:

$$A_0(0)\frac{\partial}{\partial t}\overline{b} + \sum_1^n A_j(0)\frac{\partial}{\partial x_j}\overline{b} + B(0)\overline{b} = -\frac{1}{2\pi}\int_0^{2\pi} A_0(0)R(\sigma_1, x, t)d\sigma_1 \qquad (3.11)$$

with

$$\overline{b}(x, 0) = 0.$$

Remark. The isolation of the terms $\overline{a}(x, t)$ or $\overline{b}(x, t)$ via the linearized systems such as (3.8) or (3.11) seems to have not frequently been attempted.

Now b_{11} follows from the equation

$$\frac{\partial}{\partial t}b_{11} + \sum_1^n p_1^{(j)}(0, \eta)\frac{\partial}{\partial x_j}b_{11} \qquad (3.12)$$

$$+\alpha_1(\eta)\frac{\partial}{\partial \sigma_1}(a_1 b_{11}) + \beta_1(x, t, \eta)\frac{\partial}{\partial \sigma_1}b_{11} + \gamma(\eta)b_{11}$$

$$+\epsilon_1(x, t, \eta)\frac{\partial}{\partial \sigma_1}a_1 + \rho(\sigma_1, x, t) = 0$$

with

$$b_{11}(\sigma_1, x, 0) = 0.$$

Here

$$\epsilon_1(x, t, \eta) = r_1(0, \eta)^* \cdot \sum_j \eta_j(\nabla_w(A_0(w)^{-1}A_j(w)))\Big|_{w=0} \cdot \overline{b}(x, t))r_1(0, \eta),$$

$$\rho(\sigma_1, x, t) = r_1^*(0, \eta) \cdot (R(\sigma_1, x, t) - \frac{1}{2\pi}\int_0^{2\pi} R(\sigma_1, x, t)d\sigma_1).$$

For $k \geq 2, b_{kk}(\sigma_k, x, t)$ are determined from

$$\frac{\partial}{\partial t}b_{kk} + \sum_1^n p_k^{(j)}(0, \eta)\frac{\partial}{\partial x_j}b_{kk} + \gamma_k(\eta)b_{kk} + \beta_{kk}(x, t, \eta)\frac{\partial}{\partial \sigma_k}b_{kk} = 0 \qquad (3.13)$$

with

$$b_{kk}(\sigma_k, x, 0) = -b_{1k}(\sigma_k, x, 0).$$

Here

$$\gamma_k(\eta) = r_k^*(0, \eta) \cdot A_0(0)^{-1}B(0)r_k(0, \eta)$$

$$\beta_{kk}(x, t, \eta) = r_k^*(0, \eta) \cdot \sum_1^n \eta_j(\nabla_w(A_0(w)^{-1}A_j(w)))\Big|_{w=0} \cdot \overline{a}(x, t))r_k(0, \eta).$$

Finally, $c_{kl}(\sigma_k, x, t)$ are determined from similar equations governing b_{1k} while each $c_{1kl}(\sigma_1, \sigma_k, x, t)$ satisfies a certain first order linear partial differential equation. Hypothesis (H) on the size m or the commutators $[X_k, X_l]$ of the characteristic fields allows a periodic solution c_{1kl} to such an equation (See [18] for detail). Without Hypothesis (H) we would have faced a small divisor problem.

The single pulse case: In the single pulse case, $u_1(x, t, \lambda)$, $u_2(x, t, \lambda)$ and $u_3(x, t, \lambda)$ are given as follows

$$u_1(x, t, \lambda) = a_1(\lambda S_1(x, t), x, t) r_1(0, \eta), \tag{3.14}$$

$$u_2(x, t, \lambda) = \sum_1^m b_{kk}(\lambda S_k(x, t), x, t) r_k(0, \eta) \tag{3.15}$$

$$+ \sum_2^m b_{1k}(\lambda S_1(x, t), x, t) r_k(0, \eta),$$

and

$$u_3(x, t, \lambda) = \sum_{\substack{k,l=1 \\ k \neq l}}^m c_{kl}(\lambda S_k(x, t), x, t) r_l(0, \eta) \tag{3.16}$$

$$+ \sum_{k=2}^m \sum_{l=1}^m c_{1kl}(\lambda S_1(x, t), \lambda S_k(x, t), x, t) r_l(0, \eta).$$

Here $S_j(x, t)$ are those given by (3.7). Note that they are very similar to the periodic case except that $\bar{a}_k(x, t)$ and $\bar{b}_k(x, t)$ do not appear. Actually, $a_1(\sigma, x, t)$ satisfies the same equation as (3.9) though the initial data is different. Similarly, $b_{kk}(\sigma_k, x, t)$ or $b_{1k}(\sigma_1, x, t)$ satisfy *analogous* equations to (3.10)(3.12)(3.13). To assure a nice behavior to $c_{1kl}(\sigma_1, \sigma_k, x, t)$ and $c_{kl}(\sigma_k, x, t)$, we are obliged to appeal to the requirement (3.2) but we do not need Hypothesis (H) (See [17] for detail).

4. Technicalities on Remainders

Here we indicate the precise meaning of the remainder terms $O(\lambda^{-3})$.

First of all, for $s \geq 0$, the Sobolev space of exponent s, $H^s(\mathbf{R}^n)$, is the Hilbert space (with norm $\|\cdot\|_s$) of all the (generalized) functions $g(x)$ on $\mathbf{R}^n$ that admit square integrable generalized derivatives up to "order s". More technically, in terms of Fourier transforms $\hat{g}(\xi)$ of $g(x)$,

$$\|g\|_s = \left(\int_{\mathbf{R}^n} |\hat{g}(\xi)|^2 (1 + \xi \cdot \xi)^s d\xi \right)^{1/2}.$$

Note the exponent s is not restricted to integers.

Let $n = 2$ or 3. Then our genuine solution $u(x, t, \lambda)$ is defined for $x \in \mathbf{R}^n, 0 \leq t \leq T_1, \lambda \geq \lambda_1$, with some $T_1 > 0, \lambda_1 > 0$, and $u(\cdot, t, \lambda) \in H^3(\mathbf{R}^n)^m, \frac{\partial}{\partial t} u(\cdot, t, \lambda) \in H^2(\mathbf{R}^n)^m$, $0 \leq t \leq T_1$, $\lambda \geq \lambda_1$, while for $s < 3, u(\cdot, t, \lambda)$ and $\frac{\partial}{\partial t} u(\cdot, t, \lambda)$ are respectively $H^s(\mathbf{R}^n)^m$-valued and $H^{s-1}(\mathbf{R}^n)^m$-valued continuous functions of $0 \leq t \leq T_1$.

Now the remainder estimates read as

$$\|u(\cdot, t, \lambda) - U(\cdot, t, \lambda)\|_s \leq C_s \lambda^{s-3}, \quad 0 \leq s \leq 3,$$

and

$$\|\frac{\partial}{\partial t} u(\cdot, t, \lambda) - \frac{\partial}{\partial t} U(\cdot, t, \lambda)\|_s \leq C_s \lambda^{s-2}, \quad 0 \leq s \leq 2,$$

uniformly with respect to $0 \leq t \leq T_1, \lambda \geq \lambda_1$. Here C_s stands for constants independent of t and λ.

412

Remark. $U(x,t,\lambda)$ even makes sense up to $t = \infty$, though it generally develops a shock at a certain time, say, at $t = T_0$, however smooth the initial data may be. We take the above T_1 smaller than T_0. We suspect though that T_1 might in fact be replaced by T_0.

Acknowledgement.

The present work is supported in part by Grant-in-Aid for Scientific Research, Ministry of Education, Science and Culture, Japanese Government, No. 62460005 and No. 03640169. A part of the results was obtained during the author's stay in 1991 at Université de Bordeaux II, UFR des Sciences Humaines Appliquées, as "professeur invité". We owe much to Jean-Luc Joly of Université de Bordeaux I, particularly, for informing us related results of himself and his colleagues.

References

[1] Y. Choquet-Bruhat, Ondes asymptotiques et approchées pour des systèmes d'équations aux dérivées partielles nonlinéaires, J. Math. pures et appl., **48**(1969), 117-158.

[2] R. Courant and D. Hilbert, Methods of Mathematical Physics, vol. 2, Interscience Publishers, 1962, New York.

[3] Jean-Marc Delort, Oscillations semi-linéaires multiphasées compatibles en dimension 2 ou 3 d'espace (to appear in Comm. Partial Diff. Equations).

[4] Olivier Guès, Développement asymptotique de solutions exactes de systèmes hyperboliques quasilinéaires (to appear).

[5] J. K. Hunter and J. B. Keller, Weakly nonlinear high frequency waves, Comm. Pure Appl. Math., **36**(1983), 547-569.

[6] J. K. Hunter, A. Majda and R. Rosales, Resonantly interacting, weakly nonlinear hyperbolic waves II. Several space variables, Studies in Appl. Math., **75**(1986), 187-226.

[7] Jean-Luc Joly and Jeffrey Rauch, Nonlinear resonance can create dense oscillations (to appear).

[8] Jean-Luc Joly and Jeffrey Rauch, Justification of multidimensional single phase semilinear geometric optics, preprint, Univ. Bourdeaux I, 1989.

[9] Jean-Luc Joly, Guy Métivier and Jeffrey Rauch, Remarques sur l'optique géometrique nonlinéaire multidimensionnelle, Séminaire Eq. Dérivées Partielles 1990-1991, Centre de Mathématiques, École Polytechnique, 1990.

[10] Tosio Kato, The Cauchy problem for quasi-linear symmetric hyperbolic systems, Arch. Rat. Mech. Anal., **58**(1975), 181-205.

[11] Tosio Kato, Quasi-linear equations of evolution with applications to partial differential equations, Lecture Notes in Mathematics, **448**(1975), 25-70, Springer-V., Berlin, etc.

413

[12] S. Klainerman, Global existence for nonlinear wave equation, Comm. Pure Appl. Math., **33**(1980), 43-101.

[13] P. Lax, Hyperbolic systems of conservation laws and the mathematical theory of shock waves, Reg. Conf. Ser. Appl. Math., 13(1973), SIAM.

[14] A. Majda, Compressible Fluid Flow and Systems of Conservation Laws in Several Space Variables, Springer-V., 1984, New York, etc.

[15] A. Majda, Nonlinear geometric optics for hyperbolic systems of conservation laws, IMA volumes in Math. and Appl., vol. 2, Oscillation Theory, Computation, and Methods of Compensated Compactness, 115-165, Springer-V., 1986.

[16] A. Yoshikawa, Weak asymptotic solutions to hyperbolic systems of conservation laws, Lect. Notes in Num. Appl. Anal., **10**(1989), 195-210.

[17] A. Yoshikawa, Solutions containing a large parameter of a quasi-linear hyperbolic system of equations and their nonlinear geometric optics approximation, preprint.

[18] A. Yoshikawa, Quasilinear oscillations and geometric optics, preprint.

Printing: Druckhaus Beltz, Hemsbach
Binding: Buchbinderei Schäffer, Grünstadt

Lecture Notes in Mathematics

For information about Vols. 1–1340
please contact your bookseller or Springer-Verlag

Vol. 1387: M. Petkovi´c, Iterative Methods for Simultaneous Inclusion of Polynomial Zeros. X, 263 pages. 1989.

Vol. 1388: J. Shinoda, T.A. Slaman, T. Tugué (Eds.), Mathematical Logic and Applications. Proceedings, 1987. V, 223 pages. 1989.

Vol. 1000: Second Edition. H. Hopf, Differential Geometry in the Large. VII, 184 pages. 1989.

Vol. 1389: E. Ballico, C. Ciliberto (Eds.), Algebraic Curves and Projective Geometry. Proceedings, 1988. V, 288 pages. 1989.

Vol. 1390: G. Da Prato, L. Tubaro (Eds.), Stochastic Partial Differential Equations and Applications II. Proceedings, 1988. VI, 258 pages. 1989.

Vol. 1391: S. Cambanis, A. Weron (Eds.), Probability Theory on Vector Spaces IV. Proceedings, 1987. VIII, 424 pages. 1989.

Vol. 1392: R. Silhol, Real Algebraic Surfaces. X, 215 pages. 1989.

Vol. 1393: N. Bouleau, D. Feyel, F. Hirsch, G. Mokobodzki (Eds.), Séminaire de Théorie du Potentiel Paris, No. 9. Proceedings. VI, 265 pages. 1989.

Vol. 1394: T.L. Gill, W.W. Zachary (Eds.), Nonlinear Semigroups, Partial Differential Equations and Attractors. Proceedings, 1987. IX, 233 pages. 1989.

Vol. 1395: K. Alladi (Ed.), Number Theory, Madras 1987. Proceedings. VII, 234 pages. 1989.

Vol. 1396: L. Accardi, W. von Waldenfels (Eds.), Quantum Probability and Applications IV. Proceedings, 1987. VI, 355 pages. 1989.

Vol. 1397: P.R. Turner (Ed.), Numerical Analysis and Parallel Processing. Seminar, 1987. VI, 264 pages. 1989.

Vol. 1398: A.C. Kim, B.H. Neumann (Eds.), Groups – Korea 1988. Proceedings. V, 189 pages. 1989.

Vol. 1399: W.-P. Barth, H. Lange (Eds.), Arithmetic of Complex Manifolds. Proceedings, 1988. V, 171 pages. 1989.

Vol. 1400: U. Jannsen. Mixed Motives and Algebraic K-Theory. XIII, 246 pages. 1990.

Vol. 1401: J. Steprans, S. Watson (Eds.), Set Theory and its Applications. Proceedings, 1987. V, 227 pages. 1989.

Vol. 1402: C. Carasso, P. Charrier, B. Hanouzet, J.-L. Joly (Eds.), Nonlinear Hyperbolic Problems. Proceedings, 1988. V, 249 pages. 1989.

Vol. 1403: B. Simeone (Ed.), Combinatorial Optimization. Seminar, 1986. V, 314 pages. 1989.

Vol. 1404: M.-P. Malliavin (Ed.), Séminaire d´Algèbre Paul Dubreil et Marie-Paul Malliavin. Proceedings, 1987–1988. IV, 410 pages. 1989.

Vol. 1405: S. Dolecki (Ed.), Optimization. Proceedings, 1988. V, 223 pages. 1989. Vol. 1406: L. Jacobsen (Ed.), Analytic Theory of Continued Fractions III. Proceedings, 1988. VI, 142 pages. 1989.

Vol. 1407: W. Pohlers, Proof Theory. VI, 213 pages. 1989.

Vol. 1408: W. Lück, Transformation Groups and Algebraic K-Theory. XII, 443 pages. 1989.

Vol. 1409: E. Hairer, Ch. Lubich, M. Roche. The Numerical Solution of Differential-Algebraic Systems by Runge-Kutta Methods. VII, 139 pages. 1989.

Vol. 1410: F.J. Carreras, O. Gil-Medrano, A.M. Naveira (Eds.), Differential Geometry. Proceedings, 1988. V, 308 pages. 1989.

Vol. 1411: B. Jiang (Ed.), Topological Fixed Point Theory and Applications. Proceedings. 1988. VI, 203 pages. 1989.

Vol. 1412: V.V. Kalashnikov, V.M. Zolotarev (Eds.), Stability Problems for Stochastic Models. Proceedings, 1987. X, 380 pages. 1989.

Vol. 1413: S. Wright, Uniqueness of the Injective III$_1$Factor. III, 108 pages. 1989.

Vol. 1414: E. Ramirez de Arellano (Ed.), Algebraic Geometry and Complex Analysis. Proceedings, 1987. VI, 180 pages. 1989.

Vol. 1415: M. Langevin, M. Waldschmidt (Eds.), Cinquante Ans de Polynômes. Fifty Years of Polynomials. Proceedings, 1988. IX, 235 pages.1990.

Vol. 1416: C. Albert (Ed.), Géométrie Symplectique et Mécanique. Proceedings, 1988. V, 289 pages. 1990.

Vol. 1417: A.J. Sommese, A. Biancofiore, E.L. Livorni (Eds.), Algebraic Geometry. Proceedings, 1988. V, 320 pages. 1990.

Vol. 1418: M. Mimura (Ed.), Homotopy Theory and Related Topics. Proceedings, 1988. V, 241 pages. 1990.

Vol. 1419: P.S. Bullen, P.Y. Lee, J.L. Mawhin, P. Muldowney, W.F. Pfeffer (Eds.), New Integrals. Proceedings, 1988. V, 202 pages. 1990.

Vol. 1420: M. Galbiati, A. Tognoli (Eds.), Real Analytic Geometry. Proceedings, 1988. IV, 366 pages. 1990.

Vol. 1421: H.A. Biagioni, A Nonlinear Theory of Generalized Functions, XII, 214 pages. 1990.

Vol. 1422: V. Villani (Ed.), Complex Geometry and Analysis. Proceedings, 1988. V, 109 pages. 1990.

Vol. 1423: S.O. Kochman, Stable Homotopy Groups of Spheres: A Computer-Assisted Approach. VIII, 330 pages. 1990.

Vol. 1424: F.E. Burstall, J.H. Rawnsley, Twistor Theory for Riemannian Symmetric Spaces. III, 112 pages. 1990.

Vol. 1425: R.A. Piccinini (Ed.), Groups of Self-Equivalences and Related Topics. Proceedings, 1988. V, 214 pages. 1990.

Vol. 1426: J. Azéma, P.A. Meyer, M. Yor (Eds.), Séminaire de Probabilités XXIV, 1988/89. V, 490 pages. 1990.

Vol. 1427: A. Ancona, D. Geman, N. Ikeda, École d'Eté de Probabilités de Saint Flour XVIII, 1988. Ed.: P.L. Hennequin. VII, 330 pages. 1990.

Vol. 1428: K. Erdmann, Blocks of Tame Representation Type and Related Algebras. XV. 312 pages. 1990.

Vol. 1429: S. Homer, A. Nerode, R.A. Platek, G.E. Sacks, A. Scedrov, Logic and Computer Science. Seminar, 1988. Editor: P. Odifreddi. V, 162 pages. 1990.

Vol. 1430: W. Bruns, A. Simis (Eds.), Commutative Algebra. Proceedings. 1988. V, 160 pages. 1990.

Vol. 1431: J.G. Heywood, K. Masuda, R. Rautmann, V.A. Solonnikov (Eds.), The Navier-Stokes Equations – Theory and Numerical Methods. Proceedings, 1988. VII, 238 pages. 1990.

Vol. 1432: K. Ambos-Spies, G.H. Müller, G.E. Sacks (Eds.), Recursion Theory Week. Proceedings, 1989. VI, 393 pages. 1990.

Vol. 1433: S. Lang, W. Cherry, Topics in Nevanlinna Theory. II, 174 pages.1990.

Vol. 1434: K. Nagasaka, E. Fouvry (Eds.), Analytic Number Theory. Proceedings, 1988. VI, 218 pages. 1990.

Vol. 1435: St. Ruscheweyh, E.B. Saff, L.C. Salinas, R.S. Varga (Eds.), Computational Methods and Function Theory. Proceedings, 1989. VI, 211 pages. 1990.

Vol. 1436: S. Xambó-Descamps (Ed.), Enumerative Geometry. Proceedings, 1987. V, 303 pages. 1990.

Vol. 1437: H. Inassaridze (Ed.), K-theory and Homological Algebra. Seminar, 1987–88. V, 313 pages. 1990.

Vol. 1438: P.G. Lemarié (Ed.) Les Ondelettes en 1989. Seminar. IV, 212 pages. 1990.

Vol. 1439: E. Bujalance, J.J. Etayo, J.M. Gamboa, G. Gromadzki. Automorphism Groups of Compact Bordered Klein Surfaces: A Combinatorial Approach. XIII, 201 pages. 1990.

Vol. 1440: P. Latiolais (Ed.), Topology and Combinatorial Groups Theory. Seminar, 1985–1988. VI, 207 pages. 1990.

Vol. 1441: M. Coornaert, T. Delzant, A. Papadopoulos. Géométrie et théorie des groupes. X, 165 pages. 1990.

Vol. 1442: L. Accardi, M. von Waldenfels (Eds.), Quantum Probability and Applications V. Proceedings, 1988. VI, 413 pages. 1990.

Vol. 1443: K.H. Dovermann, R. Schultz, Equivariant Surgery Theories and Their Periodicity Properties. VI, 227 pages. 1990.

Vol. 1444: H. Korezlioglu, A.S. Ustunel (Eds.), Stochastic Analysis and Related Topics VI. Proceedings, 1988. V, 268 pages. 1990.

Vol. 1445: F. Schulz, Regularity Theory for Quasilinear Elliptic Systems and – Monge Ampère Equations in Two Dimensions. XV, 123 pages. 1990.

Vol. 1446: Methods of Nonconvex Analysis. Seminar, 1989. Editor: A. Cellina. V, 206 pages. 1990.

Vol. 1447: J.-G. Labesse, J. Schwermer (Eds), Cohomology of Arithmetic Groups and Automorphic Forms. Proceedings, 1989. V, 358 pages. 1990.

Vol. 1448: S.K. Jain, S.R. López-Permouth (Eds.), Non-Commutative Ring Theory. Proceedings, 1989. V, 166 pages. 1990.

Vol. 1449: W. Odyniec, G. Lewicki, Minimal Projections in Banach Spaces. VIII, 168 pages. 1990.

Vol. 1450: H. Fujita, T. Ikebe, S.T. Kuroda (Eds.), Functional-Analytic Methods for Partial Differential Equations. Proceedings, 1989. VII, 252 pages. 1990.

Vol. 1451: L. Alvarez-Gaumé, E. Arbarello, C. De Concini, N.J. Hitchin, Global Geometry and Mathematical Physics. Montecatini Terme 1988. Seminar. Editors: M. Francaviglia, F. Gherardelli. IX, 197 pages. 1990.

Vol. 1452: E. Hlawka, R.F. Tichy (Eds.), Number-Theoretic Analysis. Seminar, 1988–89. V, 220 pages. 1990.

Vol. 1453: Yu.G. Borisovich, Yu.E. Gliklikh (Eds.), Global Analysis – Studies and Applications IV. V, 320 pages. 1990.

Vol. 1454: F. Baldassari, S. Bosch, B. Dwork (Eds.), p-adic Analysis. Proceedings, 1989. V, 382 pages. 1990.

Vol. 1455: J.-P. Françoise, R. Roussarie (Eds.), Bifurcations of Planar Vector Fields. Proceedings, 1989. VI, 396 pages. 1990.

Vol. 1456: L.G. Kovács (Ed.), Groups – Canberra 1989. Proceedings. XII, 198 pages. 1990.

Vol. 1457: O. Axelsson, L.Yu. Kolotilina (Eds.), Preconditioned Conjugate Gradient Methods. Proceedings, 1989. V, 196 pages. 1990.

Vol. 1458: R. Schaaf, Global Solution Branches of Two Point Boundary Value Problems. XIX, 141 pages. 1990.

Vol. 1459: D. Tiba, Optimal Control of Nonsmooth Distributed Parameter Systems. VII, 159 pages. 1990.

Vol. 1460: G. Toscani, V. Boffi, S. Rionero (Eds.), Mathematical Aspects of Fluid Plasma Dynamics. Proceedings, 1988. V, 221 pages. 1991.

Vol. 1461: R. Gorenflo, S. Vessella, Abel Integral Equations. VII, 215 pages. 1991.

Vol. 1462: D. Mond, J. Montaldi (Eds.), Singularity Theory and its Applications. Warwick 1989, Part I. VIII, 405 pages. 1991.

Vol. 1463: R. Roberts, I. Stewart (Eds.), Singularity Theory and its Applications. Warwick 1989, Part II. VIII, 322 pages. 1991.

Vol. 1464: D. L. Burkholder, E. Pardoux, A. Sznitman, Ecole d'Eté de Probabilités de Saint- Flour XIX-1989. Editor: P. L. Hennequin. VI, 256 pages. 1991.

Vol. 1465: G. David, Wavelets and Singular Integrals on Curves and Surfaces. X, 107 pages. 1991.

Vol. 1466: W. Banaszczyk, Additive Subgroups of Topological Vector Spaces. VII, 178 pages. 1991.

Vol. 1467: W. M. Schmidt, Diophantine Approximations and Diophantine Equations. VIII, 217 pages. 1991.

Vol. 1468: J. Noguchi, T. Ohsawa (Eds.), Prospects in Complex Geometry. Proceedings, 1989. VII, 421 pages. 1991.

Vol. 1469: J. Lindenstrauss, V. D. Milman (Eds.), Geometric Aspects of Functional Analysis. Seminar 1989-90. XI, 191 pages. 1991.

Vol. 1470: E. Odell, H. Rosenthal (Eds.), Functional Analysis. Proceedings, 1987-89. VII, 199 pages. 1991.

Vol. 1471: A. A. Panchishkin, Non-Archimedean L-Functions of Siegel and Hilbert Modular Forms. VII, 157 pages. 1991.

Vol. 1472: T. T. Nielsen, Bose Algebras: The Complex and Real Wave Representations. V, 132 pages. 1991.

Vol. 1473: Y. Hino, S. Murakami, T. Naito, Functional Differential Equations with Infinite Delay. X, 317 pages. 1991.

Vol. 1474: S. Jackowski, B. Oliver, K. Pawałowski (Eds.), Algebraic Topology, Poznań 1989. Proceedings. VIII, 397 pages. 1991.

Vol. 1475: S. Busenberg, M. Martelli (Eds.), Delay Differential Equations and Dynamical Systems. Proceedings, 1990. VIII, 249 pages. 1991.

Vol. 1476: M. Bekkali, Topics in Set Theory. VII, 120 pages. 1991.

Vol. 1477: R. Jajte, Strong Limit Theorems in Noncommutative L_2-Spaces. X, 113 pages. 1991.

Vol. 1478: M.-P. Malliavin (Ed.), Topics in Invariant Theory. Seminar 1989-1990. VI, 272 pages. 1991.

Vol. 1479: S. Bloch, I. Dolgachev, W. Fulton (Eds.), Algebraic Geometry. Proceedings, 1989. VII, 300 pages. 1991.

Vol. 1480: F. Dumortier, R. Roussarie, J. Sotomayor, H. Żoładek, Bifurcations of Planar Vector Fields: Nilpotent Singularities and Abelian Integrals. VIII, 226 pages. 1991.

Vol. 1481: D. Ferus, U. Pinkall, U. Simon, B. Wegner (Eds.), Global Differential Geometry and Global Analysis. Proceedings, 1991. VIII, 283 pages. 1991.

Vol. 1482: J. Chabrowski, The Dirichlet Problem with L^2-Boundary Data for Elliptic Linear Equations. VI, 173 pages. 1991.

Vol. 1483: E. Reithmeier, Periodic Solutions of Nonlinear Dynamical Systems. VI, 171 pages. 1991.

Vol. 1484: H. Delfs, Homology of Locally Semialgebraic Spaces. IX, 136 pages. 1991.

Vol. 1485: J. Azéma, P. A. Meyer, M. Yor (Eds.), Séminaire de Probabilités XXV. VIII, 440 pages. 1991.

Vol. 1486: L. Arnold, H. Crauel, J.-P. Eckmann (Eds.), Lyapunov Exponents. Proceedings, 1990. VIII, 365 pages. 1991.

Vol. 1487: E. Freitag, Singular Modular Forms and Theta Relations. VI, 172 pages. 1991.

Vol. 1488: A. Carboni, M. C. Pedicchio, G. Rosolini (Eds.), Category Theory. Proceedings, 1990. VII, 494 pages. 1991.

Vol. 1489: A. Mielke, Hamiltonian and Lagrangian Flows on Center Manifolds. X, 140 pages. 1991.

Vol. 1490: K. Metsch, Linear Spaces with Few Lines. XIII, 196 pages. 1991.

Vol. 1491: E. Lluis-Puebla, J.-L. Loday, H. Gillet, C. Soulé, V. Snaith, Higher Algebraic K-Theory: an overview. IX, 164 pages. 1992.

Vol. 1492: K. R. Wicks, Fractals and Hyperspaces. VIII, 168 pages. 1991.

Vol. 1493: E. Benoît (Ed.), Dynamic Bifurcations. Proceedings, Luminy 1990. VII, 219 pages. 1991.

Vol. 1494: M.-T. Cheng, X.-W. Zhou, D.-G. Deng (Eds.), Harmonic Analysis. Proceedings, 1988. IX, 226 pages. 1991.

Vol. 1495: J. M. Bony, G. Grubb, L. Hörmander, H. Komatsu, J. Sjöstrand, Microlocal Analysis and Applications. Montecatini Terme, 1989. Editors: L. Cattabriga, L. Rodino. VII, 349 pages. 1991.

Vol. 1496: C. Foias, B. Francis, J. W. Helton, H. Kwakernaak, J. B. Pearson, H$_\infty$-Control Theory. Como, 1990. Editors: E. Mosca, L. Pandolfi. VII, 336 pages. 1991.

Vol. 1497: G. T. Herman, A. K. Louis, F. Natterer (Eds.), Mathematical Methods in Tomography. Proceedings 1990. X, 268 pages. 1991.

Vol. 1498: R. Lang, Spectral Theory of Random Schrödinger Operators. X, 125 pages. 1991.

Vol. 1499: K. Taira, Boundary Value Problems and Markov Processes. IX, 132 pages. 1991.

Vol. 1500: J.-P. Serre, Lie Algebras and Lie Groups. VII, 168 pages. 1992.

Vol. 1501: A. De Masi, E. Presutti, Mathematical Methods for Hydrodynamic Limits. IX, 196 pages. 1991.

Vol. 1502: C. Simpson, Asymptotic Behavior of Monodromy. V, 139 pages. 1991.

Vol. 1503: S. Shokranian, The Selberg-Arthur Trace Formula (Lectures by J. Arthur). VII, 97 pages. 1991.

Vol. 1504: J. Cheeger, M. Gromov, C. Okonek, P. Pansu, Geometric Topology: Recent Developments. Editors: P. de Bartolomeis, F. Tricerri. VII, 197 pages. 1991.

Vol. 1505: K. Kajitani, T. Nishitani, The Hyperbolic Cauchy Problem. VII, 168 pages. 1991.

Vol. 1506: A. Buium, Differential Algebraic Groups of Finite Dimension. XV, 145 pages. 1992.

Vol. 1507: K. Hulek, T. Peternell, M. Schneider, F.-O. Schreyer (Eds.), Complex Algebraic Varieties. Proceedings, 1990. VII, 179 pages. 1992.

Vol. 1508: M. Vuorinen (Ed.), Quasiconformal Space Mappings. A Collection of Surveys 1960-1990. IX, 148 pages. 1992.

Vol. 1509: J. Aguadé, M. Castellet, F. R. Cohen (Eds.), Algebraic Topology - Homotopy and Group Cohomology. Proceedings, 1990. X, 330 pages. 1992.

Vol. 1510: P. P. Kulish (Ed.), Quantum Groups. Proceedings, 1990. XII, 398 pages. 1992.

Vol. 1511: B. S. Yadav, D. Singh (Eds.), Functional Analysis and Operator Theory. Proceedings, 1990. VIII, 223 pages. 1992.

Vol. 1512: L. M. Adleman, M.-D. A. Huang, Primality Testing and Abelian Varieties Over Finite Fields. VII, 142 pages. 1992.

Vol. 1513: L. S. Block, W. A. Coppel, Dynamics in One Dimension. VIII, 249 pages. 1992.

Vol. 1514: U. Krengel, K. Richter, V. Warstat (Eds.), Ergodic Theory and Related Topics III, Proceedings, 1990. VIII, 236 pages. 1992.

Vol. 1515: E. Ballico, F. Catanese, C. Ciliberto (Eds.), Classification of Irregular Varieties. Proceedings, 1990. VII, 149 pages. 1992.

Vol. 1516: R. A. Lorentz, Multivariate Birkhoff Interpolation. IX, 192 pages. 1992.

Vol. 1517: K. Keimel, W. Roth, Ordered Cones and Approximation. VI, 134 pages. 1992.

Vol. 1518: H. Stichtenoth, M. A. Tsfasman (Eds.), Codi[ng] Theory and Algebraic Geometry. Proceedings, 1991. VIII, 2[?] pages. 1992.

Vol. 1519: M. W. Short, The Primitive Soluble Permutati[on] Groups of Degree less than 256. IX, 145 pages. 1992.

Vol. 1520: Yu. G. Borisovich, Yu. E. Gliklikh (Eds.), Glob[al] Analysis – Studies and Applications V. VII, 284 pages. 199[2].

Vol. 1521: S. Busenberg, B. Forte, H. K. Kuiken, Mathematic[al] Modelling of Industrial Process. Bari, 1990. Editors: V. Capass[o,] A. Fasano. VII, 162 pages. 1992.

Vol. 1522: J.-M. Delort, F. B. I. Transformation. VII, 101 page[s.] 1992.

Vol. 1523: W. Xue, Rings with Morita Duality. X, 168 page[s.] 1992.

Vol. 1524: M. Coste, L. Mahé, M.-F. Roy (Eds.), Real Algebra[ic] Geometry. Proceedings, 1991. VIII, 418 pages. 1992.

Vol. 1525: C. Casacuberta, M. Castellet (Eds.), Mathematic[al] Research Today and Tomorrow. VII, 112 pages. 1992.

Vol. 1526: J. Azéma, P. A. Meyer, M. Yor (Eds.), Séminaire [de] Probabilités XXVI. X, 633 pages. 1992.

Vol. 1527: M. I. Freidlin, J.-F. Le Gall, Ecole d'Eté [de] Probabilités de Saint-Flour XX – 1990. Editor: P. L. Hennequi[n.] VIII, 244 pages. 1992.

Vol. 1528: G. Isac, Complementarity Problems. VI, 297 page[s.] 1992.

Vol. 1529: J. van Neerven, The Adjoint of a Semigroup of Line[ar] Operators. X, 195 pages. 1992.

Vol. 1530: J. G. Heywood, K. Masuda, R. Rautmann, S. [A.] Solonnikov (Eds.), The Navier-Stokes Equations II – Theo[ry] and Numerical Methods. IX, 322 pages. 1992.

Vol. 1531: M. Stoer, Design of Survivable Networks. IV, 2[06] pages. 1992.

Vol. 1532: J. F. Colombeau, Multiplication of Distributions. [X,] 184 pages. 1992.

Vol. 1533: P. Jipsen, H. Rose, Varieties of Lattices. X, 162 page[s.] 1992.

Vol. 1534: C. Greither, Cyclic Galois Extensions of Com[-] mutative Rings. X, 145 pages. 1992.

Vol. 1535: A. B. Evans, Orthomorphism Graphs of Groups. VII[,] 114 pages. 1992.

Vol. 1536: M. K. Kwong, A. Zettl, Norm Inequalities fo[r] Derivatives and Differences. VII, 150 pages. 1992.

Vol. 1537: P. Fitzpatrick, M. Martelli, J. Mawhin, R. Nussbaum[,] Topological Methods for Ordinary Differential Equation[s.] Montecatini Terme, 1991. Editors: M. Furi, P. Zecca. VII, 21[?] pages. 1993.

Vol. 1538: P.-A. Meyer, Quantum Probability for Probabilist[s.] X, 287 pages. 1993.

Vol. 1539: M. Coornaert, A. Papadopoulos, Symbolic Dynamic[s] and Hyperbolic Groups. VIII, 138 pages. 1993.

Vol. 1540: H. Komatsu (Ed.), Functional Analysis and Relate[d] Topics, 1991. Proceedings. XXI, 413 pages. 1993.